Building Knowledge for Geohazard Assessment and Management in the Caucasus and other Orogenic Regions

AF901920

NATO Science for Peace and Security Series

This Series presents the results of scientific activities supported through the framework of the NATO Science for Peace and Security (SPS) Programme.

The NATO SPS Programme enhances security-related civil science and technology to address emerging security challenges and their impacts on international security. It connects scientists, experts and officials from Alliance and Partner nations to work together to address common challenges. The SPS Programme provides funding and expert advice for security-relevant activities in the form of Multi-Year Projects (MYP), Advanced Research Workshops (ARW), Advanced Training Courses (ATC), and Advanced Study Institutes (ASI). The NATO SPS Series collects the results of practical activities and meetings, including:

Multi-Year Projects (MYP): Grants to collaborate on multi-year R&D and capacity building projects that result in new civil science advancements with practical application in the security and defence fields.

Advanced Research Workshops: Advanced-level workshops that provide a platform for experts and scientists to share their experience and knowledge of security-related topics in order to promote follow-on activities like Multi-Year Projects.

Advanced Training Courses: Designed to enable specialists in NATO countries to share their security-related expertise in one of the SPS Key Priority areas. An ATC is not intended to be lecture-driven, but to be intensive and interactive in nature.

Advanced Study Institutes: High-level tutorial courses that communicate the latest developments in subjects relevant to NATO to an advanced-level audience.

The observations and recommendations made at the meetings, as well as the contents of the volumes in the Series reflect the views of participants and contributors only, and do not necessarily reflect NATO views or policy.

The series is published by IOS Press, Amsterdam, and Springer, Dordrecht, in partnership with the NATO SPS Programme.

Sub-Series

A.	Chemistry and Biology	Springer
B.	Physics and Biophysics	Springer
C.	Environmental Security	Springer
D.	Information and Communication Security	IOS Press
E.	Human and Societal Dynamics	IOS Press

- http://www.nato.int/science
- http://www.springer.com
- http://www.iospress.nl

Series C: Environmental Security

Building Knowledge for Geohazard Assessment and Management in the Caucasus and other Orogenic Regions

edited by

Fabio Luca Bonali
Department of Earth and Environmental Sciences
University of Milano-Bicocca
Milan, Italy

Federico Pasquaré Mariotto
Department of Human and Innovation Sciences
Insubria University
Como, Italy

and

Nino Tsereteli
M. Nodia Institute of Geophysics
I. Javakhisvili Tbilisi State University
Tbilisi, Georgia

NATO OTAN *This publication is supported by:* The NATO Science for Peace and Security Programme

Springer
Published in Cooperation with NATO Emerging Security Challenges Division

Proceedings of the NATO ARW on Building Civil Protection Capacity to Mitigate Geohazards in the Caucasus: A Regional Approach
Tbilisi, Georgia
8–10 October, 2019

ISBN 978-94-024-2048-7 (PB)
ISBN 978-94-024-2045-6 (HB)
ISBN 978-94-024-2046-3 (e-book)
https://doi.org/10.1007/978-94-024-2046-3

Published by Springer,
P.O. Box 17, 3300 AA Dordrecht, The Netherlands.

www.springer.com

Printed on acid-free paper

All Rights Reserved
© Springer Nature B.V. 2021
This work is subject to copyright. All rights are reserved by the Publisher, whether the whole or part of the material is concerned, specifically the rights of translation, reprinting, reuse of illustrations, recitation, broadcasting, reproduction on microfilms or in any other physical way, and transmission or information storage and retrieval, electronic adaptation, computer software, or by similar or dissimilar methodology now known or hereafter developed.
The use of general descriptive names, registered names, trademarks, service marks, etc. in this publication does not imply, even in the absence of a specific statement, that such names are exempt from the relevant protective laws and regulations and therefore free for general use.
The publisher, the authors, and the editors are safe to assume that the advice and information in this book are believed to be true and accurate at the date of publication. Neither the publisher nor the authors or the editors give a warranty, expressed or implied, with respect to the material contained herein or for any errors or omissions that may have been made. The publisher remains neutral with regard to jurisdictional claims in published maps and institutional affiliations.

Preface

This book, displaying the outcomes of a variety of cutting-edge research efforts, is articulated in 7 parts and 25 chapters, covering a broad gamut of topics. The volume is introduced by a part aimed at emphasizing the key role played by collaboration among international organizations and scientists committed to assessing geohazard in orogenic areas. Secondly, an overall perspective on regional geological aspects, as well as seismic, volcanic, and hydrological hazard sources, is provided for the Caucasus region. The final part is devoted to examining several innovative approaches, helpful in enabling scientists to improve their research in orogenic areas prone to a wide range of natural hazards.

The research outcomes illustrated in this volume, stemming from research devoted to deepening our understanding of geological phenomena and discussing innovative tools and strategies for geohazard assessment, may spark renewed interest in studies devoted to the Caucasus and other orogenic regions.

This book represents the natural follow-up to NATO Advanced Research Workshop G5566, entitled "Building Civil Protection Capacity to Mitigate Geohazards in the Caucasus: A Regional Approach," which was held at Ivane Javakhishvili Tbilisi State University, Republic of Georgia, in 2019.

The workshop was made possible by funding from the NATO Science for Peace and Security Programme, with additional support from the International Lithosphere Program (ILP), the NATO Liaison Office (NLO) in Georgia, the European Centre on Geodynamical Hazards of High Dams (GHHD), and the LTD Arili + private company.

The Guest Editors are thankful to all contributing authors, who have provided an invaluable perspective on a comprehensive range of geological topics, from the regional to the local scale.

Milan, Italy — Fabio L. Bonali

Como, Italy — Federico Pasquaré Mariotto

Tbilisi, Georgia — Nino Tsereteli

Introduction

This book represents the natural follow-up to NATO Advanced Research Workshop (ARW) G5566, entitled "Building Civil Protection Capacity to Mitigate Geohazards in the Caucasus: A Regional Approach," which was held at Ivane Javakhishvili Tbilisi State University (Fig. 1a), between 8 and 10 October 2019 in Tbilisi, Georgia.

The Tbilisi event was aimed at sharing transboundary data and suggesting effective measures for the mitigation of natural hazards in the Caucasus region; in this area, millions of residents and several critical infrastructures and lifelines (such as hydropower plants, nuclear power plants, and oil and gas pipelines) are prone to risk. The ARW was coordinated and organized by Dr. Fabio Luca Bonali, Department of Earth and Environmental Sciences, University of Milan-Bicocca, Italy, and Dr. Otar Varazanashvili, Ivane Javakhishvili Tbilisi State University, Georgia.

More than 40 scientists (Fig. 1b), including academics and young researchers, attended the ARW, articulated in 2 days of indoor presentations and one day devoted to a field trip, aimed at providing the participants with a comprehensive look at geohazard-induced effects in the capital city and its surroundings. The indoor presentations revolved around a broad range of topics, such as the geology of orogenic regions, natural hazards and geotectonic features in the Caucasus, multidisciplinary approaches to geohazard assessment/management, microzonation, and site effects.

During the 1-day field trip, the participants took part in a guided visit to downtown Tbilisi: special attention was paid to a few selected sites, where earthquake-induced damage to buildings can be clearly observed, caused by the Ms 4.6 seismic event that struck the city in 2002 (Varazanashvili et al., 2018) (Fig. 2a–b); shortly afterwards, the group moved on to examine the impacts of hydrogeological events in Akhaldaba (41°40′34″N, 44°40′45″E), a site near Tbilisi, where local geologists illustrated ongoing mass-wasting processes, as well as the related damage and remediation measures (Fig. 2c–d).

Conference participants are from 12 different nations, subdivided in 6 NATO countries and 6 NATO Partner countries, namely Azerbaijan, Armenia, Czech

Fig. 1 (**a**) Picture of Iv. Javakhishvili Tbilisi State University, the oldest and most prominent higher educational institution in Georgia, founded in 1918. Back in those days, it was the first University in the Caucasus region. (**b**) First official pictures of the NATO-funded ARW in Tbilisi; more pictures and information are available at https://2019tbilisinatoworkshop.unimib.it/

Republic, France, Georgia, Italy, Kazakhstan, Switzerland, Turkey, the United Kingdom, Ukraine, and the United States.

This volume is structured in the following parts: (I) The Role of NATO; (II) Regional and Geological aspects; (III) Seismic Hazard Assessment; (IV) Volcanic hazard Assessment; (V) Hydrological, Landslide and Coastal Hazard Assessment; (VI) Seismic Microzonation; and (VII) Innovative Approaches to Geohazard and Risk Assessment.

Starting from Part I, Beten and Marchesoni (2020) introduce the NATO–Georgia cooperation in the framework of the Science for Peace Security (SPS) Programme,

Fig. 2 Examples of the 2002 Tbilisi earthquake-induced damage to buildings in downtown Tbilisi (**a**–**b**). Field trip at the Akhaldaba landslide (**c**–**d**)

highlighting that since the beginning of its cooperation with the Programme in 1994, Georgia has participated in 59 SPS activities. Other relevant details on the NATO–Georgia cooperation are provided by Iorio (2020): throughout its entire life as an independent nation, Georgia has developed one of the most intense and substantial partnerships with the North Atlantic Alliance. Finally, Pasquaré Mariotto et al. (2020a) document a number of multi-year projects on geohazard assessment and mitigation carried out under the umbrella of NATO, with the involvement of more than 80 scientists from across the globe.

Part II deals with regional and geological aspects: firstly, Tibaldi et al. (2020) offer insights into the active kinematics of the Greater Caucasus (territories of Georgia, Azerbaijan, and Russia) through an integrated analysis of seismological, geological-structural, and GPS-derived data. Alania et al. (2020a) illustrate a new, crustal-scale structural model aimed at displaying the recent geometry of the western Greater Caucasus; the model is built from surface and subsurface data, in agreement with the theory of fault-related folds. A second chapter by Alania et al. (2020b) shows the structural style of deformation of a convergence zone between the Lesser and Greater Caucasus, based on new data from seismic reflection profiles. The chapter by Lavecchia et al. (2020) discusses the results of a study on the seismotectonic zonation of hydrocarbon fields in active thrust belts, based on an original, structural-geological approach and a regional seismotectonic zoning of petroleum fields in Italy.

Part III is dedicated to seismic hazard assessment: firstly, Tsereteli et al. (2020a) propose a new national seismic hazard model for Georgia, based on updated techniques and datasets. The next chapter, from Abrahamson and Kuehn (2020), deals with the development of an initial non-ergodic ground-motion model for Georgia, for spectral acceleration at T=0.2 s, considering 33 local strong-motion recordings. Chelidze et al. (2020a) use complexity analysis methods to discriminate seismic events derived by human activities and background seismicity around the Enguri dam, Greater Caucasus. The work by Babayev (2020) focuses on the assessment of seismic hazard across the southern slope of the Greater Caucasus (in Azerbaijan), using parameters like moment magnitude, seismic energy, Peak Ground Acceleration (PGA), and site effects.

Regarding volcanic hazard assessment, explored in Part IV, Vashakidze et al. (2020) provide a detailed review of the eruptive history, including phases and subphases, of three major volcanoes located in the Greater Caucasus, showing evidence of quaternary activity: Elbrus, Kazbek, and Keli. Pasquaré Mariotto et al. (2020b) document the role of tectonics in controlling magma ascent at the Abuli Samsari Volcanic Ridge, Lesser Caucasus, Georgia; emphasis is also placed on volcanic hazard affecting two parallel pipelines, carrying oil and gas from the Caspian Sea to the western countries.

Part V, focused on hydrological, landslide, and coastal hazard assessment, is initiated by a work from Gaprindashvili et al. (2020), introducing a set of new hazard maps for geological hazard assessment, compiled through methods to be applied at the national level, with special emphasis on landslides and mudflows. Nichols et al. (2020) document the outcomes of their study about the significance of the Spatial Resolution of DEMs in Regional Slope Stability Analysis, focusing on an area that encompasses a portion of the Enguri Dam (and related artificial reservoir) in the Greater Caucasus, Georgia. For the same area, Ospanov and Krivchenko (2020) provide the results of a 2-year monitoring effort, aimed at assessing deformation taking place at the Khoko landslide (overlooking the Enguri reservoir), conducted through high-resolution techniques. The final chapter of this part, from Gelovani et al. (2020), examines selected examples of coastal hazard along the Georgian Black Sea Coast, highlighting the ways in which human activities have affected this unique environment over the last decades.

Part VI is designed to provide an overview of the recent advancements in seismic microzonation. The first chapter, from Albarello and Moscatelli (2020), revolves around showing how the extensive application of this technique can be a key tool for seismic risk reduction. The authors outline the methodology applied in Italy during the last 10 years, aimed at developing a common strategy to cope with future earthquakes. Tsereteli et al. (2020b) illustrate an application of seismic microzonation studies, providing the preliminary results of the first site effect assessment performed in the urban area of Mtskheta (Georgia); the outcomes of this effort represent a first step toward reducing seismic risk in Georgia. Kendzera et al. (2020) present a modelling of site amplification in the South Ukrainian nuclear power plant: three zones have been individuated and, for each, frequency responses have been separately calculated. Finally, Rushchitsky et al. (2020) present specific formulas for the evaluation of the wave number and attenuation of the harmonic SH-waves propagating in the Layered Earth Stratum.

The conclusive Part VII showcases a number of innovative approaches to geohazard and risk assessment: in the first chapter of this part, Bonali et al. (2020) touch upon the importance of Unmanned Aerial Vehicles (UAVs) and Structure from Motion (SfM) techniques for terrain and orthomosaic reconstruction as well as geological data collection in remote and inaccessible sites. Chelidze et al. (2020b) document the development of the new prototype of an original, cost-effective, multi-sensor telemetric monitoring/early warning system, suitable for forecasting the initiation of landslides and mudflows. Pasquaré Mariotto and Oppizzi (2020) provide an overview of a monitoring effort targeted to the Khoko landslide. The monitoring was conducted between 2016 and 2019, through horizontal, digital extensometers placed in trenches excavated along the Zugdidi-Jvari-Mestia-Lasdili road. The volume is concluded with an *ad-hoc* methodology, proposed by Melkadze (2020) for the assessment of building vulnerability in Georgia, to be included in a management risk approach, which will be applied at the national level.

Milan, Italy — Fabio Luca Bonali

Como, Italy — Federico Pasquaré Mariotto

Tbilisi, Georgia — Nino Tsereteli

References

Abrahamson N, Kuehn N (2020) Non-Ergodic ground-motion models for crustal earthquakes in Georgia

Alania V, Tibaldi A, Bonali FL et al (2020a) Structural architecture of the western Greater Caucasus orogen: New data from a crustal-scale structural cross-section

Alania V, Beridze T, Enukidze O et al (2020b) The geometry of the two orogens convergence and collision zones in central Georgia: new data from seismic reflection profiles

Albarello D, Moscatelli M (2020) Extensive Microzonation as a tool for seismic risk reduction: methodological and political issues

Babayev G (2020) Geohazard assessment along the southern slope of the Greater Caucasus (Azerbaijan)

Beten D, Marchesoni E (2020) NATO cooperation with Georgia in the framework of the Science for Peace Security (SPS) Programme

Bonali FL, Corti N, Russo E et al (2020) Commercial-UAV-based structure from motion for geological and geohazard studies

Chelidze T, Matcharashvili T, Abashidze V et al (2020a) Time series analysis of fault strain accumulation around large dam: the case of Enguri dam, Greater Caucasus

Chelidze Z, Varamashvili N, Chelidze T et al (2020b) Automatic telemetric monitoring/early warning systems, with multi-task sensor, applied to mass movements

Gaprindashvili M, Tsereteli E, Gaprindashvili G et al (2020) Landslide and mudflow hazard assessment in Georgia
Gelovani I, Lominadze G, Kavlashvili G et al (2020) Examples of coastal hazard along the Georgian Black Sea Coast
Iorio N (2020) Lens on NATO-Georgia cooperation: a shared engagement
Kendzera O, Rushchitsky J, Semenova Y et al (2020) Rheological properties of soils in assessing the seismic hazard of the South Ukrainian nuclear power plant
Lavecchia G, de Nardis R, Ferrarini F et al (2020) Regional seismotectonic zonation of hydrocarbon fields in active thrust belts: a case study from Italy
Melkadze T (2020) Aspects of risk management and vulnerability assessment of buildings in the Republic of Georgia
Nichols I, Gierke JS et al (2020) Significance of the spatial resolution of DEM in regional slope stability analysis Enguri Dam, Republic of Georgia
Ospanov NS, Krivchenko AA (2020) Description of a 2-year, high-resolution geodetic monitoring of the Khoko landslide, Enguri reservoir, Georgia
Pasquaré Mariotto F, Oppizzi P (2020) Extensometer-based monitoring of active deformation at the Khoko landslide (Jivari, Georgia)
Pasquaré Mariotto F, Tibaldi A, Bonali FL (2020a) An overview of the 20-year collaboration between NATO and Earth scientists to assess geohazards in the Caucasus and other critical regions
Pasquaré Mariotto F, Bonali FL, Venturini C (2020b) Tectonic control over the Abuli Samsari Volcanic Ridge, Lesser Caucasus, Georgia
Rushchitsky J, Kendzera O, Semenova Y et al (2020) On the seismic waves propagating in the layered earth stratum
Tibaldi A, Babayev G, Bonali FL et al (2020) Active kinematics of the Greater Caucasus from seismological and GPS data: a review
Tsereteli N, Danciu L, Varazanashvili O et al (2020a) The 2020 National Seismic Hazard Model for Georgia (Sakartvelo)
Tsereteli N, Moscatelli M, Albarello D et al (2020b) Preliminary results of site effects assessment in Mtskheta (Georgia)
Varazanashvili O, Tsereteli N, Bonali FL et al (2018) GeoInt: the first macroseismic intensity database for the Republic of Georgia. J Seismol 22(3):625–667
Vashakidze G, Parfenov AV, Gabarashvili K et al (2020) Quaternary volcanic activity in the Greater Caucasus: a review of Elbrus, Kazbek and Keli volcanoes

Contents

Part I
The Role of NATO

Chapter 1
NATO Cooperation with Georgia in the Framework of the Science for Peace and Security (SPS) Programme

Deniz Beten and Ettore Marchesoni

Abstract Practical cooperation in the framework of NATO's Science for Peace and Security (SPS) Programme represents a unique opportunity to build networks among scientists and experts from NATO and partner nations. By supporting the development of joint collaborative activities, the Programme contributes to promoting scientific innovation, dialogue and knowledge-sharing, while delivering tangible results that contribute to tackling emerging security challenges in key areas, such as cyber defence, counter-terrorism, and security-related advanced technology. This chapter provides an overview of Georgia's participation in SPS activities, outlining in particular the contribution of the Programmes' activities to energy and environmental security in the country and the surrounding region.

"More NATO in Georgia, more Georgia in NATO". This expression, first used by Secretary General Jens Stoltenberg on a visit to Georgia in 2015, captures the unique scope of the relationship between the North Atlantic Treaty Organization and Georgia.

Georgia enjoys a strong practical partnership and political dialogue with the Alliance, and aspires to NATO membership. Every day, NATO experts are on the ground to assist and advise Georgian authorities on the enhancement of their security and defence capabilities. This is done through the NATO Liaison Office in Tbilisi, opened in 2010, as well as a number of NATO programmes in place to help Georgia achieve its Euro-Atlantic objectives. Prominent among the opportunities for practical cooperation between Georgia and NATO is the Science for Peace and Security (SPS) Programme.

D. Beten · E. Marchesoni (✉)
NATO Emerging Security Challenges Division (ESCD) – Boulevard Leopold III, Brussels, Belgium
e-mail: Marchesoni.Ettore@hq.nato.int

© Springer Nature B.V. 2021

F. L. Bonali et al. (eds.), *Building Knowledge for Geohazard Assessment and Management in the Caucasus and other Orogenic Regions*, NATO Science for Peace and Security Series C: Environmental Security,
https://doi.org/10.1007/978-94-024-2046-3_1

1.1 The Science for Peace and Security (SPS) Programme

SPS is an established brand for NATO and has been contributing to the core goals of the Alliance for more than six decades. The Programme enhances security-related civil science and technology to address emerging security challenges and their impact, and is an integral component of NATO's Emerging Security Challenges (ESC) Division. In pursuit of NATO's Strategic Objectives, SPS activities align with a set of key priorities, which include, for instance, supporting NATO efforts to counter terrorism, further enhancing the development of security-related advanced technologies, and dealing with questions related to cyber defence, detection of explosives, and defence against chemical, biological, radiological and nuclear (CBRN) agents. Of note for this publication, SPS key priorities also focus on energy and environmental security challenges, for instance by facilitating practical cooperation dealing with the protection of critical energy infrastructure, as well as building resilience to address natural catastrophes.

SPS activities connect researchers, experts and officials from NATO and partner nations, who address shared challenges through practical scientific cooperation and capacity building. By supporting security-relevant activities in the form of grants for Multi-Year Projects (MYP), Advanced Research Workshops (ARW), Advanced Training Courses (ATC), and Advanced Study Institutes (ASI), SPS fosters the creation and expansion of international expert networks, the sharing of best practices, and the exchange of expertise and know-how among scientific communities in NATO and partner nations.

Between 2012 and 2019, SPS has launched a total of 419 new activities. Working with a balanced 360° approach, this Programme involves partners from each of NATO's partnership frameworks, engaging approximately 2000 experts every year. The Programme also invests in the development of the next generation of researchers, by actively supporting the participation and training of young scientists in its activities. As a testament to the scientific excellence supported by SPS, 21 Nobel Laureates have been involved in its activities.

SPS activities have also a high public diplomacy value for NATO, participating countries, and individual scientists. In particular, the Programme contributes to enriching the public's perception of NATO beyond the military dimension. By focusing on non-military, practical scientific cooperation and capacity building, SPS often proves essential in establishing new lines of communication with partners by addressing broad and unconventional security challenges. As demonstrated by the large number of publications and scientific articles developed in the framework of SPS activities, the Programme provides a platform for experts and scientists to contribute to the progress of the scientific disciplines they excel in.

Georgian experts' active involvement in SPS activities provides a clear example of how NATO-Georgia cooperation involves much more than hard defence, and serves as a demonstration of the various facets and benefits of practical scientific cooperation. As stressed by former Deputy Secretary General Gottemoeller during the Programme's 60th anniversary ceremony in 2018, "It is science that provides our understanding of emerging security challenges, and it is science that will underpin our solutions."

1.2 SPS Cooperation with Georgia

Since the beginning of its cooperation with the Programme in 1994, Georgia has participated in 59 SPS activities. In line with the Programme's key priorities, activities led by Georgia have addressed challenges in Environmental and Energy Security, Counter-Terrorism, Human and Social Aspects of Security, Cyber Defence, Support to NATO-led Operations, and Defence against CBRN Agents. As showcased by the examples below, activities developed with Georgian experts and institutions have not only contributed to scientific progress, but have also maintained focus on the regional dimension of security, often considering the implications of emerging security challenges on the broader Caucasus.

Under the SPS key priority of *Advanced Technologies*, Romanian and Georgian experts are developing innovative technologies for manufacturing new generation composite armor material for the protection of personnel, equipment, and vehicles. Launched in 2019, this SPS research and development Multi-Year Project aims to find new processing technologies and materials for the creation of high hardness, low cost and high capacity armor, by replacing boron carbide ceramics, which are currently in use.

Between 2017 and 2019, the SPS project on "Women, Peace and Security in the Georgian Armed Forces" strengthened the Georgian Ministry of Defence's capacity to assess its organizational climate and address personnel concerns in the Georgian Armed Forces. Led by the Ministry of Defence (MoD) of Georgia, the Slovenian and Swiss offices of the Geneva Centre for the Democratic Control of Armed Forces (DCAF), and the MoDs of the United Kingdom and Spain, this activity addressed the SPS key priority of *Human and Social Aspects of Security*. By providing training and coaching to a dedicated MoD Climate Assessment Working Group, this project enabled Georgia's first institutional climate survey in the armed forces, which was carried out between April and May 2019.

Addressing the SPS key priority of *Cyber Defence*, and following widespread cyber-attacks against Georgia in 2008, Italian and Georgian experts organized the ARW "Options for Enhanced Regional Cyber Defence Cooperation in the Southern Caucasus and Black Sea Region". The event took place in Tbilisi in 2015, and involved more than 50 participants from nine countries in the region and beyond. Participants aimed to evaluate options for enhanced information sharing, trust building and the creation of common cyber defence countermeasures, including enhanced coordination on common technical, regulatory and interoperability frameworks. By addressing shared challenges in the Southern Caucasus and Black Sea region, this activity paved the way towards improving and enhancing regional partnerships and cooperation in cyber space, and reducing the risk of duplication of efforts.

In the field of *Counter-Terrorism*, the ARW "Black Swans on the Eastern Flank" was carried out in Bucharest in April 2018 through a collaboration between Romanian and Georgian experts. The main objective of this event was to generate a tailored set of scenarios in line with the NATO Strategic Concept, and address the fundamental security challenges of a vulnerable region. With a multi-perspective approach, this activity provided an authentic bottom-up view on vulnerabilities in fields that experts identified as most likely to generate crises in this region.

Also addressing the key priority of Counter-Terrorism, the project "Protection of Underground Structures from Fuel Cloud Explosion" was completed in 2018. With this activity, researchers from the USA and Georgia designed and tested a system to protect people and infrastructure from the effects of fuel cloud explosions in underground facilities, such as shock waves and fire. The design concept of the protective system was developed based on a quantitative hazard analysis, and was deemed applicable to a variety of structures, including motorway and railway tunnels, mines and oil platforms, as well as petro-chemical plants. This initiative received broad acknowledgement among the scientific community, and its findings were presented at 12 international conferences and in 14 international journals.

1.3 Securing Georgia and the Caucasus Against Geohazards

Energy Security figures prominently among NATO's political priorities, and given Georgia's geographical location it is one of the main factors of mutual interest in Georgia-NATO relations. The relevance of this challenge is mirrored in the interest shown by Georgian scientists in the development of cooperative activities addressing the SPS key priority of *Energy Security*.

In recent years, Georgian scientists have been involved in several activities dealing with critical energy infrastructure protection, energy efficiency, as well as the mitigation of man-made and environmental risks to energy generation and transmission. In particular, with support from SPS and in cooperation with the NATO Energy Security Centre of Excellence in Vilnius, Lithuania, Georgian experts have organized several editions of the conference "Innovative Energy Solutions for Military Applications (IESMA)". As part of NATO's Smart Energy Initiative, the IESMA conference has become a recognized platform for bringing together military, academia and industry, as well as civil servants. The fourth IESMA workshop took place in November 2018 and addressed, among other topics, hybrid power generation and micro grids for dual use, using innovative energy solutions that could be especially interesting for partner countries and other international organizations, including the Comprehensive Nuclear-Test-Ban Treaty Organization and the United Nations.

Of specific relevance for this publication, Georgian scientists have been involved in two SPS activities that addressed regional security concerns at the crossroads between *Energy* and *Environmental Security*. The first such activity tackled a critical infrastructure in Georgia's Northwest, the Enguri dam. This hydroelectric facility is essential to Georgia's electric energy supply, and is built in an area prone to geohazards. Natural and man-made incidents at the dam could therefore have direct consequences for the social and geopolitical stability of Georgia and the wider Caucasus. To assess and prevent this risk, between 2015–2018 an SPS MYP evaluated the vulnerability of the Enguri dam and reservoir.

Initiated by scientists from the Institute of Geophysics (Georgia) and the University of Milan Bicocca (Italy), and with the participation of experts from Azerbaijan, Kazakhstan, United Kingdom and United States, the project studied local geological, geomorphological and geotechnical characteristics, as well as the paleoseismology and seismology of the region. Following data analysis and numerical modelling, the project team was able to assess possible risk scenarios, such as seismicity, landslides and the release of pollutants. The results of the study were used to discuss potential preventive measures and a continuous monitoring system was established. Based on these findings, a comprehensive training programme for young researchers, technicians and students from NATO and partner nations was developed and carried out. The end-users, including the society that manages the hydroelectrical facility and relevant ministries and civil protection agencies, received the results of the assessment and were trained to adopt to the proposed measures.

Building on the experience of the Enguri dam project, SPS supported the organization of the ARW 'Building Civil Protection Capacity to Mitigate Geohazards in the Caucasus: A Regional Approach', which took place in Tbilisi in October 2019. Developed by Italian and Georgian experts, the workshop aimed at exchanging knowledge about geological, seismic, volcanic, landslides and flood hazards, which could affect the lives of millions of residents in the Black Sea-Caucasus region as well as critical infrastructures such as hydropower and nuclear power plants, and gas and oil pipelines. Participants shared scientific data and methods to assess these risks and discussed effective measures for their mitigation.

40 participants from 12 nations, including professors and young researchers, gathered at the workshop, bringing together best practices and multidisciplinary perspectives. Discussions contributed to increasing knowledge about geohazards in the Caucasus, and provided a forum for sharing innovative ideas and creating new cross-border networks amongst participants.

1.4 Conclusion

This NATO Science Series publication gathers the proceedings of the SPS ARW 'Building Civil Protection Capacity to Mitigate Geohazards in the Caucasus: A Regional Approach', and includes studies from many of the international experts that contributed to the success of this activity.

The book is organized in chapters that mirror the themes addressed during the workshop, namely seismic, volcanic, hydrological, landslide and coastal hazard assessments; seismic microzonation; and innovative approaches for geohazard and risk assessment.

The chapters of this publication do not reflect necessarily views of NATO or the editors, and represent solely the perspectives of the authors.

Chapter 2
Lens on NATO-Georgia Cooperation: A Shared Engagement

Nicola Iorio

Abstract An overview of the cooperation between NATO and Georgia, encompassing a brief historical overview, recent developments and the main partnership tools. In this framework, a major role is played by Italy on several fields, including through: contribution to the Defence Institution Building School (DIB) and the Join Training Evaluation Centre (JTEC) within the Substantial NATO-Georgia Package (SNPG); presence on the Black Sea – activities and port visits of the Italian Navy under NATO flag – and further projects in the pipeline (*Carabinieri*'s expected training activities). Focus on NATO-Georgia cooperation within the Science for Peace and Security (SPS) Programme, recalling some major initiatives implemented in this framework, including the project on the responses to seismic emergencies in the Caucasus region and the project – co – directed by the University of Milan-Bicocca and the Ivane Javakhishvili Tbilisi State University – about the geological risks at the Enguri hydroelectric plant. Win-win nature of the cooperation between Tbilisi and the Alliance.

Throughout its entire life as an independent Nation, Georgia has developed one of the most intense and substantial partnerships with the North Atlantic Alliance. Joining the North Atlantic Cooperation Council – from 1997 the Euro-Atlantic Partnership Council – and the Partnership for Peace Framework in 1994, Tbilisi holds frequent political dialogue with the Alliance within the framework of the NATO-Georgia Commission, as well as ambitious practical cooperation, through a number of different tools. From joint exercises to scientific joint programmes, not to mention outstanding participation in NATO's missions and operations. Unsurprisingly, Tbilisi is one of the few NATO Partners to enjoy a privileged status

N. Iorio (✉)
Italian Embassy, Tbilisi, Georgia
e-mail: nicola.iorio@esteri.it

© Springer Nature B.V. 2021
F. L. Bonali et al. (eds.), *Building Knowledge for Geohazard Assessment and Management in the Caucasus and other Orogenic Regions*, NATO Science for Peace and Security Series C: Environmental Security,
https://doi.org/10.1007/978-94-024-2046-3_2

as an enhanced Opportunity Partner (eOP), whereby it is a recipient of special reform assistance programs.

Georgia often participates in North Atlantic Council meetings at the ministerial level. Most recently, the Minister of Foreign Affairs, Mr. Zalkaliani was in London in order to follow the works of the London Summit of NATO's Heads of State and Government (HoSG) on the occasion of the Alliance's seventieth anniversary, and to participate in part of the programme – the "*NATO Engages*" outreach event. In London NATO's Secretary General, Jens Stoltenberg, reaffirmed the importance of partnership with Georgia and the enduring relevance of the decision taken 11 years ago, at the 2008 Bucharest Summit, according to which Georgia one day [...] "will become member of NATO" when conditions allow. For the time being, the issue of NATO membership is not on the agenda of discussions within the Alliance. However, this has not hampered continuous support by all the Allies concerning the Euro-Atlantic integration process undertaken by the Caucasus Country.

NATO-Georgia Substantial Package, launched at the 2014 Wales Summit aimed at strengthening Georgia's capabilities in the field of security and defence – further enriched at the Warsaw (2016) and Bruxelles (2018) NATO HoSG Summits – is a perfect example of this common effort. And it offers to the Allies ground for cooperation in several fields of common interest. In this framework, Italy is playing a major role, praised on several occasions by the Georgian authorities. Rome contributed to the creation of a *Defence Institution Building School* (DIB) and routinely organizes capacity building and "train the trainers" courses (lastly by October 2019) performed by personnel from the Italian Army. Substantial as well is the Italian contribution to the *Joint Training and Evaluation Centre* (JTEC) one of the most important NATO-Georgian initiatives. Recently, thanks to the contributing Allies, the Georgian Minister of Defence Garibashvili hosted at JTEC's headquarters a ceremony for the delivery of transport vehicles, off-road vehicles and ambulances to the Centre (more are expected in the near future).

Further initiatives are in the pipeline right now, including a project aimed at training a battalion of Georgian Military Police, whose implementation will be carried out by Italian *Carabinieri*, under the auspices and the financing of NATO's *Defence and Capacity Building Trust Fund*. Cooperation between Italy and Georgia at the bilateral level – within the framework of the 1997 agreement on cooperation in the field of defence and related MoUs – is another key driver for common initiatives that could be further enhanced in the future.

Furthermore, Italian presence on the Black Sea through situational awareness and training initiatives contribute to foster NATO's projection of stability in the region, and the Italian role as a security provider at a global scale. The recent visit to the port of Batumi of the supply ship of the Italian Navy "*Vesuvio*" – as flagship of the Naval Group "*Standing NATO Mine Countermeasure Group 2*" – entailed not only emotional impact, but also training and capacity building activities for the Border Police of Georgia.

The cooperation between NATO and Georgia is also expressed through the *Science for Peace and Security* (SPS) *Programme*, aimed at promoting dialogue and cooperation between Allies and Partners in the field of scientific research and technological innovation. Italy has a special affection for SPS, that originated from the "three wise men report" elaborated by the Foreign Ministers of Norway (Halvard Lange), Canada (Leaster B. Pearson) and Italy (Gaetano Martino) in 1956 and focused on the need to strengthen the political dimension of the Alliance. The programme perfectly embodies the so-called "third dimension" of the Alliance, to the extent that it cannot be limited solely to military cooperation, but must have a social dimension, which includes the world of science and research. Joint programmes must be open at least to one NATO Partner Nation and include tools not only capable of "securitizing" the recipients, but also of driving their peaceful development. Among the priorities of the SPS programme, there are crucial areas of intervention such as the fight against terrorism, cyber defence, the protection of critical infrastructures in the energy sector, as well as the development of innovative technologies in the field of security.

During the last five years, the Program has initiated more than 450 cooperation activities, more than 40 of which are implemented in Partner Countries. The objective is always coupling scientific merits with addressing tangible security needs. Flexibility, adaptability to different areas and openness towards the private sector are the very strength of the programme. The above mentioned *Defence and Related Security Capacity Building* (DCB) *Initiative*, currently involving activities in Georgia, Iraq, Jordan and Moldova, can be developed through SPS projects, too.

Among the SPS projects developed in collaboration between Italy and Georgia, two are especially noteworthy. The first one has been already completed: starting in 2008, it focused on responses to seismic emergencies in the Caucasus region, with participation of experts from Georgia, Armenia and Azerbaijan and – from the NATO side – Italy, USA, Turkey, Canada and Greece, who conducted research related to capacity response of buildings and infrastructures in the face of strong seismic events. The project contributed to the creation of a *Data Acquisition Centre*, which also allowed a rapid exchange of information between the three countries and whose functions are still used.

Another key project, completed in the Spring of 2019, was co-directed by the *University of Milan – Bicocca* and the *Ivane Javakhishvili Tbilisi State University*. Its main focus was on geological risks at the Enguri hydroelectric plant, which is a fundamental infrastructure in light of the fact that it supplies around 75% of the electricity in Georgia, and has a very sensitive strategic location being very close to separatist region of Abkhazia. For these reasons, accidents at the dam could potentially have direct consequences for the social and political stability of Georgia and the entire Caucasus region. The experts, also from the United States, United Kingdom, Azerbaijan and Kazakhstan, conduct research in this regard focused on the potential implications of seismic events, landslides and the release of pollutants into the atmosphere.

Win-win initiatives, allowing Georgia to strengthen its capabilities and Allies to develop and share best practices, while both sides fostering common know how and expertise. With much room for improvement when it comes to resources devoted by NATO and Partner countries to SPS initiatives – hopefully in coordination with International Organizations such as the EU – and further engagement of the private sector. Because science and peace are the best ingredients to be secure.

Chapter 3
An Overview of the 20-Year Collaboration Between NATO and Earth Scientists to Assess Geohazards in the Caucasus and Other Critical Regions

Federico Pasquaré Mariotto, Alessandro Tibaldi, and Fabio L. Bonali

Abstract We hereby provide an overview of four multi-year projects on geohazard assessment and mitigation, carried out under the umbrella of the North Atlantic Treaty Organization (NATO), with the involvement of more than 80 scientists coming from several countries, among which the US, Georgia, Italy, Russia, Azerbaijan. The projects have been aimed at enhancing the security of people and the safety of vital infrastructures as well as facilitating cooperation between scientists from NATO and non-NATO countries. The study areas are located in the Caucasus (Georgia), in Central Asia (Kyrgyzstan) and Northeast Asia (Kamchatka).

Our work clearly demonstrates how Earth Science can contribute to improving scientific collaboration among countries that are politically in tension; moreover, geoscience can play a key role in preventing situations that may escalate into conflicts. This paper showcases the main results of the NATO-funded projects, both in terms of their scientific relevance and their geopolitical importance.

Keywords NATO · Enguri · Toktogul · Pipeline · Dam · Caucasus

3.1 Introduction

Natural hazards affecting human life, settlements and infrastructures severely affect local areas and small communities up to entire countries or regions. As far as seismic

---

F. Pasquaré Mariotto (✉)
Department of Human and Innovation Sciences, Insubria University, Como, Italy
e-mail: pas.mariotto@uninsubria.it

A. Tibaldi · F. L. Bonali
Department of Earth and Environmental Sciences, University of Milan Bicocca, Milan, Italy

CRUST-Interuniversity Center for 3D Seismotectonics with Territorial Applications, Chieti, Italy

© Springer Nature B.V. 2021

F. L. Bonali et al. (eds.), *Building Knowledge for Geohazard Assessment and Management in the Caucasus and other Orogenic Regions*, NATO Science for Peace and Security Series C: Environmental Security,
https://doi.org/10.1007/978-94-024-2046-3_3

hazard is concerned, this is a very well-known threat across the whole Caucasus region, which was hit in recent times by the M_W 6.8, 1998 earthquake in Spitak, Armenia (Gabsatarova and Baranov 2018), the M_W 6.9, 1991 (Varazanashvili et al. 2018) and by the M_W 6.0, 2009 Racha (Georgia) earthquakes (Nikolaeva and Walter 2016) as well as the M_W 6.3, 2000 Baku earthquake in Azerbaijan (Babayev et al. 2010; Babayev and Telesca 2014). Furthermore, as reported in Varazanashvili et al. (2018), damage from earthquakes in the Caucasus region usually affects more than one country, crossing political borders.

The four projects presented here involve studies on geohazard sources such as earthquakes, landslides and volcanism, and their possible relationships. The projects, conducted during the last 20 years, have been focused on assessing geohazards affecting lifelines (e.g. hydroelectrical power plants, oil and gas pipelines), which represent a mail concern as their disruption may result in major economic losses to millions of people. Another common feature shared by the hereby-showcased projects is that the selected areas are located at the border between countries where damage to infrastructures may produce international tension, which can eventually escalate and lead to turmoil and conflicts.

Over the past decades, the Science for Peace and Security (SPS) Programme, promoted by the North Atlantic Treaty Organization (NATO), has been funding several multi-year projects in Earth Sciences (Tibaldi et al. 2018), aimed at enhancing scientific cooperation and knowledge exchange among institutions from different countries. The NATO multi-year research projects presented here are the following (Fig. 3.1): *i)* G4934 SFP Project "Security against geohazards at the major Enguri hydroelectric scheme in Georgia" – in the Republic of Georgia; *ii)* CLG Project 982,957 "Volcanic hazards and countermeasures for Georgian section of Caspian pipelines" – in the Republic of Georgia; *iii)* 983,142 SFP Project "Geo-environmental security of the Toktogul hydroelectric power station region, central Asia" – in the Kyrgyz Republic (Kyrgyzstan); *iv)* CLG Project 978,989 "A multi-disciplinary approach to recent geologic catastrophes at subduction zones" – in the Kamchatka Peninsula (Russia).

Three out of four of the above research efforts involve countries that were made independent by the dissolution of the former Soviet Union and focus on geological-related threats to major energy-related infrastructures; the fourth project was especially related to regional hazards. The purpose of the first three projects was to improve our understanding of the geological hazards that might negatively affect the safety of an infrastructure of high international societal and political relevance, and suggest mitigation measures designed to prevent possible geopolitical issues.

With the present effort, we offer an overview of the projects carried out under the aegis of NATO, so as to highlight the vital role played by the cooperation between scientists and international organisations in assessing natural hazards and contributing to peacekeeping. First of all, we provide, from western to eastern regions (Fig. 3.1), a brief description of the NATO-funded Projects, along with the names of the involved institutions, the main participants, the projects' relevant outcomes, and the main suggestions for risk mitigation.

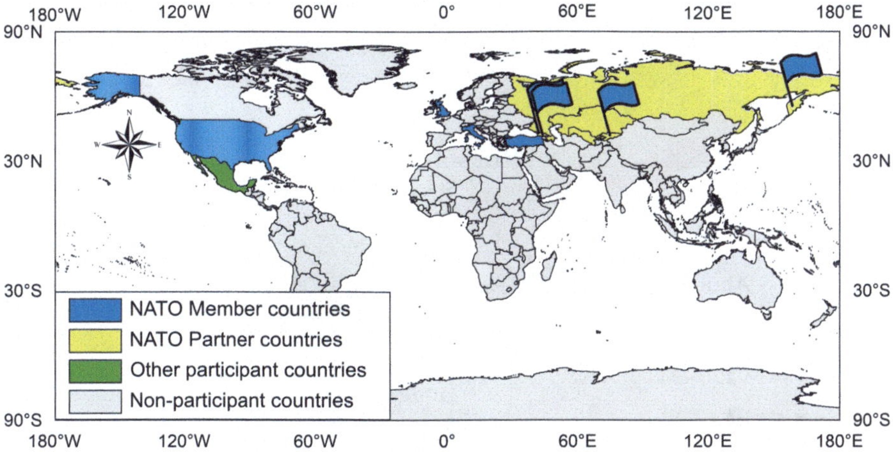

Fig. 3.1 The here described NATO-funded projects, with indication of NATO Countries (blue), NATO Partner countries (yellow) and other countries (green). Project locations are indicated by flags. Country borders are from © 2018 GADM (https://gadm.org/), version 3.6, released on 6 May 2018. NATO Member countries: Italy, the United States of America, the United Kingdom and Turkey; NATO Partner countries: Russia, Kazakhstan, Kyrgyzstan, Uzbekistan, Georgia and Azerbaijan. Others: Mexico. Reference system: WGS84 / geographic coordinates

3.2 Project CLG 982957 – Volcanic Hazards and Countermeasures for Georgian Section of Caspian Pipelines

3.2.1 Summary of the Project

This project was carried out through NATO Collaborative Linkage Grant Project 982,957, with Project Coordinator Prof. Federico Pasquaré Mariotto, Insubria University (Italy). Other Principal Investigators were Luigina Vezzoli, Insubria University; Avtandil Okrostsvaridze, Georgian Academy of Sciences, Tbilisi (Georgia); Bejan Tutberidze, Tbilisi State University, Tbilisi (Georgia); Daniel Tormey, ENTRIX Inc., Ventura, California (USA) and Salih Bayraktutan, Ataturk University, Ankara (Turkey). The project, which lasted from 2007 to 2009, was aimed at assessing volcanic and seismic hazards threatening the Georgian section of the strategic Caspian oil and gas pipelines (Fig. 3.2) and coming up with mitigation and protection measures. In the study area, the Baku-Tbilisi-Ceyhan (BTC) oil pipeline and the Baku-Tbilisi-Erzurum (BTE) gas pipelines run parallel (Aras and İşeri 2009). Along 70 km of this stretch, the pipelines lie within the northern part of the Javakheti Plateau (Figs. 3.2 and 3.3a). This 2500 km^2 region is a volcanic plateau (>1500 m a.s.l.) with up to 100 major and minor volcanic centres of Pliocene-Quaternary age (Lebedevet al. 2003; Chernyshev et al. 2006; Pasquaré et al. 2011). The youngest volcanic area in the plateau is the 40-km-long Abul-Samsari Volcanic

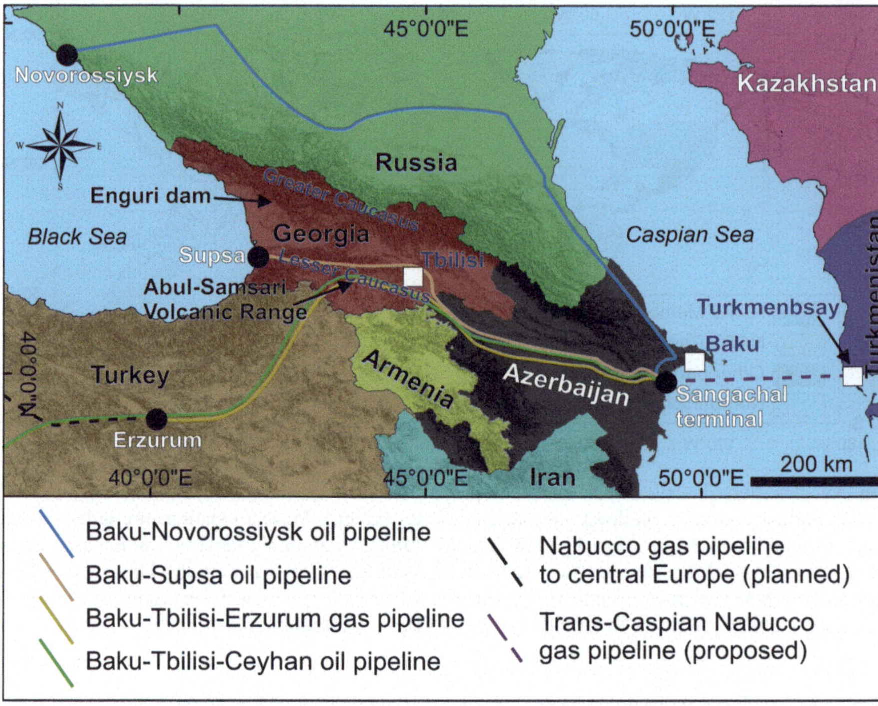

Fig. 3.2 Map showing all pipelines in the Caucasus region, country borders and study areas of two of the described NATO-supported projects. Borders are from GADM data, © 2018 GADM available at https://gadm.org/, version 3.6 (released on 6 May 2018). Reference system: WGS84 / geographic coordinates

Range (ASVR) (Fig. 3.3a), dating to the Pleistocene and Holocene times (Lebedev et al. 2003). The BTC and BTE pipelines cross the ASVR, with most of the volcanic risk posed by the Tavkvetili and Shavnabada volcanic edifices (Fig. 3.3a). The area is also seismically active, as testified by the recent, 1986, M 5.6 Paravani earthquake (Fig. 3.3a). Moreover, there are several dip-slip and strike-slip faults with Holocene displacement. The NATO-supported team, between 2007 and 2009, undertook an extensive field tectonic research aimed at finding evidence of active faults in the area, as well as defining their kinematics and identifying the possible seismic hazard affecting the pipelines.

3.2.2 *Why this Project*

When the route of the BTC and BTE was planned, the potential seismic and volcanic hazard affecting their section across the Javakheti volcanic plateau in Southern Georgia were not considered in sufficient detail. In fact, prior to the onset of these

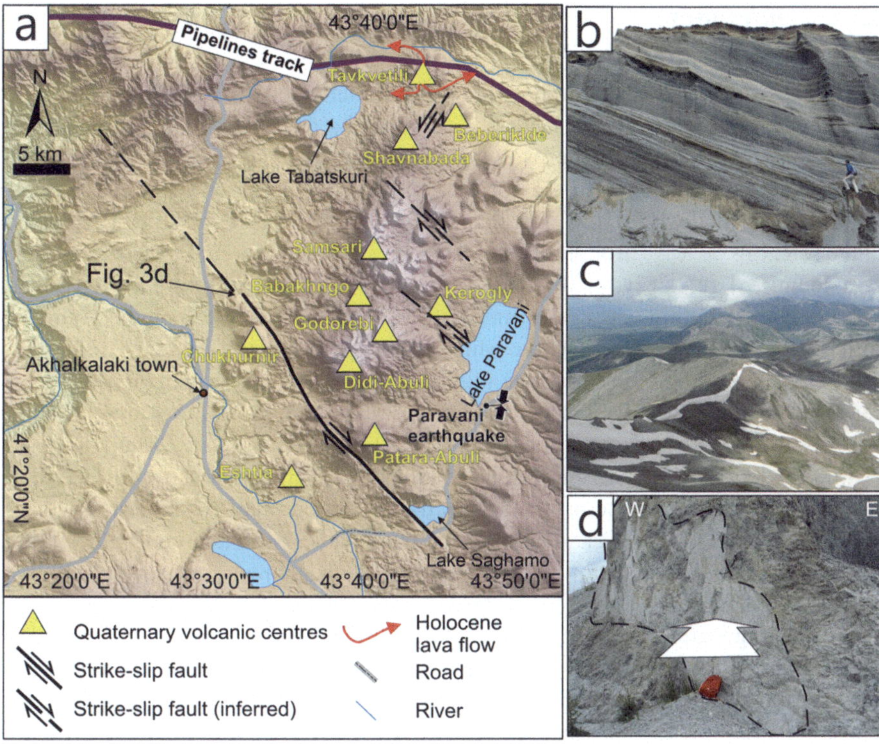

Fig. 3.3 (**a**) Shaded view of the ASVR, showing the main faults with indication of the related stress regime; P-axis from the Paravani earthquake is also reported (Heidbach et al. 2010, 2018). Post-glacial lava flows from Tavkvetili volcano are also highlighted. Modified after Pasquaré et al. (2011). Reference system: WGS84 / geographic coordinates. (**b**) The interior of a recent, monogenetic edifice. (**c**) Panoramic view of the ASVR from the top of Didi Abuli volcano, the highest in the range. (**d**) Picture of the NW-directed fault plane and indication of N-S-directed greatest principal stress, backpack for scale. A few tens of meters north of this outcrop, the fault dissects Holocene-age alluvial sediments

project, there were no publicly available scientific reports featuring an analysis of the potential geological-related extreme events that may negatively affect oil and natural gas transportation through this environmentally sensitive and geologically active area. In view of this, the main goal of the project has been to fill this critical gap, and try to come up with an overview of the geological threats that may affect these strategic lifelines, as well as individuate adequate remediation measures.

3.2.3 *Outcomes*

The main results of this research effort are summarized below; more data can be found in Pasquaré et al. (2011). Firstly, through the analysis of the aerial photos, SPOT 5 satellite images (10-m resolution), DEMs with 20 m resolution, 1: 50,000 scale topographic maps, and field surveys, the NATO-supported team of scientists performed an analysis of the overall architecture of the Abul-Samsari Volcanic Range (e.g. Fig. 3.3c). Through geological field surveys, different types of eruptions were identified (e.g. Fig. 3.3b) and evaluated for expected future activity, and their likely distribution was determined. Efforts were concentrated on the youngest edifices, Tavkvetili and Shavnabada, which are the closest to the pipelines track (Fig. 3.3a). Geomorphologic observations indicate the absence of periglacial activity on their slopes, and the well-preserved summit craters suggest that volcanic activity probably postdates the last glacial retreat (< 10 Ka BP). In terms of tectonic and seismic-related risks, the team's main result was the characterisation of a major, NW-SE trending fault system, which was tracked in the field for 25 km, as far as its NW tip (Fig. 3.3a). A great number of fault planes was analysed, all bearing outstanding evidence of right-lateral strike-slip displacement and recent activity. Fault plane striations were used to derive stress tensor orientations, with results univocally suggesting a NNE-SSW to N-S directed σ_1 (Fig. 3.3d). Moreover, the analysis of aerial photos, DEMs and satellite images indicate the NW-ward prolongation of the major fault for another 20 km (totalling 45 km in length), very close (12 km) to the pipeline track (Fig. 3.3a). The NATO-supported team concluded that the documented strike-slip stress regime (and the related trend of the horizontal σ_1) can indeed be regarded as the one presently affecting the pipeline area. In terms of volcanic hazard, the above mentioned ~NNE-SSW trending greatest principal stress (σ_1) may bear serious implications for volcanic reactivation, that might take place as fissural eruptions and emplacement of localized vents along a NNE-SSW, tectonically-controlled vector. The hazard assessment and protective strategies developed in the course of the research may be applicable to other areas of recent volcanism where oil and gas pipelines must be designed and constructed.

3.3 Project G4934 – Security Against Geohazards at the Major Enguri Hydroelectric Scheme in Georgia

3.3.1 *Summary of the Project*

This project was carried out through the NATO SPS Programme – Project G4934, the Project Coordinator was Prof. Alessandro Tibaldi, University of Milan-Bicocca (Italy). Other Principal Investigators were Dr. Nino Tsereteli, M. Nodia Institute of Geophysics, Tbilisi State University, Tbilisi (Georgia); Prof. John Gierke, Michigan Technological University, Houghton, MI (USA); Prof. Gulam Babayev,

Geology and Geophysics Institute (previously, Institute of Geology), Azerbaijan National Academy of Sciences, Baku (Azerbaijan); Prof. Aigul Sultanbekova, Kazakh Polytechnic Institute, Almaty (Kazakhstan); Prof. Derek Rust, University of Portsmouth (UK).

This research was carried out, from 2015 to 2018, by a group of 25 scientists from Italy, the United States, the United Kingdom, Switzerland, Georgia, Azerbaijan and Kazakhstan. The project was focused on assessing the vulnerability to geohazards of the largest hydroelectrical facility in the Republic of Georgia – the Enguri dam and associated reservoir (Figs. 3.4a–c), identifying possible geohazard scenarios, identifying preventive measures, and conducting training activities targeted to young scientists (Figs. 3.4d, e). The Enguri dam (the world's sixth highest) is part of the Enguri Hydroelectric power station (EHS), partially located in Abkhazia, the separatist region of Georgia. The possible collapse of a portion (or the whole) of the major "Khoko" landslide, located along one side of the reservoir (Figs. 3.4a, b), in the reservoir, might trigger a major catastrophe. Despite the very steep slopes surrounding the lake and the strong seismicity of the region surrounding the lake, no quantitative assessment of slope stability had been performed, prior to 2015, under dynamic conditions, i.e. under earthquake-induced shaking.

3.3.2 Why this Project

Problems at the Enguri reservoir would affect the energy distribution across Georgia, because most of the energy to the country is supplied by this plant. The same problem would affect Abkhazia, whose entire energy supply depends on this. Moreover, possible flooding from the reservoir would directly affect eastern Abkhazia. A stop in energy production may cause severe economic losses in both areas and increase local societal and political instability, leaving millions of residents and several tens of infrastructures such as hospitals and military bases without energy supply. This might have the effect of worsening the already tense relations between Abkhazia and the Georgian government. The situation is rooted in the status of Abkhazia, a region that was autonomous within Georgia during the Soviet Union era and claimed independence when the Soviet Union began to disintegrate. This erupted into the War of Abkhazia in 1992, which left 13,000 to 20,000 Georgians and 3000 Abkhaz killed, with about 250,000 Georgians forced into internal displacement as refugees (Tibaldi and Pasquare Mariotto 2017). This was followed by another conflict in 1998, when hundreds of Abkhaz militants entered villages of the Gali district, still populated by Georgians, leading to hundreds of casualties from both sides and further 20,000 Georgian refugees. There was another crisis in 2001, followed by the dramatic 2008 conflict, during which Russian troops coming from Abkhazia occupied the city of Zugdidi, ran raids on military bases located deeper within Georgia, and occupied the important harbour of Poti, along the Georgian Black Sea coast. The administrative line between Abkhazia and Georgia was moved eastward to the Enguri River, which still marks most of the de facto present-day border. The

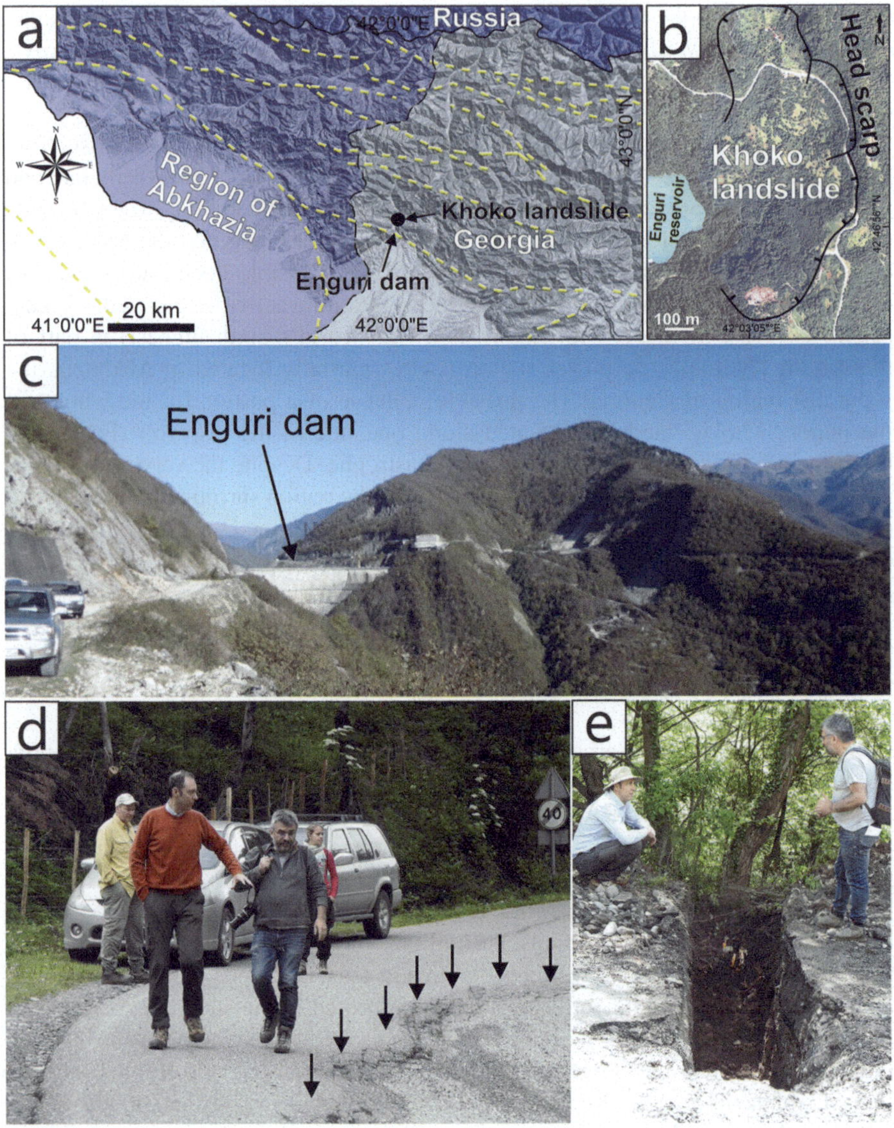

Fig. 3.4 (**a**) Location of the Enguri dam and Enguri reservoir in the Greater Caucasus; Country borders are from GADM data, © 2018 GADM available at https://gadm.org/, version 3.6 (released on 6 May 2018); yellow dashed lines represent the main active faults in the region, from the EMME Project (Gulen et al. 2011). Reference system: WGS84 / geographic coordinates. (**b**) Sketch map of the Khoko landslide (after Tibaldi et al. 2019). (**c**) Panoramic view of the Enguri dam. (**d**) Project director, Prof. Alessandro Tibaldi, left, and Dr. Paolo Oppizzi, discussing the extent of the landslide-induced road deformation, highlighted by black arrows. (**e**) Preparation of a trench for the installation of extensometers suitable for monitoring the evolution of the landslide

residents at the foothills of the Enguri dam are not only troubled by the political and economic situation, but they are also worried about the potential threat posed by earthquake hazard, as well by the possible collapse of landslides into the artificial lake.

To address the above major issues, this NATO-supported effort was intended to assessing geohazards affecting the reservoir and modelling future consequences brought about by the possible collapse of the major Khoko landslide, also in connection with possible earthquakes in this tectonically-active region. These goas were especially relevant because the Enguri hydroelectrical plant is a key Georgian infrastructure: in the event of a landslide occurrence, the hydroelectrical plant would temporarily be out of commission, with major problems to the whole country in terms of energy supply and national productivity. Moreover, the collapse of rock debris into the artificial lake would produce the partial infilling and long-term decrease in the capacity of the hydroelectric plant.

3.3.3 Outcomes

With regard to seismic hazard, through the combination of field geological, geomorphological and structural data with seismic reflection geophysical data, it has been possible to search for pre-historic surface fault ruptures in an area within a 45-km radius from the Enguri dam. As a result of this effort, it has been possible to document that the area near the dam is punctuated by strongly entrenched rivers, whose presence suggests the occurrence of major tectonic uplift. South of the dam there are widespread fluvial deposits dating from late Miocene to present times. This succession can be interpreted as evidence of a major uplift of this part of the Caucasus, followed by erosion, as proved by fission-track studies (Vincent et al. 2011). Moreover, the oldest portion of the fluvial succession shows evidence of tectonic folding, especially in correspondence of the Tsaishi fold, 45 km south of the Enguri dam, as documented by Tibaldi et al. (2017). Historical earthquake data enable locating the historic Tsaishi earthquake of 1614 AD at the south-western foot of the fold (Varazanashvili et al. 2011), with an epicentral intensity of 8–9 and magnitude (Ms) 6.0. According to Tibaldi et al. (2017), the Tsaishi is a fault-propagation fold linked with the seismic activity along a reverse fault dipping to the NE, which connects northward with a thrust system made of ramps and flats underneath the Enguri dam site.

As far as landslide hazard is concerned, the eastern slopes of the Enguri reservoir are affected by several active landslides. No study had been published on these landslides before the 2015–2018 NATO project. The quantitative evaluation of the static and dynamic behaviour of these landslides under different possible scenarios, as well as of other instability-prone slopes around the reservoir, has been performed with the aim of preventing and mitigating the effects of potential landslide-related events. In order to tackle this hazardous situation, the first numerical model of stability of the most critical slope surrounding the reservoir was developed, based on the integration of field geological surveys with geotechnical tests and drill logs from the Soviet era, which were performed in the area now covered by the reservoir. Moreover, in order to continuously monitor the evolution of instability-related processes in the area, digital extensometers were deployed in the late fall of 2016, the purpose of which has been to assess slope movements with a 0.1-mm resolution. Global Positioning System benchmarks were also installed, which have been measuring, at 6-month intervals, the displacement of the upper portion of the landslide. The results of the numerical analyses suggest that the slope is stable under normal static, gravity conditions, but that it becomes strongly unstable under dynamic conditions, such as those produced by an earthquake with a M comparable to local seismic events in the past.

3.4 Project SFP 983142 – Geo-Environmental Security of the Toktogul Hydroelectric Power Station Region, Central Asia

3.4.1 Summary of the Project

This project was carried out through NATO SPS Programme 983,142; the Project Coordinator was Prof. Derek Rust, University of Portsmouth (UK). The Other Principal Investigators were Prof. Alessandro Tibaldi, University of Milan-Bicocca (Italy) and Prof Andrey Korjenkov, Inst. Communication Information Technologies (Kyrgyzstan). This 2008–2013 NATO-funded project involved also a team from Uzbekistan, which later resigned due to country-internal political reasons.

The project had the purpose of assessing potential geo-environmental hazards affecting the Toktogul Hydro-Power Station (HPS), a key infrastructure for Kyrgyzstan and the neighbouring central Asian republics (Fig. 3.5a). The HPS dam, a 215-m-high concrete arch structure, impounds the Naryn River to produce the Toktogul reservoir, the largest of its kind in Central Asia (surface area 284 km^2, volume 19.5 km^3, production capacity 1200 MW). The well-known, NW-SE striking, regional Talas-Fergana right-lateral strike-slip fault (TFF) transects the Tien Shan orogenic belt and crosses the reservoir (Fig. 3.5). Along its ~700 km of length, the TFF is marked by the typical features associated with paleoseismic deformation, such as scarps, offset streams and drainage divides, as well as aligned landslides (Burtman et al. 1987, 1996; Trifonov et al. 1992, 2015; Korjenkov et al. 2012, 2013, 2014; Tibaldi et al. 2015; Rust et al. 2018).

The NATO-supported team conducted extensive fieldwork, as well as paleoseismic and slope stability analyses (e.g. Figs. 3.5b–e), which enabled documenting major clues to historical and Holocene ground-rupturing earthquakes. Further studies comprised detailed interpretations of aerial photos and new detailed (50-cm resolution) digital elevation models. The team also carried out a landslide hazard

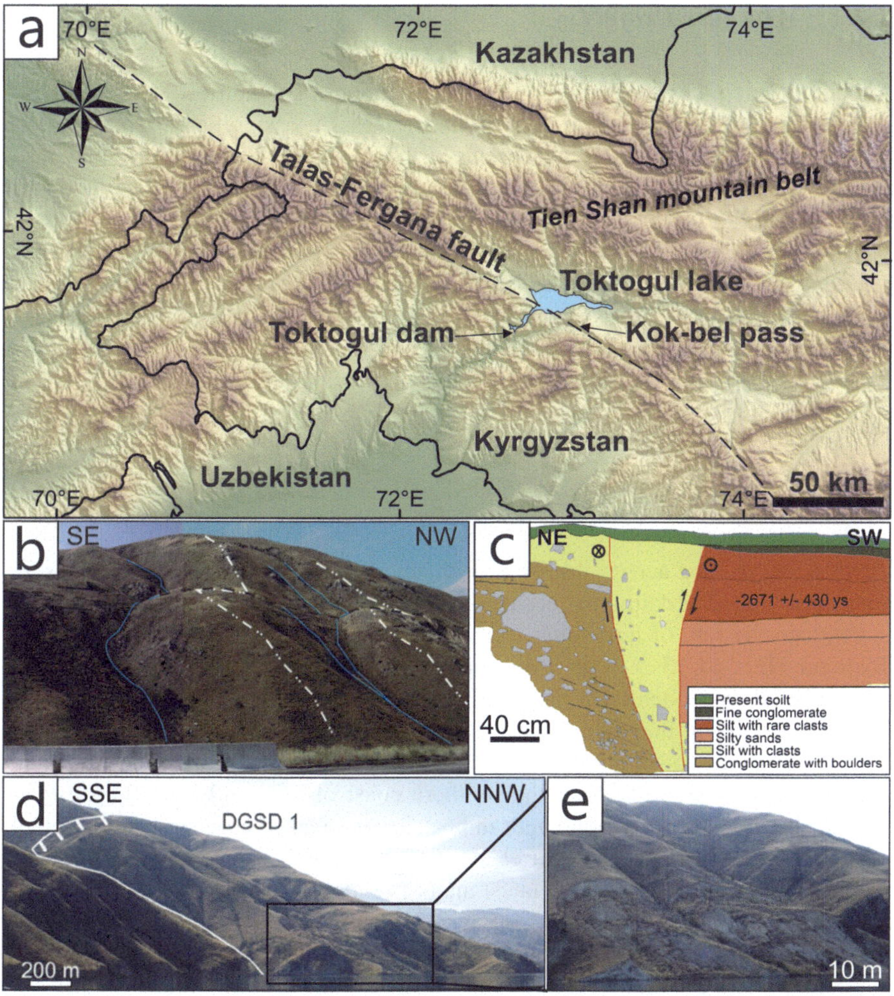

Fig. 3.5 (**a**) Location of the studied area and of the Toktogul reservoir and dam. The TFF trace is recognizable, highlighted in this shaded view of western Tien Shan mountain belt. Country borders are from © 2018 GADM (https://gadm.org/), version 3.6, released on 6 May 2018. Reference system: WGS84 / geographic coordinates. (**b**) Example of the TFF segment located southeast of the Toktogul water reservoir, where its continuous surface trace (segmented black line) and the systematic right-lateral strike-slip offset of gullies (blue lines) and water divides (white point-and-dash lines) can be clearly observed. (**c**) Interpreted log of one of the several trenches excavated along the TFF trace (here cutting a Quaternary alluvial fan). Age obtained by Optically-Stimulated Luminescence. (**d**) Photo of a major deformed slope (DGSD 1) facing the Toktogul reservoir. (**e**) Photo of an active landslide located at the foot of the DGSD 1

assessment near the HPS. Both geohazard sources (earthquakes and landslides) were recognized as significant potential threats to the Toktogul region.

3.4.2 Why this Project

Already existing tensions between the two neighbouring countries, Kyrgyzstan and Uzbekistan, may escalate into a crisis, in the case of a significant landslide into the water reservoir, triggered by a strong earthquake. The above scenario may have a major impact on operations at the Toktogul HPS, and may lead to variations in water availability for the Uzbek region and local population downstream of the dam. This, in turn, may lead to the activation of local religious and nationalistic elements. The Fergana Valley is a "powder keg" that could be ignited by a natural disaster affecting the HPS, possibly triggering a regional war (Chellaney 2013). Therefore, the assessment of earthquake and landslide hazard in the Toktogul region has a major scientific interest, as well as crucial geopolitical implications.

3.4.3 Outcomes

Trenches were dug across the TFF, with the purpose of applying paleoseismic techniques aimed at dating the occurrence of prehistoric earthquakes. This enabled to estimate the rate of horizontal tectonic displacements along the fault, near the Toktogul HPS, at 13 ± 1 mm/yr. Further paleoseismological studies along the fault allowed us to identify a fault segment that was reactivated during two major earthquakes that hit the area in the 14th and 16th centuries A.D. (Korjenkov et al. 2013). The magnitude of the sixteenth century event was at least 7 and the associated seismic intensity was IX. Near the southern coast of Toktogul lake, the offset of several alluvial fans of different ages indicates that repeated coseismic faulting took place in Holocene times (Tibaldi et al. 2015). A trench, opened across this fault segment, has offered clues to the latest major pre-historic earthquake that struck the area after 2617 ± 430 year ago (e.g. Fig. 3.5c), with a moment Magnitude as much as 6.9. Therefore, it is worth pointing out that other strong earthquakes might occur in the future along the studied segment of the regional strike-slip fault.

Moreover, several huge rockslides were recognized along the TFF, near the Toktogul reservoir (Korjenkov et al. 2014; Tibaldi et al. 2015). Although significant landslides aligned along an active fault are not necessarily always coeval with earthquakes, the presence of a major fault might represent a mechanical weakness zone, responsible for new slope failures, as documented along major faults in the Alps (Tibaldi and Pasquaré 2008; Pasquaré Mariotto and Tibaldi 2016) and elsewhere (Tibaldi and Ferrari 1992; Tibaldi et al. 1995).

Finally, limit equilibrium methods were applied, focused on the south-western bank of the Toktogul reservoir, and a numerical analysis was performed, intended to assess possible instability factors and slope failure mechanisms (Tibaldi et al. 2015). The numerical analyses suggest possible slope instability in the case of seismic shaking, and underscore the necessity to carry out further studies in the area.

3.5 CLG Project 978989 – A Multidisciplinary Approach to Recent Geologic Catastrophes at Subduction Zones

3.5.1 Summary of the Project

This project was carried out through NATO Collaborative Linkage Grant – Project CLG 978989; the Project Coordinator was Prof. Alessandro Tibaldi, University of Milan-Bicocca (Italy). The Other Principal Investigators were: Vera Ponomareva, Institute of Volcanic Geology and Geochemistry, Far East Division, Russian Academy of Sciences (Russia); Philip Kyle, New Mexico Institute of Mining and Technology (USA); Jonathan M. Lees, Dept. of Geological Sciences, University of North Carolina (USA); Andrei Ivan Kozhurin, Geological Institute of Moscow (Russia).

This multidisciplinary, Russian-US-Italian effort, between 2003 and 2005, was aimed at understanding the relationship of eruptive activity with major tectonic earthquakes in subduction-related volcanic arcs, with a focus on Kamchatka (Russia). The central part of this Peninsula is characterized by a well-developed depression associated with active volcanism, aligned in a NE–SW direction. On the east, the depression is bounded by a prominent system of active faults known as the East Kamchatka Fault Zone (EKFZ) (Tibaldi 2004; Kozhurin et al. 2006). In order to improve understanding of the behaviour and kinematic role of this fault zone a fieldwork programme, including paleoseismological study of trenches, was conducted in the north-central part of this system (e.g. Fig. 3.6b).

3.5.2 Why this Project

The Kamchatka Peninsula is one of the most geologically active regions in the world, as it lies at the northwestern margin of the Pacific plate (Gorbatov et al. 1997). In Kamchatka there is a major seismicity, associated with the subduction of the Pacific plate at a rate of 9–10 cm/year (Geist and Scholl 1994). In historical times, three major earthquakes struck off the coast of the Kamchatka Peninsula, in 1737 (Mw 9.0), 1923 (Mw 8.5) and 1952 (Mw 9.0). Each can be classified as a megathrust earthquake and triggered tsunamis. The 1952 earthquake, for example, produced a large tsunami that brought about destruction and loss of life around the Kamchatka peninsula and the Kuril Islands, reaching as far as the Hawaiian Islands, where damage was estimated at one million USD. Beside seismic hazard, Kamchatka is home to about 30 active volcanoes and hundreds of monogenetic vents. Recent volcanism is highly explosive and is marked by a major magma output rate (Fedotov and Masurenkov 1991).

This NATO-supported project focused on this geohazard-prone region for both scientific and geopolitical reasons. From a scientific standpoint, research aimed at individuating, mapping and characterising active faults above subduction zones, especially in relation to Holocene volcanic arcs, are key to shedding light on tectonic processes taking place along plate boundaries. Moreover, as major earthquakes can induce volcanic eruptions (Bonali et al. 2013, 2015), this type of studies is also helpful in assessing the relations between seismicity and volcanic activity at volcanic arcs. Moreover, at the surface, faulting and dyking deform volcanic edifices, favouring sector collapses (e.g. Tibaldi et al. 2006; Tibaldi et al. 2008; Vezzoli et al. 2008) and controlling the alignment of vents (e.g. Bonali et al. 2011).

From a geopolitical point of view, the capital of Kamchatka – Petropavlovsk-Kamčatskij – is situated along the coastline, near the subduction trench and at the foothill of the active Avača volcano. The city hosts 181,181 residents and is home to one of the most important nuclear submarine bases in Russia. A major earthquake or volcanic eruption may hit these infrastructures and possibly trigger devastating tsunamis, which may affect several other countries in the Pacific region.

3.5.3 Outcomes

The most relevant scientific outcomes of this project (related to the reconstruction of the paleoseismic events that induced major faulting affecting post-glacial alluvial terraces along a major late Quaternary fault system (with the most notable Kumroch fault) in north-central Kamchatka (Tibaldi 2004; Kozhurin et al. 2006). Trenching across the fault provided evidence for five faulting events, which were found in a Holocene succession made of several distinct and dated tephra layers (Fig. 3.6d). The oldest earthquake was dated to about 10.5 ka BP, and was followed by another event at about 6 ka BP and then another one at about 4.5 ka BP. The two final events were closely spaced in time about 3.2–3.3 ka ago; no further events are recorded thereafter, despite the presence of a complete tephra succession encompassing the most recent, historical eruptions. A possible interpretation is that this fault zone is subjected to pulses of activity: therefore, renewed faulting cannot be ruled out and may be imminent, potentially triggering significant regional, shallow earthquakes (~M 6.5). Such seismic events represent a hazard source that needs to be considered in seismic hazard assessment, in addition to the earthquakes associated with the subduction of the Pacific Plate beneath Kamchatka.

At a regional level, two major faults, including the Kumroch fault, are marked by combined dip-slip and right-lateral displacements (Fig. 3.6c). This suggests that the entire East Kamchatka Fault Zone (EKFZ) may be moving as an active, dextral transtensive zone, with the normal component particularly significant along the segment forming the easternmost margin of the Central Kamchatka Depression (CKD). Movements along this fault zone, together with slip on the Nachiki Transverse Zone further south, may define a regional-scale block moving towards the ESE, separately from the rest of Kamchatka.

It is worth noting that there may be some correlations between faulting events and the activity at the nearest volcanoes. In fact, all faulting events are accompanied by eruptions (although the opposite has never been observed). This may reflect changes

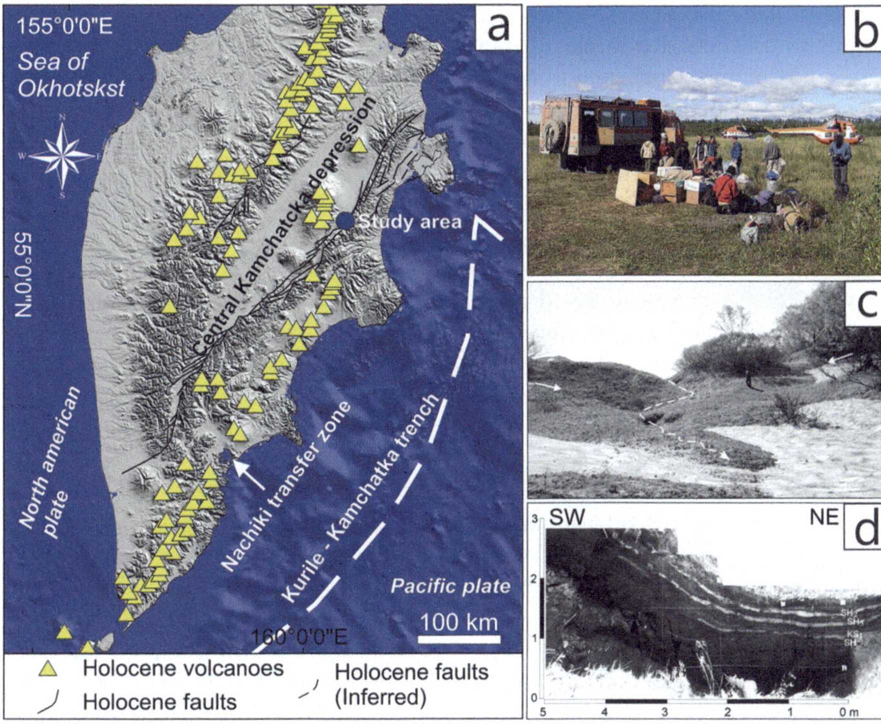

Fig. 3.6 (**a**) Shaded view of the Kamchatka peninsula showing main Holocene faults, the central Kamchatka depression and the main Holocene volcanoes. Blue dot locates the area studied during the NATO-funded Project. Reference system: WGS84 / geographic coordinates. (**b**) The international team during base camp setup. (**c**) Combined, dip-slip and right-lateral offsets of a shallow gully along the major active fault, person for scale. (**d**) One of the trenches dug by hand across the Kumroch fault; notice the vertically offset volcanic horizons and three major faulting events

in the stress field at depth, that have the effect of pushing magma to the surface and induce brittle deformation in the crust.

3.6 Discussion and Final Remarks

By showcasing four NATO-funded projects, which involved 81 scientists, this paper has highlighted the role that Earth Science can play in taking action aimed at reducing the possibility of societal and political crises due to geological hazards and related natural disasters. Definitely, research devoted to assessing natural hazards is not a novelty: Earth scientists have been working on these topics, in cooperation with national civil protection agencies, for decades. Nevertheless, it is certainly worth highlighting that Earth Science can be a key factor in preventing to the effects of natural disasters, which may exacerbate geopolitical tensions, possibly leading to conflicts. Researchers from different scientific backgrounds have been very active in areas subjected to economic, societal or political tensions; however, the potential risks linked with geohazards may be overlooked if Earth scientists are not involved. An example of the above is the Khoko landslide in Georgia, which has been threatening the Enguri reservoir for tens of years, causing extensive damage to the Jvari-Mestia road. Although this hazard source had been recognized for a long time, no monitoring and remediation measures had been put in place. In this case, thanks to the financial support provided by NATO, it has been possible to come up with countermeasures aimed at enhancing preparedness and minimizing the consequences of possible extreme events in the future. This has been made possible by training young local researchers and setting up instruments aimed at monitoring a possible escalation in slope movements. This will enable Georgian public officials to take the most appropriate civil protection actions, in the unfortunate case of an impending landslide event.

Another key aspect is represented by the necessity of establishing close relations with end-users and the relevant stakeholders, from the beginning of the research project, so as to make joint decisions on follow-up activities.

In spite of the major advancements in studying geohazards made over the last two decades, many regions at the worldwide level are still exposed to unrecognized geological-related threats. As in these regions poverty is still a major issue, this greatly reduces the preparedness of local populations toward the possible occurrence of natural, extreme events. Our work emphasizes the importance of building up research teams spanning as many Earth Science disciplines as possible, and conducting research in areas affected by a combination of geohazards and critical geopolitical conditions. In doing so, Earth scientists will show that they can valuably contribute to a safer and peaceful global society.

Acknowledgements The work at Enguri has been carried out with funding from NATO SFP project G4934; The work at Toktogul was developed with funding from NATO SFP project 983142; The studies on the pipeline were conducted with funding from NATO project CLG 982957; The studies in Kamchatka were funded by the NATO CLG Project 978989. Several colleagues from different countries, who collaborated in these research efforts, are acknowledged.

References

Aras B, Iseri E (2009) The Nabucco natural gas pipeline: from opera to reality. SETA Policy Brief 34:1–11

Babayev G, Telesca L (2014) Strong motion scenario of 25th November 2000 earthquake for Absheron peninsula (Azerbaijan). Nat Hazards 73(3):1647–1661

Babayev G, Ismail-Zadeh A, Le Mouel J-L (2010) Scenario-based earthquake hazard and risk assessment for Baku (Azerbaijan). Nat Hazards Earth Syst 10:2697–2712

Bonali FL, Corazzato C, Tibaldi A (2011) Identifying rift zones on volcanoes: an example from La Réunion island, Indian Ocean. Bull Volcanol 73(3):347–366

Bonali FL, Tibaldi A, Corazzato C et al (2013) Quantifying the effect of large earthquakes in promoting eruptions due to stress changes on magma pathway: the Chile case. Tectonophysics 583:54–67
Bonali FL, Tibaldi A, Corazzato C (2015) Sensitivity analysis of earthquake-induced static stress changes on volcanoes: the 2010 M w 8.8 Chile earthquake. Geophys J Int 201(3):1868–1890
Burtman VS, Skobelev SF, Sulerzhitsky LD (1987) The Talas-Fergana fault: recent offsets in the Chatkal district of the Tien Shan. Dokl Akad Nauk SSSR 296:1173–1176
Burtman VS, Skobelev SF, Molnar P (1996) Late Cenozoic slip on the Talas-Fergana fault, the Tien Shan, Central Asia. Bull Geol Soc Am 108:1004–1021
Chellaney B (2013) Water, peace, and war: confronting the global water crisis. Rowman & Littlefield, Lanham
Chernyshev IV, Lebedev VA, Arakelyants MM (2006) K-Ar dating of quaternary volcanics: methodology and interpretation of results. Petrology 14:62–80
Fedotov SA, Masurenkov Yu P (1991) Active volcanoes of Kamchatka. Nauka, Moscow, 2 volumes (in Russian)
Gabsatarova IP, Baranov SV (2018) New data on aftershocks of the December 7, 1988, Spitak earthquake. Seism Instrum 54(2):144–157
Geist EL, Scholl DW (1994) Large-scale deformation related to the collision of the Aleutian arc with Kamchatka. Tectonics 13(3):538–560
Gorbatov A, Kostoglodov V, Suarez G et al (1997) Seismicity and structure of the Kamchatka subduction zone. J Geophys Res B 102(8):17883–17898
Gulen L, EMME WP2 Team (2011) Active faults and seismic sources of the Middle East region: earthquake model of the Middle East (EMME) project. In: Abstracts of the AGU fall meeting, San Francisco, California, 5–9 December 2011
Heidbach O, Tingay M, Barth A et al (2010) Global crustal stress pattern based on the world stress map database release 2008. Tectonophysics 482(1–4):3–15
Heidbach O, Rajabi M, Cui X et al (2018) The world stress map database release 2016: crustal stress pattern across scales. Tectonophysics 744:484–498
Korjenkov AM, Rust D, Tibaldi A et al (2012) Parameters of the strong aleoearthquakes along the Talas-Fergana fault, the Kyrgyz Tien-Shan. In: D'Amico S (ed) Earthquake research and analysis – seismology, Seismotectonics and earthquake geology. In Tech publishers, Rijeka, Croatia, pp 33–84
Korjenkov AM, Abdieva SV, Burtman VS et al (2013) Indications of late medieval earthquakes in the Talas-Fergana fault zone, Tien Shan. Geotectonics 47:444–453
Korjenkov AM, Abdieva SV, Belousov TP et al (2014) An age of rockslides and palaeoearthquakes in the Karasu River valley (Talas-Fergana Fault, Kyrgyzstan). Seism Instrum 50:97–108
Kozhurin A, Acocella V, Kyle PR et al (2006) Trenching studies of active faults in Kamchatka, eastern Russia: Palaeoseismic, tectonic and hazard implications. Tectonophysics 417(3–4):285–304
Lebedev VA, Chernishev IV, Dudauri OZ et al (2003) The Samsari volcanic Centre as an example of recent volcanism in the lesser Caucasus: K-Ar geochronological and Sr-Nd isotopic data. Dokl Earth Sci. 393A:1323–1328
Nikolaeva E, Walter TR (2016) InSAR observations of the 2009 Racha earthquake, Georgia. Nat Hazards Earth Syst. Sci16:2137–2144
Pasquaré Mariotto F, Tibaldi A (2016) Inversion kinematics at deep-seated gravity slope deformations revealed by trenching techniques. Nat Hazards Earth Syst Sci 16:663–674
Pasquaré F, Tormey D, Vezzoli L et al (2011) Mitigating the consequences of extreme events on strategic facilities: evaluation of volcanic and seismic risk affecting the Caspian oil and gas pipelines in the republic of Georgia. J Environ Manag 92:1774–1782
Rust D, Korjenkov A, Tibaldi A (2018) Geologic slip-rate determinations on the Talas Fergana fault: mismatch with geodetic slip rate. Geophys Res Lett 45:3880–3888
Tibaldi A (2004) Recent surface faulting investigated in Kamchatka volcanic arc. EOS, Transact Am Geophys Union 85(14):133–141

Tibaldi A, Ferrari L (1992) Latest Pleistocene-Holocene tectonics of the Ecuadorian Andes. Tectonophysics 205(1–3):109–125
Tibaldi A, Pasquaré F (2008) Quaternary deformations along the 'Engadine–Gruf tectonic system', Swiss–Italian border. J Quat Sci 23:475–487
Tibaldi A, Pasquarè Mariotto F (2017) The struggle between Abkhazia and Georgia, as observed from NATO-supported scientists. http://www.asrie.org/2017/06/the-struggle-between-abkhazia-and-georgia-as-observed-from-nato-supported-scientists
Tibaldi A, Ferrari L, Pasquaré G (1995) Landslide triggered by earthquakes and their relationships with faults and mountain slope geometry: an example from Ecuador. Geomorphology 11:215–226
Tibaldi A, Bistacchi A, Pasquaré F et al (2006) Extensional tectonics and volcano lateral collapses: insights from Ollagüe volcano (Chile-Bolivia) and analogue modelling. Terra Nova 18(4):282–289
Tibaldi A, Pasquaré Mariotto F, Papanikolaou D et al (2008) Discovery of a huge sector collapse at the Nisyros volcano, Greece, by on-land and offshore geological-structural data. J Volcanol Geotherm Res 177:485–499
Tibaldi A, Corazzato C, Rust D et al (2015) Tectonic and gravity-induced deformation along the active Talas-Fergana fault, Tien Shan, Kyrgyzstan. Tectonophysics 657:38–62
Tibaldi A, Alania V, Bonali FL et al (2017) Active inversion tectonics, simple shear folding and back-thrusting at Rioni Basin, Georgia. J Struct Geol 96:35–53
Tibaldi A, Korjenkov A, Pasquaré Mariotto F et al (2018) NATO and earth scientists: an ongoing collaboration to assess geohazards and contribute to societal security in Central Asia and the Caucasus. Episodes 41(3) 193:205
Tibaldi A, Oppizzi P, Gierke JS et al (2019) Landslides near Enguri dam (Caucasus, Georgia) and possible seismotectonic effects. Nat Hazards Earth Syst Sci 19:71
Trifonov VG, Makarov VI, Skobelev SF (1992) The Talas-Fergana active right-lateral fault. Annales Tectonicae VI:224–237
Trifonov VG, Korjenkov AM, Khaled OM (2015) Recent geodynamics of major strike-slip zones. Geod Geodyn:10):1–10)23
Varazanashvili O, Tsereteli N, Tsereteli E (2011) Historical earthquakes in Georgia (up to 1900): source analysis and catalogue compilation. Monograph, M. Javakhishvili Tbilisi State University, Georgia
Varazanashvili O, Tsereteli N, Bonali FL et al (2018) GeoInt: the first macroseismic intensity database for the Republic of Georgia. J Seismol 22(3):625–667
Vezzoli L, Tibaldi A, Renzulli A (2008) Faulting-assisted lateral collapses and influence on shallow magma feeding system at Ollagüe volcano (central volcanic zone, Chile-Bolivia Andes). J Volcanol Geotherm Res 171(1–2):137–159
Vincent SJ, Carter A, Lavrishchev VA et al (2011) The exhumation of the western greater Caucasus: a thermochronometric study. Geol Mag 148:1–21

Part II
Key Studies Focused on Regional and Geological Aspects

Chapter 4
Active Kinematics of the Greater Caucasus from Seismological and GPS Data: A Review

Alessandro Tibaldi, Gulam Babayev, Fabio L. Bonali, Federico Pasquaré Mariotto, Elena Russo, Nino Tsereteli, and Noemi Corti

Abstract In this paper, we describe the active kinematics of the Greater Caucasus (territories of Georgia, Azerbaijan and Russia) through an integrated analysis of seismological, geological-structural and GPS data. Alignments of crustal earthquake epicentres indicate that most seismic areas are located along the southern margin of the mountain belt and in its north-eastern sector, in correspondence of major, active WNW-ESE faults, parallel to the mountain range. Focal Mechanism Solutions (FMS) delineate dominant reverse fault kinematics in most sectors of the mountain belt, though swarms of strike-slip FMS indicate the presence of active transcurrent faulting, especially along the southeastern border of the Greater Caucasus. The mountain belt is characterized by dominant NNE-SSW-oriented P-axes. In the central-southern sector, in correspondence of the local collision between the Lesser and Greater Caucasus, P-axes are mainly NNW-SSE oriented. GPS data show dominant motions to the NNW, with rates increasing in eastward direction. All observations are consistent with a component of eastward escape of the central-eastern part of the Greater Caucasus.

Keywords Caucasus · Earthquakes · Active fault · Focal mechanisms · GPS data

---

A. Tibaldi (✉) · F. L. Bonali · E. Russo · N. Corti
University of Milan Bicocca, Milan, Italy
e-mail: alessandro.tibaldi@unimib.it

G. Babayev
Seismology and geomagnetism Department, Geology and Geophysics Institute, Azerbaijan National Academy of Sciences, Baku, Azerbaijan

F. Pasquaré Mariotto
Department of Human and Innovation Sciences, Insubria University, Como, Italy

N. Tsereteli
M. Nodia Institute of Geophysics, M. Javakhishvili Tbilisi State University, Tbilisi, Georgia

© Springer Nature B.V. 2021

F. L. Bonali et al. (eds.), *Building Knowledge for Geohazard Assessment and Management in the Caucasus and other Orogenic Regions*, NATO Science for Peace and Security Series C: Environmental Security,
https://doi.org/10.1007/978-94-024-2046-3_4

4.1 Introduction

The convergence between the Arabian and the Eurasian plates resulted in the development of the Greater Caucasus (GC) and the Lesser Caucasus fold-and-thrust belts, separated across most of their length by the Transcaucasian depression (Fig. 4.1) (Adamia et al. 1977, 2010; Banks et al. 1997; Mosar et al. 2010; Sosson et al. 2010). The whole sub-horizontal shortening of the Caucasus was quantified at hundreds of kilometers (Barrier and Vrielynck 2008; Meijers et al. 2015) and, according to several studies, reached its maximum rate in the Miocene-Pliocene (Avdeev 2011; Avdeev and Niemi 2011; Vincent et al. 2007, 2011).

At present, convergence between the Eurasian and African-Arabian plates is still active, producing widespread deformation within the mountain belt and in surrounding regions, as testified to by seismological, paleoseismological and GPS data (Rebai et al. 1993; Jackson and Ambraseys 1997; Koçyiğit et al. 2001; Allen et al. 2004; Reilinger et al. 2006; Tan and Taymaz 2006; Kadirov et al. 2008, 2012; Avagyan et al. 2010; Pasquarè et al. 2011; Kadirov et al. 2015; Alizadeh et al. 2016; Tsereteli et al. 2016; Adamia et al. 2017; Sokhadze et al. 2018; Tibaldi et al. 2017a, 2017b, 2019a).

Focal mechanism solutions (FMS) indicate dominant thrust faulting along planes striking parallel to the mountain belt, in agreement with field data (Jackson and McKenzie 1984; Philip et al. 1989; Tan and Taymaz 2006; Tsereteli et al. 2016).

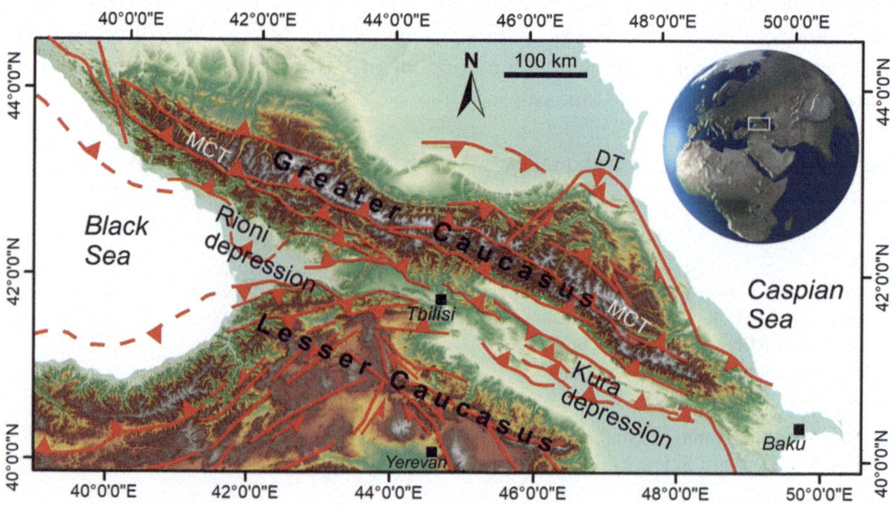

Fig. 4.1 Location of the Greater Caucasus and main active faults (faults after Tan and Taymaz 2006; Forte et al. 2014; Tsereteli et al. 2016). Triangles point to the hanging-wall block of reverse faults. Fault planes dip mainly to the north in the western Greater Caucasus, whereas in the central-eastern Greater Caucasus, thrust faults located at the two opposite mountain sides (northern and southern) have converging dips. The tip line trace of the Dagestan Blind Thrust (DT) is projected to the surface. MCT = Main Caucasus Thrust

Other FMS, instead, show more complex relations with field data and show also other types of kinematics, comprising reverse, transcurrent and normal faulting (Tsereteli et al. 2016; Tibaldi et al. 2019b). The situation is even more complicated because, based on some authors, the escape of crustal blocks to the west, towards the Black Sea, and to the east, towards the South Caspian basin, is still taking place (e.g. Kazmin 2002), in a much similar way as in Central Asia, as a consequence of the collision of India against Eurasia (e.g. Tibaldi et al. 2015).

Understanding the active tectonics of the Caucasus is of paramount importance for a better assessment of geohazards, especially seismic hazard. Moreover, there is a major concentration of residents in Tbilisi, the capital of Georgia, hosting 1.2 million citizens, and in Baku, the capital of the Azerbaijan Republic, with over 2.3 million citizens; both cities are located in active tectonic basins at the southern foothills of the GC. Hundreds of rural villages are scattered in the mountain regions, and all were built without taking into consideration antiseismic criteria. All the above shows that the Caucasus and Transcaucasus regions are subject to an extreme seismic hazard and risk.

In spite of the societal and scientific relevance of these regions in terms of the geohazards affecting them, seismogenetic structures are still poorly documented. After the Soviet era, several major research efforts have been conducted, but a greater exchange of data and a wider regional perspective would be necessary in order to gain a holistic insight into the ongoing tectonic processes. Above all, several long faults extend from Georgia into Azerbaijan, and their activation can have severe impacts on both countries. An integration of data from the two countries is thus necessary. In this perspective, through the present paper we wish to contribute to the integration of international data, which is needed for better understanding seismic hazard in the GC and the Transcaucasian regions. To this end, we collected all relevant data on seismicity, FMS, field structural geology, and GPS of the GC encompassing Georgia, Russia and Azerbaijan territories. By doing so, we intend not only to share key data useful for seismic hazard/risk assessment with a trans-boundaries perspective, but also to contribute to the following challenging scientific questions: i) Are FMS always coherent with field structural data? ii) Is there a consistency between active fault kinematics and GPS data? iii) Are these data coherent with previous hypotheses on the lateral escape of parts of the Caucasus belt?

4.2 Main Tectonic Features of Greater Caucasus

During the Alpine orogeny, an inversion of the previous extensional tectonic regime resulted in a series of ramp-and-flat fault structures in the GC that piled up Mesozoic-Cenozoic deposits (Adamia et al. 1977, 2010, Adamia et al. 2011a; Gamkrelidze 1986; Philip et al. 1989; Banks et al. 1997; Saintot et al. 2006; Mosar et al. 2010; Sosson et al. 2010, 2013; Forte et al. 2014). While in the Transcaucasian region, the Rioni and Kura basins developed as foreland basins during the Alpine orogeny, in the more external areas, part of the pre-Alpine basins survived in the

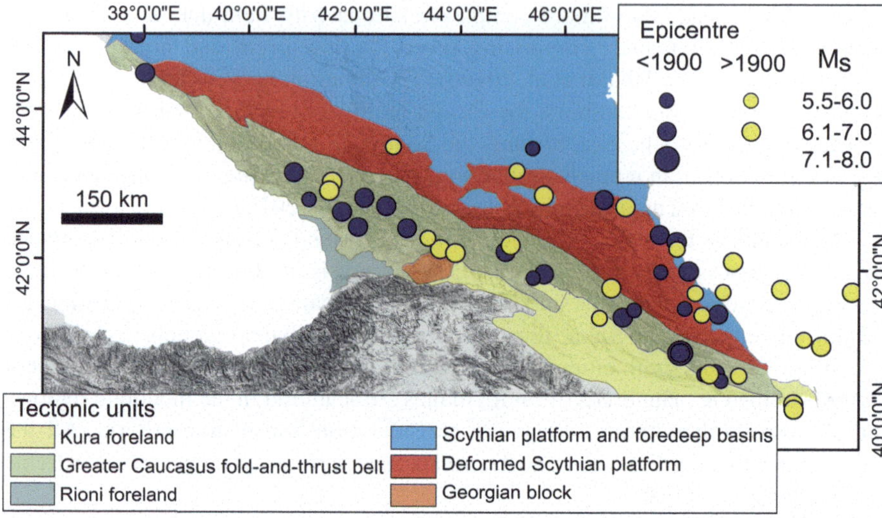

Fig. 4.2 Map of main historical and instrumental earthquakes ($M_S > 5.5$, $M_W > 5.7$) superimposed on the main tectonic units (units after Tibaldi et al. 2019b)

form of the present-day Caspian and Black Sea depressions (Forte et al. 2010, 2014; Mosar et al. 2010; Adamia et al. 2011a; Alania et al. 2016) (Figs. 4.1 and 4.2).

The pre-Mesozoic basement and Jurassic sedimentary rocks in the axial zone of the GC, and Cretaceous to Cenozoic sedimentary rocks along the northern and southern sides of the belt, are affected mainly by WNW-ESE-striking reverse faults (Adamia et al. 2011a; Mosar et al. 2010). Along the southwestern margin of the GC, a main structural salient affects the Rioni Basin, where part of the rock succession has been uplifted during Miocene-Pliocene-Quaternary. This gave rise to an arcuate fold-and-thrust belt with the convex side pointing southward (Tibaldi et al. 2017a, 2017b). In the northeastern margin of the GC, another salient is represented by an arcuate blind fault (Sobornov 1994), known as Dagestan Thrust (DT in Fig. 4.1), Dagestan salient or North Caucasus Thrust (Kadirov et al. 2008). This blind fault dips southward, allowing propagation of the shortening further northward than the remaining northern Caucasus front (Forte et al. 2014). The external, northern part of the deformed zone belongs to the Scythian Platform, which further north is essentially undeformed together with the foredeep basins (Fig. 4.2).

According to geodetic data, the convergence rate between the Arabian and the Eurasian plate is ca. 20–30 mm/year, with a decrease towards the western Caucasus (De Mets et al. 1990, 1994; Triep et al. 1995; Reilinger et al. 2006). Based on GPS stations, the northern Arabian plate moves at a rate of 18 ± 2 mm/year towards N25° ± 5°W relative to Eurasia. GPS data also indicate that the total convergence in the GC is about 4 mm/year, and in the Lesser Caucasus about 11 mm/year (McClusky et al. 2000). In the Transcaucasian depression, GPS stations move at about 6 mm/year towards NE in a Eurasia-fixed reference frame (McClusky et al.

2000). Repeated GPS measurements in Azerbaijan in 1998–2015 defined active convergence between the Lesser Caucasus/Kura Depression and the GC with strain concentrated along the Main Caucasus Thrust (Fig. 4.1) (Philip et al. 1989; Reilinger et al. 2006; Kadirov et al. 2008, 2012; Telesca et al. 2013, 2017). Present day slip rates on the Main Caucasus Thrust decrease from 10 ± 1 mm/year in eastern Azerbaijan to 4 ± 1 mm/year in western Azerbaijan (Kadirov et al. 2008, 2015).

4.3 Seismicity of the Greater Caucasus

4.3.1 Source of Data

In order to get a complete picture of seismicity in the GC, we merged several databases, comprising the New Catalogue of Strong Earthquakes in the USSR from Ancient Times through 1977 (Kondorskaya and Shebalin 1982), the Catalogue of Large Historical Earthquakes of the Caucasus (Shebalin and Tatevossian 1997), the Catalogue of Historical earthquakes in Georgia (up to 1900) (Varazanashvili et al. 2011), the Recent Developments of the Middle East Catalogue (Zare et al. 2014), Earthquakes of Azerbaijan for 1983–2002 (Gasanov 2003), and the Catalogue for GC (Mumladze et al. 2015). In addition, the catalogue of earthquakes in the GC was updated for the period 2006–2017, in Georgia from Varazanashvili et al. (2018) and in Azerbaijan from Telesca et al. (2012, 2017). The catalogue for the northern sector of the Caucasus belonging to the Russian territory was updated with data from the International Seismological Centre (ISC). We also combined these catalogues with a dataset of 1000 main seismic events in Georgia and Azerbaijan that were relocated by Tibaldi et al. (2019b). Finally, for the adjacent regions we used: the Catalogue for the territory of Turkey near the Georgia border provided by the National Earthquake Monitoring Center of Bogazici University (Turkey), the Catalogue for the territory of Iran provided by the International Institute of Earthquake Engineering and Seismology (IIEES) (Shahvar et al. 2013; Karimiparidari et al. 2013), the Catalogue of Armenia, taken from Sargsyan et al. (2017).

Since, during the former Soviet era, another type of magnitude was under use (energy class K, Rautian 1964) in the studied territories, we had to unify the earthquake catalogues in terms of magnitudes. With this purpose, we used the equation of Zare et al. (2014) for Georgia and the regional equation of Yetirmishli et al. (2015) for Azerbaijan (more details in Tibaldi et al. 2019b).

4.3.2 Epicentre Distribution

Figure 4.3 shows the plot of instrumental epicentres with $M_s > 3.1$ of the Caucasus and adjacent regions, covering the time window between 1900 and 2017. It is

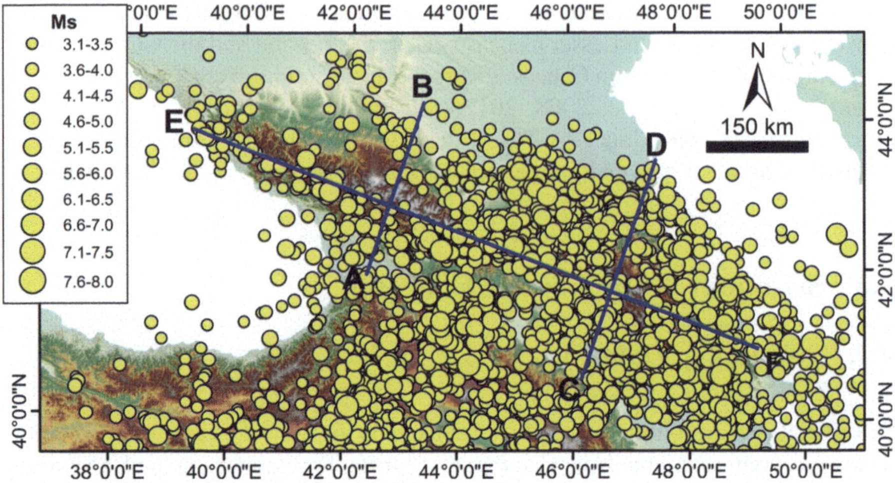

Fig. 4.3 Map of earthquakes in the Caucasus and adjacent regions, scaled with magnitudes. The three blue lines locate sections of Fig. 4.4

possible to observe the elongation of the seismic zone along the mountain belts, and its prolongation towards the East into the Caspian Sea. There is also a strong decrease in epicentre frequency approaching the platform north of the GC. South of the GC, on the other hand, there is a broad zone marked by major seismicity, which involves the Transcaucasian depression, and which extends westward into the Black Sea and eastward into the Caspian Sea. Widespread seismicity is also present in the Lesser Caucasus and in the regions located further south.

By focusing only on the strongest earthquakes with $M_s > 5.5$ ($M_W > 5.7$) within the GC (Fig. 4.2) that occurred both in historic and instrumental times, it is possible to observe that their spatial distribution is not uniform. In fact, the westernmost part of the mountain belt is seismically less active (Tsereteli et al. 2016; Mumladze et al. 2015): Fig. 4.2 shows events only along the southwestern coastal segment of the GC. Only approaching the easternmost coast of the Black Sea, a denser cluster of WNW-ESE-elongated earthquakes is present. This cluster comprises pre- and post-1900 main events, all of which are located along the southern side of the mountain belt, except for one event, which is located along its northern side, north of Elbrus volcano.

Starting from the longitude of Kazbek volcano and moving eastward, there is a northward shift in the seismic activity with $M_s > 5.5$ (Fig. 4.2). Along the southern side of the GC, there is still a clear WNW-ESE alignment of earthquakes, but they are located in a more axial position within the mountain belt with respect to the earthquakes plotted west of Kazbek volcano. Besides, in the central-eastern GC, strong earthquakes increase in number and distribution, and they are spread also along the northern side of the mountain range. Along the northeastern foothills, the epicentres of the largest events are located mostly towards the foreland.

The eastern sector of the mountain range is the seismically most active. Here, the strongest seismic events in the whole Caucasus were recorded, ranging from 742 AD to 2012 AD, with M_s from 5.7 to 7.8. A NNW-SSE-elongated cluster of events is present along the Caspian Sea coast, mostly onshore. This cluster approaches another WNW-ESE alignment of earthquakes present along the southern foothills of the GC.

The underwater section of the active compressional belt extends down into the Caspian Sea, as attested by the occurrence of 7 earthquakes (from 1902 AD to 2000 AD) with $M_s = 5.6–6.8$.

4.3.3 Hypocentres Distribution

About 95% of the total earthquakes in the GC occurred within a depth range of 1 to 20 km, and the remaining events at depths from 21 to 160 km. A section across the western part of the GC, with events with $M_s \geq 3$ only, shows a homogenous distribution down to about 22 km (Fig. 4.4, Sec. A-B). Events with larger M values are located in the southern side of the GC, from 4 to 21 km, whereas such earthquakes disappear south of the Rioni salient. Below the axis of the western GC, the seismicity is very low in terms of number of events and M values, and is characterized by low M and scattered events also below the northern side of the mountain belt.

A section across the eastern part of the GC, again with events of $M_s \geq 3$ only, shows a deeper distribution of earthquakes (Fig. 4.4, Sec. C-D): below the Kura fold-and-thrust belt, there are essentially earthquakes of medium M (4–5) at depth mostly between 5 and 30 km. Along the southern side of the GC, earthquakes increase in size and number, and depths increase up to 40 km. Below the axis of the GC, earthquakes extend again at depths greater than 40 km. The axis coincides with an abrupt decrease in earthquake magnitudes, which remain low northward, although the number of events does not decrease. Below the axis, hypocentres extend down to a depth of 100 km. Below the Dagestan salient, M increases again and earthquake depths range from very shallow to 38 km.

A section parallel to the GC (Fig. 4.4, Sec. E-F), with events of $M_s \geq 3$ only, shows clearly the gradual increase in hypocentre depths from west to east, in agreement with previous findings by Mumladze et al. 2015. This section shows that most of the seismicity is concentrated at depths less than 20 km and extends down to a depth of 80 km. In this case, magnitude values are homogeneously distributed. Below other parts of the eastern GC, hypocentres extend down to about 150 km of depth (Mumladze et al. 2015).

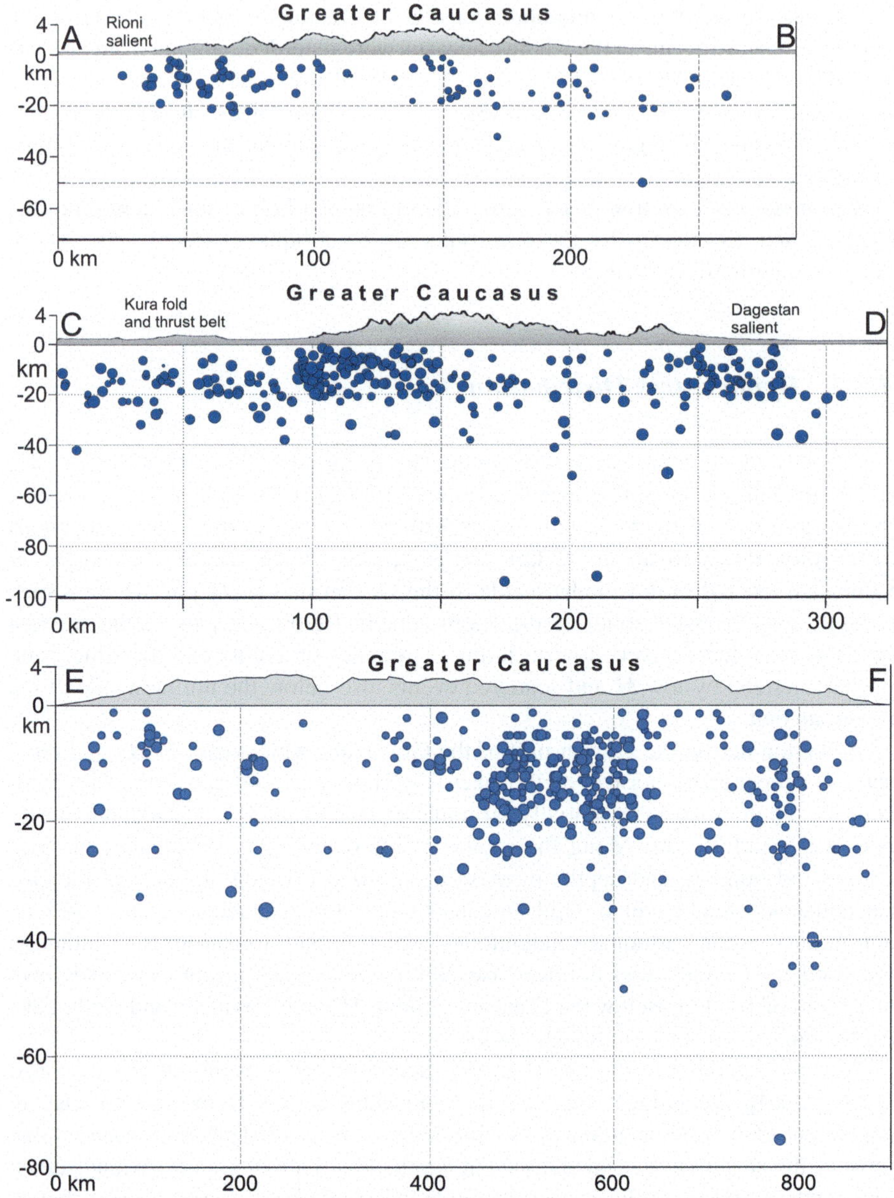

Fig. 4.4 Sections showing hypocentres of earthquakes with $M_S \geq 3$. Section traces are located in Fig. 4.3. Note that sections A-B and C-D are at the same scale, whereas section E-F has been plotted at another scale to fit in the figure

4.3.4 *Focal Mechanism Solutions*

4.3.4.1 Data Source

Several previous studies investigated FMS of various sectors of the Caucasus region, using different methodologies (Jackson and McKenzie 1988; Fuenzalida et al. 1997; Ekström and England 1989; Pondrelli et al. 1995; Ambraseys and Jackson 1998; Tan and Taymaz 2006; Tsereteli et al. 2016; Adamia et al. 2017; Tibaldi et al. 2018). In the present study, we chose to review and interpret the newest FMS of major earthquakes that were presented by Tibaldi et al. (2019b), because this represents the most complete database of selected FMS. In this database, only earthquakes for which the number of polarities is $\geq$8, and azimuthal gaps are less than 180°, were included in the analysis. The database can be considered homogenous and FMS from the various regions can be directly compared; however, due to the type of available data source, we used two different methods: for earthquakes in Georgia before 2003, the first motion polarity technique (Lander 2004), whereas for post-2003 events, the frequency sensitive moment tensor inversion technique (Barth et al. 2007). Regarding Azerbaijan, the whole data set of earthquakes that occurred between 1953 and 2017 is based on the first motion polarity technique (Lander 2004). The FMS of 1953–2003 are from Agayeva and Babayev (2009) and Babayev et al. (2017).

4.3.4.2 FMS Results

In general, 30% of FMS show reverse fault slip along planes striking WNW-ESE, consistently with the dominant orientation of known recent/active reverse faults based on field data (Fig. 4.1). 34% of FMS show strike-slip kinematics, whereas 6% transpression, 4% transtension and 4% normal slip. The remaining 22% do not present clear solutions.

More in detail (for depth < 10 km), Figs. 4.5a–c show that most earthquakes have reverse kinematics in the western and central part of the GC, whereas, in the eastern part, strike-slip deformation dominates. The beach balls of reverse events indicate a dominant WNW-ESE strike of the slip planes all over the GC. Along the southern side of the GC, most of the earthquakes with strike-slip FMS present one slip plane striking WNW-ESE as well. This means that most FMS are compatible with fault slip along planes parallel to the mountain belt.

At depths of 10–15 km (Fig. 4.6a), reverse FMS increase in number while strike-slip FMS decrease, accompanied by very few normal motions. In most cases, the slip planes of strike-slip FMS are oblique to the WNW-elongation of the GC, whereas a few strike-slip events show one slip plane parallel to the mountain belt. The slip planes of normal FMS are from parallel to perpendicular to the mountain belt.

At depths of 15–30 km (Fig. 4.6b), normal faulting events decrease in frequency, while other earthquake kinematics show similar characteristics to the events

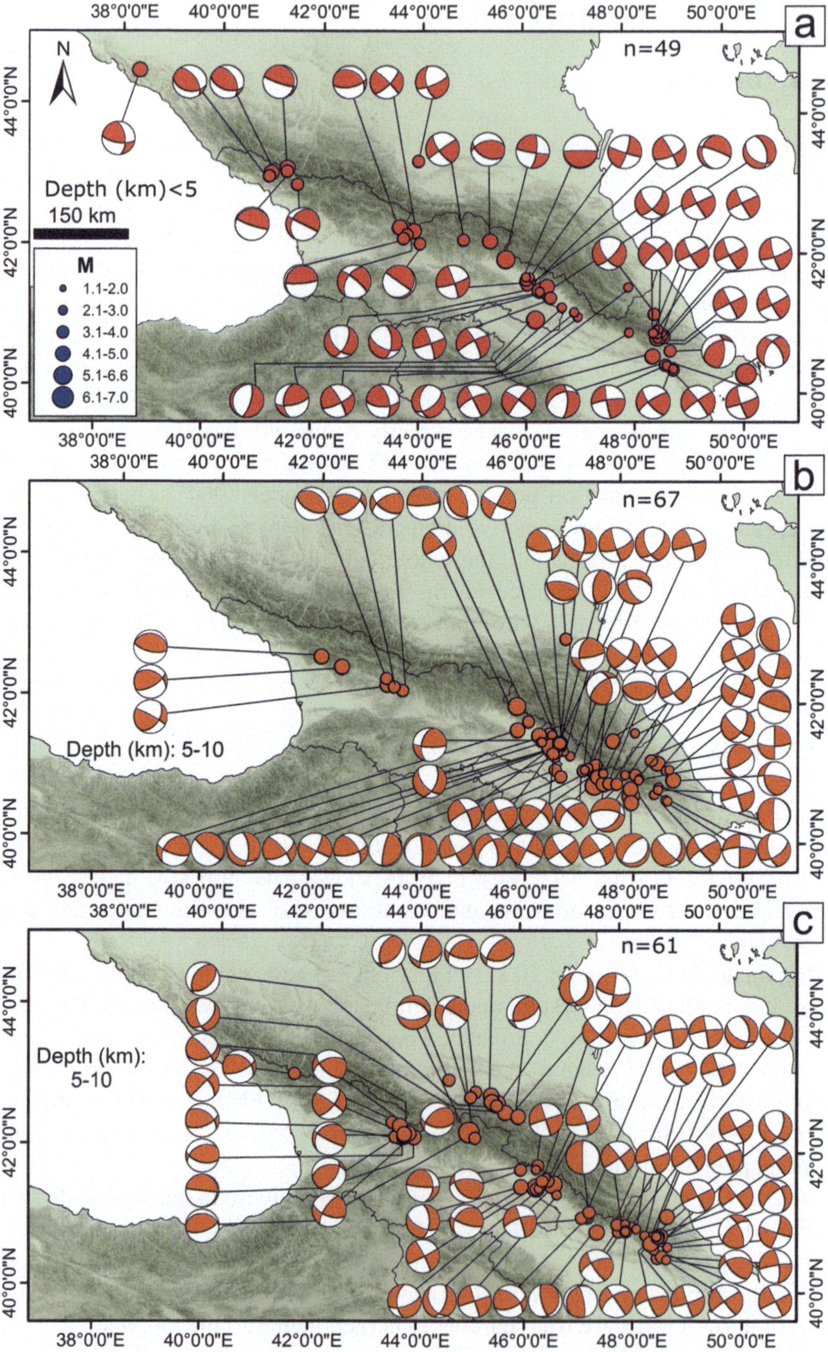

Fig. 4.5 Focal mechanism solutions of the Greater Caucasus, subdivided by M and depth. (**a**) Hypocentre depth < 5 km. (**b**) and (**c**) Hypocentre depth of 5–10 km. Redrawn after Tibaldi et al. (2017b)

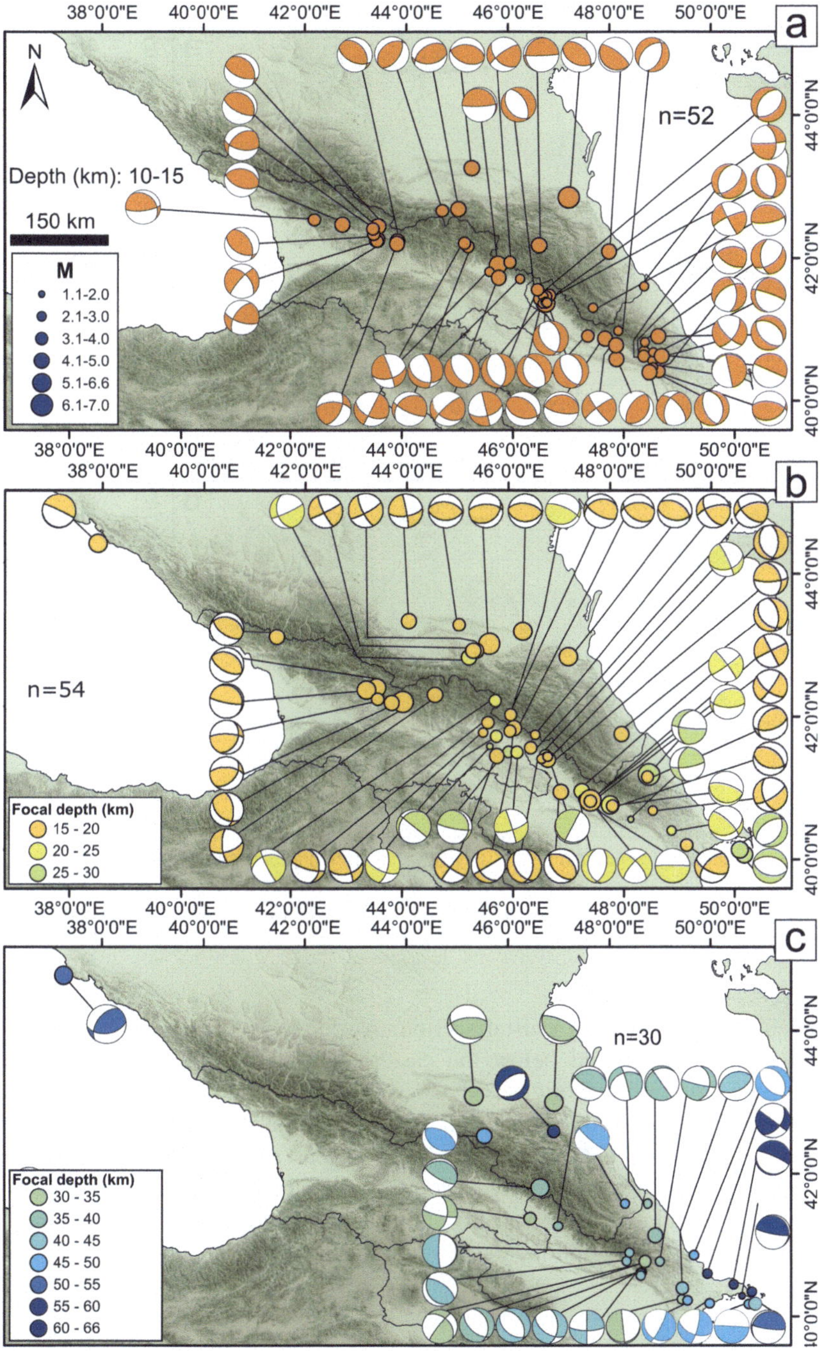

Fig. 4.6 Focal mechanism solutions of the Greater Caucasus subdivided by M and depth. (**a**) Hypocentre depth of 10–15 km. (**b**) Hypocentre depth of 15–30 km. (**c**) Hypocentre depth ≥ 30 km, colour of beach balls is referred to depth intervals. Redrawn after Tibaldi et al. (2019b)

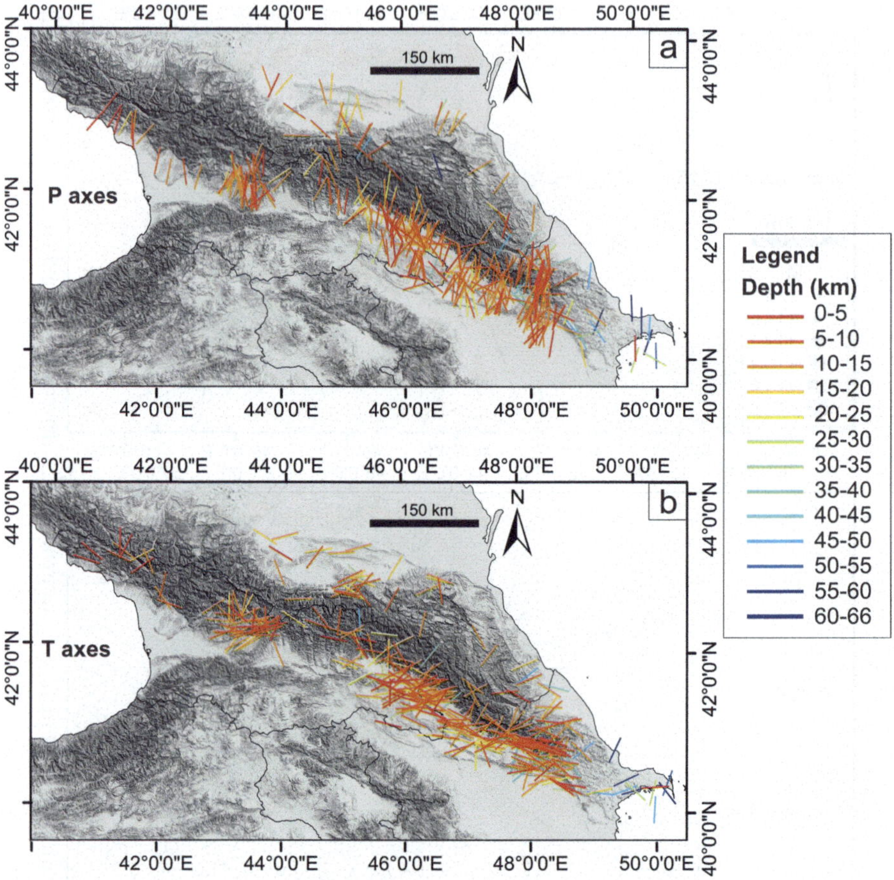

Fig. 4.7 (**a**) P axes and (**b**) T axes derived from focal mechanism solutions calculated in Tibaldi et al. (2019b). Colours refer to earthquake depth

recorded in the 10–15 km crustal section. At depths ≥ 30 km (Fig. 4.6c), most seismicity is concentrated below the eastern part of the GC, with earthquakes characterised by reverse motions.

In Fig. 4.7 we have plotted the P and T axes of all the aforementioned FMS. Most P axes trend from N-S to NNE-SSW, and a minor part rotates to NE-SW. The T axes are mostly horizontal, consistently with the dominance of strike-slip motions. Most T axes trend from E-W to WNW-ESE. In the southeastern part of the mountain belt, most T axes are slightly oblique to the southern slope range.

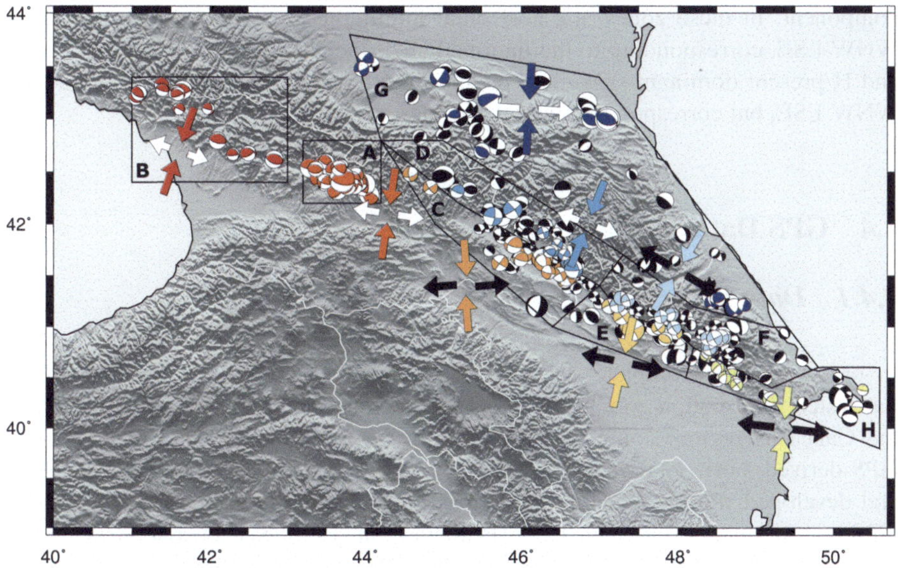

Fig. 4.8 Formal stress inversion of focal mechanisms, redrawn after Tibaldi et al. (2017b). Colours indicate the eight separate inversion zones as described in the text. Black beach balls represent weak earthquakes not used for inversion. Coloured, converging arrows give the orientations of the greatest principal stress (σ_1). Diverging arrows give least principal stress (σ_3) for strike-slip regime (black) and intermediate principal stress (σ_2) for predominantly thrust faulting regime (white). Arrow lengths are scaled by stress ratio R

4.3.5 Stress Distribution

Tibaldi et al. (2019b) used the above described 366 FMS to perform a formal stress inversion. For this purpose, the authors used the main tectonic zones shown in Fig. 4.2, simplified from Adamia et al. (2011a, b). Tectonic units were chosen based on the fact that they have experienced different tectonic histories, giving rise to distinct morphostructural patterns and being nowadays characterized by clustered seismicity. The boxes shown in Fig. 4.8 delimitate the areas where a stress field inversion was performed. The boxes do not coincide perfectly with the tectonic units of Fig. 4.2 because they were adjusted in order to take into consideration the distribution of seismicity. Tibaldi et al. (2019b) applied the methodology of Michael (1984), updated with the contributions of Lund and Townend (2007) and Barth and Wenzel (2010).

Results show that in zones A, E, H and G, the σH_{max} trends NNE-SSW (Fig. 4.8). In zones B, D and F, the σH_{max} is slightly rotated clockwise, trending more NE-SW. In zone C, the σH_{max} is rotated anticlockwise and trends NNW-SSE. All regions show dominant compressional tectonics, thus everywhere the greatest principal stress (σ_1) is horizontal, corresponding to the σH_{max}. Zones A, B, D and G are characterised by a thrust faulting regime with a very subordinate strike-slip component. In these zones, the horizontal minimum stress (σH_{min}) trends about WNW-ESE, corresponding to the intermediate principal stress (σ_2). Zones C, E, F and H present dominant strike-slip tectonics with a σH_{min} always trending about WNW-ESE, but corresponding here to the least principal stress (σ_3).

4.4 GPS Data

4.4.1 Data Source

In order to get a complete dataset of GPS horizontal motions in the GC, we integrated information coming from several regional and local investigations. For the Caucasus region, comprising the territories of Azerbaijan and Georgia, the GPS-derived, horizontal velocity map were started in 1988 (Reilinger et al. 1997) and developed further (Ahadov and Jin 2017, and references therein). In Russia, GPS measurements started in 1991 (Prilepin et al. 1997; Shevchenko et al. 1999; Reilinger et al. 1997, 2006). Later on, the GPS stations network was gradually improved (e.g. Milyukov et al. 2010, 2015a, 2018). In the Ossetian segment of the GC, GPS observations started in 2010 (Milyukov et al. 2015b).

Data presented here were integrated from the works of Reilinger et al. (1997, 2006), Shevchenko et al. (1999), McClusky et al. (2000), Guliev et al. (2002), Kadirov et al. (2012, 2013), Aktuğ et al. (2013) and Sokhadze et al. (2018). As a whole, the period covered by measurements ranges from 1988 to 2017, though in some areas the period is shorter. We selected the GPS data referred to at least 4 years of observations and we plotted the GPS vectors of the Russian, Georgian and Azerbaijan GC and Transcaucasian depression, in order to make a detailed comparison between FMS, structural field data and GPS data. GPS horizontal velocities are shown in the Eurasian plate fixed reference frame.

4.4.2 Results

GPS data are described hereunder from west to east. The easternmost GPS station (KRPO) shows motions of 3 mm/year towards NE, consistent with the station ZECK (Fig. 4.8). At the southwestern foothills of the GC, in correspondence of the northern part of the Rioni uplifted area, the station INGU indicates motions towards NNW at a velocity of 2.5 mm/year. North of the Main Caucasus Thrust (MCT), in the core of the GC, station MATS indicates motions towards NNE at rate of 2 mm/year. More to the west, station ULKA shows motions to the WNW of 2 mm/year. In the SP, in the northern slope of the mountain belt, stations SHAT and BEUG, very close each other, give 2 mm/year to the NNE and N respectively.

More to the east, in correspondence of the Georgian block, three out of four stations show motions towards NNE, whereas station LESO has motions towards ENE. Among the first three stations, KHOT and SACC, located in the southern part of the Georgian block, have velocity in the order of 4–5 mm/year. Stations KHUR and LESO, which correspond to the northern part of the Georgian block, have velocity of 2 mm/year.

In the western part of the Kura basin zone, GPS stations KRES and KUDI indicate northward motions at a rate of 3–4 mm/year. North of the MCT, only stations ZURA and DUBK are present and are located in the SP in correspondence of the Dagestan salient. They provide motions to about N ad NNE respectively, at 2 mm/year. In the central part of the Kura basin zone, four GPS stations (KTYX, SHEK, KEBE, YEVL) show northward motions, and another three (KURD, QURD, MLAY) indicate motions towards NNE. Rates are in the order of 3 mm/year apart from MLAY that shows 5 mm/year. The stations located along the northern slopes of the Lesser Caucasus, here show motions of 5–7 mm/year to the NNE.

In the eastern part of the Kura basin zone, stations KIZA, MEDR and KHID show motions towards NNE, with rate increasing southwards from 3 mm/year to 7.7 mm/year in the Lesser Caucasus. Other stations (SIYE, BAKU, SANG, SHIK) instead, have motions towards NE at rates of 3–4 mm/year. Finally, station SAMU, located along the northeastern slope of GC, has motions of 2 mm/year directed to the NNW.

As a comparison, in the Lesser Caucasus, the great majority of GPS vectors point to NNE at velocity rates higher than in the GC.

4.5 Discussion

4.5.1 Seismicity

Seismicity increases in frequency from west to east (Figs. 4.2 and 4.3), and hypocentres depth increases in the same direction (Fig. 4.4). The strong seismicity of this mountain belt is due to lithospheric shortening in consequence of its position between the converging Eurasian and Africa-Arabian plates (Nemčok et al. 2011; Alizadeh et al. 2016). In regard to instrumental seismicity with $M_s > 3$, several earthquakes are arranged in WNW-ESE elongated clusters, consistently with the trend of the GC; most seismic events are concentrated along the whole southern side of the mountain belt. Clusters with the same elongation appear in the central and eastern part of the GC axis. Along the northeastern side, instead, a dense distribution of epicentres is present, most of which are arranged along an arc, with a northward convex side. This arc is in correspondence of the faults that border the Dagestan salient, and this attests to its important seismic activity. Several seismic events are also present in the Caspian Sea, suggesting the prosecution of the GC active structures below sea level. A diffuse seismicity is also present in the Kura basin and in correspondence of the Rioni salient, where recent and active structures were recognized by Tibaldi et al. (2017a, 2017b). A similar pattern characterizes the historical seismicity with $M_s > 5.5$, with the difference that only a few of such events are present in the core of the GC and no events were recorded in the Kura and Rioni basins (Fig. 4.2).

The above observations suggest that a strong seismic activity is located in proximity of the main reverse and strike-slip faults that bound the northeastern part of the mountain belt, especially those related to the Dagestan salient. Major seismicity is also linked to the WNW-ESE reverse faults that bound the mountain belt to the south. The dip of the Dagestan main reverse fault is toward SSW, whereas in the same area only minor faults dip toward NNE (Sobornov 1994). Furthermore, the dip of the main faults situated along the southern border of the mountain belt is towards NNE, thus indicating a single main southward vergence in the western part of the orogen, and a double-vergence in the eastern part. This has already been claimed by Mosar et al. (2010) and Forte et al. (2014) who noticed that the eastern GC has a pro-wedge and a retro-wedge actively propagating into the foreland sedimentary basin of Kura to the south and Terek to the north. The pronounced seismicity in the Rioni salient and in the Kura basin indicates that parts of these areas are subject to uplift and shortening, as suggested also by the presence of several folds involving the Tertiary strata (Forte et al. 2010; Tibaldi et al. 2017a, 2017b).

4.5.2 Active Fault Kinematics

Previous works based on field geological-structural data or geophysical prospecting documented that active faults of the GC fold-and-thrust belt, Kura and Rioni basins, and the deformed SP have essentially reverse motions (Sobornov 1994; Banks et al. 1997; Agayeva and Babayev 2009; Adamia et al. 2011a, 2011b, 2017; Kadirov et al. 2012; Babayev et al. 2014; Alizadeh et al. 2016; Tsereteli et al. 2016; Tibaldi et al. 2017a, 2017b).

Strike-slip faults are reported for the eastern sector of the mountain belt, corresponding to the lateral boundaries of the actively indenting wedge that culminates with the Dagestan thrust. The western boundary may correspond in part to the Borjomi-Kazbegi left-lateral strike-slip regional fault (Philip et al. 1989; Ruppel and McNutt 1990). However, our field and seismological data do not suggest the presence of late Quaternary activity along this structure, in agreement with seismological data from an improved seismic network by O'Connor et al. (2005), who indicated a lack of seismicity along the proposed Borjomi-Kazbegi fault, as well as no offset of Early-Middle Eocene sediments across the supposed fault trace. Similarly, another large NE-striking left-lateral strike-slip fault was hypothesized as running from Tbilisi towards NNE, with a right-stepping geometry with respect to the Borjomi-Kazbegi fault (Philip et al. 1989). However, this fault remains highly speculative too, at least regarding Quaternary activity along the southernmost segment.

Seismic activity may be present along the northern segment of the NE-SW, left-lateral strike-slip fault, visible from about 100 km NE of Tbilisi. This fault offsets the reverse faults of the northern Caucasus slope as a transfer fault. West of this fault, other SSW-dipping reverse faults may be active and linked with fold growth in the Russian external part of the deformed SP.

The eastern boundary of the Dagestan salient was claimed to correspond to a NW-SE-striking right-lateral strike-slip main fault (Philip et al. 1989). Our seismological data do confirm the existence of a corridor of earthquake epicentres elongated with the same direction and running southward from the eastern tip of the Dagestan thrust. This is especially clear from the distribution of historical and instrumental earthquakes with $M_s > 5.5$ (Fig. 4.2). Moreover, FMS along this corridor show several right-lateral strike-slip events, if we assume the NW-SE planes as slip planes.

Other FMS with strike-slip kinematics characterize the southern margin of the eastern GC and are mostly shallower than 30 km. If we assume that slip planes coincide with field-mapped faults, i.e. striking WNW-ESE, it turns out that this sector of the GC is separated from the Kura basin by a swarm of right-lateral transpressional faults. This indicates an eastward escape of the eastern GC with respect to the tectonic blocks located further south, as suggested by Tibaldi et al. (2019b). This tectonic escape may be the consequence of the collision between the Lesser Caucasus and the structurally thickened crust of the GC immediately to the west, i.e. in correspondence of the Georgian block.

4.5.3 Stress Field and GPS Deformation Field

The great majority of FMS have a horizontal P-axis (Fig. 4.7a); similarly, the stress field based on formal inversion of FMS shows a dominant horizontal σ_1 across the whole GC, coherently with an active orogenic belt (Fig. 4.8). More interestingly, especially in the eastern part of the belt, the T-axis is vertical or horizontal, showing the coexistence of reverse and strike-slip faulting, respectively. The formal inversion of FMS, instead, indicates that some main tectonic zones are characterized by horizontal σ_1 and σ_2, suggesting the dominance of reverse faulting. This is the case of the western, central and northeastern parts of the mountain belt. In the southeastern portion of the orogen, stress inversion results in horizontal σ_1 and σ_3, suggesting the dominance of strike-slip faulting.

In spite of the permutations of σ_2 and σ_3, the σ_1 trend is more stable across the GC, being mostly NNE-SSW and secondarily NE-SW. This is coherent with previous works that documented the σ_1 trending about normally to the mountain chain in the northern GC (Babayev 2009) and in the western GC (Tsereteli et al. 2016). The only different orientation of σ_1 is observed in zone C (Fig. 4.8), east of the Georgian block, where it trends NNW-SSE, corresponding to an anticlockwise rotation of compression. Zone C is located immediately east of the sector of collision between the GC and the Lesser Caucasus. Moreover, in this sector, also the mapped faults, based on field geological data and on crustal seismic exploration, strike WNW-ESE. The stress state in zone C is thus consistent with right-lateral strike-slip motions along faults parallel to the GC. To the east, the stress inversion results in a σ_1 slightly oblique to WNW-ESE faults, thus producing right-lateral transpressional displacements. These data are all consistent with the eastward tectonic escape of the eastern GC, as already suggested by Tibaldi et al. (2019b).

Present-day rates of deformation accumulation along the main faults of Caucasus, evidenced by GPS data, suggest that the main fault activity should be located along the southern and northern borders of the mountain chain (Reilinger et al. 2006). Among the active faults, continuous GPS measurements from 1998 to 2016 indicate stronger activity of the reverse faults along the southern border (Kadirov et al. 2008, 2012, 2015). Moreover, the shortening along the southern border of the GC increases from west to east, reaching the maximum convergence between the Lesser Caucasus, the Kura basin and the eastern GC (Kadirov et al. 2015). The rate of convergence between the Lesser and Greater Caucasus, in fact, increases from ~4 mm/year in the Rioni basin, to ~14 mm/year in the Kura Basin (Reilinger et al. 2006).

The concentration of epicentres along the northern and southern sides of the GC, and the scarcity of earthquakes in the core of the GC, especially for those with $M_s > 5.5$, as explained in this paper, are consistent with GPS data and other types of seismological data. In fact, GPS measurements indicate fast shortening rates, coherently with the studies by McClusky et al. (2000), Reilinger and Barka (1997), and Reilinger et al. (1997, 2004), who showed a total of ~10 mm/year of NE-SW shortening across the whole Caucasus, divided into ~6 mm/year in the GC and ~ 4 mm/year in the Lesser Caucasus. Moreover, Vernant et al. (2004) reported 14 ± 2 mm/year of N-S shortening across the eastern GC. These GPS data are apparently at odds with the convergence rates obtained from the magnitudes of historic and recent earthquakes, which indicate that the Caucasus is shortening at 2–3 mm/year (maximum rate = 6 mm/year) (Westaway 1990). This contradictory data can be explained assuming that the GC and the Lesser Caucasus are subject to only minimal deformation at their core; this implies that the ongoing contraction is taking place especially across the Transcaucasian region (Reilinger et al. 2006; Forte et al. 2010) and, according to our view, also along the northeastern slope of the GC.

In Fig. 4.10 we compare the stress field, obtained from the formal stress inversion, with the average GPS vector computed within each main tectonic unit. In unit B, the average GPS vector trends NW-SE, rotated anticlockwise to the local orientation of the σ_1 based on formal stress inversion. This might be consistent with a left-lateral component of motions along the WNW-striking faults. Here, the various GPS stations indicate a decrease of the slip rate northward, confirming the existence of active reverse faults striking WNW-ESE.

In the Georgian block (unit A), average GPS vector trends NNE-SSW, slightly rotated in a clockwise manner with respect to the local orientation of the σ_1 based

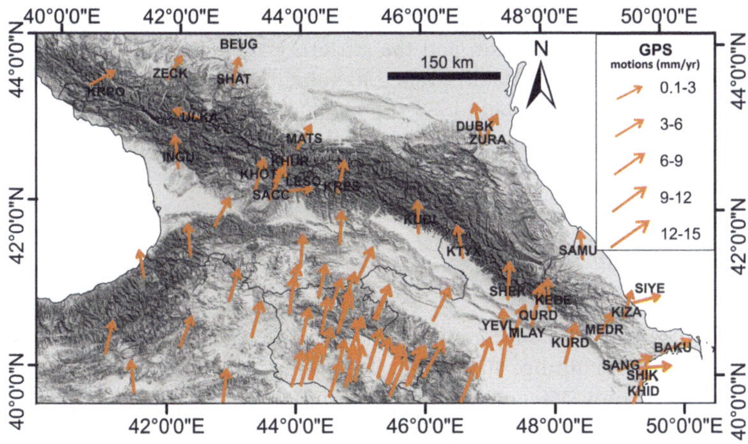

Fig. 4.9 Map of GPS site velocities with respect to Eurasia. Data were integrated from Reilinger et al. (1997, 2006), Shevchenko et al. (1999), McClusky et al. (2000), Guliev et al. (2002), Kadirov et al. (2012, 2013), Aktuğ et al. (2013) and Sokhadze et al. (2018). As a whole, the period covered by measurements is from 1988 to 2017. Names of GPS stations discussed in the text are shown

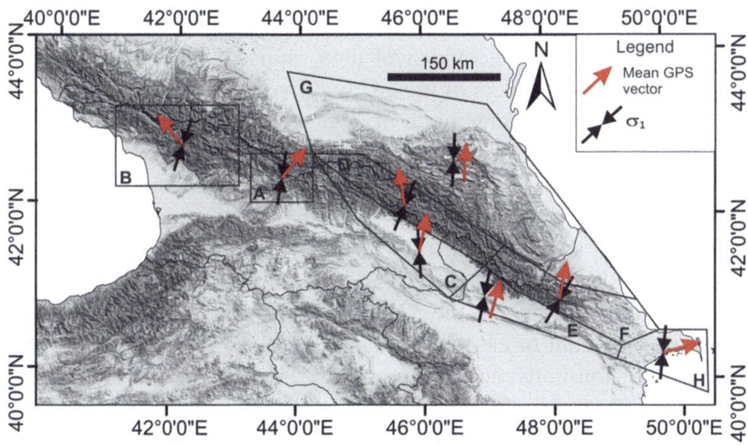

Fig. 4.10 Comparison of the stress field (black arrows), obtained from formal inversion of FMS of Tibaldi et al. (2019b), with the GPS field (red arrows) obtained from averaging vector data shown in Fig. 4.9. Calculation was performed within each seismotectonic unit (boxes)

on formal stress inversion. Further north, also station MATS shows motions to the NNE, confirming that this part of the GC, located in correspondence of the zone of collision with the Lesser Caucasus, reacts with pure contraction. In the same Georgian block, the various GPS vectors delineate a gradual northward decrease of shortening (Fig. 4.9), and this suggests the conditions for the development of slip along number of reverse faults.

To the east, at the Kura basin (units C and E, Fig. 4.10), average GPS vectors trend N12–14°, slightly rotated or coinciding with the σ_1. The clockwise rotation of average GPS at units C and H suggests the presence of right-lateral transpressional motions between the Kura basin and the eastern GC, as already suggested by the results of stress inversion. The difference in velocity between the Lesser Caucasus vectors and the stations located along the southern slopes of the GC indicates the presence of contraction across the Transcaucasian depression.

4.6 Conclusions

In the GC, focal mechanism solutions indicate that most seismic events are represented by reverse faulting, strike-slip faulting and transpressional/transtensional faulting, whereas only 4% correspond to normal faulting. This pattern is coherent with a stable horizontal σ_1 and a local permutation of the σ_2 and σ_3 axes.

At depths <30 km, reverse kinematics dominates along the northern and southwestern flanks of the GC, whereas strike-slip kinematics dominates along the southeastern flank. Both kinematics can develop along faults striking parallel to the mountain belt, i.e. WNW-ESE, in response to the NNE-SSW to NE-SW trend of the σ_1.

GPS data, the distribution of epicentres, and the location of the main faults based on geological-structural and geophysical data, indicate that most seismotectonic activity occurs along the borders of the GC, with very poor deformation at the core of the mountain belt. The stronger activity occurs in correspondence of the northeastern and southeastern border of the GC: the northeastern sector corresponds to the Dagestan thrust system and its lateral strike-slip transfer faults; the southeastern sector corresponds to a swarm of WNW-ESE faults with transpressional motions. Transpression here is suggested by the oblique orientation of the local σ_1 to fault strike, and by the GPS vectors that trend northward along the northern Kura basin, and trend NNE to NE further to the northeast. This transpression is interpreted as the effect of a component of eastward escape of the central-eastern part of the GC, following the collision between the structurally thickened crust of the central GC and the Lesser Caucasus.

The central part of the GC is characterised by a gradual decrease of GPS vectors towards NNE, compatible with the possible development of slip along WNW-ESE faults. Here, the collision between the GC and the Lesser Caucasus is consistent with GPS motions, directed northward to the NNE, and with the trend of σ_1. This suggests the transmission of the effects of NNE-directed collision across the central core of the GC.

Acknowledgments We acknowledge the reviews of Victor Alania and Ali Ismet Kanli on a previous version of the manuscript. This study has been conducted in the framework of: NATO project SfP G4934 "Georgia Hydropower Security", International Lithosphere Program - Task Force II, European Space Agency project n. 32309 "Active tectonics and seismic hazard of southwest Caucasus by remotely-sensed and seismological data" (Leader A. Tibaldi), and project 216758 of the Shota Rustaveli National Science Foundation.

References

Adamia S, Lordkipanidze MB, Zakariadze GS (1977) Evolution of an active continental margin exemplified by the alpine history of the Caucasus. Tectonophysics 40:183–199

Adamia S, Alania V, Chabukiani A et al (2010) Evolution of the late Cenozoic basins of Georgia (SW Caucasus): a review. In: Sosson M, Kaymakçı N, Stephenson R, Bergerat F (eds) Sedimentary basin tectonics from the Black Sea and Caucasus to the Arabian platform, Geological Society, vol 340. Special Publication, London, pp 239–259

Adamia S, Zakariadze G, Chkhotua T et al (2011a) Geology of the Caucasus: a review. Turk J Earth Sci 20:489–544

Adamia Sh, Alania V, Chagelishvili R et al (2011b) Tectonic setting of Georgia (Caucasus). In: Abstracts of the 3rd International Symposium on the Geology of the Black Sea Region, Bucharest, Romania, 1–10 October 2011. Abstracts Supplement to Geo Eco Marina 17, pp 11–13

Adamia Sh, Alania V, Tsereteli N et al (2017) Post-collisional tectonics and seismicity of Georgia. Tectonic evolution, collision, and seismicity of Southwest Asia. In: Honor of Manuel Berberian's forty-five years of research contributions. Geological Society of America Special Paper 525, pp 535–572

Agayeva ST, Babayev GR (2009) Analysis of earthquake focal mechanisms for greater and lesser Caucasus applying the method of world stress map. In: Proceedings of Geology Institute, Azerbaijan National Academy of Sciences, vol 2. Nafta-Press, Baku, pp 40–44

Ahadov B, Jin S (2017) Present-day kinematics in the eastern Mediterranean and Caucasus from dense GPS observations. Phys Earth Planet Inter 268:54–64

Aktuğ B, Meherremov E, Kurt M et al (2013) GPS constraints on the deformation of Azerbaijan and surrounding regions. J Geodyn 67:40–45

Alania V, Chabukiani A, Chagelishvili R et al (2016) Growth structures, piggyback basins and growth strata of Georgian part of Kura foreland fold and thrust belt: implication for late alpine kinematic evolution. In: Sosson M, Stephenson R, Sh A (eds) Tectonic evolution of the eastern Black Sea and Caucasus, Geological Society, vol 428. Special Publications, London. https://doi.org/10.1144/SP428.5

Alizadeh AA, Guliyev IS, Kadirov FA et al (2016) Geosciences of Azerbaijan volume I: geology. Springer International Publishing

Allen M, Jackson J, Walker R (2004) Late Cenozoic reorganization of the Arabia-Eurasia collision and the comparison of short-term and long-term deformation rates. Tectonics 23:TC2008. https://doi.org/10.1029/2003TC001530

Ambraseys NN, Jackson JA (1998) Faulting associated with historical and recent earthquakes in the eastern Mediterranean region. Geophys J Int 133(2):390–406. https://doi.org/10.1046/j.1365-246X.1998.00508.x

Avagyan A, Sosson M, Karakhanian A et al (2010) Recent tectonic stress evolution in the lesser Caucasus and adjacent regions. In: Sosson M, Kaymakci N, Stephenson RA, Bergerat F, Starostenko V (eds) Sedimentary Basin tectonics from the Black Sea and Caucasus to the Arabian platform, Geological Society, vol 340. Special Publications, London, pp 393–408. https://doi.org/10.1144/SP340.170305-8719/10

Avdeev B (2011) Tectonics of the greater Caucasus and the Arabia-Eurasia Orogen. Dissertation. University of Michigan

Avdeev B, Niemi NA (2011) Rapid Pliocene exhumation of the central greater Caucasus constrained by low-temperature thermochronometry. Tectonics 30:TC2009. https://doi.org/10.1029/2010TC002808

Babayev G (2009) Analysis of earthquake focal mechanisms for greater and lesser Caucasus applying the method of world stress map. Azerbaijan National Academy of Science. Catalogue of Azerbaijan Republican Seismological Center, Baku, pp 67–74

Babayev G, Tibaldi A, Bonali F et al (2014) Evaluation of earthquake-induced strain in promoting mud eruptions: the case of Shamakhi–Gobustan–Absheron areas, Azerbaijan. Nat Hazards 72(2):789–808

Babayev G, Akhmedova EV, Kadirov FA (2017) Analysis of stress-strain state of Caucasus region (Azerbaijan) on the basis of maximum horizontal stress vectors and world stress map. Application technique. Geophys J 3(39) (in Russian):26–39. https://doi.org/10.24028/gzh.0203-3100.v39i3.2017.104026

Banks C, Robinson A, Williams M (1997) Structure and regional tectonics of the Achara-Trialeti fold belt and the adjacent Rioni and Kartli foreland basins, Republic of Georgia. In: Robinson AG (ed) Regional and petroleum geology of the Black Sea and surrounding region, vol 68. American Association of Petroleum Geologists, Tulsa, pp 331–336

Barrier E, Vrielynck B (2008) Palaeotectonic map of the Middle East, Atlas of 14 maps, tectonosedimentary-palinspastic maps from Late Norian to Pliocene. Comm for the Geol Map of the World (CCMW, CCGM), Paris, France

Barth A, Wenzel F (2010) New constraints on the intraplate stress field of the Amurian plate deduced from light earthquake focal mechanisms. Tectonophysics 482:160–169. https://doi.org/10.1016/j.tecto.2009.01.029

Barth A, Wenzel F, Giardini D (2007) Frequency sensitive moment tensor inversion for light to moderate magnitude earthquakes in eastern Africa. Geophys Res Lett 34:L15302. https://doi.org/10.1029/2007GL030359

De Mets C, Gordon RG, Argus DF et al (1990) Current plate motion. Geophys J Int 101:425–478

De Mets C, Gordon RG, Argus DF et al (1994) Effects of recent revision to the geomagnetic reversal time scale on estimates of current plate motion. Geophys Res Lett 21:2191–2194. https://doi.org/10.1029/94GL02118

Ekström G, England P (1989) Seismic strain rates in regions of distributed continental deformation. J Geophys Res 94(10). https://doi.org/10.1029/JB094iB08p10231tb06847.x

Forte A, Cowgill E, Bernardin T et al (2010) Late Cenozoic deformation of Kura fold-thrust belt, southern greater Caucasus. Geol Soc Am Bull 122:465–486

Forte A, Cowgill E, Whipple KX (2014) Transition from a singly vergent to doubly vergent wedge in a young orogen: the greater Caucasus. Tectonics 33:2077–2101

Fuenzalida H, Rivera L, Haessler H et al (1997) Seismic source study of the Racha-Dzhava (Georgia) earthquake from aftershocks and broad-band teleseismic body-wave records: an example of active nappe tectonics. Geophys J Int 130:29–46

Gamkrelidze I (1986) Geodynamic evolution of the Caucasus and adjacent areas in alpine time. Tectonophysics 127:261–267

Gasanov A (2003) Earthquakes of Azerbaijan for 1983–2002. Elm, Baku. (in Russian)

Guliev IS, Kadirov FA, Reilinger RE et al (2002) Active tectonics in Azerbaijan based on geodetic, gravimetric, and seismic data. In: Doklady earth sciences c/c of Doklady-Akademiia Nauk, vol 383. Interperiodica Publishing, pp 174–177

Jackson J, Ambraseys NN (1997) Convergence between Eurasia and Arabia in eastern Turkey and the Caucasus. In: Giardini D, Balassanian S (eds), Historical and prehistorical earthquakes in the Caucasus, NATO ASI. Ser. Environ 28, pp 79–90

Jackson J, McKenzie D (1984) Active tectonics of the alpine-Himalayan Belt between western Turkey and Pakistan. Geophys J R Astron Soc 77:185–264

Jackson J, McKenzie D (1988) The relationship between plate motions and seismic moment tensors, and the rates of active deformation in the Mediterranean and Middle East. Geophysical Geophys J R Astron Soc 93(1):45–73. https://doi.org/10.1111/j.1365-246X.1988.tb01387.x

Kadirov F, Mammadov S, Reilinger R et al (2008) Some new data on modern tectonic deformation and active faulting in Azerbaijan (according to global positioning system measurements). In: Proceedings of Azerbaijan National Academy of Sciences. Earth Sci 1, pp 82–88

Kadirov FA, Floyd MA, Alizadeh A et al (2012) Kinematics of the Caucasus near Baku, Azerbaijan. J Nat Hazards 63:997–1006

Kadirov FA, Gadirov AG, Babayev GR et al (2013) Seismic zoning of the southern slope of greater Caucasus from the fractal parameters of the earthquakes, stress state, and GPS velocities. Izv, Phys Solid Earth 49(4):554–562

Kadirov FA, Floyd M, Reilinger R et al (2015) Active geodynamics of the Caucasus region: implications for earthquake hazards in Azerbaijan. In: Proceedings of Azerbaijan National Academy of Sciences. Sci Earth 3, pp 3–17

Karimiparidari S, Zaré M, Memarian H et al (2013) Iranian earthquakes, a uniform catalog with moment magnitudes. J Seismol 17(3):897–911

Kazmin VG (2002) Expulsion tectonics in the Black Sea–Caucasus area. In: 1st international symposium of Istanbul Technical University, the Faculty of Mines on Earth Sciences and Engineering, Istanbul, 16–18 May 2002

Koçyiğit A, Yılmaz A, Adamia S et al (2001) Neotectonics of East Anatolia plateau (Turkey) and lesser Caucasus: implication for transition from thrusting to strike-slip faulting. Geodin Acta 14:177–195

Kondorskaya N, Shebalin N (eds) (1982) New catalogue of strong earthquakes in the USSR from ancient times through 1977. NOAA, USA

Lander AV (2004) The FA2002 program system to determine the focal mechanisms of earthquakes in Kamchatka, the Commander Islands and the northern Kuriles. In: Report KEMSD GS RAS, Petropavlovsk-Kamchatsky

Lund B, Townend J (2007) Calculating horizontal stress orientations with full or partial knowledge of the tectonic stress tensor. Geophys J Int 170(3):1328–1335

McClusky S, Balassanian S, Barka A et al (2000) Global positioning system constraints on plate kinematics and dynamics in the eastern Mediterranean and Caucasus. J Geophys Res 105:5695–5719. https://doi.org/10.1029/1999JB900351

Meijers MJ, Smith B, Kirscher U et al (2015) A paleolatitude reconstruction of the South Armenian block (lesser Caucasus) for the late cretaceous: constraints on the Tethyan realm. Tectonophysics 644:197–219

Michael AJ (1984) Determination of stress from slip data: faults and folds. J Geophys Res 89:11,517–11,526

Milyukov V, Kopaev A, Zharov V et al (2010) Monitoring crustal deformations in the northern Caucasus using a high precision long base laser strainmeter and the GPS/GLONASS network. J Geodyn 49(3–4):216–223

Milyukov VK, Mironov AP, Rogozhin EA et al (2015a) Velocities of contemporary movements of the northern Caucasus estimated from GPS observations. Geotectonics 49(3):210–218

Milyukov VK, Mironov AP, Steblov GM et al (2015b) The contemporary GPS-derived horizontal motions of the main elements of tectonic structure in the Ossetian segment of greater Caucasus. Izv, Phys Solid Earth 51(4):522–534

Milyukov VK, Mironov AP, Ovsyuchenko AN et al (2018) Velocities of present-day horizontal movements in the central sector of the greater Caucasus according to GPS observations and their relation to tectonics and the deep structure of the Earth's crust. In: Doklady earth sciences, vol 481. Pleiades Publishing, pp 879–882

Mosar J, Kangarli T, Bochud M et al (2010) Cenozoic–recent tectonics and uplift in the greater Caucasus: a perspective from Azerbaijan. In: Sosson M, Kaymakçı N, Stephenson R, Bergerat F (eds) Sedimentary basin tectonics from the Black Sea and Caucasus to the Arabian platform, Geological Society, vol 340. Special Publication, London, pp 261–279

Mumladze T, Forte AM, Cowgill ES et al (2015) Subducted, detached, and torn slabs beneath the greater Caucasus. Geo Res J 5:36–46

Nemčok M, Feyzullayev A, Kadirov A et al (2011) Neotectonics of the Caucasus and Kura valley, Azerbaijan. Global Eng Technol Rev 1:1):1–1)14

O'Connor TM, Szymanski E, Krasovec M et al (2005) Rethinking the tectonic model of the Caucasus: an investigation of the southern section of the proposed Borjomi-Kazbegi fault. In: Abstracts of the AGU Fall Meeting, December 2005

Pasquarè FA, Tormey D, Vezzoli L et al (2011) Mitigating the consequences of extreme events on strategic facilities: evaluation of volcanic and seismic risk affecting the Caspian oil and gas pipelines in the republic of Georgia. J Environ Manag 92:1774–1782

Philip H, Cisternas A, Gvishiani A et al (1989) The Caucasus: an actual example of the initial stages of continental collision. Tectonophysics 161:1–21

Pondrelli S, Morelli A, Boschi E (1995) Seismic deformation in the Mediterranean area estimated by moment tensor summation. Geophys J Int 122:938–952. https://doi.org/10.1111/j.1365-246X.1995.tb06847.x

Prilepin MT, Balasanyan S, Baranova SM et al (1997) The kinematic study of the Caucasus region using GPS techniques. Izv, Phys Solid Earth 33(6):486–493

Rautian T (1964) Ob opredelenii energii zemletryaseniy nrasstoyaniyakh do 3000 km. Akademiya Nauk SSSRTrudy Instituta Fiziki Zemli 32:88–93. (in Russian)

Rebai S, Philip H, Dorbath L et al (1993) Active tectonics in the lesser Caucasus: coexistence of compressive and extensional structures. Tectonics 12(5):1089–1114

Reilinger R, Barka A (1997) In: Giardini D, Balassanian S (eds) GPS constraints on fault slip rates in the Arabia-Africa-Eurasia plate collision zone: implications for earthquake recurrence times, Historical and prehistorical earthquakes in the Caucasus. Kluwer Academic Publishers, pp 91–108

Reilinger RE, McClusky S, Oral MB et al (1997) Global positioning system measurements of present-day crustal movements in the Arabia-Africa-Eurasia plate collision zone. J Geophys Res 102(5):9983–9999

Reilinger RE, McClusky S, Lawrence S (2004) GPS constraints on continental deformation in the eastern Mediterranean and Caucasus region: evidence for plate/block-like behavior of the continental lithosphere. In: Abstracts of the AGU fall meeting, 2004

Reilinger RE, McClusky S, Vernant P et al (2006) GPS constraints on continental deformation in the Africa-Arabia-Eurasia continental collision zone and implications for the dynamics of plate interactions. J Geophys Res Solid Earth 111(B5). https://doi.org/10.1029/2005JB004051

Ruppel C, McNutt M (1990) Regional compensation of the greater Caucasus mountains based on an analysis of Bouguer gravity data. Earth Planet Sci Lett 98(3–4):360–379

Saintot A, Brunet MF, al YF (2006) The Mesozoic-Cenozoic tectonic evolution of the greater Caucasus. In: Gee DG, Stephenson RA (eds) European lithosphere dynamics, Geological Society, vol 32. Special Publication, London, pp 277–289

Sargsyan GV, Abgaryan GR, Mkhitaryan KA et al (2017) Seismic conditions of the territory of Armenia and adjacent areas after the 1988 Spitak earthquake. Publishing house of Eldorado, Gyumri, p 279. (in Russian)

Shahvar MP, Zare M, Castellaro S (2013) A unified seismic catalog for the Iranian plateau (1900–2011). Seismol Res Lett 84:233–249. https://doi.org/10.1785/0220120144

Shebalin NV, Tatevossian RE (1997) Catalogue of large historical earthquakes of the Caucasus. In: Historical and Prehistorical earthquakes in the Caucasus, vol 28. Kluwer Academic Publishers, Dordrecht, pp 201–232

Shevchenko VI, Guseva TV, Lukk AA et al (1999) Recent geodynamics of the Caucasus Mountains from GPS and seismological evidence. Izv, Phys Solid Earth 35(9):691–704

Sobornov KO (1994) Structure and petroleum potential of the Dagestan thrust belt, northeastern Caucasus, Russia. Bull Can Pet Geol 42(3):352–364

Sokhadze G, Floyd M, Godoladze T et al (2018) Active convergence between the lesser and greater Caucasus in Georgia: constraints on the tectonic evolution of the lesser–greater Caucasus continental collision. Earth Planet Sci Lett 481:154–161. https://doi.org/10.1016/j.epsl.2017.10.007

Sosson M, Rolland Y, Danelian T et al (2010) Subductions, obduction and collision in the lesser Caucasus (Armenia Azerbaijan, Georgia), new insights. In: Sosson M, Kaymakci N, Stephanson R, Bergarat F, Storatchenoko V (eds) Sedimentary Basin tectonics from the Black Sea and Caucasus to the Arabian platform, Geological Society, vol 340. Special Publication, London, pp 329–352

Sosson M, Adamia SH, Muller C et al (2013) From greater to lesser Caucasus: new insights from surface and subsurface data along a N-S trending transect (Georgia): thick-skin versus thin-skin tectonics. Darius News 3:5–7

Tan O, Taymaz T (2006) Active tectonics of the Caucasus: earthquake source mechanisms and rupture histories obtained from inversion of teleseismic body waveforms. In: Dilek Y, Pavlides S (eds) Postcollisional tectonics and magmatism in the Mediterranean region and Asia, Special Paper, vol 409. Geological Society of America, pp 531–578

Telesca L, Babayev G, Kadirov F (2012) Temporal clustering of the seismicity of the Absheron-Prebalkhan region in the Caspian Sea area. Nat Hazards Earth Syst Sci 12:3279–3285

Telesca L, Lovallo M, Babayev G, Kadirov F (2013) Spectral and informational analysis of seismicity: an application to the 1996–2012 seismicity of the northern Caucasus–Azerbaijan part of the greater Caucasus–Kopet dag region. Physica A: Statistical Mechanics and its Applications 392(23):6064–6078

Telesca L, Kadirov F, Yetirmishli G, Safarov R, Babayev G, Ismaylova S (2017) Statistical analysis of the 2003-2016 seismicity of Azerbaijan and surrounding areas. J Seismol 21:1467–1485. https://doi.org/10.1007/s10950-017-9677-x

Tibaldi A, Corazzato C, Rust D et al (2015) Tectonic and gravity-induced deformation along the active Talas–Fergana fault, Tien Shan, Kyrgyzstan. Tectonophysics 657:38–62

Tibaldi A, Alania V, Bonali FL et al (2017a) Active inversion tectonics, simple shear folding and back-thrusting at Rioni Basin, Georgia. J Struct Geol 96:35–53

Tibaldi A, Russo E, Bonali FL et al (2017b) 3-D anatomy of an active fault-propagation fold: a multidisciplinary case study from Tsaishi, western Caucasus (Georgia). Tectonophysics 717:253–269

Tibaldi A, Bonali FL, Russo E et al (2018) Structural development and stress evolution of an arcuate fold-and-thrust system, southwestern greater Caucasus, Republic of Georgia. J Asian Earth Sci 156:226–245

Tibaldi A, Oppizzi P, Gierke J et al (2019a) Landslides near Enguri dam (Caucasus, Georgia) and possible seismotectonic effects. Nat Hazards Earth Syst 19(1):71–91

Tibaldi A, Tsereteli N, Varazanashvili O et al (2019b) Active stress field and fault kinematics of the greater Caucasus. J Asian Earth Sci 188:104108. https://doi.org/10.1016/j.jseaes.2019.104108

Triep EG, Abets GA, Lerner-Lam AL et al (1995) Active thrust front of the greater Caucasus: the April 29, 1991, Racha earthquake sequence and its tectonic implications. J Geophys Res 100(B3):4011–4033

Tsereteli N, Tibaldi A, Alania V et al (2016) Active tectonics of Central-Western Caucasus, Georgia. Tectonophysics 691:328–344. https://doi.org/10.1016/j.tecto.2016.10.025

Varazanashvili O, Tsereteli N, Tsereteli E (2011) Historical earthquakes in Georgia (up to 1900): source analysis and catalogue compilation. Monograph, Pub Hause MVP, Tbilisi

Varazanashvili O, Tsereteli N, Bonali FL et al (2018) GeoInt: the first macroseismic intensity database for the republic of Georgia. J Seismol 22(3):625–667

Vernant P, Nilforoushan F, Hatzfeld D et al (2004) Present-day crustal deformation and plate kinematics in the Middle East constrained by GPS measurements in Iran and northern Oman. Geophys J Int 157:381–398. https://doi.org/10.1111/j.1365-246X.2004.02222.x

Vincent SJ, Morton AC, Carter A et al (2007) Oligocene uplift of the Western greater Caucasus: an effect of initial Arabia-Eurasia collision. Terra Nova 19:160–166

Vincent SJ, Carter A, Lavrishchev VA et al (2011) The exhumation of the western greater Caucasus: a thermochronometric study. Geol Mag 148:1–21

Westaway R (1990) Seismicity and tectonic deformation rate in soviet Armenia: implications for local earthquake hazard and evolution of adjacent regions. Tectonics 9(3):477–503. https://doi.org/10.1029/TC009i003p00477

Yetirmishli GJ, Abdullayeva RR, Kazimova ES et al (2015) Strong earthquakes on the territory of Azerbaijan for the period of 2012–2014. Seismoprog Obs Territ Azerb 12(1):19–26

Zare M, Amini H, Yazdi P et al (2014) Recent developments of the Middle East catalog. J Seismol 18(N4):749–772

Chapter 5
Structural Architecture of the Western Greater Caucasus Orogen: New Data from a Crustal-Scale Structural Cross-Section

Victor Alania, Alessandro Tibaldi, Fabio L. Bonali, Onise Enukidze, and Elena Russo

Abstract A new, crustal-scale structural model aimed at illustrating the recent geometry of the western Greater Caucasus, has been built based on surface and subsurface data, in agreement with the theory of fault-related folds. The crustal-scale, structural cross-section shows that the western Greater Caucasus has a double vergence with a pro- and a retro-wedge actively propagating towards the Rioni foreland basins to the south and the Terek basin to the north. The retro-wedge and the Terek foreland basin developed on the upper plate, whereas the pro-wedge and the Rioni foreland basin developed on the lower, subducting plate. The formation of an antiformal stack in the axial zone of the western Greater Caucasus is related to reactivation of pre-existing normal faults during the orogeny. The main style of deformation within the pro-and retro-wedge is a series of fault-propagation and fault-bend folds, duplexes and structural wedges. Within the frontal part of the pro- and retro-wedge, a series of thrust-top basins developed.

Keywords Western greater Caucasus · Structural cross-section · Pro-wedge · Retro-wedge · Double wedge orogen

---

V. Alania (✉) · O. Enukidze
M. Nodia Institute of Geophysics, Iv. Javakhishvili Tbilisi State University, Tbilisi, Georgia
e-mail: victor.alania@tsu.ge

A. Tibaldi · F. L. Bonali · E. Russo
University of Milan Bicocca, Milan, Italy

CRUST-Interuniversity Center for 3D Seismotectonics with Territorial Applications, Chieti Scalo, Italy

© Springer Nature B.V. 2021
F. L. Bonali et al. (eds.), *Building Knowledge for Geohazard Assessment and Management in the Caucasus and other Orogenic Regions*, NATO Science for Peace and Security Series C: Environmental Security,
https://doi.org/10.1007/978-94-024-2046-3_5

5.1 Introduction

The Greater Caucasus (GC) is a more than 1100 km long NW–SE directed mountain range located between the Crimea and Caspian Sea (Fig. 5.1). The GC was developed as an intracontinental mountain belt resulting from the Late Cenozoic structural inversion of a Mesozoic–Early Cenozoic back-arc basin associated with the Arabia/Eurasia convergence (e.g. Adamia et al. 1977, 1981; Dercourt et al. 1986; Khain 1974; Mosar et al. 2010; Nikishin et al. 2010; Sosson et al. 2016).

Many tectonic models of the western GC have been developed over the past 40 years, based on geological and topographical data, seismic profiles and seismicity

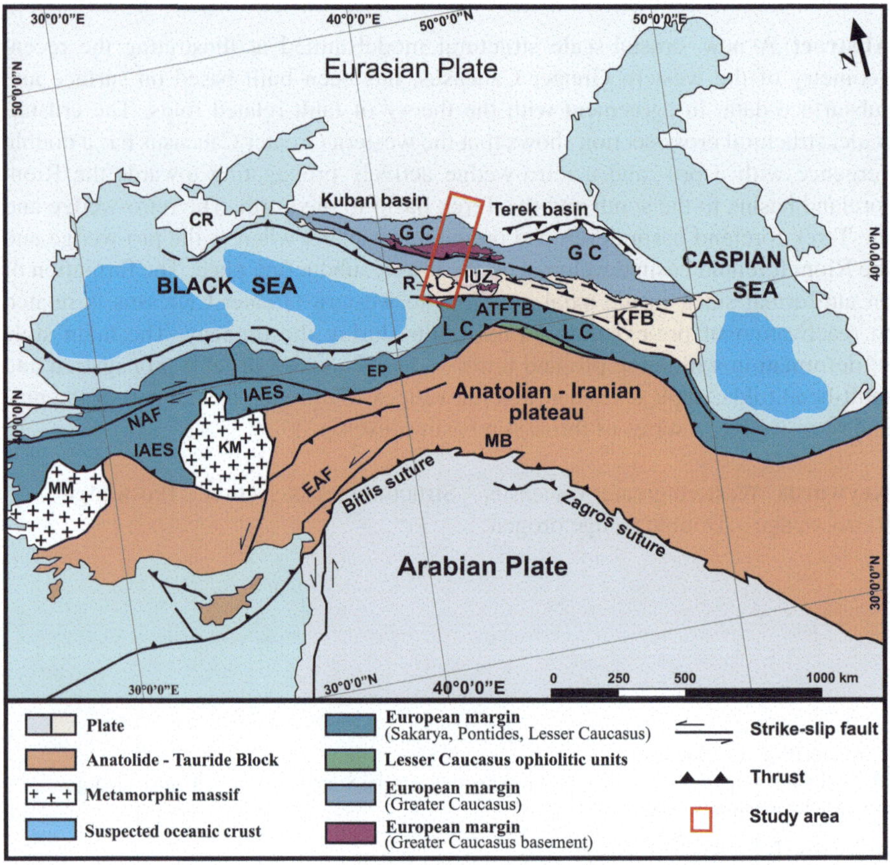

Fig. 5.1 Tectonic map of the Arabia-Eurasia collision zone. (Modified from Sosson et al. 2016). Abbreviations: *CR* Crimea, *GC* Greater Caucasus, *LC* Lesser Caucasus, *R* Rioni, *IUZ* Imereti uplift zone, *Dz* Dzirula, *KFB* Kura Foreland Basin, *ATFTB* Achara-Trialeti fold-and-thrust belt, *MB* Mus Basin, *EP* Eastern Pontides, *KM* Kirsehir Massif, *EAF* Eastern Anatolian Fault, *NAF* North Anatolian Fault, *IAES* Izmir-Ankara-Erzincan Suture, *MM* Menderes Massif

(e.g. Adamia et al. 2011a, 2011b; Banks et al. 1997; Cowgill et al. 2016; Dotduyev 1986; Ershov et al. 1999, 2003; Forte et al. 2014; Gamkrelidze 1986; Giorgobiani and Zakaraya 1989; Mumladze et al. 2015; Nikishin et al. 2010, 2012; Robinson et al. 1996; Saintot et al. 2006; Tan and Taymaz 2006; Tari et al. 2018; Tibaldi et al. 2018, 2020; Varazanashvili et al. 2018; Vincent et al. 2016). Several structural cross-sections through the western GC have been interpreted as a single-vergent orogenic wedge developed above the top of the north-dipping subducting slab (e.g. Cowgill et al. 2016; Forte et al. 2014; Mumladze et al. 2015). According to Forte et al. (2014), the GC is an important natural example of the transition from single- to double-vergent orogen. In contrast to the central and eastern parts, which represent a double-vergent orogenic wedge (Cowgill et al. 2016; Forte et al. 2014; Mosar et al. 2010; Sobornov 1994, 1996; Sosson et al. 2016), the western GC is predominantly a single-vergent orogen (Cowgill et al. 2016; Forte et al. 2014).

The onset of compressional deformation in the GC took place in Late Eocene-Oligocene times, after the end of the Eocene rifting events within the GC basin (Adamia et al. 2011a; Khain 1974; Mosar et al. 2010; Nikishin et al. 2010; Vincent et al. 2016, 2018). The growth of the GC was mainly influenced by the reactivation of former extensional faults at the crustal scale (e.g. Mosar et al. 2010; Nikishin et al. 2010; Sosson et al. 2016; Vincent et al. 2016, 2018). Based on fission-track data, the exhumation process in the GC started at the end of the Late Eocene-Oligocene and reached its peak in the Pliocene-Pleistocene (Avdeev and Niemi 2011; Vincent et al. 2007, 2011). Recent GPS and earthquake data indicate that the western GC is still tectonically active (e.g. Adamia et al. 2017; Reilinger et al. 2006; Tan and Taymaz 2006; Tsereteli et al. 2016; Varazanashvili et al. 2018; Tibaldi et al. 2020).

In this paper, we present a crustal-scale structural cross-section across the western GC orogenic system, in order to constrain the subsurface geometries and deformation structural style. Our analysis uses subsurface structural geologic techniques to interpret the structural cross-section according to the theory of fault-related folding (Shaw et al. 2006; Suppe 1983; Suppe and Medwedeff 1990) and orogenic wedge models (e.g. Naylor and Sinclair 2008).

5.2 Tectonic Setting

The western GC, from south to north, is divided into five structural domains (Fig. 5.2): the Rioni Basin (RB) (western part of the Transcaucasian intermountain depression), the Southern Slope of the GC (SSGC), the Main Range of the GC (MRGC), the Northern Slope of the GC (NSGC) and the Scythian Young platform – Terek basin (Adamia et al. 2011a, 2011b). The RB is located between the western GC and Lesser Caucasus (LC) orogens and can be divided into the Rioni foreland basin (RFB) and the Rioni foreland fold-and-thrust belt (RFFTB) (Adamia et al. 2010; Adamia et al. 2011c; Banks et al. 1997; Cowgill et al. 2016; Tibaldi et al. 2017a, 2017b).

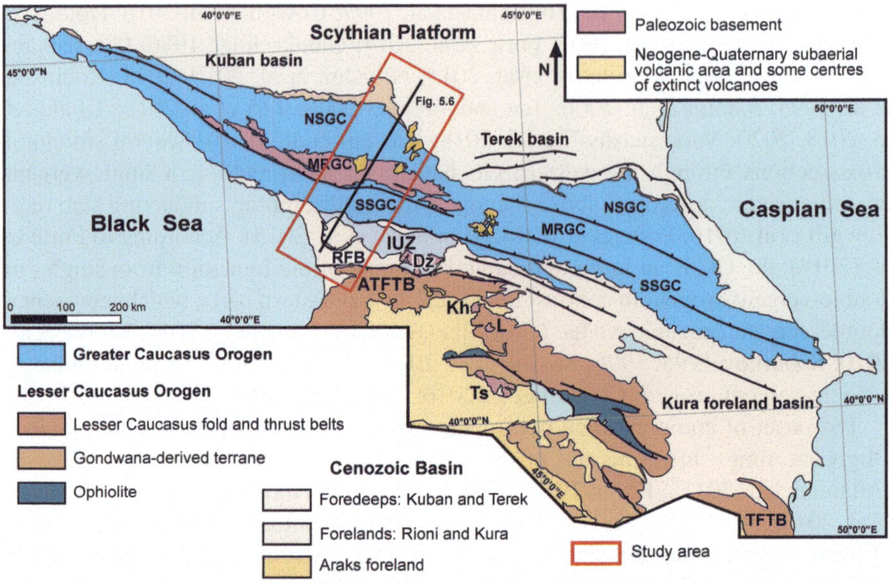

Fig. 5.2 Simplified tectonic map of the Caucasus. (modified after Adamia et al. 2011a, 2011b; Mosar et al. 2010; Sosson et al. 2016). Abbreviations: *SSGC* Southern Slope of Greater Caucasus, *MRGC* Main Range of Greater Caucasus, *NSGC* Northern Slope of Greater Caucasus, *RFB* Rioni foreland basin, *IUZ* Imereti uplift zone, *ATFTB* Achara-Trialeti fold-and-thrust belt, *TFTB* Talysh fold-and-thrust belt, *DZ* Dzirula, *Kh* Khrami, *L* Loki, *Ts* Tsakhkuniats

The RFFTB is characterized by curve-shaped thrust trajectories and is dominated by a thin-skinned tectonic style (Adamia et al. 2010, 2017; Banks et al. 1997; Tari et al. 2018; Tibaldi et al. 2017a, 2017b). According to Tibaldi et al. (2018), the structural development of the arcuate RFFTB system is related to oblique, asymmetric indentation of an upper crustal block moving to the SSW. The sedimentary cover of the RB is commonly >5 km thick and is composed of Jurassic, Cretaceous, Paleogene and Miocene-Pleistocene strata. The MRGC is represented by the Variscan basement, whereas the more external zones (SSGC and NSGC) comprise mainly Jurassic, Cretaceous and Cenozoic sedimentary rocks (e.g. Adamia et al. 2011a; Vincent et al. 2016).

South-vergent fault-related folds constitute the structure of the SSGC. The Variscan basement of the MR of western GC is made of granitoids, metamorphic rocks and their Paleozoic cover (e.g. Adamia et al. 2011a). Upper Jurassic to Eocene, mainly shallow water siliciclastic sediments ($\sim$ 4 km) are preserved along the NSGC and Scythian platforms of the pre-Mesozoic basement (Vincent et al. 2016). Within the RB, the Oligocene–Lower Miocene (the Maykopian series) is represented mainly by an alternation of clays and sandstones; their thickness is about 800-900 m. The Middle-Upper Miocene, Pliocene and Pleistocene units are represented by shallow marine and continental sediments. The thickness of these units is about 1500-2000 m. The Holocene is represented by undifferentiated continental strata and river-bed alluvial deposits (e.g. Adamia et al. 2010).

5.3 The Crustal-Scale Structural Cross-Section

In order to determine the structural style of deformation, we have constructed a crustal-scale structural cross-section through the western GC, elaborated by using surface geology from 1:200.000 scale geological maps and geological cross-sections from the northern part of the study area (Djanelidze and Kandelaki 1956; Luparev and Prokhorov 1957; Potapenko 1961) (Fig. 5.3), as well as published, interpreted seismic reflection profiles (Tibaldi et al. 2018a) (Figs. 5.4a, b) and original field data. The lithostratigraphy for the western GC is based on Adamia et al. (2011a) and Vincent et al. (2016).

Within the RFFTB, two levels of detachment have been distinguished (Tibaldi et al. 2017a, 2017b). The upper detachment has been identified in the Upper Jurassic evaporites whereas the lower detachment occurs within Lower Jurassic shales (Fig. 5.4b). The seismic reflection profile (R-15) reveals the presence of upper and lower structural complexes. The upper structural complex is represented by the shallow, south-vergent Tsaishi fault-propagation fold. The back-limb of the Tsaishi anticline is complicated by three back-thrusts. The lower structural complex is represented by a duplex and is made of Jurassic strata (Fig. 5.4b).

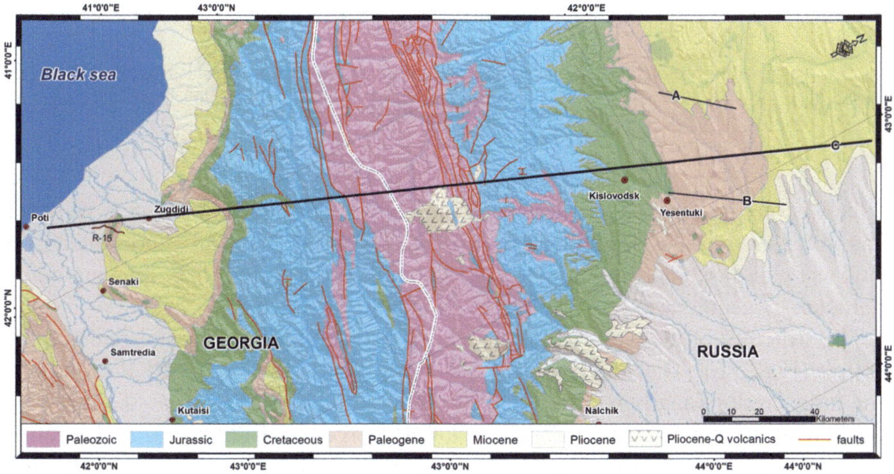

Fig. 5.3 Geological map of study area. Modified after the geological map of the western Caucasus (Djanelidze and Kandelaki 1956, K-38-XIII; Luparev and Prokhorov 1957, L-38-XXXI; Potapenko 1961, K-38-I. Scale 1:200,000. Ministry of Geology and Mineral Protection, USSR)

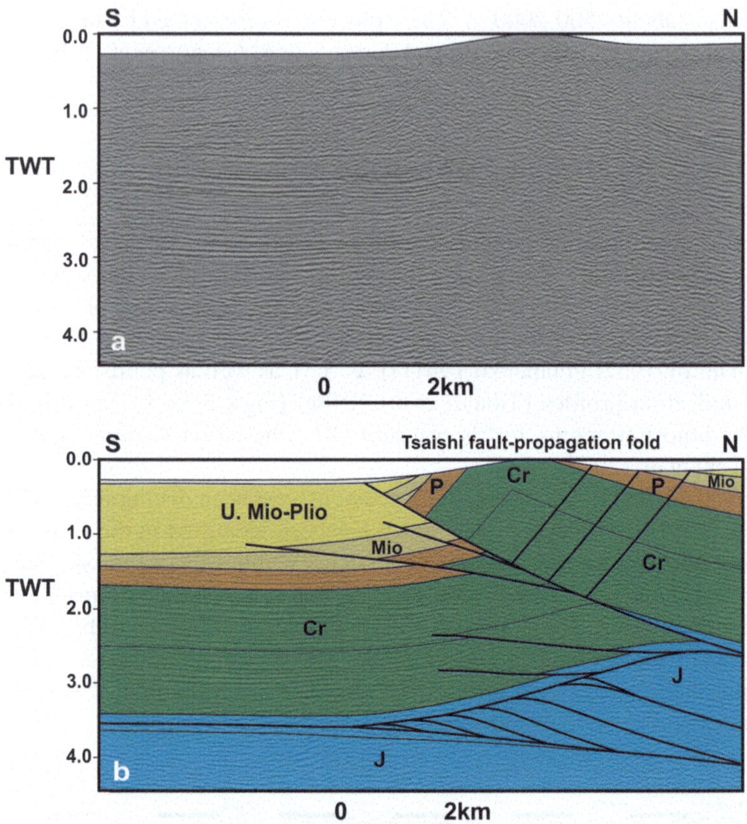

Fig. 5.4 (**a**) Uninterpreted and (**b**) interpreted seismic reflection profiles R-15 (Tibaldi et al. 2017a). Location is shown in Fig. 5.3. Abbreviations: *J* Jurassic, *Cr* Cretaceous, *P* Paleogene, *Mio* Miocene, *U. Mio-Plio* Upper Miocene-Pliocene

Two geological cross-sections through the NSGC, provided in Fig. 5.5, were re-interpreted on the basis of published data (Potapenko 1961) and are represented by a north-vergent basement structural wedge and fault-propagation folds showing many characteristics similar to retro-wedge model features (Fig. 5.5).

The structure of the western GC is here described along the crustal-scale cross-section (Fig. 5.6). The length of the structural cross-section is 298 km and the structures predominantly record a N-S shortening direction. The cross-section (Fig. 5.6) shows that the western GC has a double vergence, with a pro- and a retro-wedge actively propagating into the foreland basins of Rioni to the south and Terek to the north. On the basis of crustal-scale structural cross-section, the western GC could be divided, from south to north, into five main structural units: (1) the RFB, (2) the southwestern GC, forming the pro-wedge of the orogen (SSGC and RFFTB) where Jurassic, Cretaceous, and Tertiary rocks are exposed, (3) the Main Range (or Axial Zone), made by an antiformal stack duplex involving the basement, (4) the

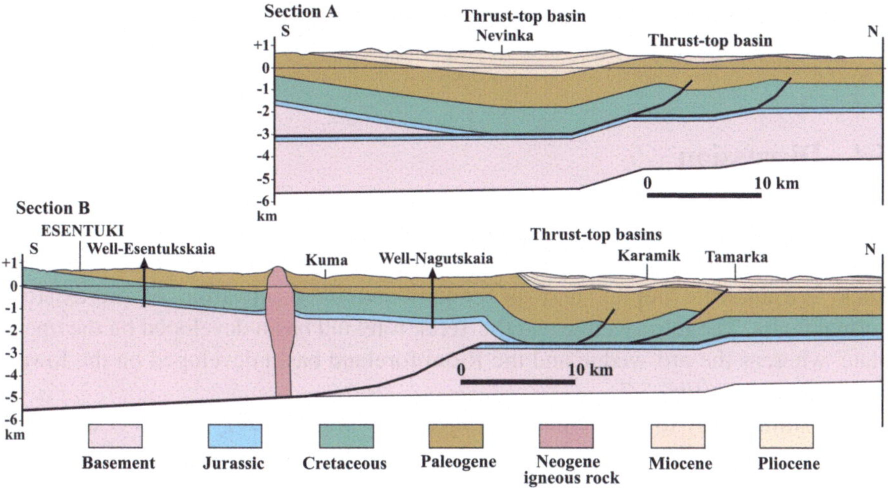

Fig. 5.5 Geological cross sections (A, B) through the Northern Slope of the Greater Caucasus. (Modified after Potapenko 1961, K-38-I. Scale 1:200.000. Ministry of Geology and Mineral Protection USSR). Location is shown in Fig. 5.3

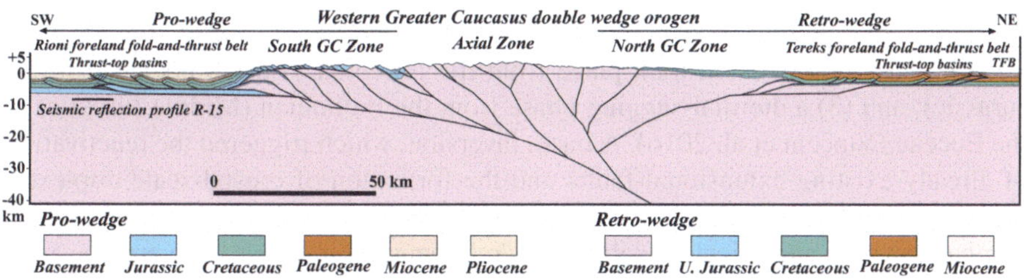

Fig. 5.6 Crustal-scale, structural cross-section (C). Location is shown in Fig. 5.2

northeastern GC, which represents the retro-wedge of the orogeny, where Upper Jurassic, Cretaceous and Tertiary rocks are exposed, and (5) the Terek Foreland Basin (TFB).

The structure of the southwestern and northeastern GC is similar to a basement-involved thin-skinned fold and thrust belt (e.g. Pfiffner 2006) and is represented by south- and north-vergent structural wedges and fault-bend and fault-propagation folds. The frontal structure of the NW GC Zone is represented by two south-vergent fault-propagation folds. The formation of these structures is related to the northward propagating duplex (Fig. 5.6).

The crustal-scale structural cross-section (Fig. 5.6) and the two geological cross-sections across the northern part of the western GC (Fig. 5.5) show that pro-and retro wedge synclines are filled by the Miocene shallow marine and continental syn-tectonic sediments, forming a typical thrust-top basin (e.g. Butler and Grasso 1993). The growth anticline is mainly represented by fault-propagation folds (Fig. 5.6).

5.4 Discussion

The tectonic model across the western GC (Fig. 5.7) demonstrates that basement uplift is related to deep-seated thrusting; moreover, the formation of an antiformal stack, crustal-scale duplex may be conditioned by reactivation of pre-existing normal faults. The retro-wedge and the Terek foreland basin developed on the upper plate, whereas the pro-wedge and the Rioni foreland basin developed on the lower, subducting plate (Fig. 5.7).

In a similar way to the Pyrenean orogen (e.g. Muñoz 1992; Verges et al. 1995) and numerical models of Alpine-type orogens (e.g. Erdos et al. 2014; Grool et al. 2019), it is possible to identify three main factors that control the deformational structural style of an orogenic system, such as the western GC: (1) a rifting inheritance, (2) the efficiency of decoupling between thin-skinned and thick-skinned deformation, and (3) the syn-orogenic sedimentation.

The tectono-stratigraphic chart provided in Fig. 5.8 shows the evolution of the western GC. It is possible to distinguish several phases of extension Within the GC back-arc basin: (1) the first phase started in the Sinemurian (Early Jurassic); (2) a main phase of extension took place from the Aalenian to the Bajocian (Middle Jurassic); and (3) a thermal-sagging phase from the Bathonian (Middle Jurassic) to the Eocene (Vincent et al. 2016). A basin inversion, which triggered the reactivation of already-existing extensional faults and the formation of crustal-scale duplexes, began in Late Eocene-Oligocene times (e.g. Vincent et al. 2016). As a result of a comparison with numerical modeling research focused on Alpine-type orogens (e.g. Erdos et al. 2014; Grool et al. 2019), we suggest that build-up of the western GC is related to a two-phase compressive deformation: (1) an early, symmetric inversion

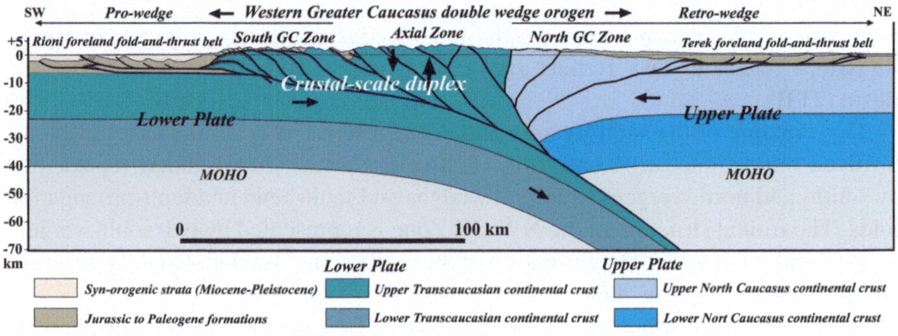

Fig. 5.7 Tectonic model of the wGC. Location is the same as shown in Fig. 5.6

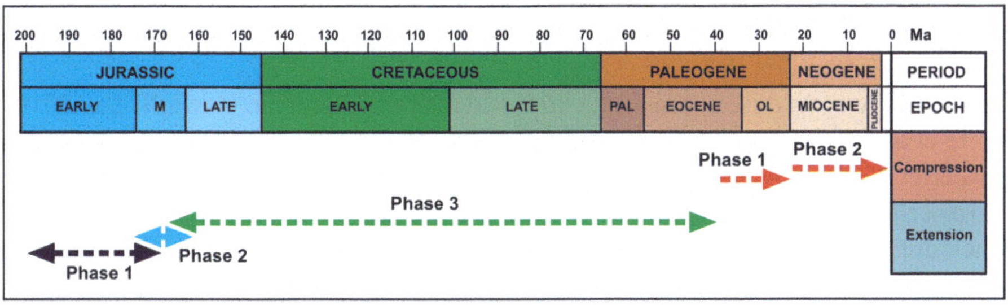

Fig. 5.8 Tectonostratigraphic chart. (Modified after Vincent et al. 2016)

during the Late Eocene-Oligocene, and (2) the formation of an asymmetric, double vergent structure during the Miocene-Pliocene. The asymmetric shortening may indicate the onset of the subduction of continental crust as well. From Miocene times onward, the compressional deformation within pro-and retro-wedge foreland basins and the onset of Miocene syn-orogenic sedimentation took place. A similar picture can be observed in the Kura foreland fold-and-thrust belt, where compressional deformation began from the Middle Miocene (Alania et al. 2017). The geometry of syn-orogenic sedimentation, associated with footwall synclines and the sedimentary infill of thrust-top basins, provides information on the thrusting activity within retro- and pro-wedges of the western GC orogen. The evolution of the thrust-top basins was mainly controlled by the kinematics of competing growth fault-propagation folds (Fig. 5.6).

The build-up of north- and south-vergent structures in the southwestern and northeastern GC was governed by the basement wedge(s) propagation to the south and to the north along detachment horizons within the cover-generating thin-skinned structures made of Mesozoic and Tertiary strata. A major detachment in the retro-wedge can be observed within the Upper Jurassic-Cretaceous boundary: the structure is mainly represented by north-vergent fault-propagation folds and duplexes. The frontal part of the Terek foreland fold-and-thrust belt is introduced by a triangle zone (Fig. 5.6). The formation of south-vergent fault-related folds within the pro-wedge is related to two detachment levels, and can be interpreted mainly in terms of asymmetric fault-propagation folding with a steep and narrow forelimb and a gentle and wide backlimb (Fig. 5.6).

In summary, the presented structural model shows that the western GC is an asymmetric, double-wedge orogen. The inversion of a GC back-arc rift basin during Oligocene-Neogene time generated an upper crustal, antiformal stack duplex, an outer pro-wedge fold-and-thrust belt, and a retro-wedge fold-and-thrust belt.

The above illustrated, new results on the structural architecture of the western GC, certainly need to be integrated by further structural studies, leading to new, balanced and reconstructed cross-sections, as well as kinematic models and the assessment of the degree of shortening.

5.5 Conclusions

In a similar way as the eastern and central part of the GC orogen (Cowgill et al. 2016; Forte et al. 2014; Mosar et al. 2010; Sobornov 1994, 1996; Sosson et al. 2016), the western GC is a typical, double-vergent orogenic structure represented by pro- and retro-wedges and the associated foreland basins. The formation of an antiformal stack duplex involving the Variscan basement in the axial zone of the western GC is related to reactivation of pre-existing normal faults during the orogeny. Thick-skinned deformation style is dominant within the NE and SW areas of this part of the GC, whereas in the RFB and TFB, a thin-skinned tectonic style of deformation dominates. The main style of deformation within the pro-and retro-wedges is represented by a set of fault-related folds (fault-bend and fault-propagation folds) and their duplexes. Within the frontal part of pro- and retro-wedges of the western GC, a series of thrust-top basins developed.

Acknowledgments This study has been carried out in the framework of the NATO project SfP G4934 "Georgia Hydropower Security". We acknowledge the useful review and suggestions of Federico Pasquaré Mariotto, Alexander Chabukiani and Nino Sadradze on a previous version of the manuscript.

References

Adamia S, Lordkipanidze MB, Zakariadze GS (1977) Evolution of an active continental margins exemplified by the alpine history of the Caucasus. Tectonophysics 40:183–199

Adamia S, Chkhotua T, Kekelia M et al (1981) Tectonics of Caucasus and adjoining regions: implications for the evolution of the Tethys Ocean. J Struct Geol 3:437–447

Adamia S, Alania V, Chabukiani A et al (2010) Evolution of the late Cenozoic basins of Georgia (SW Caucasus): a review. Geol Soc Lond Spec Publ 340:239–259. https://doi.org/10.1144/SP340.11

Adamia S, Alania V, Chabukiani A et al (2011a) Great Caucasus (Cavcasioni): Longliving Northtethyan Back-Arc Basin. Turk J Earth Sci 20:611–628. https://doi.org/10.3906/yer-1005-12

Adamia S, Zakariadze G, Chkhotua T et al (2011b) Geology of the Caucasus: a review. Turk J Earth Sci 20:489–544. https://doi.org/10.3906/yer-1005-11

Adamia S, Alania V, Chagelishvili R et al (2011c) Tectonic setting of Georgia (Caucasus). Geo-Eco-Marina 17:11–13

Adamia S, Alania V, Tsereteli N et al (2017) Postcollisional tectonics and seismicity of Georgia. Geol Soc Am Spec Pap 525:535–573. https://doi.org/10.1130/2017.2525(17

Alania V, Chabukiani A, Chagelishvili R et al (2017) Growth structures, piggyback basins and growth strata of Georgian part of Kura foreland fold and thrust belt: implication for late alpine kinematic evolution. Geol Soc Lond Spec Publ 428:171–185. https://doi.org/10.1144/SP428.5

Avdeev B, Niemi NA (2011) Rapid Pliocene exhumation of the central greater Caucasus constrained by low-temperature thermochronometry. Tectonics 30:TC2009. https://doi.org/10.1029/2010TC002808

Banks C, Robinson A, Williams M (1997) Structure and regional tectonics of the Achara-Trialeti fold belt and the adjacent Rioni and Kartli foreland basins. Republic of Georgia. In: Robinson

A (ed) Regional and petroleum geology of the Black Sea and surrounding region. AAPG, Tulsa, pp 331–346
Butler RWH, Grasso M (1993) Tectonic controls on base-level variations and depositional sequences within thrust-top and foredeep basins: examples from the Neogene thrust belt of Central Sicily. Basin Res 5:137–151
Cowgill E, Forte A, Niemi N et al (2016) Relict basin closure and crustal shortening budgets during continental collision: an example from Caucasus sediment provenance. Tectonics 35:2918–2947. https://doi.org/10.1002/2016TC004295
Dercourt J, Zonenschain LP, Ricou LE et al (1986) Geological evolution of the Tethys belt from the Atlantic to the Pamirs since the Lias. Tectonophysics 123:241–315
Djanelidze A, Kandelaki N (1956) Geological map of the USSR, Caucasus series sheet K-38-XIII. Ministry of Geology, Moscow
Dotduyev SI (1986) Nappe structure of the greater Caucasus range. Geotectonics 20:420–430
Erdos Z, Huismans RS, Van der Beek P, Hieulot C (2014) Extensional inheritance and surface processes as controlling factors of mountain belt structure. J Geophys Res 119:9042–9061. https://doi.org/10.1002/2014JB011408
Ershov AV, Brunet MF, Korotaev MV et al (1999) Late Cenozoic burial history and dynamics of the northern Caucasus molasse basin: implications for foreland basin modelling. Tectonophysics 313:219–241
Ershov AV, Brunet MF, Nikishin AM et al (2003) Northern Caucasus basin: thermal history and synthesis of subsidence models. Sediment Geol 156:95–118. https://doi.org/10.1016/S0037-0738(02)00284-1
Forte A, Cowgill E, Whipple KX (2014) Transition from a singly vergent to doubly vergent wedge in a young orogen: the greater Caucasus. Tectonics 33:2077-101. https://doi.org/10.1002/2014tc003651
Gamkrelidze IP (1986) Geodynamic evolution of the Caucasus and adjacent areas in alpine time. Tectonophysics 127:261–277
Giorgobiani T, Zakaraya D (1989) The folded structure of the North-Western Caucasus and the mechanism of its formation. Metsniereba, Tbilisi
Grool AR, Huismans RS, Ford M (2019) Salt décollement and rift inheritance controls on crustal deformation in orogens. Terra Nova 31:562–568. https://doi.org/10.1111/ter.12428
Khain VY (1974) Structure and main stages in the tectono-magmatic development of the Caucasus: an attempt at geodynamic interpretation. Am J Sci 274:16–23
Luparev N, Prokhorov N (1957) Geological map of the USSR, Caucasus series sheet L-38-XXXI. Ministry of Geology, Moscow
Mosar J, Kangarli T, Bochud M et al (2010) Cenozoic–recent tectonics and uplift in the greater Caucasus: a perspective from Azerbaijan. Geol Soc Lond Spec Publ 340:261–279. https://doi.org/10.1144/SP340.12
Mumladze T, Forte AM, Cowgill ES et al (2015) Subducted, detached, and torn slabs beneath the greater Caucasus. Geo Res J 5:36–46. https://doi.org/10.1016/j.grj.2014.09.004
Muñoz JA (1992) Evolution of a continental collision belt: ECORS-Pyrenees crustal balanced cross-section. In: KR MC (ed) Thrust tectonics. Springer, pp 235–246. https://doi.org/10.1007/978-94-011-3066-0_21
Naylor M, Sinclair HD (2008) Pro- versus retro-foreland basins. Basin Res 20:285–303. https://doi.org/10.1111/j.1365-2117.2008.00366.x
Nikishin AM, Ershov AV, Nikishin VA (2010) Geological history of Western Caucasus and adjacent foredeeps based on analysis of the regional balanced section. Dokl Earth Sci 430:155–157. https://doi.org/10.1134/S1028334X10020017
Nikishin AM, Ziegler PA, Bolotov SN et al (2012) Late Paleozoic to Cenozoic evolution of the Black Sea-southern eastern European region: a view from the Russian platform. Turk J Earth Sci 21(5):571–634. https://doi.org/10.3906/yer-1005-22
Pfiffner OA (2006) Thick-skinned and thin-skinned styles of continental contraction. Geol Soc Am Spec Pap 414:153–177. https://doi.org/10.1130/2006.2414(09)

Potapenko U (1961) Geological map of the USSR, Caucasus series sheet K-38-I. Ministry of Geology, Moscow

Reilinger R et al (2006) GPS constraints on continental deformation in the Africa–Arabia–Eurasia continental collision zone and implications for the dynamics of plate interactions. J Geophys Res:BO5411. https://doi.org/10.1029/2005JB004051

Robinson AG, Rudat JH, Banks CJ et al (1996) Petroleum geology of the Black Sea. Mar Pet Geol 13:195–223

Saintot A, Brunet MF et al (2006) The Mesozoic-Cenozoic tectonic evolution of the greater Caucasus. Geol Soc Lond Spec Publ 32:277–289

Shaw JH, Connors CD, Suppe J (2006) Seismic interpretation of contractional fault-related folds. AAPG

Sobornov K (1994) Structure and petroleum potential of the Dagestan thrust belt, northeastern Caucasus, Russia. Bull Can Petrol Geol 42:352–364

Sobornov K (1996) Lateral variations in structural styles of tectonic wedging in the northeastern Caucasus, Russia. Bull Can Petrol Geol 44:385–399

Sosson M, Stephenson R, Sheremet Y et al (2016) The eastern Black Sea-Caucasus region during cretaceous: new evidence to constrain its tectonic evolution. Compte-Rendus Geosci 348:23–32. https://doi.org/10.1016/j.crte.2015.11.002

Suppe J (1983) Geometry and kinematics of fault-bend folding. Am J Sci 283:684–721

Suppe J, Medwedeff DA (1990) Geometry and kinematics of fault-propagation folding. Eclogae Geol Helv 83:409–454

Tan O, Taymaz T (2006) Active tectonic of the Caucasus: earthquake source mechanisms and rupture histories obtained from inversion of teleseismic body waves. Geol Soc Am Spec Pap 409:531–578. https://doi.org/10.1130/2006.2409(25)

Tari G, Vakhania D, Tatishvili G et al (2018) Stratigraphy, structure and petroleum exploration play types of the Rioni Basin, Georgia. Geol Soc Lond Spec Publ 464:403–438. https://doi.org/10.1144/SP464.14

Tibaldi A, Alania V, Bonali FL et al (2017a) Active inversion tectonics, simple shear folding and back-thrusting at Rioni Basin. Georgia. J Struct Geol 96:35–53. https://doi.org/10.1016/j.jsg.2017.01.005

Tibaldi A, Russo E, Bonali FL et al (2017b) 3-D anatomy of an active fault-propagation fold: a multidisciplinary case study from Tsaishi, western Caucasus (Georgia). Tectonophysics 717:253–269. https://doi.org/10.1016/j.tecto.2017.08.006

Tibaldi A, Bonali FL, Russo E et al (2018) Structural development and stress evolution of an arcuate fold-and-thrust system, southwestern greater Caucasus, Republic of Georgia. J Asian Earth Sci 156:226–245. https://doi.org/10.1016/j.jseaes.2018.01.025

Tibaldi A, Tsereteli N, Varazanashvili O et al (2020) Active stress field and fault kinematics of the greater Caucasus. J Asian Earth Sci 188:104108. https://doi.org/10.1016/j.jseaes.2019.104108

Tsereteli N, Tibaldi A, Alania V et al (2016) Active tectonics of Central-Western Caucasus, Georgia. Tectonophysics 691:328–344. https://doi.org/10.1016/j.tecto.2016.10.025S0040-1951(16)30478-4

Varazanashvili O, Tsereteli N, Bonali F et al (2018) GeoInt: the first macroseismic intensity database for the Republic of Georgia. J Seismol 22(3):625–667. https://doi.org/10.1007/s10950-017-9726-5

Verges J, Millan H, Roca E et al (1995) Eastern Pyrenees and related foreland basins: pre-, syn- and post-collisional crustal-scale cross-sections. Mar Pet Geol 12:903–915. https://doi.org/10.1016/0264-8172(95)98854-X

Vincent SJ, Morton AC, Carter A et al (2007) Oligocene uplift of the Western greater Caucasus: an effect of initial Arabia-Eurasia collision. Terra Nova 19:160–166. https://doi.org/10.1111/j.1365-3121.2007.00731.x

Vincent SJ, Carter A, Lavrishchev VA et al (2011) The exhumation of the western Greater Caucasus: a thermochronometric study. Geol Mag 148:1–21. https://doi.org/10.1017/S0016756810000257

Vincent SJ, Braham W, Lavrishchev VA et al (2016) The formation and inversion of the western greater Caucasus Basin and the uplift of the western greater Caucasus: implications for the wider Black Sea region. Tectonics 35:2948–2962. https://doi.org/10.1002/2016TC004204
Vincent SJ, Saintot A, Mosar J et al (2018) Comment on "relict basin closure and crustal shortening budgets during continental collision: an example from Caucasus sediment provenance" by Cowgill et al. (2016). Tectonics 37(3):1006–1016. https://doi.org/10.1002/2017TC004515

Chapter 6
The Geometry of the Two Orogens Convergence and Collision Zones in Central Georgia: New Data from Seismic Reflection Profiles

Victor Alania, Tamar Beridze, Onise Enukidze, Rusudan Chagelishvili, Zurab Lebanidze, Davit Maqadze, Alexander Razmadze, Nino Sadradze, and Niko Tevzadze

Abstract The thrust front of the central Lesser Caucasus and southern part of the Greater Caucasus orogens is one of the key regions of an apparent convergence zone between two orogens. The formation of the complex structure of the Lesser Caucasus-Greater Caucasus convergence zone is governed by the northward and southward-directed thrusting. Here we show the structural style of deformation of a convergence zone between Lesser Caucasus retro-wedge and Greater Caucasus pro-wedge based on seismic reflection profiles. Seismic profiles reveal the transition between convergence and the initial collision zone within the study area. The frontal part of the Lesser Caucasus retro-wedge is represented by a shallow triangle zone and north-vergent structural wedge. The thrust front of Greater Caucasus pro-wedge is represented by south-vergent fault-related folds. Between the Lesser Caucasus and Greater Caucasus frontal parts, the undeformed Kura foreland basin is located.

V. Alania (✉) · O. Enukidze · A. Razmadze · N. Sadradze
M. Nodia Institute of Geophysics, I. Javakhishvili Tbilisi State University, Tbilisi, Georgia
e-mail: victor.alania@tsu.ge

T. Beridze
A. Janelidze Geological Institute, I. Javakhishvili Tbilisi State University, Tbilisi, Georgia

R. Chagelishvili
Georgian National Museum, Tbilisi, Georgia

Z. Lebanidze
Faculty of Exact and Natural Sciences, I. Javakhishvili Tbilisi State University, Tbilisi, Georgia

D. Maqadze
A. Tvalchrelidze Caucasus Institute of Mineral Materials, I. Javakhishvili Tbilisi State University, Tbilisi, Georgia

N. Tevzadze
Georgian Oil and Gas Ltd., Tbilisi, Georgia

© Springer Nature B.V. 2021

F. L. Bonali et al. (eds.), *Building Knowledge for Geohazard Assessment and Management in the Caucasus and other Orogenic Regions*, NATO Science for Peace and Security Series C: Environmental Security,
https://doi.org/10.1007/978-94-024-2046-3_6

The structure of the Lesser Caucasus-Greater Caucasus initial collision zone is fairly complex. The orogens collision zone is represented by south-vergent imbricate thrusts. Under the triangle zone, there are two structural wedges. The frontal part of the eastern Achara-Trialeti fold-and-thrust belt exhibit structural variations along strike within the study area. Near the Lesser Caucasus-Greater Caucasus convergence zone is located Tbilisi city, where the north-and-south-vergent thrusts of this collision zone represent a significant seismic hazard.

Keywords Lesser Caucasus · Greater Caucasus · Seismic reflection profile · Convergence zone · Collision zone

6.1 Introduction

The Lesser Caucasus (LC) and Greater Caucasus (GC) orogens, which are among the best examples of collision-driven far-field deformations, are located in the northernmost part of the Arabia/Eurasia collision zone (Fig. 6.1). The geodynamic evolution of LC and GC is the result of the Arabia/Eurasia collision during Alpine times, which led to the inversion of back-arc basins and formation of two orogens with the intermontane basin (or foreland basin) in between (Fig. 6.2) (e.g. Adamia et al. 1981; Banks et al. 1997; Khain 1974; Mosar et al. 2010; Nemcok et al. 2013; Sosson et al. 2016). During the late Cenozoic, in response to collision between Arabia and Eurasian plates and their subsequent deformations, convergence between the LC and GC orogens took place in far-field zone (e.g. Adamia et al. 2010; Alania et al. 2017a; Banks et al. 1997; Nemcok et al., 2013; Sokhadze et al. 2018). Between the two opposite verging chains of the LC and the GC, the Rioni and Kura foreland basins (KFB) are located (Adamia et al. 2010, 2011a; Alania et al. 2017a; Banks et al. 1997; Cowgill et al. 2016; Forte et al. 2014; Sokhadze et al. 2018; Tari et al. 2018; Tibaldi et al. 2017).

The thrust front of the central LC and southern part of the western KFB is our study area and is one of the key regions of apparent convergence between the two orogens (Fig. 6.3). The complex structures of the LC-GC convergence zone are formed by northward and southward-directed thrusting (e.g. Banks et al. 1997). Structural studies of the frontal part of the central LC and the western KFB were mainly based on surface geology and drilling data. Currently available structural models considerably differ from each other (e.g. Adamia et al. 2010; Alania et al. 2017a; Banks et al. 1997; Basheleishvili 1985; Gamkrelidze 1976; Nemcok et al. 2013; Papava 1967; Sokhadze et al. 2018; Sosson et al. 2013, 2016). According to one group of authors, the frontal part of the western KFB in the study area is overthrusted on the frontal part of the LC (Basheleishvili 1984; Gamkrelidze 1976; Papava 1967), while another group believes that frontal part of LC is represented by a triangle zone, and the south-vergent thrust is a passive-backthrust formed as a result of the northward displacement of a duplex (Alania et al. 2017a; Banks et al. 1997; Nemcok et al. 2013; Sosson et al. 2013, 2016). Between the triangle zone and

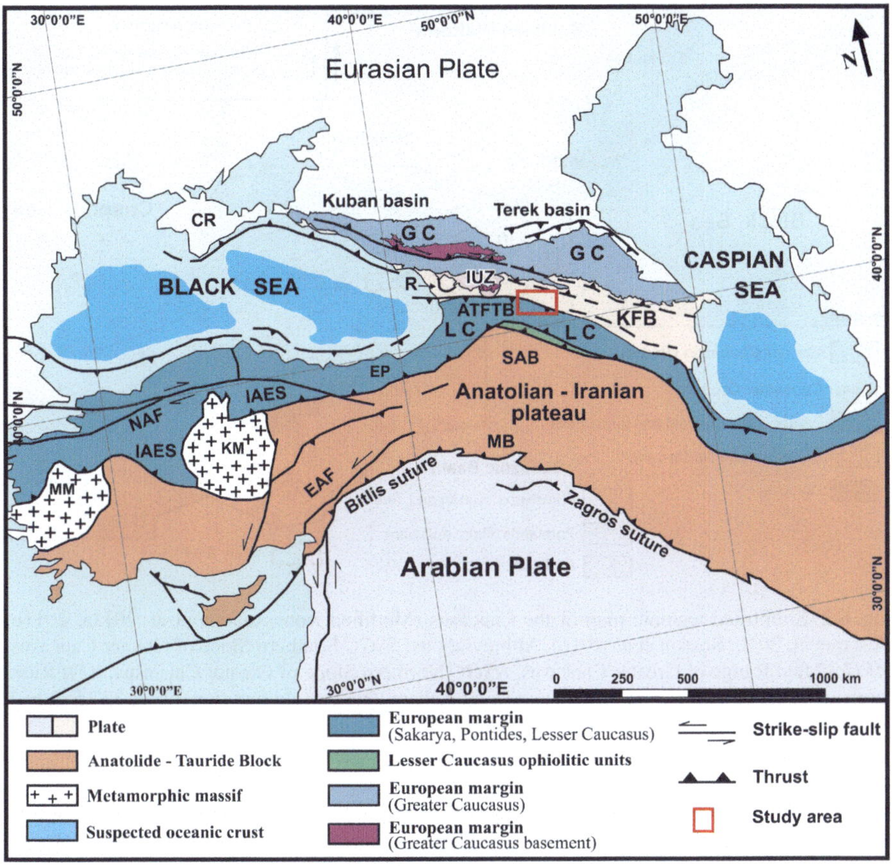

Fig. 6.1 Tectonic map of the Arabia - Eurasia collision zone (modified from Sosson et al. 2016). Abbreviations: *CR* Crimea, *GC* Greater Caucasus, *LC* Lesser Caucasus, *R* Rioni, *IUZ* Imereti uplift zone, *Dz* Dzirula, *KFB* Kura Foreland Basin, *ATFTB* Achara-Trialeti fold-and-thrust belt, *SAB* South Armenian Block, *MB* Mus Basin, *EP* Eastern Pontides, *KM* Kirsehir Massif, *EAF* Eastern Anatolian Fault, *NAF* North Anatolian Fault, *IAES* Izmir-Ankara-Erzincan Suture, *MM* Menderes Massif

Kura foreland fold-and-thrust belt (KFFTB) is located KFB (Alania et al. 2017a; Banks et al. 1997; Nemcok et al. 2013). According to Nemcok et al. (2013), east of Mtskheta there is an ongoing collision between the LC and GC orogens. Sokhadze et al. (2018) suggest that the southward offset of the convergence zone near Tbilisi is related to the incipient collision between the LC and GC orogens.

Historical Mtskheta earthquake (1275, Ms = 6.5), recent GPS, earthquakes and paleoseismic data indicate that the LC-GC convergence zone is tectonically fairly active (Adamia et al. 2004, 2017; Boichenko et al. 2016; Pasquarè et al. 2011; Reilinger et al. 2006; Sokhadze et al. 2018; Tibaldi et al. 2020; Tsereteli et al. 2016; Varazanashvili et al. 2011, 2018).

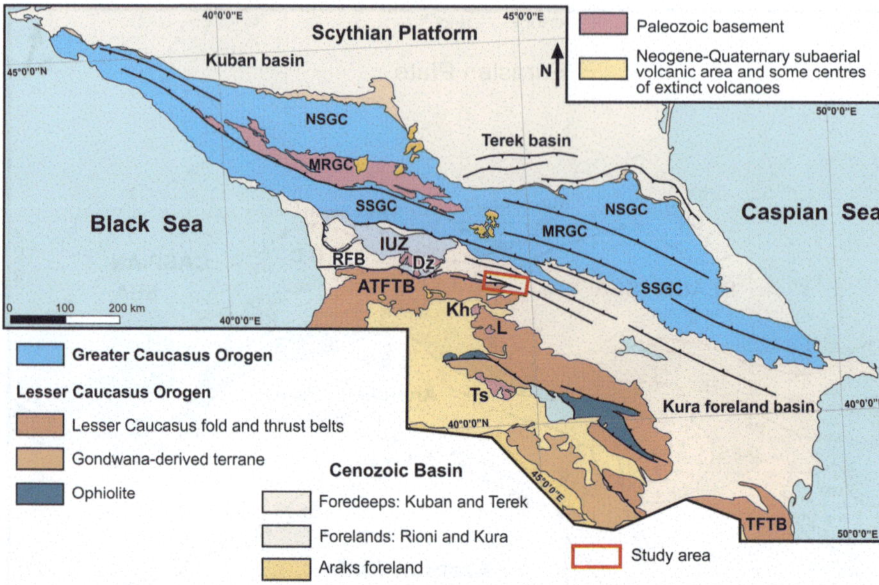

Fig. 6.2 Simplified tectonic map of the Caucasus (Modified from Adamia et al. 2011a, 2011b; Mosar et al. 2010; Sosson et al. 2016). Abbreviations: *SSGC* Southern Slope of Greater Caucasus, *MRGC* Main Range of Greater Caucasus, *NSGC* Northern Slope of Greater Caucasus, *RFB* Rioni foreland basin, *IUZ* Imereti uplift zone, *ATFTB* Achara-Trialeti fold-and-thrust belt, *TFTB* Talysh fold-and-thrust belt, *DZ* Dzirula, *Kh* Khrami, *L* Loki, *Ts* Tsakhkuniats

To determine the deformation structural style of the LC-GC convergence zone, in this study we have extended previous interpretations (e.g. Adamia et al. 2010; Banks et al. 1997; Alania et al. 2017a, 2020a, 2020b; Forte et al. 2014; Nemcok et al. 2013; Sokhadze et al. 2018; Sosson et al. 2013, 2016; Tibaldi et al. 2020) and presented the structural evidence for the presence of convergence and initial collision zone transition on the ground of interpreted seismic reflection profiles within the study area.

6.2 Geological Setting

Like the GC orogen, the LC is a double wedge orogen and forms a W-E-trending, bivergent thrust system (Alania et al. 2019). The Achara-Trialeti fold-and-thrust belt (ATFTB) is located in the frontal part of the LC. Our study area includes the frontal part of the eastern ATFTB and the southern part of the western KFB (Fig. 6.3).

The general structure of the eastern ATFTB and KFB was described by Buleishvili (1964), Gamkrelidze (1949), Gamkrelidze (1976), Milanovski and Khain (1963), Papava (1967), and others. The structure within the study area (Fig. 6.3) is represented by W-E trending folds and north- and south-vergent reverse-

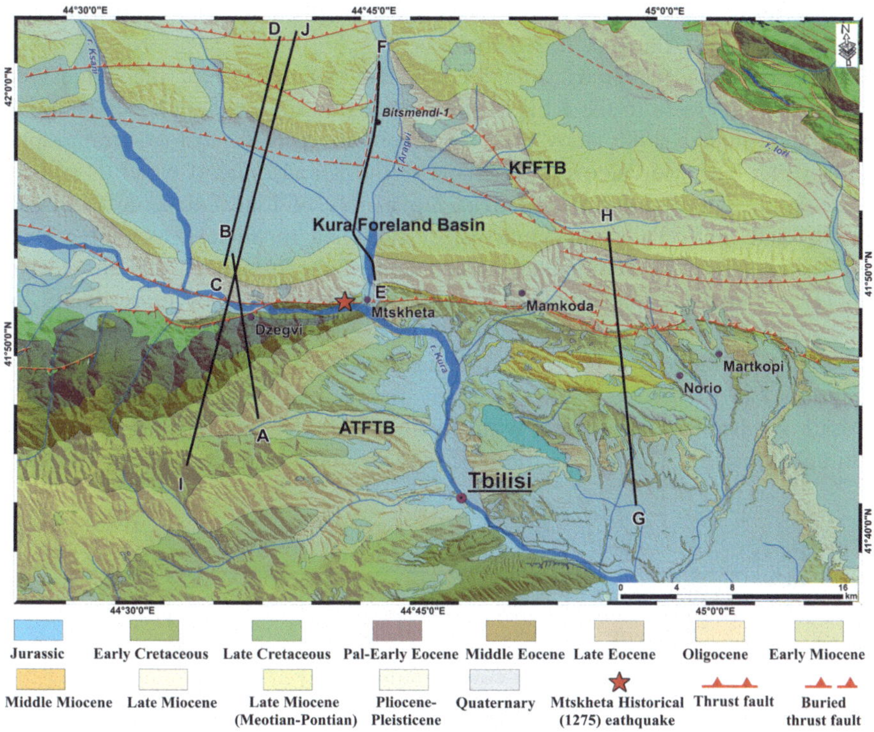

Fig. 6.3 Geological map of the study area. (Modified from Papava 1967)

faults and thrusts (e.g. Gamkrelidze 1976; Papava 1967). The eastern ATFTB plunges east in the southern part of Tbilisi area, and continues plunging into the eastern part of the KFFTB (e.g. Adamia et al. 2002). The eastern ATFTB can be divided into three major structural trends (from south to north): (1) the backthrust zone, (2) the forethrust zone, and (3) triangle zone. These structural trends are formed by fault-bend, fault-propagation folds and duplex that transfer shortening from the mountain belt towards the north into the KFB (Alania et al. 2017b). The frontal structure of the zone extending from Gori to Norio is characterized by a well-developed frontal (north-dipping) monocline involving the stratigraphy down to Lower Miocene. According to Banks et al. (1997), formation of the ATFTB started in the Late Miocene and was related to northward displacement of the Lesser Caucasus basement wedge. According to Sosson et al. (2013), the northern part of the eastern ATFTB evidences a triangle zone as the result of the northeastward motion of the inverted Achara-Trialeti Basin (by thick-skin tectonics) in front of the southward motion of the KFB (by thin-skin tectonics).

The Kura basin which developed formerly as a foreland basin (Oligocene-Lower Miocene) continued its evolution as a fold-and-thrust belt during Middle Miocene-Pleistocene (Alania et al. 2008, 2017a, 2018). Western KFFTB is mainly represented by the series of NEN dipping thrust faults, most of which are associated with fault-propagation folds (Alania et al. 2020b).

Stratigraphy in the study area records the evolution from the extensional Achara-Trialeti Basin to the KFB of the Arabia-Eurasia collision zone. The main stratigraphic units of the study area are schematized in the tectono-stratigraphic chart of Fig. 6.4. The Paleozoic basement of the Khrami massif which is situated to the south of the study area is made up mainly of Late Paleozoic (Variscan) granites and Early Paleozoic gneisses, diorites and tonalites (Zakariadze et al. 2007) (Fig. 6.2). The sedimentary succession of the frontal part of the eastern ATFTB and southern part of KFB is commonly represented by more than 7 km thick Jurassic, Cretaceous, Paleogene and Neogene deep marine, shallow marine and thick continental strata (e.g. Adamia et al. 2011b). The oldest formation exposed south of the study area (Khrami massif) and introduced by the Lower Jurassic sandstones and shales, with conglomerates covers the basement rocks with unconformity. The Middle Jurassic is represented by Bajocian volcanogenic-

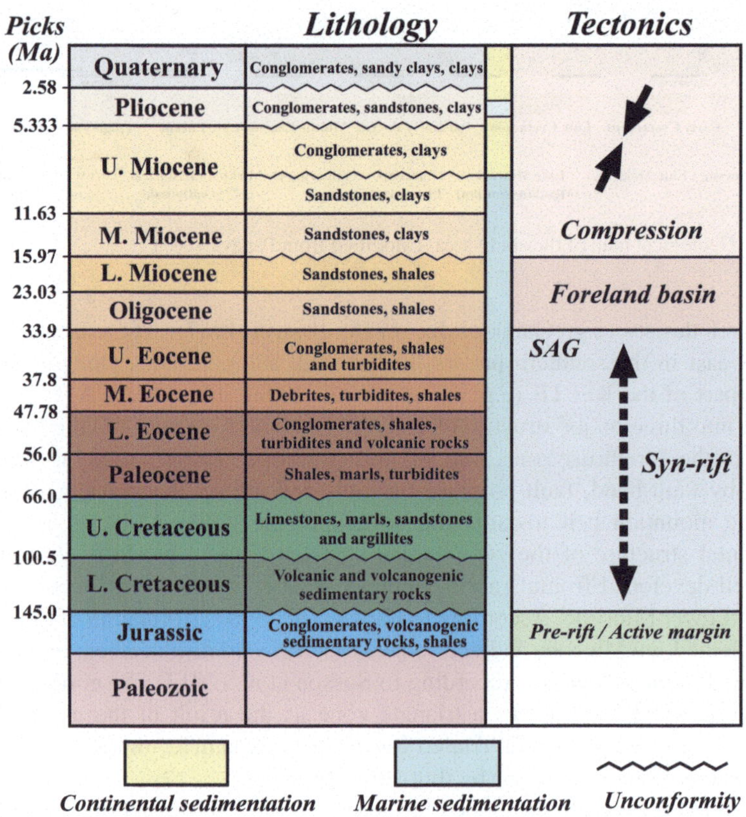

Fig. 6.4 Tectono-stratigraphic chart of the study area. (Modified from Adamia et al. 2002, 2010; Adamia et al. 2017)

sedimentary rocks whereas the Upper Jurassic is represented by limestones and volcanogenic-sedimentary rocks. The thickness of Jurassic strata is about 200-600 m (Papava 1967). During the Cretaceous and Paleogene, Achara-Trialeti extensional basin was filled with approximately 3500-4000 m of sediments (Adamia et al. 2011b). The Lower Cretaceous strata crop-out west of the study area and is mainly composed of Aptian-Albian volcanic and volcanogenic-sedimentary rocks. The Upper Cretaceous is composed of limestones, marls, sandstones and argillites (thickness is about 600-1800 m). Paleogene is represented by Paleocene (shales, marls, turbidites - 50-450 m), Lower Eocene (conglomerates, shales, turbidites, dacite volcanics - 500-1000 m), Middle Eocene (debrites, turbidites, shales - 300-800 m) and Upper Eocene (conglomerates, shales, turbidites - total thickness of the unit is 400-800 m) strata (Adamia et al. 2002, 2011b, Banks et al. 1997; Beridze 2019; Papava 1967). In the frontal part of the LC and KFB the Oligocene–Early Miocene rock formations in some places conformably follow the Late Eocene and are composed of shales and sandstones. Maximum thickness of the Oligocene–Lower Miocene deposits are reported within the basin of the river Kura (2.0-2.5 km) (Buleishvili 1964). In the vicinities of Tbilisi and further westward, the Lower Miocene is represented by cross-bedded sandstones and shales (Robinson et al. 1997). Syn-tectonic Middle-Upper Miocene and Pliocene strata are represented by about 1500-2000-m-thick shallow marine and continental sediments (Alania et al. 2017a).

6.3 Interpretation of Seismic Reflection Profiles

To illustrate the structural architecture within the LC-GC convergence zone, we present one previously published (A-B) (Alania et al. 2020a) and three new interpreted post-stack time-migrated seismic profiles (C-D, E-F, G-H) (Figs. 6.5, 6.6, 6.7 and 6.8). The location of the seismic reflection profiles and structural cross-section is shown in Fig. 6.3.

Fault-related folding and wedge thrust folding theories were used in the interpretation of seismic reflection profiles and construction of the structural cross-section (Shaw et al. 2006; Medwedeff 1990; Suppe and Medwedeff 1990). The identification of stratigraphic units at depth for seismic profiles was based on outcrop data correlations. The surface geological information is obtained from 1:50,000 geological map (Papava 1967) and our personal field data (Fig. 6.2).

The seismic reflection profile A-B reveals the presence of upper and lower structural complexes (Fig. 6.5b). The upper structural complex is represented by a shallow triangle zone which includes a north-vergent Armazi fault-propagation fold, north-vergent duplex, and south-vergent passive-backthrust at the mountain front. The lower structural complex is represented by a structural wedge and is made up of Lower Cretaceous-Jurassic strata. Two north-vergent blind thrusts are developed above the structural wedge (Fig. 6.5b).

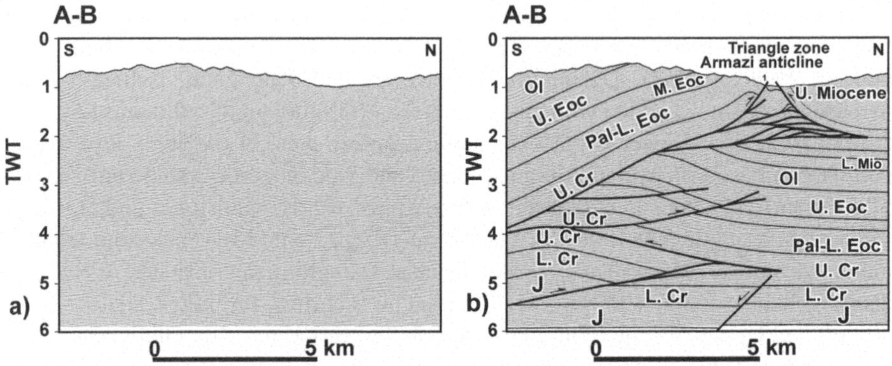

Fig. 6.5 (**a**) Uninterpreted and (**b**) interpreted seismic reflection profiles A-B (Alania et al. 2020a). Location is shown in Fig. 6.3. Abbreviations: *J* Jurassic, *L. Cr* Lower Cretaceous, *U. Cr* Upper Cretaceous, *Pal-L. Eoc* Paleocene-Lower Eocene, *M. Eoc* Middle Eocene, *U. Eoc* Upper Eocene, *Ol* Oligocene, *L. Mio* Lower Miocene, *M. Mio* Middle Miocene

Seismic interpretation of the C-D profile shows four shallow south-vergent thrusts (Fig. 6.6b). One is a passive-backthrust which is a constituent part of the shallow triangle zone. Upper Miocene growth strata are developed within the backlimb of the passive-backthrust. Deformed growth strata folds have scarps inclined towards the axis of the syncline (Fig. 6.6c). In the seismic profiles (C-D, E-F) the frontal structure of western KFFTB is defined by a Bitsmendi fault-propagation fold which involves the Late Miocene unit (Figs. 6.6b and 6.7b). The backlimb of the Bitsmendi anticline dips gently (8°–3°) to the north, whereas the forelimb dips steeply (13°–11°) to the south, and steepens with increasing depth (Figs. 6.6b and 6.7b). Normal faults and positive structural inversion in the foreland domain under the Bitsmendi anticline are well-observed in the seismic profiles C-D, E-F (Figs. 6.6b and 6.7b).

The seismic reflection profile G-H reveals the presence of north-and south-vergent fault-related folds, south-and north-vergent thrusts, north-vergent duplexes and structural wedges (Fig. 6.8b).

6.4 Discussion

The results obtained after seismic profile interpretation, enable to establish the general picture of the LC-GC convergence zone. Like the Apennines-Alps and Pamir-Tien Shan convergence zones (e.g. Fu et al., 2010; Toscani et al., 2014), the LC–GC convergence zone is a unique example to understand ongoing intracontinental mountain building within the Arabia/Eurasia collision system.

Fission-track cooling ages indicate that the exhumation process in the GC started at the end of Upper Eocene-Oligocene (Vincent et al. 2007, 2011) and reached

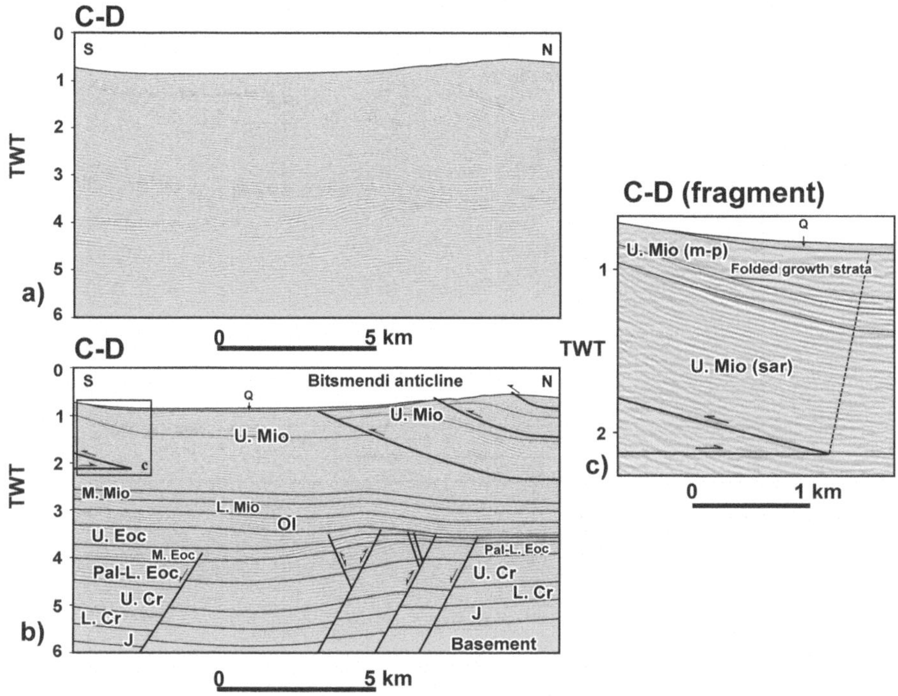

Fig. 6.6 (**a**) Uninterpreted, (**b**) interpreted seismic reflection profiles C-D, and (**c**) fragment of seismic profile C-D. Location is shown in Fig. 6.3. Abbreviations: *J* Jurassic, *L. Cr* Lower Cretaceous, *U. Cr* Upper Cretaceous, *Pal-L. Eoc* Paleocene-Lower Eocene, *M. Eoc* Middle Eocene, *U. Eoc* Upper Eocene, *Ol* Oligocene, *L. Mio* Lower Miocene, *M. Mio* Middle Miocene, *U. Mio* Upper Miocene, *U. Mio (sar)* Upper Miocene (Sarmatian), *U. Mio (m-p)* Upper Miocene (Meotis-Pontian), *Q* Quaternary

its maximum rate in the Pliocene (Avdeev and Niemi 2011). Apatite fission-track data from the central part of the LC orogen record two successive stages of compressional exhumation in Neogene: (1) Early to Middle Miocene, and (2) Middle Miocene to present (Cavazza et al. 2017, 2019). Syn-orogenic (Miocene-Pliocene) sediments within KFB preserve important information for understanding the tectonic process of convergence between the LC and the GC orogens. In the KFB, the transition from marine to continental deposition occurred between the late Miocene and Pleistocene. During the Upper Miocene and Pliocene, the LC and GC was the source area of continental, coarsening upwards deposits in the KFB (Adamia et al. 2010). The geometry and unconformity surfaces of the Miocene growth strata in the study area are observed in outcrops (Alania et al. 2017a), as well as in seismic profiles (Figs. 6.6c and 6.7b). Fission track data and comparison of growth strata ages enable us to suppose that the formation of the LC frontal part took place during the middle-upper Miocene. Later in the Pliocene deformation of the frontal part of the western KFTFB as well as convergence and collision of two orogens occurred.

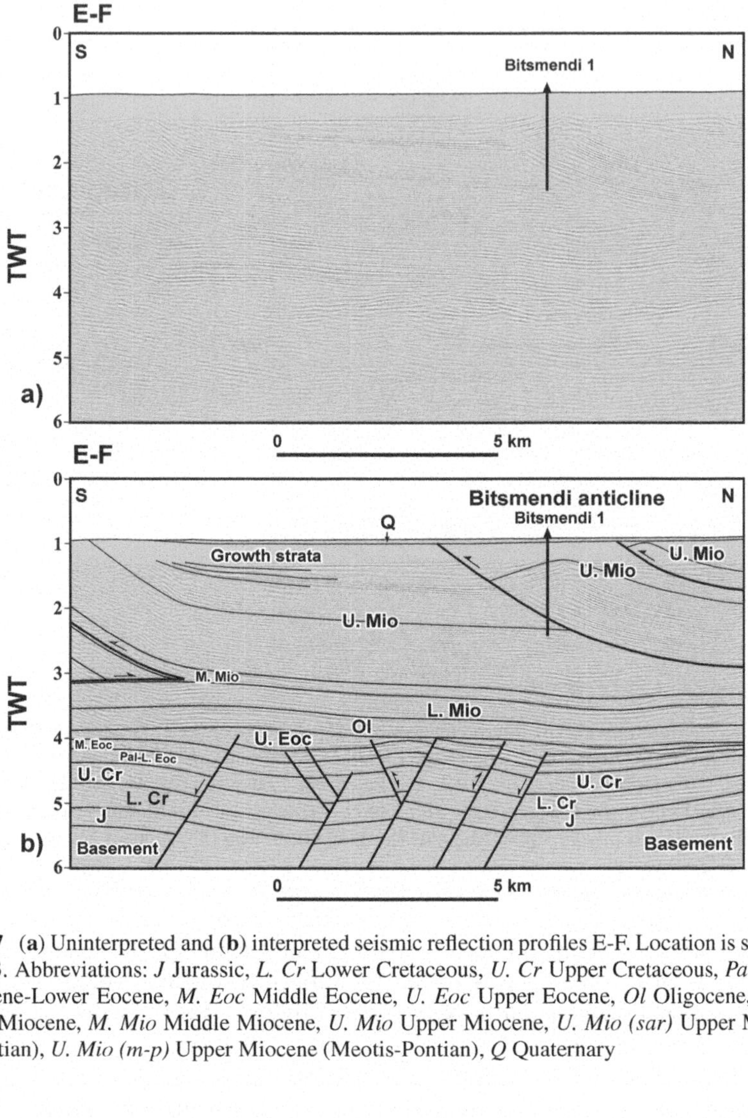

Fig. 6.7 (**a**) Uninterpreted and (**b**) interpreted seismic reflection profiles E-F. Location is shown in Fig. 6.3. Abbreviations: *J* Jurassic, *L. Cr* Lower Cretaceous, *U. Cr* Upper Cretaceous, *Pal-L. Eoc* Paleocene-Lower Eocene, *M. Eoc* Middle Eocene, *U. Eoc* Upper Eocene, *Ol* Oligocene, *L. Mio* Lower Miocene, *M. Mio* Middle Miocene, *U. Mio* Upper Miocene, *U. Mio (sar)* Upper Miocene (Sarmatian), *U. Mio (m-p)* Upper Miocene (Meotis-Pontian), *Q* Quaternary

In the western part of the study area, the convergence zone of two orogens is introduced by frontal parts of the LC retro-wedge and the GC pro-wedge with KFB in between (Fig. 6.9). The frontal part of the LC retro-wedge is represented by two different thrust systems: a shallow triangle zone and a lower structural wedge. Within the triangle zone, the duplex sequence comprises Cretaceous-Neogene strata that have been formed due to passive-roof duplex style of deformation. The southern part of the Kura foreland basin is represented by Miocene strata that were deformed and uplifted by passive back thrusting at the triangle zone. The kinematic evolution of the south-vergent backthrust is related to the northward propagating duplex and

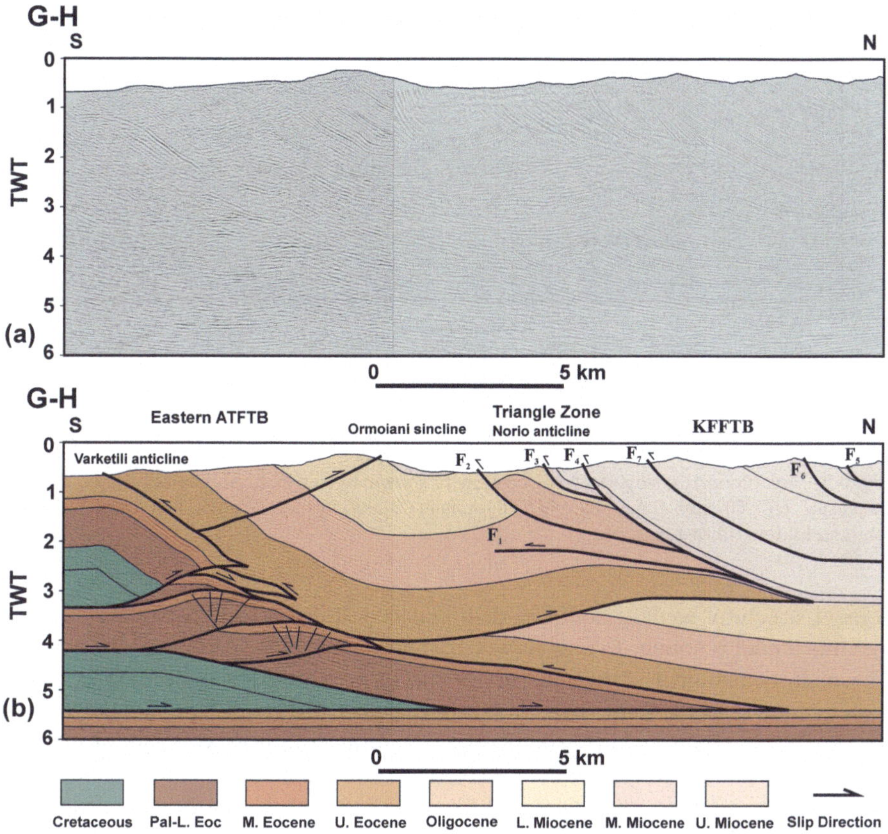

Fig. 6.8 (**a**) Uninterpreted and (**b**) interpreted seismic reflection profiles G-H. Location is shown in Fig. 6.3. Abbreviations: *ATFTB* Achara-Trialeti fold-and-thrust belt, *KFFTB* Kura foreland fold-and-thrust belt

triangle tip is located in middle Miocene. The lower structural wedge is composed of Jurassic-Cretaceous strata and wedge tip is located in upper Cretaceous strata (Fig. 6.9). The interpreted seismic profile and structural cross-section have revealed that Armazi anticline is a breakthrough fault-propagation fold developed above the thrust sheet. The imbricated structure below the Armazi anticline is characterized by break-backward thrusting (Figs. 6.5b and 6.9). The frontal part of the GC pro-wedge is represented by south-vergent thrusts composed of Miocene deposits (Fig. 6.9). Similar to Nemchok et al. (2013) transitional zone of convergence and collision of two orogens could be observed in the study area. The distance between them gradually decreases eastwards and the orogens collision takes place in the eastern part of the study area.

The deep structure of the LC-GC initial collision zone is fairly well-observed in the interpreted seismic profile (Figs. 6.8b and 6.10). The structure of the initial collision zone of the LC-GC is fairly complex. Under the triangle zone, two north-

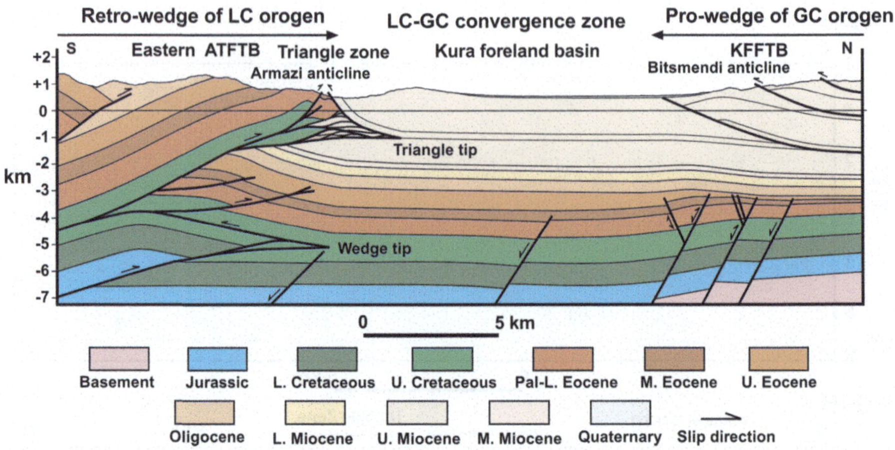

Fig. 6.9 Structural cross-section I-J. Location is shown in Fig. 6.3. Abbreviations: *LC* Lesser Caucasus, *GC* Greater Caucasus, *ATFTB* Achara-Trialeti fold-and-thrust belt, *KFFTB* Kura foreland fold-and-thrust belt

vergent structural wedges could be distinguished at depth (Fig. 6.10). We suggest that this model is similar to double-wedge fault systems models (McQuarrie 2002). The main detachment of frontal part of the LC retro-wedge is developed in the shales of the Upper Eocene lower horizons (Fig. 6.10). The initial collision zone of the orogens is represented by south-vergent imbricate thrusts. We suggest that F_1, F_2, F_3, and F_4 thrusts might be the passive-backthrust and their formation should be kinematically related to the displaced from the south to the north upper structural wedge and wedge tip (or triangle tip) is located at the base of Middle Miocene. The lower structural wedge is composed by Paleocene-Middle Eocene strata and wedge tip is located in the lower part of Upper Miocene shales (Fig. 6.10). The frontal part of the KFFTB is represented by three south-vergent thrusts (F_5, F_6, F_7). The main detachment within the KFFTB is located in Upper Miocene shales. As for the F_7 thrust, it represents the deformation front of the GC pro-wedge (Fig. 6.10). The interpreted seismic profiles and structural cross-section (Figs. 6.5b, 6.6b, 6.7b, 6.8b and 6.9) show along-strike structural variations within the frontal part of the eastern ATFTB, mainly due to changes in the detachment level and strength and variations in the wedge thickness.

Based on published information about historical and recent earthquake data (Tibaldi et al. 2020; Tsereteli et al. 2016; Varazanashvili et al. 2011, 2018), fission-track data from LC (Cavazza et al. 2017, 2019), age of syn-tectonic units from frontal part of eastern ATFTB and KFFTB (Alania et al. 2017a), and interpreted seismic profiles, indicates that compressive deformation within LC-GC convergence zone started in the Middle-Upper Miocene and continues today. Near the LC-GC convergence zone is located Tbilisi city (Fig. 6.3), where the north-and-south-vergent thrusts of this zone pose significant seismic hazard.

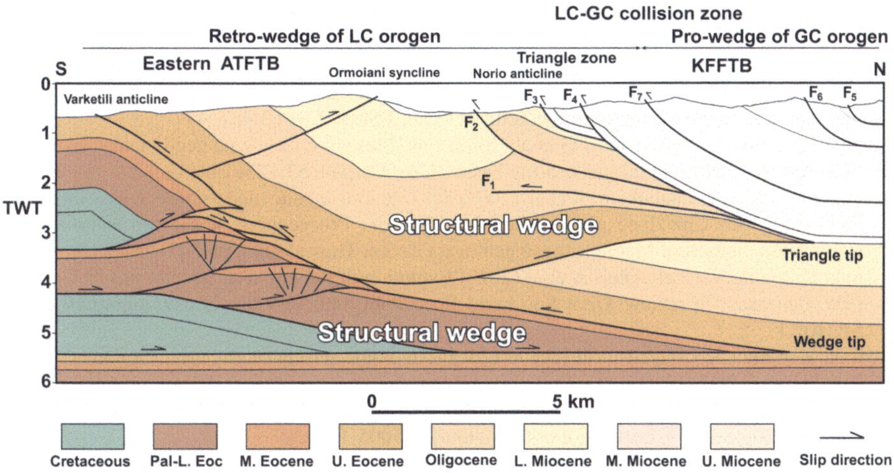

Fig. 6.10 Structural model of interpreted seismic profile I-J (LC-GC collision zone). Location is shown in Fig. 6.3. Abbreviations: *LC* Lesser Caucasus, *GC* Greater Caucasus, *ATFTB* Achara-Trialeti fold-and-thrust belt, *KFFTB* Kura foreland fold-and-thrust belt

6.5 Conclusions

The interpreted seismic profiles and structural cross-section illustrate the deformation structural style and the deep structure of the LC-GC convergence zone.

- Within our study area, the formation of the complex structure of the LC-GC convergence zone is governed by the northward and southward-directed thrusting and is characterized by the transition between convergence and collision zone.
- The frontal part of the LC retro-wedge is represented by a shallow triangle zone and a north-vergent structural wedge. The thrust front of GC pro-wedge is represented by south-vergent fault-related folds and between the LC and GC frontal parts, the undeformed Kura foreland basin is located.
- The structure of the LC-GC initial collision zone is fairly complex. The collision zone between the two orogens is represented by south-vergent imbricate thrusts. Under the south-vergent thrusts, there are two structural wedges.

Acknowledgments The work was supported by Shota Rustaveli National Science Foundation of Georgia (SRNSF #217942 - Structure, time and magnitude of deformation of the eastern Achara-Trialeti fold and thrust belt, Georgia). We are very grateful to Georgian State Agency of Oil and Gas and Georgian Oil & Gas Limited for providing the seismic profiles for our seismic interpretation. We acknowledge the useful review and suggestions of Federico Pasquaré Mariotto, Fabio Luca Bonali and Ara Avagyan on a previous version of the manuscript.

References

Adamia S, Chkhotua T, Kekelia M et al (1981) Tectonics of Caucasus and adjoining regions: implications for the evolution of the Tethys Ocean. J Struct Geol 3:437–447

Adamia S, Alania V, Ananiashvili G et al (2002) Late Mesozoic-Cenozoic geodynamic evolution of the eastern Georgian oil-gas bearing basins. Geol Carpath 53:155–159

Adamia Sh, Alania V, Chabukiani A et al. (2004) Seismic source zone models of recent earthquakes of Georgia. In: Chatzipetros AA, Pavlides SB (eds) Proceedings of the 5th international symposium on eastern Mediterranean geology, Greece, Thessaloniki, pp 545–551

Adamia S, Alania V, Chabukiani A et al (2010) Evolution of the late Cenozoic basins of Georgia (SW Caucasus): a review. Geol Soc Lond Spec Publ 340:239–259. https://doi.org/10.1144/SP340.11

Adamia S, Alania V, Chagelishvili R et al (2011a) Tectonic setting of Georgia (Caucasus). Geo-Eco-Marina 17:11–13

Adamia S, Zakariadze G, Chkhotua T et al (2011b) Geology of the Caucasus: a review. Turk J Earth Sci 20:489–544. https://doi.org/10.3906/yer-1005-11

Adamia S, Alania V, Tsereteli N et al (2017) Postcollisional tectonics and seismicity of Georgia. Geol Soc Am Spec Pap 525:535–573. https://doi.org/10.1130/2017.2525(17)

Alania V, Enukidze O, Koiava K et al (2008) Thrust systems and time of deformation of the Gare Kakheti foothills, eastern Georgia (Georgia). Boll Geofis Teor Appl 49:207–208

Alania V, Chabukiani A, Chagelishvili R et al (2017a) Growth structures, piggyback basins and growth strata of Georgian part of Kura foreland fold and thrust belt: implication for late alpine kinematic evolution. Geol Soc Lond Spec Publ 428:171–185. https://doi.org/10.1144/SP428.5

Alania V, Sosson M, Enukidze O et al (2017b) Structural architecture of the eastern Achara-Trialeti fold and thrust belt, Georgia: implications for kinematic evolution. Geophys J 4:81–82

Alania V, Enukidze O, Glonti N et al (2018) Structural architecture of the Kura foreland fold-and-thrust belt using seismic reflection profile, Georgia. Universal J Geosci 6:184–190. https://doi.org/10.13189/ujg.2018.060602

Alania V, Enukidze O, Tevzadze N (2019) Some problems of Caucasus tectonics: new insight and data from the structural cross-sections. In: Abstracts of the AAPG GTW conference exploration and production in the Black Sea, Caucasus, and Caspian Region, Batumi, Georgia, 18–19 September 2019

Alania V, Beridze T, Enukidze O et al (2020a) Structure of frontal part of the eastern Achara-Trialeti fold-and-thrust belt using seismic profile, Georgia. Bull Georg Nat Acad Sci. (in press)

Alania V, Enukidze O, Sadradze N et al (2020b) 3D structural model of the Bitsmendi anticline using seismic profiles, Georgia. Bull Georg Nat Acad Sci 14(1):61–67

Avdeev B, Niemi NA (2011) Rapid Pliocene exhumation of the central greater Caucasus constrained by low-temperature thermochronometry. Tectonics 30:TC2009. https://doi.org/10.1029/2010TC002808

Banks C, Robinson A, Williams M (1997) Structure and regional tectonics of the Achara-Trialeti fold belt and the adjacent Rioni and Kartli foreland basins. Republic of Georgia. In: Robinson A (ed) Regional and petroleum geology of the Black Sea and surrounding region. AAPG, Tulsa, pp 331–346

Basheleishvili LV (1985) The northern periphery of the eastern segment of the Adzharian–Trialeti fold zone (between the towns of Gori and Kaspi): peculiarities of geological structure. Bull Georg Acad Sci 119:457–459

Beridze T (2019) Middle Eocene submarine mass-transport deposits of Tbilisi environs (eastern Achara-Trialeti fold and thrust belt, Georgia). In: Abstracts of the IAS 34-th meeting on sedimentology, Rome, Italy, 10–13 September 2019

Boichenko G, Cowgill E, Stahl T et al (2016) A paleoseismic investigation of a frontal foreland thrust in the Greater Caucasus, Georgia. In: Abstracts of the AGU fall meeting, Denver, USA, 12–16 December 2016

Buleishvili D (1964) Lower Miocene. In: Gamkrelidze P (ed) Geology of the Georgian SSR. Nedra, pp 238–264. (in Russian)
Cavazza W, Albino I, Zattin M et al (2017) Thermochronometric evidence for Miocene tectonic reactivation of the Sevan-Akera suture zone (lesser Caucasus): a far-field tectonic effect of the Arabia-Eurasia collision? Geol Soc Lond Spec Publ 428:187–198. https://doi.org/10.1144/SP428.4
Cavazza W, Albino I, Galoyan G et al (2019) Continental accretion and incremental deformation in the thermochronologic evolution of the lesser Caucasus. Geosci Front 10:2189–2202. https://doi.org/10.1016/j.gsf.2019.02.007
Cowgill E, Forte A, Niemi N et al (2016) Relict basin closure and crustal shortening budgets during continental collision: an example from Caucasus sediment provenance. Tectonics 35:2918–2947. https://doi.org/10.1002/2016TC004295
Forte A, Cowgill E, Whipple KX (2014) Transition from a singly vergent to doubly vergent wedge in a young orogen: the greater Caucasus. Tectonics 33:2077–2101. https://doi.org/10.1002/2014tc003651
Fu B, Ninomiya Y, Guo J (2010) Slip partitioning in the northeast Pamir–Tian Shan convergence zone. Tectonophysics 483:344–364. https://doi.org/10.1016/j.tecto.2009.11.003
Gamkrelidze P (1949) Geological structure of Adjara-Trialeti folded system. Metsniereba, Tbilisi. (in Russian)
Gamkrelidze I (1976) Mechanism of tectonic structure formation and some general problems of tectogenesis. Metsniereba, Tbilisi. (in Russian)
Khain VY (1974) Structure and main stages in the tectono-magmatic development of the Caucasus: an attempt at geodynamic interpretation. Am J Sci 274:16–23
McQuarrie N (2002) The kinematic history of the central Andean fold-thrust belt, Bolivia: implications for building a high plateau. GSA Bull 114(8):950–963
Medwedeff DA (1990) Geometry and kinematics of an active, laterally propagating wedge thrust, wheeler ridge, California. Structural geology of fold and thrust belts. In: Mitra S, Fisher G (eds) Geometry and kinematics of an active, laterally propagating wedge thrust, wheeler ridge. The Johns Hopkins University Press, California, pp 3–28
Milanovski E, Khain V (1963) Geological structure of the Caucasus. Moscow State University. (in Russian)
Mosar J, Kangarli T, Bochud M et al (2010) Cenozoic–recent tectonics and uplift in the greater Caucasus: a perspective from Azerbaijan. Geol Soc Lond Spec Publ 340:261–279. https://doi.org/10.1144/SP340.12
Nemcok M, Glonti B, Yukler A, Marton B (2013) Development history of the foreland plate trapped between two converging orogens: Kura Valley, Georgia, case study. Geol Soc Lond Spec Publ 377:159–188. https://doi.org/10.1144/SP377.9
Papava P (1967) Some question of eastern part of the Trialeti Ridge and perspective of oil-bearing of Cretaceous and Paleogene deposits. In: Gamkrelidze P (ed) Geology of the Georgian SSR. NEDRA, pp 188–204. (in Russian)
Pasquarè FA, Tormey D, Vezzoli L et al (2011) Mitigating the consequences of extreme events on strategic facilities: evaluation of volcanic and seismic risk affecting the Caspian oil and gas pipelines in the Republic of Georgia. J Environ Manag 92:1774–1782. https://doi.org/10.1016/j.jenvman.2011.02.003
Reilinger R et al (2006) GPS constraints on continental deformation in the Africa–Arabia–Eurasia continental collision zone and implications for the dynamics of plate interactions. J Geophys Res:BO5411. https://doi.org/10.1029/2005JB004051
Robinson AG, Griffith ET, Gardiner AR, Home AK (1997) Petroleum geology of the Georgian fold and thrust belts and foreland basins. In: Robinson A (ed) Regional and petroleum geology of the Black Sea and surrounding region. AAPG, Tulsa, pp 347–367
Shaw JH, Connors CD, Suppe J (2006) Seismic interpretation of contractional fault-related folds. AAPG, Tulsa

Sokhadze G, Floyd M, Godoladze T et al (2018) Active convergence between the lesser and greater Caucasus in Georgia: constraints on the tectonic evolution of the lesser–greater Caucasus continental collision. EPSL 481:154–161. https://doi.org/10.1016/j.epsl.2017.10.007

Sosson M, Adamia S, Muller C et al (2013) From greater to lesser Caucasus: new insights from surface and subsurface data along a N-S trending transect (Georgia): thick-skin versus thin-skin tectonics. Darius News 3:5–7

Sosson M, Stephenson R, Sheremet Y et al (2016) The eastern Black Sea-Caucasus region during cretaceous: new evidence to constrain its tectonic evolution. Compte-Rendus Geosci 348:23–32. https://doi.org/10.1016/j.crte.2015.11.002

Suppe J, Medwedeff DA (1990) Geometry and kinematics of fault-propagation folding. Eclogae Geol Helv 83:409–454

Tari G, Vakhania D, Tatishvili G et al (2018) Stratigraphy, structure and petroleum exploration play types of the Rioni Basin, Georgia. Geol Soc Lond Spec Publ 464:403–438. https://doi.org/10.1144/SP464.14

Tibaldi A, Alania V, Bonali FL et al (2017) Active inversion tectonics, simple shear folding and back-thrusting at Rioni Basin. Georgia. J Struct Geol 96:35–53. https://doi.org/10.1016/j.jsg.2017.01.005

Tibaldi A, Tsereteli N, Varazanashvili O et al (2020) Active stress field and fault kinematics of the greater Caucasus. J Asian Earth Sci 188:104108. https://doi.org/10.1016/j.jseaes.2019.104108

Toscani G, Bonini L, Ahmad M et al (2014) Opposite verging chains sharing the same foreland: kinematics and interactions through analogue models (Central Po plain, Italy). Tectonophysics 633:268–282. https://doi.org/10.1016/j.tecto.2014.07.019

Tsereteli N, Tibaldi A, Alania V et al (2016) Active tectonics of Central-Western Caucasus, Georgia. Tectonophysics 691:328–344. https://doi.org/10.1016/j.tecto.2016.10.025S0040-1951(16)30478-4

Varazanashvili O, Tsereteli N, Tsereteli E (2011) Historical earthquakes in Georgia (up to 1900): source analysis and catalog compilation. LTD MVP, Tbilisi

Varazanashvili O, Tsereteli N, Bonali FL et al (2018) GeoInt: the first macroseismic intensity database for the Republic of Georgia. J Seismol 22(3):625–667. https://doi.org/10.1007/s10950-017-9726-5

Vincent SJ, Morton AC, Carter A et al (2007) Oligocene uplift of the Western greater Caucasus: an effect of initial Arabia-Eurasia collision. Terra Nova 19:160–166. https://doi.org/10.1111/j.1365-3121.2007.00731.x

Vincent SJ, Carter A, Lavrishchev VA et al (2011) The exhumation of the western Greater Caucasus: a thermochronometric study. Geol Mag 148:1–21. https://doi.org/10.1017/S0016756810000257

Zakariadze G, Dilek Y, Adamia S et al (2007) Geochemistry and geochronology of the Neoproterozoic Pan-African Transcaucasian Massif (Republic of Georgia) and implications for island-arc evolution of the late Precambrian Arabian Nubian shield. Gondwana Res 11:97–108. https://doi.org/10.1016/j.gr.2006.05.012

Chapter 7
Regional Seismotectonic Zonation of Hydrocarbon Fields in Active Thrust Belts: A Case Study from Italy

Giusy Lavecchia, Rita de Nardis, Federica Ferrarini, Daniele Cirillo, Simone Bello, and Francesco Brozzetti

Abstract The study of seismic activity triggered by transient stress changes associated with hydrocarbon reservoir exploitation and deep fluid injection is especially important, in a seismic risk perspective, in regions, where relevant natural seismicity is coupled with widespread and regionally distributed petroleum fields. Hydrocarbon occurrences in Italy derive from a variety of petroleum systems characterized by source/reservoirs distributed in age from Mesozoic to Quaternary. Several of them are located along and in the proximity of the Apennine Outer Thrust System which started to develop in Late Pliocene times and is still active, as well as seismogenic, along most of its extent (especially northern-central Italy and Sicily). Other occurrences are sited along the front of the more internal Pede-Apennine compressional belt, which was mainly active in Early Pliocene times. Still, a large percentage of hydrocarbon fields are observable within the Adriatic-Pelagian foreland, whereas a small number of fields, highly relevant in terms of oil and/or gas extraction, are sited within the intermountain Apennine active extensional domain. In this paper, after grouping the Italian oil and gas fields in stratigraphically and kinematically homogenous Hydrocarbon Field Assemblages (HFAs), we build a regional seismotectonic zonation of the HFAs, taking into account their surface and depth location with respect to the 3D boundary of regional seismogenic provinces. The provinces differ in kinematics (extensional, compressional, strike-slip) and in characterizing seismogenic thickness and depth, which may coincide or not with the depth range of the hydrocarbon recovery, as well as with the depth of fluid withdrawal and injection activities. The HFAs zonation is a new conceptual product intended as a regional-scale tool that contains base 3D geometric-kinematic information on the active and potentially seismogenic

G. Lavecchia · R. de Nardis · F. Ferrarini · D. Cirillo · S. Bello · F. Brozzetti
DiSPUTer-Dipartimento di Scienze Psicologiche, della Salute e del Territorio, Università G. d'Annunzio Chieti-Pescara, Chieti, Italy

CRUST- Centro InteRUniversitario per l'Analisi SismoTettonica tridimensionale, Chieti, Italy
e-mail: glavecchia@unich.it

© Springer Nature B.V. 2021
F. L. Bonali et al. (eds.), *Building Knowledge for Geohazard Assessment and Management in the Caucasus and other Orogenic Regions*, NATO Science for Peace and Security Series C: Environmental Security,
https://doi.org/10.1007/978-94-024-2046-3_7

deformation in correspondence of all large Italian hydrocarbon fields. In case of a seismic sequence ongoing near an HFA (epicentral distance <~10 km), it could provide information on the Quaternary fault pattern and structural style characterizing the area, as well as on the associated crustal state of stress, as input constraints to the discussion on the likelihood of a direct or indirect link, controlled by tectonics, between the hypocentral area and the hydrocarbon field.

Keywords Hydrocarbon fields · Earthquakes · Quaternary faults · Seismotectonic provinces · Triggered seismicity · Italy

7.1 Introduction

Italy is the most hydrocarbon-rich country of southern Europe, with petroleum exploration ongoing since the second half of the XIX century and now spread from the Alps to Sicily (Mattavelli et al. 1993; Anelli et al. 1996; Pieri 2001; Casero 2004; Fantoni et al. 2008; Bertello et al. 2010; Cirilli et al. 2015, 2018) (Fig. 7.1). Italy is also seismically the most active among western European countries, with moderate to destructive (Mw 7.0–7.5) earthquakes occurring from northern Italy to Sicily (ISIDe Working Group 2016; Rovida et al. 2019; Guidoboni et al. 2018).

The overlapping of petroleum/gas exploited reservoirs and relevant well-known seismic activity areas makes Italy a key area to investigate the possibility that human activities associated with hydrocarbon-related activities might trigger the tectonic earthquake activity beyond its natural threshold.

This topic was dramatically brought to the general attention of Italian researchers after the Emilia 2012 seismic sequence (Mw 6.1, northern Italy) that occurred at the front of the Apennine thrust belt, in an area partially occupied by an important producing oil field (Lavecchia et al. 2012b; Malagnini et al. 2012; Burrato et al. 2012; de Nardis et al. 2014; Lavecchia et al. 2015). Two international commissions (ICHESE 2014; Astiz et al. 2014) analyzed in depth and modelled the temporal, geometric, mechanical and dynamic relationships between the seismogenic sources and the injection/extraction activities. As also confirmed by subsequent studies (Juanes et al. 2016; Albano et al. 2017), they both concluded that neither induced nor triggered effects could be proved in this specific case.

At a regional scale, the question of a possible increase in seismic hazard due to oil and gas exploitation in Italy, that might anticipate the natural occurrence of destructive earthquakes (Mw $\geq$ 5.5), remains open (Mucciarelli et al. 2015). To be fully answered, it would require detailed studies and modelling site by site. As preliminary to such deep investigations, the availability of a regional seismotectonic zonation of the hydrocarbon occurrences might be useful to provide, whenever necessary (e.g. during a seismic crisis close or also eventually far from an exploitation site), fast answers on the likely geometric-kinematic interlinks between active faults/seismogenic sources and hydrocarbon reservoirs and on the related natural seismogenic potential.

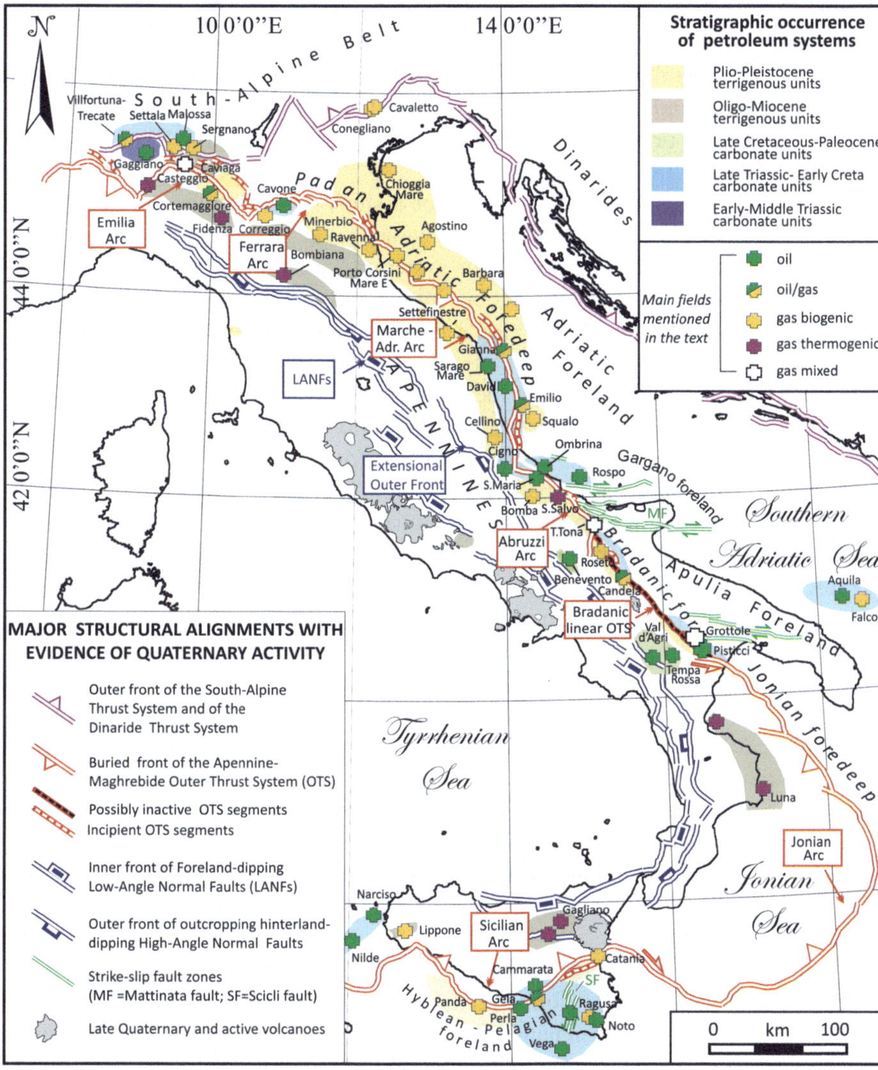

Fig. 7.1 Overview of the stratigraphic and tectonic location of the major petroleum systems along the Apennine-Maghrebide and the south-Alpine compressional belts (after Bertello et al. 2010). The main hydrocarbon fields mentioned in the text are also reported with their prevailing accumulation type (oil and/or gas) and fluid (thermogenic or biogenic gas) origin as derived from the literature (Pieri 2001; Casero 2004; Fantoni et al. 2008; Casero and Bigi 2013; Zappaterra 2013). The schematized surface traces of the structural elements with evidence of Quaternary activity extending across the national territory are also drawn (see text for details). This map, as well as all the others in the paper, is projected using a Lambert Conformal Conic projection (13.0, 42.0, 39.0 and 45.0 are the central meridian, the latitude of origin and the standard parallels, respectively; GCS: WGS84)

The goal of this paper is to build such a regional-scale seismotectonic framework for the Italian hydrocarbon field occurrences. The adopted methodological approach is necessarily multi-disciplinary. Relevant information from petroleum geology, structural geology and seismology are integrated and unified through the construction of a number of original thematic maps, originally built on a GIS platform at a common unified 1:2,000,000 scale.

The paper is organized as follows. After a short specification on the meaning of the terms "induced" and "triggered" seismicity, the main tectono-stratigraphic characteristics of the Italian petroleum systems are summarized and the major structural elements with evidence of activity in Quaternary times are outlined (Fig. 7.1). Subsequently, an updated regional dataset of oil and gas fields, grouped in kinematically homogeneous hydrocarbon field assemblages (HFAs) is compiled (Fig. 7.3). The HFA boundaries are identified by integrating the online information on the individual drilling sites for petroleum exploration in Italy since 1895 (Fig. 7.2) with the geographic and compositional information from literature on the hydrocarbon fields and related petroleum-systems. These are then integrated with the geometry and the time-space evolution of the major structural elements of the Apennine fold-and-thrust belt, possibly controlling the trap formations (Fig. 7.3 and references therein). Afterwards, based on the geometric-kinematic setting of distinct active deformation domains highlighted from earthquakes and tectonic data (Figs. 7.4, 7.5 and 7.6) and on a conceptual model on the structural style and long-term history of progressive deformation in peninsular Italy and Sicily (Fig. 7.7), a seismotectonic zonation of the HFAs is built (Figs. 7.8 and 7.9). It is intended as a regional-scale tool to provide information on the prevailing geometry, kinematics and depth range of the seismogenic deformation field in correspondence of each HFAs. To finish, the vertical distance between the HFAs well bottoms and the underlying seismogenic shear zone bounding the 3D seismotectonic province is introduced as spatial parameter useful to preliminary discuss the likelihood of a structural connection between an exploited field and an ongoing seismic sequence (Figs. 7.10 and 7.11).

The definition of the spatial and tectonic correlation between hydrocarbon fields and earthquake occurrence in Italy, contained in such a database, is certainly not meant to discriminate between natural and anthropogenic seismicity, which requires detailed seismicity, pore pressure diffusion analyses and stress modelling. In our aim, the HFAs zonation can help to contextualize any seismic sequence of doubtful origin (e.g. tectonic versus induced and triggered) in the regional seismotectonic framework. It could be useful to quickly identify, in the light of substantially structural-geological criteria, "critical" areas of exploitation that might, on the one hand, produce triggering effects on natural seismicity and on the other hand suffer a real increase in seismic risk during a natural seismic crisis.

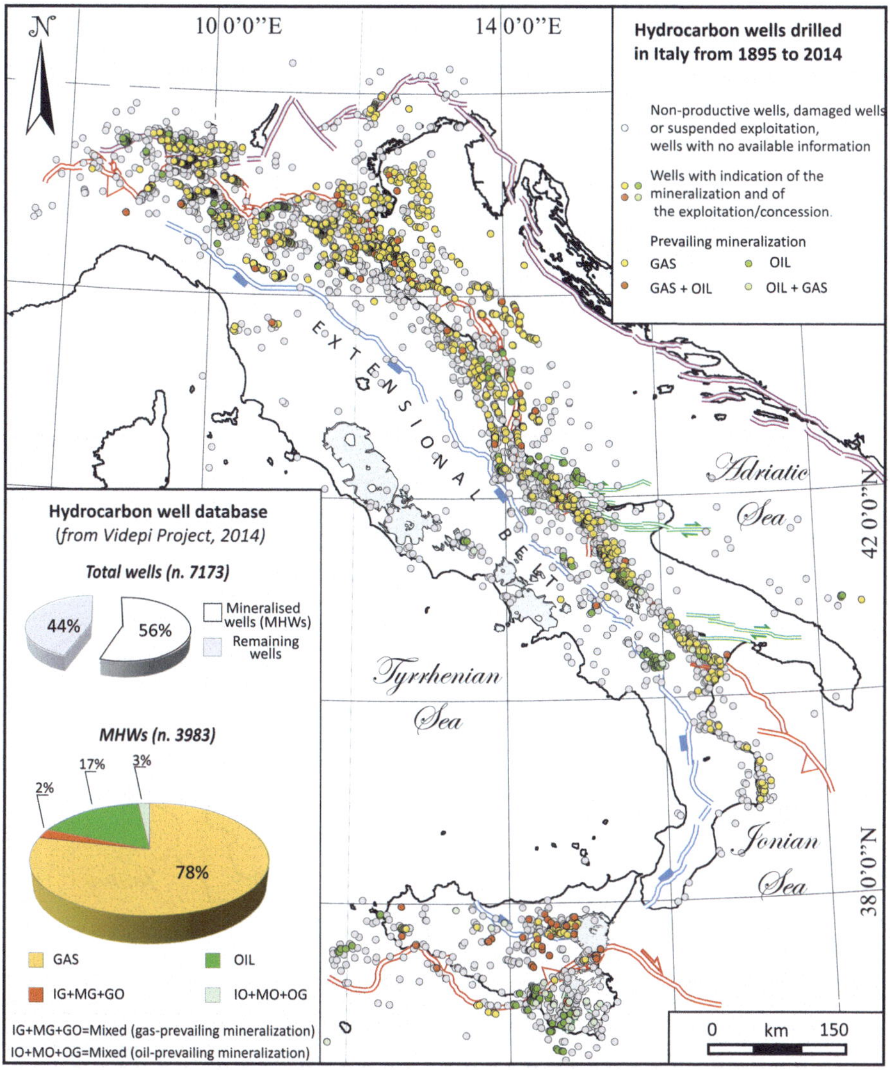

Fig. 7.2 Map of the Italian hydrocarbon-wells given in the Videpi database (VIDEPI Project 2014) and referred to the time interval 1895–2014 (last access April 2014). The productive and potentially productive wells (coloured dot symbols), carrying indication on the mineralization and on the exploitation/concession (56% of the total dataset, upper pie chart in the left inside inset of the figure) are distinguished within an overall dataset of 7173 data and referred to as Mineralized Hydrocarbon Wells (MHWs). The MHWs are coloured based on the prevailing oil versus gas mineralization; the different percentages are provided in the lower pie chart of the inset. The schematized surface trace of the major Late Pliocene-Quaternary structural elements extending across the national territory is also drawn (see text for details); key as in Fig. 7.1

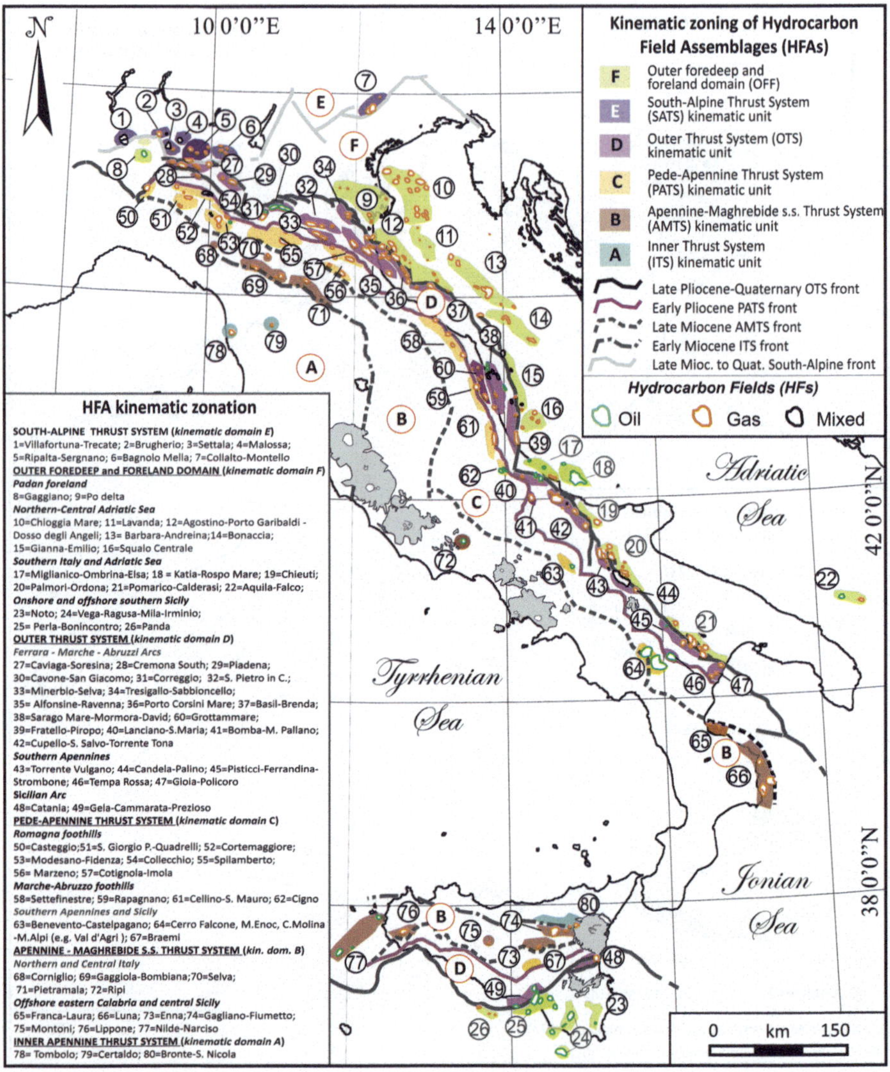

Fig. 7.3 Map of the Italian Hydrocarbon Fields (HFs) grouped into Hydrocarbon Field Assemblages (HFAs). The HFs boundaries and subdivision into gas, oil and mixed mineralization were obtained by integrating the Hydrocarbon Field Map by Casero (2004) with the Mineralized Hydrocarbon Well database (MWHs in Fig. 7.2) extracted from VIDEPI Project (2014). The HFAs boundaries were traced in order to enclose neighbouring HFs similar in both the reservoir tectono-stratigraphic characteristics and the trap tectonic location. The HFAs names, corresponding to the numbers from 1 to 80 in the map, are given in the left side inset within the figure. The different HFAs colours highlight their belonging to different kinematic domains

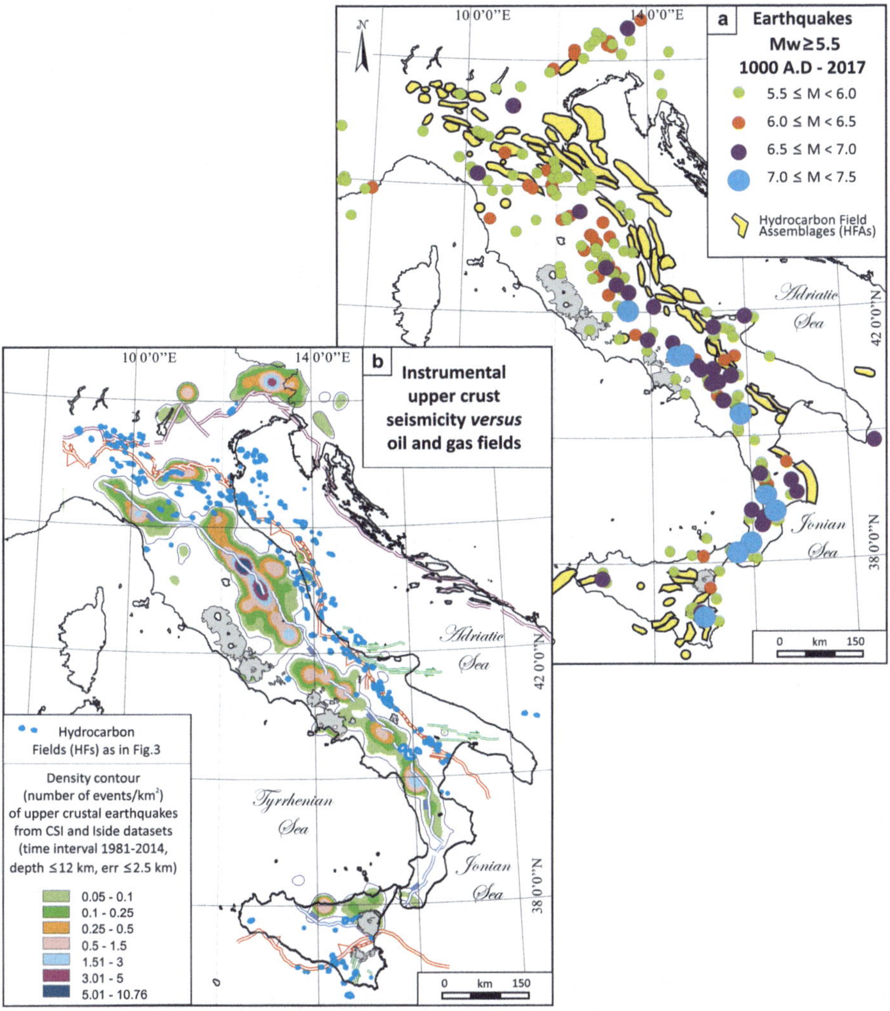

Fig. 7.4 Schematic view of the spatial relationships between earthquakes and hydrocarbon fields. (**a**) Hydrocarbon Field Assemblages (HFAs) as defined in Fig. 7.3 versus major historical events (Mw ≥ 5.5) from CPTI15v.2 dataset for the time interval 1000 A.D. to December 2017 (Rovida et al. 2019). (**b**) Hydrocarbon Fields (HFs) as defined in Fig. 7.3 versus a density contour map of the upper crust (≤12 km) instrumental seismicity ($M_L > 2.5$) for the time interval 1981–2014; data are extracted from the CSI1.1 seismic catalogue for the time interval 1981–2002 (Castello et al. 2006) and from the ISIDe database for the time interval 2003–2014 (ISIDe Working Group 2016)

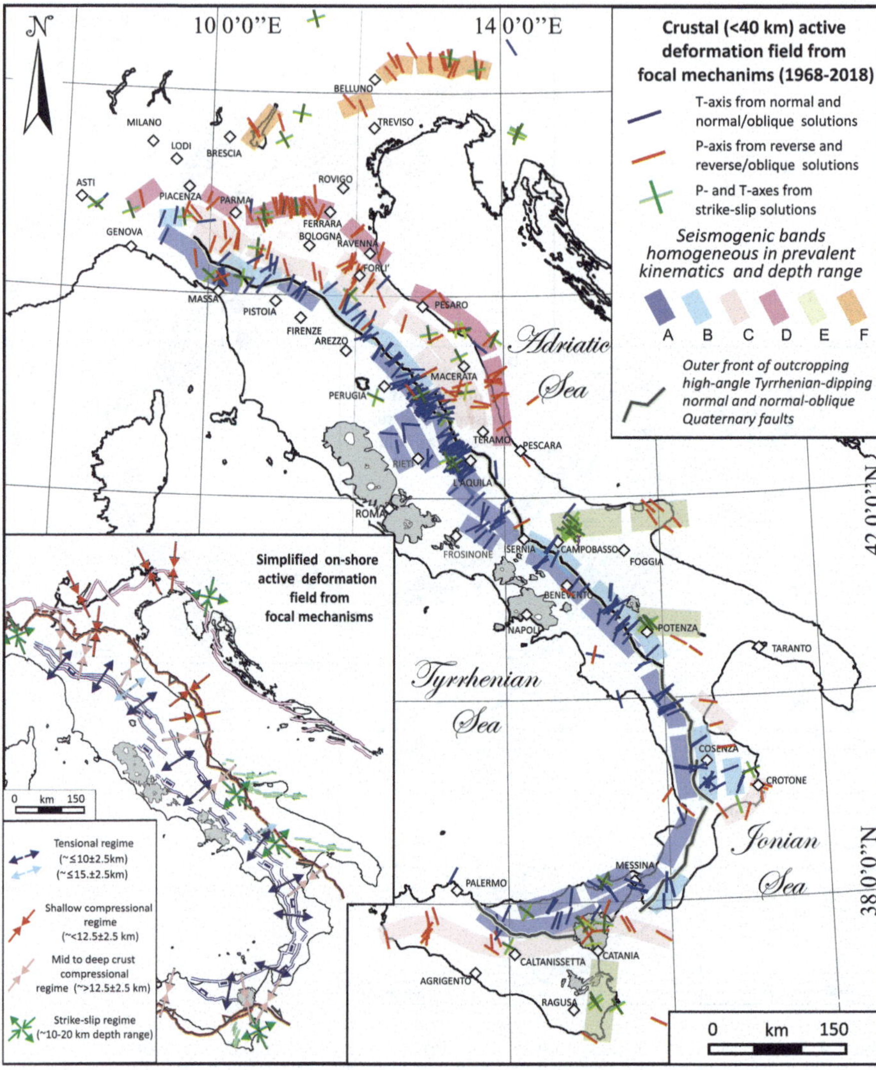

Fig. 7.5 Active deformation field from a compilation of focal mechanisms taken from the RCMT (Pondrelli and Salimbeni 2006) and TDMT catalogues (Scognamiglio et al. 2006) plus other focal solutions deduced from specific papers (references in Lavecchia et al. 2007a) for the Italian earthquakes with Mw > 3.5 located onshore at depths <40 km and occurred in the interval 1968–2018. The blue and red bars represent the surface projections of T- and P-axes from focal mechanisms classified as normal/normal-oblique and reverse/reverse-oblique, respectively. The light and dark green bars represent the T-and P-axis, respectively, of strike-slip type focal mechanisms. The kinematic classification has been obtained using the Frohlich (1992) triangular diagram. The coloured bands envelope deformation zones with clear prevailing kinematics, along the major structural trends: band A = intermountain Apennine and northern Sicily upper crust extension; band B = mid-crust extension; band C = pede-Apennine compression; band D = outer front compression; band E = foreland strike-slip; band F = south-Alpine compression. The average prevailing attitude of the P- and T-axes, subdivided for depth ranges, is schematized within the inset together with the major active structures

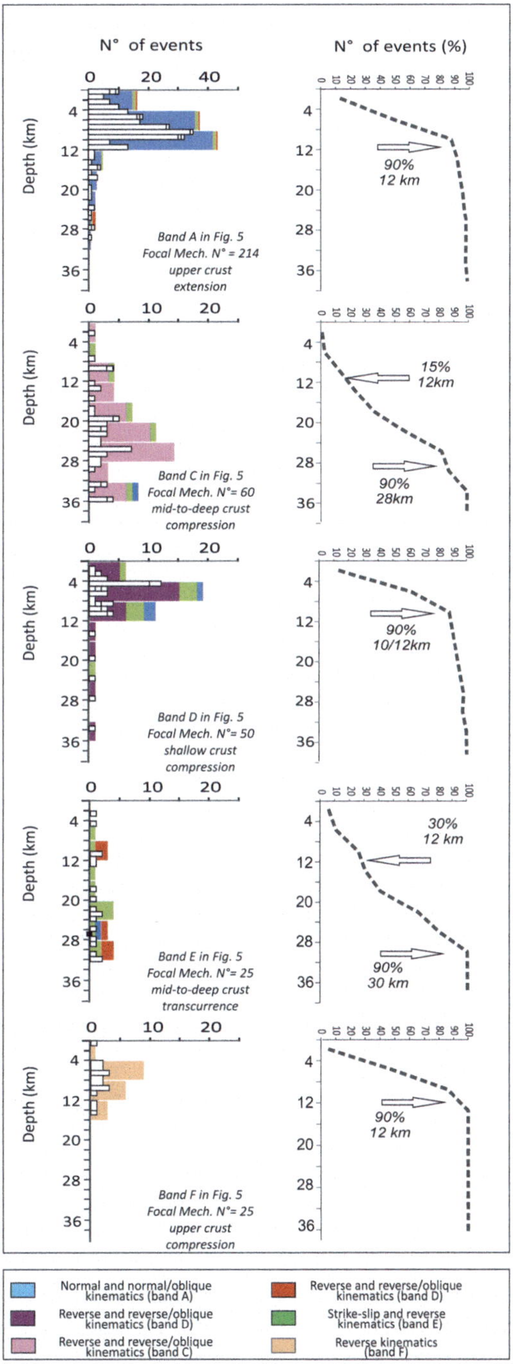

Fig. 7.6 Depth distribution and kinematics of the focal mechanisms (FM) falling within the boundaries of five structural-kinematic bands (A to F) identified in Fig. 7.5 (Mw > 3.5, depth <40 km, time interval 1968–2018). The FM hypocentral coordinates are derived from the CSI1.1 seismic catalogue for the events older than 2002 (Castello et al. 2006), from the ISIDe database for the more recent events and from specific papers for some minor sequences and for the most important seismic sequences (references in the main text). Left column: number of events for kinematics versus depth; right column: cumulate number of events for kinematics (4 km intervals) versus depth

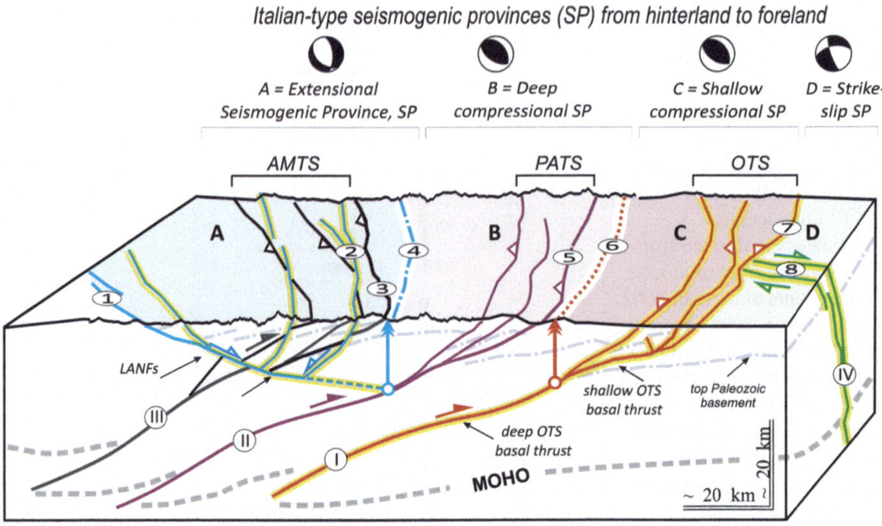

Fig. 7.7 Conceptual structural/seismotectonic block diagram transversal to the arcuate trends of the compressional and extensional structures across the peninsular Italy, illustrating the inferred 3D geometry of the active and possibly seismogenic compressive, extensional and strike-slip structures and also highlighting the major Late Miocene to Quaternary compressional kinematic units across the Apennine-Maghrebide system as identified in Fig. 7.3. Key: A, B, C, D = seismogenic provinces as in Fig. 7.8; 1 = surface trace of the foreland-dipping Quaternary low-angle normal faults (e.g. LANFs trace in Fig. 7.1); 2 = surface trace of the outermost outcropping hinterland-dipping Quaternary high-angle normal faults (e.g. Extensional Front trace in Fig. 7.1); 3 = outer front of the Late Miocene Apennine-Maghrebide Thrust System (AMTS in Fig. 7.3); 4 = surface projection of the LANF basal detachment depth tip line (~12–15 km); 5 = outer front of the Early Pliocene Pede-Apennine System (PATS in Fig. 7.3); 6 = surface projection of the 10–12 km depth contour line of the Outer Thrust System (OTS in Fig. 7.3) basal thrust; 7 = outer front of the buried OTS; 8 = Intra-foreland active strike-slip faults; I = OTS basal thrust; II = PATS basal thrust; III = AMTS basal thrust; IV = active foreland strike-slip zones. The active and possibly seismogenic structures discussed in the paper are highlighted in yellow

7.2 Triggered Versus Induced Seismicity

Seismicity is defined as "induced" or "triggered" when slip along a fault is promoted by transient changes in stress or pore pressure (Cesca et al. 2013). The adjective "induced" refers to a process of seismic stress release that is entirely controlled by a transient non-tectonic phenomenon and occurs in an aseismic area. The adjective "triggered" refers to seismic events that are mainly controlled by the background tectonic stress and, only subordinately, by the transient phenomenon. Therefore, triggered earthquakes represent the premature release of tectonic stress (Gibson and Sandiford 2013). Transient phenomena responsible for stress changes triggering earthquake activity may have a natural origin, being driven by either exogenous forces (e.g. sun and moon tides) or endogenous forces (e.g. magmatic, volcanic and geothermal activities). For instance, earthquake-earthquake interactions may be

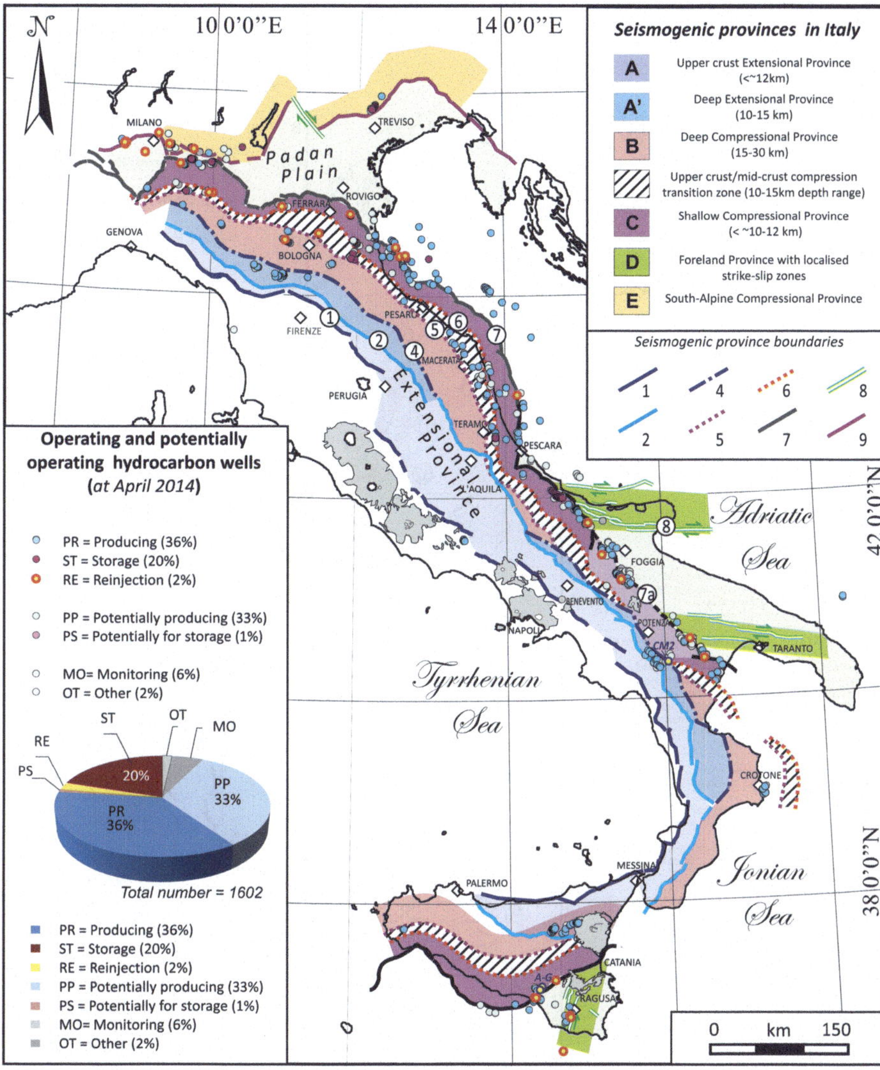

Fig. 7.8 Regional-scale seismogenic provinces of Italy with Italian hydrocarbon-wells operating and potentially-operating at 2014, as extracted from Videpi Project database; the pie chart diagram highlights in percentage the different wells destination. A-G = Armatella and Giaurone Enhanced Oil Recovery fields (close to Gela, southern Sicily); CM2 = Costa Molina reinjection well (wastewater disposal in Val d'Agri, Southern Apennines) (from Andrei et al. 2010). The structural elements, highlighted with a number within a circle, are the same as in the block diagram of Fig. 7.7, except for number 5. Key: 1 = surface trace of the foreland-dipping Quaternary low-angle normal faults (LANFs); 2 = surface trace of the hinterland-dipping Quaternary high-angle normal faults; 4 = surface projection of the LANF basal detachment depth tip line (~12–15 km); 5 = inferred surface projection of the ~15 km depth contour line of the OTS Basal Thrust; 6 = inferred surface projection of the ~10 km OTS depth contour line; 7 = buried OTS front; 8 = Intra-foreland active strike-slip faults; 9 = south-Alpine thrust front

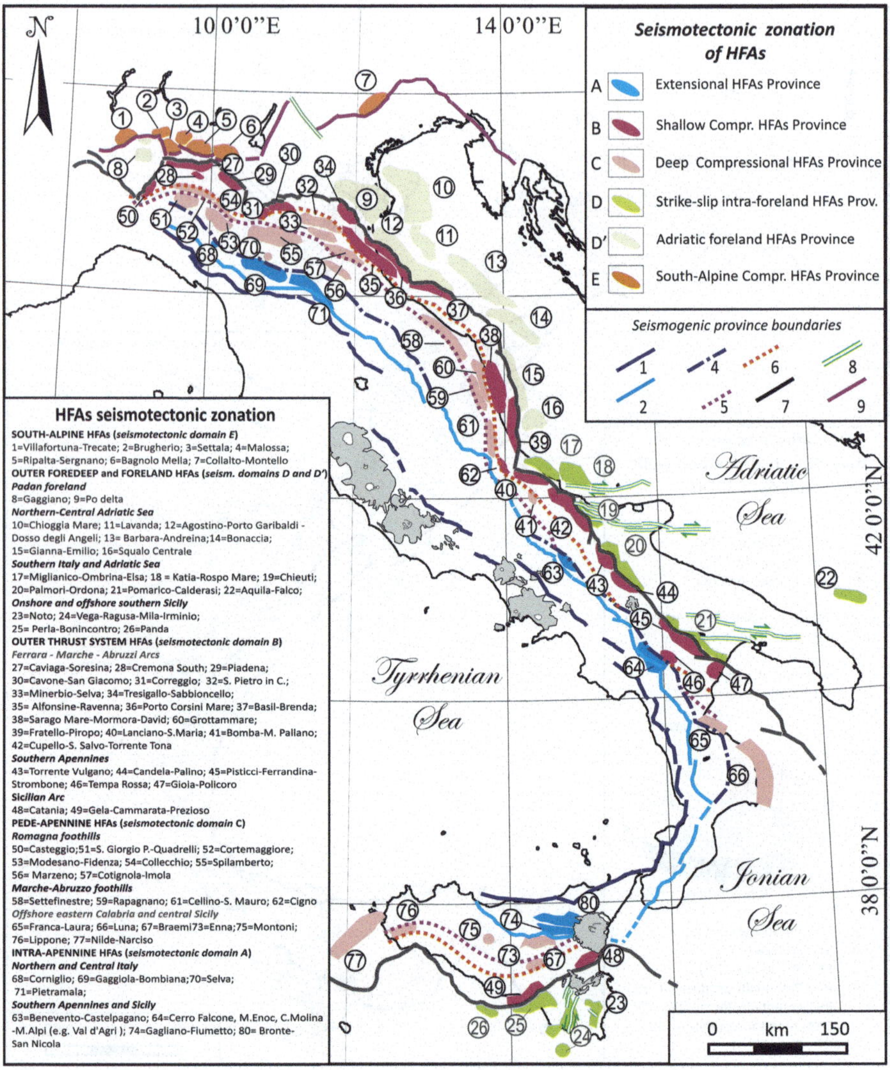

Fig. 7.9 Regional seismotectonic zonation of the hydrocarbon field assemblages (HFAs) of Italy. The boundaries and the key of the Hydrocarbon Field Assemblages (HFAs) are those defined in Fig. 7.3. The different HFA's colour highlights their belonging to various seismotectonic provinces. The boundaries of the seismotectonic provinces and the key is the same as those in Fig. 7.8

considered as a specific case of naturally-triggered seismicity (Stein et al. 2012). Modifications of the state of stress at the point to trigger earthquakes may also be due to human activities associated with oil and gas extraction, geothermal energy production and mining, as well as with water reservoir impoundments, wastewater disposals and fluid injections (Cesca et al. 2013; Ellsworth 2013 and references

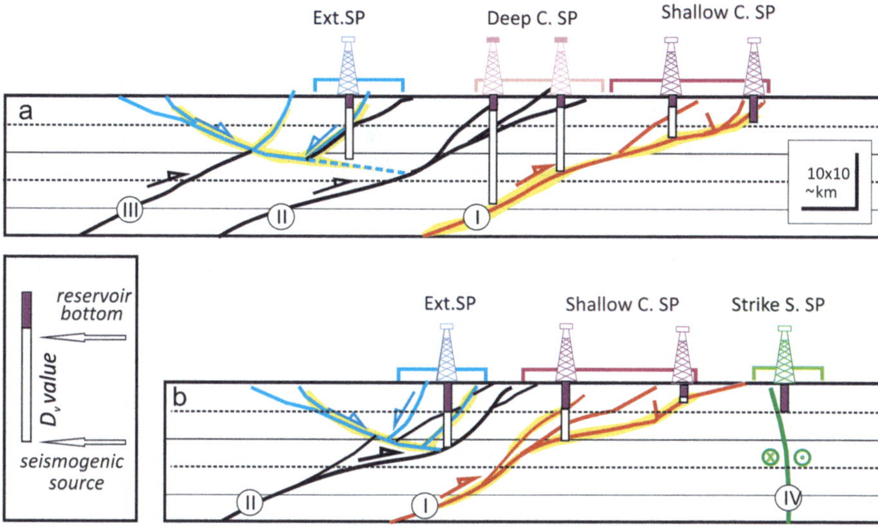

Fig. 7.10 Conceptual representation of the "D-value" intended in this paper as the vertical distance between the considered reservoir (potentially subject to extraction/injection activities) and the inferred underlying seismogenic source. Note that within the same seismogenic province, moving from one HFA to the other, the D-value may have a relevant range of variability, due to the local depth of the reservoir and to the punctual depth of the underlying fault source. Sketches A and B are not in scale and highlight different oil field structural position (see the derricks) along two conceptual WSW-ENE transects across northern-central Italy and southern Italy, respectively. Key: Ext.SP = oil field within the extensional Seismogenic Province; C.SP. = oil fields within the shallow and /or deep Compressional Seismogenic Province, S.SP = oil field within the Strike-slip Seismogenic Province. The active and possibly seismogenic structures discussed in the paper are highlighted in yellow. The roman numbers in the circle symbol mark the same structural elements as in Fig. 7.7, key: I = Outer Thrust System basal thrust; II = Pede-Apennine Thrust System basal thrust; III = Apennine-Maghrebide Thrust System basal thrust; IV = active foreland strike-slip zones

therein). These activities can trigger seismic instabilities of various sizes (from microseismicity to moderate/large earthquakes) and at various scales (from local to regional) by either reducing the effective normal stress or rising the shear stress. The stress change from all the different possible mechanisms is small (i.e. 0.01–0.05 MPa), but may be sufficient to trigger earthquake activity when the state of stress is near to failure (Grasso 1992). Therefore, activity on deep reservoirs in tectonically active zones can carry a real risk of triggering damaging earthquakes. What remains especially undetermined is the spatial and temporal distance between industrial activity and the associated seismic event. In fact, a high-volume fluid injection can propagate far away from the injection point, activating fault structures several kilometres far away (Mulargia and Bizzarri 2014). As well, triggered events

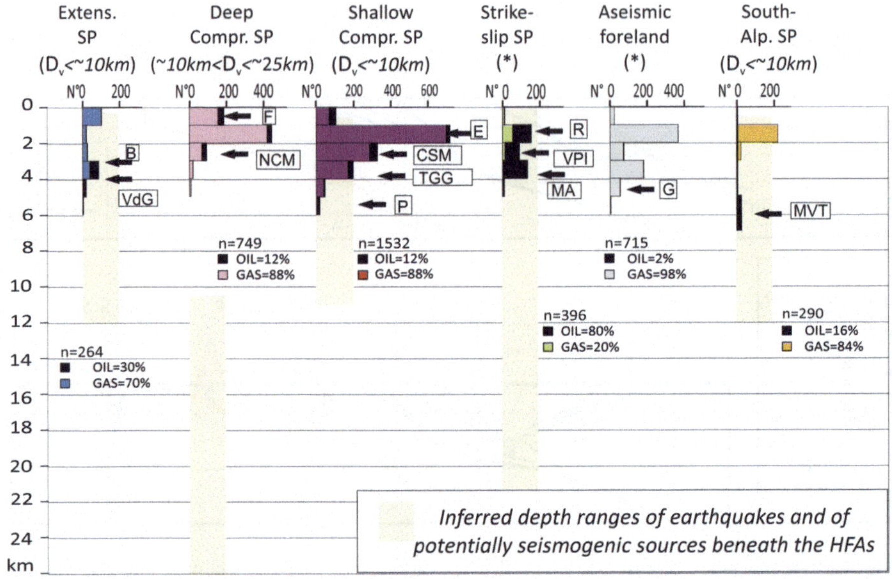

Fig. 7.11 Depth range of the Mineralized Hydrocarbon Wells (MHWs in Fig. 7.2, falling within the boundaries of the Hydrocarbon Field Assemblages of Fig. 7.9, for any of the Seismogenic Provinces identified in Fig. 7.8. Each hydrocarbon depth histogram highlights the different percentage of the oil (black) versus gas (coloured). The histogram vertical scale is in kilometres (km); the horizontal one reports the NHWs number (N°). The well depth distribution is compared with the earthquake depth range within the provinces (yellow vertical bar), as derived from the earthquake depth histograms of Fig. 7.6. The D-value range characterizing each province (e.g. the vertical distance Dv between the possible reservoirs and the underlying seismogenic fault) is also indicated. (*) = not inferred D-value due to sub-vertical geometry of the seismogenic faults in the Strike-Slip province and to the lack of known seismogenic sources in the northern Adriatic foreland domain. Some of the most relevant Italian oil fields are highlighted. Key (in alphabetic order): B = Benevento (Campania); CSM = Cavone (Ferrara Arc), S.Maria Mare; E = Emma (Marche-Adriatic Arc); F = Fidenza (Emilia-Romagna); G = Gaggiano (south-Alpine foreland); MA = Miglianico (Abruzzi), Aquila (south Adriatic); MVT = Malossa, Villafortuna-Trecate (western Southern Alps); NCM = Nilde (Sicily), Cortemaggiore (Romagna); P = Prezioso (Sicily); R = Rospo (Pescara offshore); TGG = Tempa Rossa-Gorgoglione (Lucania), Gela (Sicily); VdG = Val D'Agri (Campania-Lucania); VPI = Vega, Perla, Irminio (Sicily)

can occur up to several years after well activities have begun or even several years after all well activities have stopped (Nicholson and Wesson 1992). The November 2007, Pohang earthquake (Mw 5.5, South Korea), which occurred in proximity to an enhanced geothermal system site where high-pressure hydraulic injection had been performed during the previous 2 years, was located at depths of 4–5 km, nearly 10–12 km eastward of the active Yangsan regional strike-slip fault (Grigoli et al. 2018). Combining seismological and geodetic analyses, Grigoli et al. (2018) concluded that the Pohang earthquake was an induced event, which might, in turn, transfer static stress to the larger nearby faults, in such a way as to potentially increase the seismic hazard in the area. The November 2011, Oklahoma earthquake sequence (Pennsylvania-USA), consisting of three major events (one Mw 5.7 and two Mw 5.0) associated with a large structurally controlled oil field, started near the bottom of the active injection well and spread along a linear zone for nearly 10 km, activating in sequence three pre-existing steeply dipping strike-slip fault segments at depths between 1 and 6 km (McGarr et al. 2002; Keranen et al. 2013; Block et al. 2014). These events, which were located within the continental interior of the United States, ~180 km from the nearest known Quaternary fault, and are a classic example of induced seismicity, had a delay of nearly 20 years from the initiation of the injection activity.

7.3 Stratigraphic and Structural Framework of the Italian Petroleum Systems

The complex stratigraphic and structural context of the Italian hydrocarbon fields have been largely reviewed in the literature and schematized in a number of tectono-stratigraphic cycles, schematic maps and cross-sections that have clearly highlighted: (1) the age, stratigraphy and paleogeographic setting of the petroleum system; (2) the age and kinematics of the tectonic event determining the trap formation; (3) the hydrocarbon accumulation type (oil and gas) and the fluid origin (thermogenic or biogenic gas); (4) the geographic location of the reservoirs within different kinematic domains (Pieri and Groppi 1981; Mattavelli et al. 1993; Zappaterra 1994; Anelli et al. 1996; Holton 1999; Pieri 2001; Casero 2004; Fantoni et al. 2008; Bertello et al. 2010; Casero and Bigi 2013; Zappaterra 2013; Cirilli et al. 2015, 2018). In our meaning of the term, oil petroleum systems include not only the reservoirs (mostly carbonate) but also the source rocks (that are mostly organic-rich shales) and seals (both carbonates and marls/marly limestones). This is valid at least for most of the Italian Triassic/Jurassic petroleum systems and, to our knowledge, also for most of the onshore and offshore petroleum systems from the Alps to the Apennines. Italian oil petroleum systems are mainly ascribed to early-middle Triassic, late Triassic-early Cretaceous and late Cretaceous-early Eocene carbonates deposited within the subsiding continental margin during the Tethyan extensional phase and/or involved in the south-Alpine and Apennine compressional phases (Bertello et al. 2010). The largest accumulations are found in the western and central Po-Plain of northern Italy (Gaggiano, Middle Triassic; Villafortuna, Late Triassic; Cavone, Early Cretaceous), in the Adriatic offshore of the central Italy (Rospo field, Paleocene-Eocene), in southern Italy (Val d'Agri – Tempa Rossa, Paleocene-Early Eocene) and in southern Sicily (Gela, Late Triassic-Early Jurassic) (Fig. 7.1, and references therein). Gas reservoirs mainly coincide with Cenozoic terrigenous units infilling the Alps-Apennine foredeep basins. Thermogenic gas occurrences are sited within the inner Oligo-Miocene flysch deposits, later tectonized and thrusted, and now residing along the northern Apennines margin (e.g. Fidenza field), in Calabria (e.g. Luna field) and Sicily (e.g. Gagliano field). Biogenic gas fields are abundant, instead, in the outer Plio-Pleistocene foredeep successions, mainly in the western and northern Padan Plain (e.g. Settala, Sergnano, Caviaga) and in the northern Adriatic Sea (e.g. Agostino-P. Garibaldi, Barbara), but also in southern Italy (e.g. Roseto, Candela) and Sicily (e.g. Ragusa-Noto, Panda). In Fig. 7.1, we outlined the above information, compiling an introductory map of the major hydrocarbon fields with their prevailing accumulation type (oil and/or gas) and fluid origin (thermogenic or biogenic gas) and with an indication of the stratigraphic occurrence of the petroleum systems. In the same map, we also drew the major alignments of tectonic structures that will be mentioned in the text and which carry, although not always with continuity, evidence of activity in the Quaternary (~2.6 My, following the IUGS Sub-commission on Quaternary Stratigraphy, Cohen and Gibbard 2011). The depicted structural elements, although schematic, derive from the integration of the Structural and the Neotectonic Maps of Italy, scale 1:500,000 (CNR – P.F. Geodinamica 1987, 1990) with detail maps and specific studies from the literature. Fundamental pieces of information derive from the overview on the anatomy of the Apennine-Alpine orogen by Vai and Martini (2001 and reference therein) and from the overview on the geophysical exploration in Italy by Finetti et al. (2005 and reference therein). Our own field data and structural interpretations, related to the Late Pliocene-Quaternary extensional and compressional structures in peninsular Italy and Sicily, were also considered (Lavecchia 1988; Lavecchia et al. 1994, 2003, 2007a, 2011a, b, 2017; Boncio et al. 2000, 2007; Bell et al. 2005; Brozzetti et al. 2009b; Visini et al. 2010; Brozzetti 2011; Lavecchia and Bell 2012; Sgroi et al. 2012; Di Naccio et al. 2013; Romano et al. 2013; Di Giulio et al. 2016; Brozzetti et al. 2017a; Ferrarini et al. 2017).

Specifically, in the map of Fig. 7.1, we represent: (1) the Outer Thrust Front of the South-Alpine thrust system and of the Dinaride thrust system; (2) the buried thrust front of the Apennine-Maghrebide compressional belt, hereinafter referred to as Outer Thrust System, OTS; (3) the inner boundary of the Intermountain Apennine-northern Sicily extensional belt, mainly consisting in foreland-dipping low-angle normal faults, hereinafter referred to as LANFs; (4) the outer outcropping front of Intermountain Apennine-northern Sicily extensional belt, mainly consisting in hinterland-dipping high-angle normal faults, antithetic to the LANFs; (5) – localized zones of active strike-slip deformation, such as the Mattinata (MF) and the Scicli (SF) faults (in the Gargano and Hyblean forelands, respectively). The extensional, compressional and strike-slip structures of peninsular Italy and Sicily drawn in the map represent the last stage of a deformation history that, starting from late Miocene times, led to the progressive nucleation, and outward migration, of a radial foredeep-thrust system (e.g. Apennine-Maghrebide belt) surrounding a coeval and substantially-coaxial extensional domain in the rear (e.g. the Tyrrhenian basin) (Alvarez 1972; Malinverno and Ryan 1986; Lavecchia 1988; Patacca et al. 1990 among many others).

A summary of the acronyms used across the paper may be found in Table 7.1.

Table 7.1 Summary of most common terms used in this paper with their acronyms

Acronyms used in this paper		
Hydrocarbon Well	HW	Well for hydrocarbon exploration in the VIDEPI 2014 database (*)
Mineralized Hydrocarbon Well	MHW	Oil and/or gas hydrocarbon wells from VIDEPI database carrying indications on the found mineralization and on the exploitation/concession (*)
Hydrocarbon Field	HF	Area devoted to exploration/exploitation of hydrocarbons including one or more reservoirs in the underground with common sourcing and trapping mechanisms (**)
Hydrocarbon Field Assemblage	HFA	Area containing neighboring HFs lying within the boundary of a same kinematic province and showing substantially similar source and reservoir (***)
Outer Thrust System	OTS	Most external compressional kinematic block of the Apennine-Maghrebide system (late Pliocene-quaternary in age)
Low-angle normal faults	LANFs	Active fault-system dipping toward the Adriatic foreland and bordering inward the quaternary intra-Apennine extensional belt
High-angle normal faults	HANFs	Quaternary fault-system developed within the LANF hanging-wall rock-volume

Key: (*) oil and gas well database projected in Fig. 7.2; (**) oil and gas fields projected in Fig. 7.3; (***) HFAs projected in Figs. 7.3, 7.4a and 7.9

7.4 From Hydrocarbon Wells to Hydrocarbon Fields (HFs) and Hydrocarbon Field Assemblages (HFAs)

A dataset of on-shore and off-shore hydrocarbon wells drilled in Italy from 1895 to 2014 has been made available online by VIDEPI Project (2014) (last access April 2014). It provides information on the well coordinates (latitude, longitude, depth), on the mineralization of the productive wells (gas versus oil) and on the drilling purposes and results. The total dataset consisted of 7173 wells. From such a total dataset, we extracted a sub-set of 3983 wells (56%) that are those carrying indication on the mineralization and on the exploitation/concession (hereinafter, they will be referred to as Mineralized Hydrocarbon Wells, MHWs) (Fig. 7.2). The MHWs are mainly concentrated close and along the OTS front and within the Adriatic – Hyblean – Pelagian foreland, as well as along portions of the South-Alpine front. The well total depths cover a wide range, with a prevailing concentration in the 1–2 km interval (50% of total), and with depths >5000 km in a few cases (2%), mainly in the northern Alps and southern Sicily. In large prevalence, the MHWs show gas-type mineralization (78%); subordinately they are oil-rich (17%) or mixed (5%) (pie chart in Fig. 7.2). Nearly 40% wells, among the MHWs drilled between 1948 to 2012, were operating or potentially operating at April 2014. They are more or less uniformly distributed over the MHW total and include 36% of producing wells,

33% of potentially productive wells not operating at present, 20% used for storage and 2% for reinjection.

By superimposing in map view the distribution of the punctual MHWs extracted from the VIDEPI database with the surface boundary of the major hydrocarbon fields redrawn from georeferenced regional maps in Casero (2004) and Casero and Bigi (2013) (Fig. 7.2), we compiled a GIS database of oil and gas fields in Italy, hereinafter referred to as Hydrocarbon Fields (HFs). To follow, neighboring HFs (along-strike distance less than about 30 km), characterized by a common trap tectonic setting and source/reservoir chrono-stratigraphy, were grouped into Hydrocarbon Field Assemblages (HFAs) (Zappaterra 1994; Anelli et al. 1996; Pieri 2001; Casero 2004; Fantoni et al. 2008; Bertello et al. 2010; Casero and Bigi 2013; Zappaterra 2013). The HFA boundaries were traced also taking into account the HFAs location within distinct kinematic domains. The latter are meant to correspond to fold-and-thrust systems (distinguished in space and age) and to regional peaks of tectonic activity marked by the age of corresponding foredeep deposits (Waschbusch and Royden 1992). The HFA data, stored as ESRI shapefiles, are mapped in Fig. 7.3.

7.5 Kinematic Zonation of HFAs

The reconstruction of the kinematic domains that we propose (Fig. 7.3) has been carried out by drawing the outer compressional fronts at each identified kinematic stage. It is based on the critical analysis of a considerable amount of papers (Patacca et al. 1990; Boccaletti et al. 1990; Lavecchia et al. 1994, 2007b; Casero 2004; Patacca and Scandone 2004; Cosentino et al. 2010; Santilano et al. 2019; Ghielmi et al. 2010 among many others). Specifically, from the hinterland to the foreland of the Apennine-Maghrebide system, we distinguish four fold-and-thrust kinematic blocks which rejuvenate north-to NNE-ward in the northern Apennines, eastward in central-southern Italy and southward in Sicily. The Inner Thrust System (ITS) is bordered outward by a frontal thrust system of prevailing Early Miocene age (kinematic unit A). The Apennine-Maghrebide s.s. Thrust System (AMTS) (kinematic unit B), marked in peninsular Italy by the main topographic bulge, is bordered outward by a frontal thrust system, Late Miocene in age (kinematic unit B). The Pede-Apennine Thrust System (PATS) is bordered by a frontal thrust system of Early Pliocene age (kinematic unit C). The Outer Thrust System (herein referred to as OTS – kinematic unit D) is bordered outward by a frontal thrust system (mostly buried) of Late Pliocene-Quaternary age and develops both onshore and offshore. Kinematic units A and B involved in the deformation both the Mesozoic-to-Miocene sedimentary cover and the rocks pertaining to the underlying crystalline basement. Kinematic units C and D involved in the deformation also the Pliocene foredeep units. The OTS was active along its entire length from Late Pliocene times to Early Pleistocene times. Its activity, and/or the activity of its hanging-wall splays, also during Late Quaternary times is taken into consideration for most of the belt but remains especially controversial for the Campania-Lucania segment (Bradanic linear OTS in southern Italy, Fig. 7.1). A well distinct compressional domain is represented by the South-Alpine Thrust System (kinematic unit E) which is bounded southward by a late Miocene frontal thrust system, reactivated in late Pliocene-Quaternary times. The outer foredeep and foreland domain (OFF) is represented by the Padan-Adriatic-Pelagian kinematic unit (F) consisting of Late Pliocene-Quaternary undeformed foredeep terrains and of autochthonous Mesozoic carbonate units. Discrete narrow zones of strike-slip deformation, also active in Quaternary times, locally dissect the foreland terrains. The geographic position of the HFAs within the boundary of the kinematic units reflects the age of the fold-and-thrust structure that hosts the reservoirs and the traps and, therefore, it does imply a kinematic zonation of the HFAs. The attribution of each HFA to a well distinct kinematic domain is represented in the map of Fig. 7.3, built on a GIS platform on an original scale of 1:2,000,000.

7.6 Seismicity Versus HFAs

As first step to analyze the spatial relations between hydrocarbon fields and potentially seismogenic faults in Italy, we examined the HFAs distribution with respect to the epicentral distribution of the major historical and instrumental events (Mw $\geq$ 5.5) (Fig. 7.4a). The earthquake data were extracted from the CPTI15v2.0 catalogue (time interval 1000–2017) (Rovida et al. 2019). An overlap between earthquakes and HFAs may be observed along the southern Alpine front, along a wider extent of the Ferrara Arc and of the Northern Adriatic arc, along some sectors of the Emilia-Romagna and Marche foothills, in Sicily and very close to some foreland strike-slip faults in the Gargano and Hyblean forelands. In Fig. 7.4b, we compared the distribution of the oil and gas fields (Hydrocarbon Fields, HFs in Fig. 7.3) with a density contour map (number of earthquakes/km^2) of the upper crust ($\leq$12 km, M_L > 2.5) seismicity in the period 1981–2014, extracted from the CSI1.1 seismic catalogue (Castello et al. 2006) and from the ISIDe database (vertical error $\leq$2.5 km) (ISIDe Working Group 2016). We observe a prevailing lack of spatial overlap between the HFs and the instrumental earthquakes, with a few exceptions such as the Cavone field (HFA 30) in the northwestern Ferrara Arc, the Sarago Mare field in the central Adriatic arc (HFA 38) and the Val d'Agri field within the extensional belt of southern Italy (HFA 64).

Successively, we compiled an updated collection of available focal mechanisms (FMs) from regional databases (Regional Centroid Moment Tensor, RCMT catalogue, Pondrelli and Salimbeni 2006; Time Domain Moment Tensor catalogue TDMT, Scognamiglio et al. 2006; Montone and Mariucci 2016) and from some papers on specific crustal seismic sequences with events Mw > 3.5, located at depth <40 km, that occurred onshore in the time interval 1968–2018. Following previous Authors (Gulia and Weimer 2010; Macchiavelli et al. 2012; Sgroi et al. 2012; Montone and Mariucci 2016), we classified the FMs into kinematic classes (strike-slip, reverse and reverse/oblique, normal and normal/oblique) and represented them in map view as T-axes for the normal and normal oblique events and as P-axes for the others (Fig. 7.5). Starting from the T- and P-axes spatial distribution, and also taking into account the prevailing structural trends highlighted by the active extensional, compressional and strike-slip fault alignments (see inset of Fig. 7.5), we identified a number of narrow (~20 km wide) structural bands with an homogenous FM kinematics. From the Tyrrhenian hinterland to the Adriatic-Pelagian foreland we identified: A and B) – two extensional bands running across the Apennine mountain chain from northern Italy to Sicily; C) – one, and locally two, compressional bands extending with an arcuate shape across the Romagna-Marche external foothills of central-northern Italy and across the mainland region of Sicily; D) – a more segmented compressional band running along the arcuate Padan-Adriatic front of the OTS; E) – two strike-slip bands within the Apulian and Hyblean foreland and F) – two nearly E-W striking narrow compressional bands along discrete sectors of the south-Alpine front.

In order to analyze the earthquake depth distribution within these bands, we associated to each FM the hypocentral coordinates derived from the CSI1.1 seismic catalogue (Castello et al. 2006) for older events, from the ISIDe database for recent events, and from specific papers for the most important seismic sequences and for some of the minor instrumental events (Gasparini et al. 1985; Anderson and Jackson 1987; Frepoli and Amato 1997; Neri et al. 2005; Piccinini et al. 2006; Scafidi et al. 2015; Brozzetti et al. 2009a, 2017b; Lavecchia et al. 2007a, 2016, 2017; Sgroi et al. 2012; Presti et al. 2013; Ferrarini et al. 2014). We have obtained a total dataset of nearly 400 FM, ranging in magnitude from $3.5 < Mw < 7$ in the time interval from 1968 to 2018. We summarized the results in the depth histograms in Fig. 7.6, which refer to the structural/kinematic band-types of Fig. 7.5; the data are presented as number of events and percentage (%) versus depth. The Intermountain extensional band (A and B in Fig. 7.5) contains about 250 events with $3.5 < Mw \leq 6.9$, including the large Irpinia 1980 (Mw 6.9), L'Aquila 2012 (Mw 6.3) and the Norcia 2016 (Mw 6.5) normal fault earthquakes; ninety per cent of the analyzed events are shallower than 12 km. The Pede-Apennine and Mainland Sicily compressional bands contain about 60 events, with $3.5 < Mw \leq 5.5$, including the south-verging Belice 1968 thrust earthquake in Sicily and the north-verging Monghidoro 2003 (Mw 5.3) and Canossa 2008 (Mw 5.5) events in northern Italy; ninety per cent of the earthquakes are shallower than 28 km and eighty-five per cent of the total are clustered below 10–12 km (C in Fig. 7.5). The Padan-Adriatic compressional band (D in Fig. 7.5) contains about 50 events with $3.5 < Mw \leq 6.0$, including the Emilia 2012 north-verging thrust earthquake (Mw 6.1); ninety per cent of the events are located at depths shallower than 12 km. The foreland strike-slip band (E in Fig. 7.5) contains about 25 events, with $3.5 < Mw \leq 5.8$, including the Potenza 1990 (Mw 5.8) and the twin earthquakes of San Giuliano 2002 seismic sequence (Mw 5.7); ninety per cent of the events fall at depths shallower than 30 km with a significant increase of seismicity below 12 km. The South-Alpine compressional bands (F in Fig. 7.5) contains about 25 events, with $3.5 < Mw \leq 6.5$, including the south-verging Friuli 1976 thrust earthquake (Mw 6.5); ninety per cent of the events are located at depths shallower than 12 km. One topic to be highlighted is the inward deepening of the compressional seismic activity, pointed out by the FM spatial distribution, which is prevailingly shallower than 12 km close and along the OTS front and deepens to depth ranges of 12 to 30 km toward the foothills regions. This configuration fits well with the deepening of the compressional seismicity observed by previous authors at a regional scale (Chiarabba et al. 2005; Chiarabba and De Gori 2016; Carannante et al. 2013) and along discrete segments of the belt (Parolai et al. 2001; Finetti et al. 2001; Picotti and Pazzaglia 2008; Visini et al. 2010). It also fits well with the deepening of the OTS basal thrust, as reconstructed across central Italy and Sicily in some of our previous papers (Lavecchia et al. 2003, 2007a; Sgroi et al. 2012). The present activity of the compressional-extensional front along the entire system is also well testified by geodetic data information (Devoti et al. 2017) and borehole breakout data (Montone and Mariucci 2016), as well as by regional stress pattern maps (Pierdominici and Heidbach 2012; Macchiavelli et al. 2012; Carafa and Barba 2013). The reconstructed prevailing kinematics of the seismogenic deformation at different depth ranges is schematized within the inset of Fig. 7.5. The P- and T-axes from compressional and extensional solutions, respectively, remain substantially perpendicular to the arcuate structural trend of the active geologic structures, either compressional or extensional, from north to south Italy including Sicily. Different is the case of the strike-slip solution within the Adriatic-Pelagian foreland, whose P-axes remain SE-NW oriented and do not rotate from Sicily to the central Adriatic Sea. This configuration might suggest the co-existence of two different active geodynamic motors responsible for the observed deformation pattern, such as the Africa-Europa plates relative convergence on one side and mantle expansion in the western Mediterranean on the other, that lead to the progressive development and outward-migration of an arcuate extension-compression pair (Cuffaro et al. 2011; Bell et al. 2013).

7.7 The Seismogenic Provinces

The analyzed FM distribution highlights well-distinguished depth layers with homogeneous kinematics, located in different geographic areas (Figs. 7.5 and 7.6). The identified volumes fit, at a regional scale, the spatial distribution and the possible extrapolation at depth of the outcropping and/or buried active structures derived from integrated geological, geophysical and seismological information. From this observation, a 3D-seismotectonic model, articulated in first-order Seismogenic Provinces (after in referred to as SP), may be built. In this paper, the model will be developed for peninsular Italy and Sicily. We consider the SP as the surface projections of crustal volumes active in Late Pliocene-Quaternary times, homogeneous from the structural-kinematic and seismological point of view and with uniform earthquake potential and/or maximum predictable earthquakes (Reiter 1991). The SP surface geometry is controlled by the 3D shape of the first-order fault structures (master faults alignments), that are active on a regional scale (Lavecchia et al. 2002, 2011a, b; Brozzetti et al. 2017a). The methodological approach adopted to define the SP and to construct the corresponding seismotectonic map is the kinematic one, which analyses the active tectonics from the long-term geological constraints point of view (Barbano et al. 1982; Meletti et al. 2000, 2008). Specifically, we interpret the active tectonic structural style in peninsular Italy and Sicily in the framework of the Late Miocene-Quaternary progressive history of deformation, leading to the nucleation of the Tyrrhenian-Apennine system which is characterized by the outward migration (with time) of crustal slices. This resulted in synchronous and co-axial extension, in the rear, and compression on the front (Lavecchia 1988; Lavecchia et al. 1994; Visini et al. 2010). The shear planes bounding the crustal slices possibly reach and dislocate the crust bottom, according to a thick-skinned structural style, as hypothesized by some authors on the basis of geologic constraints and seismic line interpretations system (e.g. Barchi et al. 1998; Speranza and Chiappini 2002; Lavecchia et al. 2003; Butler et al. 2004). The adopted kinematic model is schematized in the conceptual block diagram of Fig. 7.7, which is intended as representative of any transversal section across the arcuate Apennine-Maghrebide system from the intra-mountain extensional hinterland to the foreland. The diagram highlights the crust-scale seismogenic master faults (yellow lines) and their role in delimiting at the surface the boundary between the seismogenic provinces. The diagram also shows pre-existing and presently inactive shear planes. The identified provinces are named from west to east as Extensional (A), Deep Compressional (B), Shallow Compressional (C) and Strike-Slip (D). The schematic seismotectonic zonation of the Italian territory provided in Fig. 7.8 is based on such a conceptual model, taking into account the available geological-seismological information (DISS Working Group 2018 and previously cited papers).

7.7.1 The Extensional Intermountain Province

The extensional Province (A and A' in Fig. 7.8) extends with an average NW-SE direction along the whole axis of the Apennine mountain chain and continues southward with a N-S strike in northern Calabria, a NNE-SSW strike in southern Calabria and northeastern Sicily, and finally with a WNW-ESE strike in northern Sicily. The inner boundary of the Province schematically coincides with the trace envelope of the eastward-dipping low-angle extensional master faults (LANFs in Figs. 7.1 and 7.7) active since Late Pliocene times and extends from northern Italy to southern Italy and Calabria (Brozzetti et al. 2009b; Brozzetti 2011; Di Naccio et al. 2013; Brozzetti et al. 2017a, b; Lavecchia et al. 2017; Castaldo et al. 2018). Its eastern boundary (point-and-dash blue line in Figs. 7.7 and 7.8) has been chosen to coincide with the surface projection of the LANFs tip line at depth, which is within the depth range of 12–15 km (Boncio et al. 2000; Lavecchia et al. 2011a, b; Sgroi et al. 2012; Ferrarini et al. 2017; DISS Working Group 2018, and references therein). The FMs analyzed in Fig. 7.5 show normal and normal-oblique kinematics, with a nearly WSW-ENE trending T-axis in central and southern Italy and a nearly N-S striking axis in northern Sicily. The Extensional Province is highly seismogenic and is characterized, in historical times, by widespread, frequent and rather strong (M_{max} up to 7) earthquakes (Michetti et al. 2000; Rovida et al. 2019). Recent major earthquakes, which mainly concentrated in the central Apennine segment of the province (e.g. Colfiorito 1997, Mw 6.0, L'Aquila 2009, Mw 6.3; Norcia 2016, Mw 6.5), highlight the activity of the westward-dipping fault sources at depths shallower than 12 km (Boncio et al. 2010; Lavecchia et al. 2011a, b, 2012a, 2016, 2017; Brozzetti et al. 2019). Conversely, the catastrophic Irpinia 1980 (southern Italy) earthquake (Mw 6.8), activated the east-dipping extensional system, at depths $\leq$ ~12 km (Westaway 1993). Only a microseismic activity, although well constrained from local seismic networks, marks the activity of the deeper LANF portions (10–15 km depth), leaving open the question on the low-angle normal fault seismogenic release style (creep versus discrete seismogenic shear) (Chiaraluce et al. 2007; Castaldo et al. 2018).

The values of maximum expected magnitude within areas corresponding to the northern and the southern sectors of this Province are equal to Mw 7.4 for the northern Apennines and to Mw 7.5 for the southern Apennines. These values have been recently computed by Santulin et al. (2017), starting from the highest estimates given by the CPTI15v2.0 earthquake catalogue (Rovida et al. 2019) and the database of the composite seismogenic sources (DISS Working Group 2018) and increasing them by their standard deviation.

7.7.2 The Deep and Shallow Compressional Province

The Compressional Province extends with an arcuate shape from the Po Plain to Sicily. It is characterized by two first-order, outward-convex arcs: the NNE-to-ENE verging, northern Padan-Adriatic arc and the southern NE-to-S verging Ionian-Sicilian arc (Visini et al. 2010). The eastern boundary of the Province corresponds to the OTS front (e.g. the outer front of the Apennine-Maghrebide compressional belt active in Late Pliocene-Quaternary times). The inner boundary, as depicted in Fig. 7.8, simply corresponds to the outer limit of the Extensional Province. The deformation field from focal mechanisms (Fig. 7.5), as well as borehole break-out data (Montone and Mariucci 2016), is prevailingly characterized by sub-horizontal P-axes, striking nearly N-S in the Ferrara arc (Lavecchia et al. 2012b), nearly WSW-ENE in eastern-central Italy and nearly N-S in Sicily (Fig. 7.6). The Compressional Province, as shown in the conceptual diagram of Fig. 7.7, substantially coincides with the crust-scale rock volume in the hanging-wall of the OTS basal thrust. As interpreted on the basis of seismic reflection profiles (Barchi et al. 1998; Lavecchia et al. 2003), this rock volume moves along a relatively low-angle west-dipping fault cutting the entire crust, with a thick-skinned deformation style. Crustal strength envelopes estimated for the Marche Pede-Apennine area (Lavecchia et al. 2003) and for the central Padan region (Viganò et al. 2012) show rheological stratification with a compressional brittle stress regime within the sedimentary crust in the first 8–10 km and within the uppermost lower crust at depths between 15/18 km and 25 km.

The spatial distribution of the available reverse and reverse-oblique FM (Mw > 3.5 dataset in Figs. 7.5 and 7.6) shows as the earthquake release along the OTS basal thrust and its hanging-wall splays is also focused in two depth ranges: a shallower one within the upper sedimentary crust and a deeper one other within the crystalline basement. Therefore, following the approach used in our previous papers (Lavecchia et al. 2003, 2007a, b and references therein), we schematically subdivided the compressional province into two sub-provinces, called Shallow Compressional and Deep Compressional (B and C in Fig. 7.8), which are roughly the surface projections of the shallow crust OTS basal thrust segment (depth < 10/12 km) and of the mid-to-deep OTS basal thrust segment (depth between 10/12 and 30 km), respectively. In the map of Fig. 7.8, the transition between the shallow and the deep province is represented as a white-ruled band, more or less coinciding with the surface projection of the OTS basal thrust depth contour lines 10 and 15 km. The Shallow Compressional province is characterized by a normal-to-thick crust (average 35 km) (Cassinis et al. 2003) and by a rather relevant number of historical and instrumental, small to moderate magnitude, earthquakes (Mw $\leq$ ~ 6.0) (Rovida et al. 2019), which mainly occur close to the Adriatic coastline and may be associated with seismic shear on both the OTS upper basal thrust crust segment (<~12 km) and its frontal and lateral hanging-wall splays (DISS Working Group 2018 and references therein). The Emilia 2012 (Mw 6.1) earthquake, as well as the Ancona 2013 (Mw 4.9) sequence, belongs to such sub-province. Only minor seismicity characterizes the shallow OTS basal thrust in Sicily (Sgroi et al. 2012), although evidence of active fold-and-thrusting is pointed out by geological and morphological analyses (Barreca et al. 2014).

The Deep Compressional province is characterized by a slightly thickened crust (about 35–40 km thick) (Cassinis et al. 2003) and by a significant number of moderate historical and instrumental earthquakes which mainly occur within the Apennine and Mainland Sicily foothill regions, and in general do not reach magnitude higher than Mw ~6.0 except for the Maiella 1706 earthquake (Mw 6.8) (Rovida et al. 2019; DISS Working Group 2018) (Fig. 7.8). The OTS basal thrust of northern-central Italy and/or its splays were activated at mid-crustal depths (15–25 km) by some instrumental compressional earthquakes with Mw up to 5.5 (e.g. Reggio Emilia 1996, Mw 5.4; Faenza 2000, Mw 4.8; Monghidoro 2003, Mw 5.3; Canossa 2008, Mw 5.5) and by frequent background seismicity in the Marche foothills (Parolai et al. 2001; Piccinini et al. 2006; Lavecchia et al. 2007a; Carannante et al. 2013). Reverse and transpressive instrumental events with moderate magnitude also occur at mid-crustal depth in mainland Sicily (Lavecchia et al. 2007b). The Belice 1968 seismic sequence, in central-eastern Sicily, reached an equivalent moment magnitude derived from macroseismic intensities of Mw 6 (Monaco et al. 1996; Rovida et al. 2019). The maximum values of expected magnitude recently calculated by Santulin et al. (2017) for areas lying within the boundary of the compressional province range from a minimum of Mw 6.7 in

Sicily, to Mw 7.1 in northern Peninsular Italy, to a maximum of Mw 7.4 in offshore Calabria.

7.7.3 The Strike-Slip Province

The Strike-slip Province (D in Fig. 7.8), as here defined, extends outwards from the OTS front, south of the Pescara town latitude. From a geological point of view, it consists of the undeformed Plio-Pleistocene foredeep at the OTS basal thrust footwall and of the contiguous Adriatic-Pelagian foreland, cross-cut by discrete zones of near-vertical strike-slip deformation (Di Bucci et al. 2010; Adinolfi et al. 2015). The southern Italy 1990–1991 Potenza M_W 5.7 (Boncio et al. 2007) and San Giuliano 2002 M_W 5.7 (RCMT) earthquakes well testify the seismogenic deformation typical of this province. The strike-slip faults are characterized by E-W right-lateral kinematics in the Adriatic foreland and by N-S left lateral kinematics across the Pelagian foreland in Sicily, being all compatible with a regional NW-SE compression, evidently independent from the stress field associated to the structures of the arcuate compressional province, which is radial along the belt (Fig. 7.5a). The observed prevailing seismogenic depth in the province is in the 10–20 km range, as testified by earthquake activity (Fig. 7.6) and by strength profiles (Boncio et al. 2007). Outcropping active strike-slip faults (e.g. the Mattinata fault across the Gargano foreland and the Scicli fault, across the Hyblean foreland), are located within the boundary of the province, thus implying the possibility of a seismogenic release also at upper crustal depths. The values of maximum expected magnitude recently calculated by Santulin et al. (2017) is equal to Mw 7.0 offshore Gargano and Mw 7.6 across the Hyblean foreland.

7.8 Seismotectonic Map of Hydrocarbon Field Assemblages (HFAs)

A seismotectonic zonation of the Hydrocarbon Field Assemblages (HFAs) (Fig. 7.9) is here obtained by projecting, on a GIS platform, the HFAs boundaries as drawn in Fig. 7.3, on the map of the seismotectonic provinces of Fig. 7.8. The spatial overlap between the HFAs and the provinces gives information on the geometry, active kinematics and depth range of the potentially seismogenic source possibly interacting with the HFAs or neighbouring to them. The HFAs zonation may also be used to measure the proximity between an ongoing sequence and the closest HFAs in terms of horizontal distance range. As stressed by Dahm et al. (2015) and Grigoli et al. (2017), the definition of spatial "closeness" or "proximity" is subjective and needs to be specified in the different contexts also at the light of the local size of the potentially seismogenic structure. In the Italian case, we suggest distinguishing three different "proximity" ranges.

1. A sequence whose aftershock epicentral area substantially overlaps with an HFA or whose main event is distant less than 10 km from the injection point, may be defined as "close sequence".
2. As most the potentially seismogenic fault has a horizontal length that is less than 25 km (Trippetta et al. 2019), the term "close enough sequence" may be used when the main event is distant between 10 and 25 km from the reference injection site, i.e. when it substantially lies within the surface boundaries of the fault box representing the associated individual seismogenic source.
3. "Distant" sequences are those located at a distance >25 km. In this reasoning, we are not taking fault orientation into account.

An analogous proximity criteria can be also used when considering the vertical distance Dv, hereunder referred to as Dv-value, between the depth of the extraction/injection wells characterizing each HFAs and the inferred depth of the underlying closest seismogenic fault. A simple graphic representation of the Dv-value is given in Fig. 7.10, along the trace of two schematic regional transects, extending in W-E direction across northern-central and southern Italy. The Dv-value has a great range of variability, depending on the seismotectonic province hosting the HFA and on the specific site of the HFA within the province boundaries. At the regional scale, we simply distinguish two Dv-value classes: Dv < 5–10 km and Dv > 10–15 km.

The above horizontal and vertical spatial parameters, together with the overall structural-geological and seismological constraints, can contribute the discussion on the likelihood or not of a structural connection between a fault and a reservoir, but the evaluation of the probability of a fault being triggered by hydrocarbon activities requires deep analysis of fluid pressure at well bottom and at hypocentral depth and necessarily rests on the evaluations of pore pressure changes and stress transfer mechanisms. We ought not to forget, that, in case of waste disposals, the subsurface fluid injection may propagate far from the injection point and activate fault structures several kilometres away (Nicholson and Wesson 1992; Keranen et al. 2013; Mulargia and Bizzarri 2014).

The different percentage of mineralized hydrocarbon wells (MHWs in Fig. 7.2) falling within the HFAs boundaries in the different Italian seismogenic provinces is schematized in the depth histograms of Fig. 7.11. Obviously, the depth of the well bottoms is not directly revealing the reservoir depth but can give general information on the oil/gas exploring research strategy in the various provinces. Moreover, it can offer a rough estimate of the maximum possible depth where direct contact between potentially seismogenic sources and extraction/injection activities might occur. A preliminary evaluation of the Dv-value range characterizing the HFAs in the different seismogenic provinces might be obtained from a comparison between the depths of the MHWs and the depth range of the associated analyzed earthquakes. Therefore, in Fig. 7.11, we have drawn the depth range of the seismogenic sources (yellow bar), as inferred both from the FMs analysis (Fig. 7.6) and from the general seismotectonic considerations (Fig. 7.8).

Based on the HFAs distribution within the seismotectonic provinces, we distinguish six homogeneous groups of assemblages with different kinematics and earthquake potential and different chance to interact with a potentially seismogenic fault (A, B, C, D, D' and E).

A – The HFAs within the boundary of the Intermountain Extensional Province represent a small percentage, accounting for 7% of the total MHWs (Figs. 7.9 and 7.11). The bottom of the wells are prevailingly located in two depth ranges: 0–1 and 3–4 km. The reservoirs are mainly gas (70%) and subordinately oil (30%). Among these latter, a large number of MHWs fall in the large Val d'Agri oil field (Monte Alpi, Costa Molina and Cerro Falcone, HFA 64 in Fig. 7.9), that ranks among the largest onshore field of western Europe (Bertello et al. 2010). The Val d'Agri field, whose reservoirs are located in the Mesozoic carbonate substratum at depths between 2 and 4 km (Zappaterra 2013), is associated to a fold-and-thrust structure deformed during Early Pliocene times and now inactive. The field is partially dissected by the normal faults of the Quaternary extensional belt (Valoroso et al. 2009 and references therein). The extensional faulting produce a typical horst-and-graben structure, above an east-dipping basal detachment (Maschio et al. 2005; Brozzetti 2011). In the south-eastern sector of the Val d'Agri field, in June 2006, wastewater injections into a high-rate disposal well induced, at a depth between 1 and 5 km, a microseismic sequence ($M_L \leq 1.8$) with extensional kinematics focused along a previously unknown NE-dipping fault (Stabile et al. 2014). The microseismicity started 1 km beneath the injection well, since only 3 h after the initiation of injection (Improta et al. 2015). Over the following 7.5 years, the seismicity rate correlates with short-term increases in injection pressure.

In instrumental times, the Val d'Agri area has remained substantially aseismic, but in historical times it was struck by the destructive 1857 (Mw 6.7) event. Not far from the Val d'Agri field, both the regional east-dipping faults and the antithetic west-dipping ones have released extensional normal fault earthquakes at upper crustal depths. This is the case of the 9 September 1998 Mercure earthquake (Mw 5.6, Brozzetti et al. 2009a) and of the strong Ms 6.9 Irpinia 1980 event (e.g. Westaway 1993; Pantosti and Valensise 1990). Therefore, the possibility of a connection between the Val d'Agri reservoir and a seismogenic fault plane, capable of triggering earthquake activity, cannot be excluded and should be considered in order to analyze possible seismic risk increments in the area.

The HFAs within the boundary of this Province are characterized by a Dv-value not greater than 10/12 km and most commonly less than ~5 km (Figs. 7.10 and 7.11).

B – The HFAs within the surface boundaries of the Deep Compressional Province contain 19% of the total MHWs (Figs. 7.9 and 7.11). The wells are mainly drilled across the Early Pliocene fault-related anticlines in the hanging wall block of the Pede-Apennine basal thrust. The well bottoms are prevailingly located at depths shallower than 2 km and only a minor percentage exceeds 3 km. The wells are mainly drilled through gas reservoirs (88%) and subordinately through oil ones

(12%). Among the representative producing fields are S. Pietro in Casale and Fidenza (northern Italy, HFAs 32 and 53), Cellino-S.Mauro (Central Italy, HFA 61) and Luna (Calabria offshore, HFA 66). The reservoirs falling within such a Province might appear to be less prone to favour triggered seismicity effects on active faults. In fact, the thrust structures at the reservoir depths (1–4 km) are inactive since Late Pliocene times, although we cannot exclude that they might be reactivated in an extensional regime. The seismogenic portion of the OTS basal thrust located beneath the reservoirs is found at mid-to deep crust depths (~12–25 km) and therefore a minimum vertical distance of at least 8–10 km exists between the extraction activities and the earthquake fault sources. Specifically, the typical Dv-value is ~10–15 km for the HFAs located close to the boundary with the Shallow Compressional Province, above the ruled transition zone of Fig. 7.8 (e.g. HFAs 52, 54, 32, 33 and 57 in northern Italy; 58, 59, 60, 61 in Central Italy and 67, 77 in mainland Sicily).

The HFAs within the boundary of this Province are characterized by a Dv-value > ~ 10 km. For the HFAs of this province located in a more internal position within the Emilia-Romagna foothills (e.g. HFAs 51, 53, 55, 56 and 70 in northern Italy and 65–66 in Calabria) it would be > ~ 15 km (Figs. 7.10 and 7.11).

C – The HFAs within the surface boundaries of the Shallow Compressional Province contain the highest MHW percentage, accounting for 38% of the total (Figs. 7.9 and 7.11). In fact, in this Province, large hydrocarbon reservoirs are hosted by growing anticlines driven by faults that extend to seismogenic depth (Mucciarelli et al. 2015).

The well bottoms are prevailingly located in the 1–4 km depth range, but may also reach 6 km. Exploration wells in this province are mainly drilled through the Late Pliocene-Pleistocene fault-related anticlines, in the hanging wall block of the OTS basal thrust. Those sited just along or very close to the OTS front may actually intersect reservoirs within the undeformed foredeep succession at the basal thrust footwall. The MWHs are prevalently drilled through gas reservoirs (88%). The oil ones are mainly sited within the Mesozoic carbonates at depths of 2–3 km; the gas ones, prevailingly of biogenic nature, are located within the uppermost Plio-Pleistocene terrigenous foredeep deposits within the first 2 km depth. Among the most representative Mesozoic reservoirs, there are the Cavone one in the Ferrara Arc (HFA 30), Sarago Mare and Mormora Mare in the Marche-Adriatic arc (HFA 38) and Prezioso and Cammarata in Sicily (HFA 49). Among the most representative fields in the terrigenous deposits, we recall Ravenna and Porto Corsini Mare (HFAs 35 and 36) in northern Italy and Candela-Palino (HFA 44) and Accettura-Pisticci (HFA 45) in the southern Apennines.

We recall also the large Caviaga-Soresina gas field (HFA 27), located at the outermost front of the north-verging Apennine thrust-system. In a pioneer work on the induced seismicity process (Caloi et al. 1956), the extraction of large quantities of methane from this field was considered as responsible for inducing the 1951–1952 Caviaga thrust sequence at a depth of about 5 km (Mw 5.4, Rovida et al. 2019). Such an interpretation was strongly conditioned by the lack of knowledge of a possible active thrust source in the area. Caciagli et al. (2015), owing to a deeper understanding of the regional seimotectonics and of crustal structure, revisited this interpretation. By integrating relocation of the instrumental events, with the analysis of the historical seismicity and with an estimation of stress field variation in proximity to wells, they concluded that the event had a tectonic origin.

The hydrocarbon reservoirs within the Shallow Compressional Provinces may be cross-cut or delimited at the base by upper crustal seismogenic segments belonging to the OTS basal thrust and/or by its hanging wall splays, as well shown by available geologic and seismic sections (Casero 2004; Bertello et al. 2010). The Dv-value within the boundary of this province is on average in the order of few kilometres (Figs. 7.10 and 7.11). Therefore, the HFAs attributed to this province appear to be the most suitable for having a direct geometric link between the hydrocarbon field and the upper crust seismogenic segment of the OTS basal thrust. Nevertheless, the Emilia 2012 thrust earthquake, which occurred at upper crustal depths in the proximity of the Cavone oil field, where both extraction and fluid injection activities were ongoing (Lavecchia et al. 2015), did not show any evidence of induced or even triggered seismic activity. Albano et al. (2017), based on Coulomb stress evaluation and computed stress changes induced from 20 years of fluid injection in the Cavone oil field, evaluated that it represents less than 10% of that generated by an Mw 4.5 earthquake occurred in the same area one year before the onset of the May–June 2012 sequence. Therefore, they concluded that natural stress increase was responsible for the sequence onset. The same conclusions were also drawn by other Authors (Dahm et al. 2015; Juanes et al. 2016).

D – The HFAs within the surface boundaries of the Strike-Slip Province account for 10% of the total MHWs (Figs. 7.9 and 7.11). The well bottoms are prevailingly located in the 1–4 km depth range. Such wells are mainly drilled through the carbonate shelf terrains of the undeformed southern Adriatic-Pelagian foreland and across the undeformed Plio-Peistocene foredeep at the OTS basal thrust footwall. The wells are prevalently oil (80%), and subordinately gas (20%). The Rospo Mare potentially producing field, in the central Adriatic offshore (HFA 18 in Fig. 7.4), and the Aquila producing one (HFA 22), in the Puglia offshore, are huge oil accumulations within Early Cretaceous and Late Jurassic/Cretaceous reservoirs, respectively, at depths of about 1300 m and 3850 m, respectively (Zappaterra 2013). Large oil fields are also located in the onshore and offshore south-eastern Sicily (e.g. HFAs 23 and 24), with reservoirs in the Late Triassic-Liassic platform at average depths of ~2500. The typical Dv-value within the boundary of this province is not easy to evaluate given the near sub-vertical geometry of the seismogenic faults. In this case, it would be more suitable to consider the horizontal distance between the hydrocarbon occurrences and the seismogenic fault. A horizontal distance <~5 km might be, for example, the case of the Ombrina-Rospo Mare Fields in the Abruzzo offshore. These might interact with an offshore E-W strike-slip system parallel to the on-shore Mattinata one. Another case is that of the Ragusa-Vega fields (HFA 24), in Sicily, which might interact with the N-S system of strike-slip faults associated to the Scicli fault system.

D' – The MHWs drilled within the northern Adriatic foreland domain, which account for 18% of the total (Figs. 7.9 and 7.11), are located within an aseismic domain, which at the moment is not undergoing any relevant tectonic activity.

Therefore, in general, they should be the less relevant ones from the seismic hazard point of view, if we exclude the presence of still unrecognized earthquake sources in the area. Specifically, the HFAs sited within the Late Pliocene-Quaternary foredeep deposits just close to the off-shore Romagna and Adriatic thrust front (e.g. Agostino-P. Garibaldi – Dosso degli Angeli, HFA 12), might be more prone to compressional deformation, possibly associated to the progressive outward growth of the fold-and-thrust system.

E – The MHWs drilled in the hanging wall of the South-Alpine thrust system account for 7% of the total (Figs. 7.9 and 7.11). They are located in two distinct areas: one in the south-western Alps, west of the Garda Lake (HFAs 1–6) and the other in the zone of the south-eastern Alps (HFA 7). The HFAs defined in this province are all sited within anticline structures in the hanging-wall of south-verging active and seismogenic thrust segments (DISS Working Group 2018) with a Dv-value being reasonably <10 km. A more detailed study of the possible spatial connection between the HFAs falling in this province and the seismogenic sources province is not afforded in this paper.

Concluding, on the seismotectonic zonation of Fig. 7.8, we have projected the Italian hydrocarbon-wells classified as operating and potentially-operating in 2014, in the Videpi database (VIDEPI Project 2014). They are classified as producing (36%) and potentially producing (33%), storage (20%) and potentially for storage (1%), reinjection (2%), monitoring (6%) and other (2%). The reinjection wells are mostly located along the OTS front from the northern Apennines to Sicily; a small percentage is located along the South-Alpine Thrust front in northwestern Italy and within the Vega-Ragusa strike-slip zone within the Hyblean foreland in Sicily. The only known Italian reinjection well used as wastewater disposal causing induced seismicity is the Costa Molina 2 (CM2) one, located in the southeastern sector of the Val d'Agri (Improta et al. 2015). This well is available in the literature (Stabile et al. 2014), although not yet available within the Videpi database.

In Italy, we also have two cases of enhanced oil recovery (EOR) with CO_2 capture and sequestration, applied to carbonate heavy-oil reservoirs in the Giaurone and Armatella Fields (Gela-Cammarata HFA in Sicily) (Andrei et al. 2010; Olea 2017) (Fig. 7.8). This technique is aimed at improving the efficiency of oil recovery and extending the oil fields production life, through the use of carbon dioxide (CO_2) as a tertiary recovery agent. In the Italian case, the CO_2 was captured from the flue gas stream of the Gela Refinery, located 15 km from the oil field (Olea 2017); the initial year of CO_2 flooding was 2015.

7.9 Final Remarks

In this paper, based on an original structural-geological approach, a regional seismotectonic zoning of the Italian petroleum fields, grouped into homogenous Hydrocarbon Field Assemblages (HFAs), is built on a GIS platform, as an informative tool to gain fast information on the HFAs 3D state of active strain and on their proximity or not with Quaternary and potentially seismogenic fault structure.

Such information may be useful to highlight local situations that might need deeper investigation on the possible connection between individual seismogenic sources and hydrocarbon production-related activities, for an improvement of risk evaluation. In particular, it could be useful to preliminary answer to some of the questions that, according to Davis and Frohlich (1993), should be advanced to start discriminating natural versus anthropogenic seismic events.

1. Is the ongoing sequence unexpected in the region or is it closely correlated with the seismotectonic setting and active stress field?
2. How far is the sequence from the nearest exploited HFAs?
3. How deep is the sequence beneath the well bottoms?
4. Does the sequence occur near injection depths or at expected seismogenic depths?
5. Are there Quaternary faults that may potentially channel flow to the site of earthquakes?

Evidently, the zonation itself, and more in general the structural-tectonic constraints, cannot be used to answer the remaining questions posed by Davis and Frohlich (1993), concerning the presence (or absence) of fluid pressure at well bottom and at hypocentral depth enough to encourage seismicity. The answer to this question requires independent analyses and different approaches, such as a combination of high-resolution seismicity methodologies with Coulomb stress analysis and with empirical and/or modelled seismicity response due to well-monitored injection volumes (McNamara et al. 2015).

The injection of fluids, which may be used to enhance hydrocarbon production, causes a pore pressure increase with consequent reduction of the effective stress. In cases of wastewater injection into deep wells, the fluid pressure increase can cause weakening processes on a preexisting fault, thus triggering also relatively large (up to M_L 5.5) earthquakes (Ellsworth 2013). Otherwise, fluid extraction activities decrease the pore pressure within the reservoir, thus increasing the effective stress. The accommodation of the reservoir depletion originates poroelastic mechanisms of stress transfer from the reservoir to the surrounding rock mass. As a consequence, seismic instabilities ($M_L \leq 5$) may occur on a local scale, generally restricted to a finite fault close to the reservoir.

On the other hand, massive hydrocarbon extraction and mass removal imply significant unloading at upper-crust levels, which could induce large earthquakes (M 6–7), also at large distances ($\geq$15 km) from the hydrocarbon field (Grasso 1992). This was the case discussed for two large reverse-oblique events occurred in the seismic region of California (1985 M 6.1 Kettleman North Dome and 1987 M 6 Whittier Narrows earthquakes), as well as for a remarkable sequence of strong thrust earthquakes (two M 7 events in 1976 and another M 7 one in 1984) occurred at depths between 15 and 20 km within an almost aseismic desert area, close to the town of Gazli (Uzbekistan) (Bossu et al. 1996). In both the California and the Uzbekistan cases, a striking geometric correspondence was observed between the hydrocarbon field extents and the respective underlying aftershock sequence, supporting the hypothesis that fault zones buried under reservoir-containing anticlines were possibly stimulated by the unloading process.

Combining the seismotectonic zoning of the Hydrocarbon Field Assemblages (HFAs) elaborated in this paper (Fig. 7.9) with the above general observations, we note that in Italy conditions favourable to triggered earthquake activity cannot be a priori excluded. Most of the Italian Hydrocarbon Field Assemblages (HFAs) are located within the boundary of seismogenic provinces undergoing relevant historical and instrumental seismicity (Figs. 7.4 and 7.5). At a regional scale, the horizontal distance between any HFA and the closest active and potentially seismogenic fault may be deduced from the HFAs zonation in Fig. 7.9. The elaborated seismotectonic map (Fig. 7.8) represents the surface projection of inferred 3D-fault sources. Therefore, it contains rough additional information on the likely vertical distance between the reservoir and/or the depth of the extraction/injection activities with the inferred underlying seismogenic source. In this paper, we referred to such a vertical distance as Dv-value (Figs. 7.10 and 7.11).

In the Extensional Province, where the inferred Dv-value is <8–10 km, the exploration wells might drill the seismogenic fault source. In such cases, very large volumes of water and/or pressure perturbation might leak directly to basement faults, thus favouring the process of weakening of a pre-existing fault by elevating the fluid pressure and raising the shear stress above the strength of the contact. Furthermore, some of the hydrocarbon wells presently operating in Italy, are used for storage and, although in a very small percentage, also for reinjection activities (Fig. 7.8), and might be thus responsible for increasing the existing hazard. This may be just the case of the Costa Molina 2 well in Val d'Agri (Stabile et al. 2014).

The HFAs within the Deep Compressional Province, where a large vertical distance between the reservoir and the fault-source is observed (Dv-value > 10–12 km) would be less suitable to trigger earthquake activity.

The HFAs within the Shallow Compressional Province (Dv-value < 8–10 km) could present, during periods of relevant extraction activities, an increase in the probability of occurrence of triggered damaging reverse and reverse-oblique earthquake associated with unloading processes. As well, the associated secondary fluid injections could be another factor inducing seismicity. Nevertheless, all the authors studying the real case study offered by the Emilia 2012 thrust sequence (Mw 6.1) which occurred close to one of the important operating oil field of in northern Italy, concluded that the earthquake was not induced and "not even human triggered", but was exclusively controlled by tectonic stress (ICHESE 2014; Astiz et al. 2014; Lavecchia et al. 2015; Dahm et al. 2015; Juanes et al. 2016; Albano et al. 2017).

In any case, an evaluation of eventual additional ground shakings, related to the variation of the earthquake occurrences triggered by hydrocarbon production activities, is a necessary step for seismic hazard assessment purposes, especially in the Italian region characterized by a high seismic risk (Romeo et al. 2000). The understanding of how and why a seismicity increase over the natural threshold might derive from hydrocarbon-related human activities and if and how it might affect seismic hazard is starting to become a leading line of research also in Italy (Mucciarelli et al. 2015). A full interdisciplinary and detailed approach to the triggered phenomena, such as those performed for the Emilia 2012 earthquakee will be necessary, but certainly a prerequisite is represented by the full knowledge of the natural tectonic hazard. The proposed HFAs seismotectonic zonation (Figs. 7.8 and 7.9) is just a first step in such a direction. We are aware that to make it actually useful for the scientific community and possibly for civil protection purposes, an online product is needed, with the related geological and seismological database and with statistical treatment of data. This product is in preparation and will be available shortly.

Acknowledgments This work was carried out with Ud'A University funding to Giusy Lavecchia. We wish to thank Werter Paltrinieri, two anonymous reviewers and the guest editors Fabio Bonali and Federico Pasquaré Mariotto for valuable reviews and criticisms of the document in various stages of its development.

References

Adinolfi GM, De Matteis R, Orefice A, Festa G, Zollo A, de Nardis R, Lavecchia G (2015) The September 27, 2012, ML 4.1, Benevento earthquake: a case of strike-slip faulting in Southern Apennines (Italy). Tectonophysics 660:35–46

Albano M, Barba S, Tarabusi G, Saroli M, Stramondo S (2017) Discriminating between natural and anthropogenic earthquakes: insights from the Emilia Romagna (Italy) 2012 seismic sequence. Sci Rep 7:282

Alvarez W (1972) Rotation of the Corsica–Sardinia microplate. Nature 248:309–314

Anderson H, Jackson J (1987) Active tectonics of the Adriatic region. Geophys JR Astr Soc 91:937–983

Andrei M, De Simoni M, Delbianco A, Cazzani P, Zanibelli L (2010) Enhanced oil recovery with CO2 capture and sequestration. In: World Energy Congress 2010 (WEC Montreal 2010), Montreal, Quebec, Canada, 12–16 September 2010, vol 3. World Energy Council, London, pp 2015–2034

Anelli L, Mattavelli L, Pieri M (1996) Structural–stratigraphic evolution of Italy and its petroleum systems. In: Ziegler PA, Horvath F (eds) Peri-Tethys Memoir 2, Memoirs of the Museum of Natural History and Materials, vol 170. French National Museum of Natural History, Paris, pp 455–483

Astiz L, Dieterich JH, Frohlich C, Hager BH, Juanes R, Shaw JH (2014) On the potential for induced seismicity at the Cavone oilfield: analysis of geological and geophysical data, and geo-mechanical modeling. Report for Laboratorio di Monitoraggio Cavone, p 139

Barbano MS, Coli M, Ghisetti F, Lavecchia G, Riuscetti M, Scandone P, Slejko D, Valpreda E, Vezzani L (1982) (Gruppo Redazionale della Carta Sismotettonica del P.F.G.): Carta sismotettonica d'Italia. Mem Soc Geol It 24:491–496

Barchi MR, Minelli G, Pialli G (1998) The CROP 03 Profile: a synthesis of results on deep structures of the Northern Apennines. Mem Soc Geol It 52:383–400

Barreca G, Bruno V, Cocorullo C, Cultrera F, Ferranti L, Guglielmino F, Guzzetta L, Mattia M, Monaco C, Pepe F (2014) Geodetic and geological evidence of active tectonics in south-western Sicily (Italy). J Geod 82:138–149

Bell K, Lavecchia G, Stoppa F (2005) Reasoning and beliefs about Italian geodynamics. Boll Soc Geol It 5:119–127

Bell K, Lavecchia G, Rosatelli G (2013) Cenozoic Italian magmatism – isotope constraints for possible plume-related activity. J S Am Earth Sci 41:22–40

Bertello F, Fantoni R, Franciosi R, Gatti V, Ghielmi M, Pugliese A (2010) From thrust-and-fold belt to foreland: hydrocarbon occurrences in Italy. Geol Soc, London, Pet Geol Conf Series 7:113–126

Block LV, Wood CK, William L, Yeck WL, King VM (2014) The 24 January 2013 ML 4.4 Earthquake near Paradox, Colorado, and its relation to deep well injection. Seismol Res Lett 85(3):609–624

Boccaletti M, Calamita F, Deiana G, Gelati R, Massari F, Moratti G, Ricci Lucchi F (1990) Migrating foredeep-thrust belt system in the northern Apennines and Southern Alps. Paleogeogr Paleoclimatol Paleoecol 77:3–14

Boncio P, Brozzetti F, Lavecchia G (2000) Architecture and seismotectonics of a regional low-angle normal fault zone in central Italy. Tectonics 19:1038–1055

Boncio P, Mancini T, Lavecchia G, Selvaggi G (2007) Seismotectonics of strike–slip earthquakes within the deep crust of southern Italy: geometry, kinematics, stress field and crustal rheology of the Potenza 1990–1991 seismic sequences (M_{max} 5.7). Tectonophysics 445:281–300

Boncio P, Pizzi A, Brozzetti F, Pomposo G, Lavecchia G, Di Naccio D, Ferrarini F (2010) Coseismic ground deformation of the 6 April 2009 L'Aquila earthquake (central Italy, Mw 6.3). Geophys Res Lett. https://doi.org/10.1029/2010GL042807

Bossu R, Grasso JR, Plotnikova LM, Nurtaev B, Fréchete J, Moisy M (1996) Complexity of intracontinental seismic faultings: the Gazli, Uzbekistan, sequence. Bull Seismol Soc Am 86(4):959–971

Brozzetti F (2011) The Campania-Lucania Extensional Fault System, southern Italy: a suggestion for a uniform model of active extension in the Italian Apennines. Tectonics 30:TC5009. https://doi.org/10.1029/2010TC002794

Brozzetti F, Boncio P, Lavecchia G, Pace B (2009a) Present activity and seismogenic potential of a low-angle normal fault system (Citta di Castello, Italy): constraints from surface geology, seismic reflection data and seismicity. Tectonophysics 463:31–46

Brozzetti F, Lavecchia G, Mancini G, Milana G, Cardinali M (2009b) Analysis of the 9 September 1998 M_w5.6 Mercure earthquake sequence (Southern Apennines, Italy): a multidisciplinary approach. Tectonophysics 476:210–225

Brozzetti F, Cirillo D, Liberi F, Piluso E, Faraca E, de Nardis R, Lavecchia G (2017a) Structural style of Quaternary extension in the Crati Valley (Calabrian Arc): evidence in support of an east-dipping detachment fault. Ital J Geosci 136:434–453

Brozzetti F, Cirillo D, de Nardis R, Cardinali M, Lavecchia G, Orecchio B, Presti D, Totaro C (2017b) Newly identified active faults in the Pollino seismic gap, southern Italy, and their seismotectonic significance. J Struct Geol 94:13–31. https://doi.org/10.1016/j.jsg.2016.10.005

Brozzetti F, Boncio P, Cirillo D, Ferrarini F, de Nardis R, Testa A, Liberi F, Lavecchia G (2019) High-resolution field mapping and analysis of the August–October 2016 Coseismic surface faulting (Central Italy earthquakes): slip distribution, parameterization, and comparison with global earthquakes. Tectonics 38(2):417–439

Burrato P, Vannoli P, Fracassi U, Basili R, Valensise G (2012) Is blind faulting truly invisible? Tectonic-controlled drainage evolution in the epicentral area of the May 2012, Emilia-Romagna earthquake sequence (northern Italy). Ann Geophys. https://doi.org/10.4401/ag-6182

Butler RWH, Mazzoli S, Corrado S, De Donatis M, Di Bucci D, Gambini R, Naso G, Nicolai C, Scrocca D, Shiner P, Zucconi V (2004) Applying thick-skinned tectonic model to the Apennine thrust belt of Italy – limitations and implications. In: Mc Clay KR (ed) Thrust tectonics and hydrocarbon systems, AAPG Memoir 82. American Association of Petroleum Geologists, Tulsa, pp 647–667

Caciagli M, Camassi R, Danesi S, Pondrelli S, Salimbeni S (2015) Can we consider the 1951 Caviaga (northern Italy) earthquakes as noninduced events? Seism Res Lett 86:1335–1344

Caloi P, Depanfilis M, Di Fillipo D, Marcelli I, Spadea MC (1956) Terremoti della val Padana del 15-16 Maggio 1951. Ann Geofis 9:63–105

Carafa M, Barba S (2013) The stress field in Europe: optimal orientations with confidence limits. Geophys J Int 193(2):531–548. https://doi.org/10.1093/gji/ggt024

Carannante S, Monachesi G, Cattaneo M, Amato A, Chiarabba C (2013) Deep structure and tectonics of the northern-central Apennines as seen by regional-scale tomography and 3-D located earthquakes. JGR Solid Earth 118:5391–5403

Casero P (2004) Structural setting of petroleum exploration plays in Italy. In: Crescenti V, D'Offizi S, Merlino S, Sacchi L (eds) Geology of Italy, Special volume of the Italian Geological Society for the IGC 32 Florence. Italian Geological Society, Bologna, pp 189–199

Casero P, Bigi S (2013) Structural setting of the Adriatic basin and the main related petroleum exploration plays. Mar Petrol Geol 42:135–147

Cassinis R, Scarascia S, Lozej A (2003) The deep crustal structure of Italy and surrounding areas from seismic refraction data. Boll Soc Geol It 122:365–376

Castaldo R, de Nardis R, DeNovellis V, Ferrarini F, Lanari R, Lavecchia G, Pepe S, Solaro G, Tizzani P (2018) Coseismic stress and strain field changes investigation through 3-D finite element modeling of DInSAR and GPS measurements and geological/seismological data: the L'Aquila (Italy) 2009 earthquake case study. J Geophys Res. https://doi.org/10.1002/2017JB014453

Castello B, Selvaggi G, Chiarabba C, Amato A (2006) CSI Catalogo della sismicità italiana 1981–2002, versione 1.1. INGV-CNT, Roma. http://csi.rm.ingv.it/

Cesca S, Dost B, Oth A (2013) Preface to the special issue "Triggered and induced seismicity: probabilities and discrimination" J Seismoll7:1–4. https://doi.org/10.1007/s10950-012-9338-z

Chiarabba C, De Gori P (2016) The seismogenic thickness in Italy: constraints on potential magnitude and seismic hazard. Terra Nova 28:402–408

Chiarabba C, Jovane L, Di Stefano R (2005) A new view of Italian seismicity using 20 years of instrumental recordings. Tectonophysics 395:251–268

Chiaraluce L, Chiarabba C, Collettini C, Piccinini D, Cocco M (2007) Architecture and mechanics of an active low angle normal fault: Alto Tiberina Fault, northern Apennines, Italy. J Geophys Res. https://doi.org/10.1029/2007JB005015

Cirilli S, Buratti N, Gugliotti L, Frixa A (2015) Palynostratigraphy and palynofacies of the Upper Triassic Streppenosa Fm. (SE Sicily, Italy) and inference on the main controlling factors in the organic rich shale deposition. Rev Palaeobot Palynol 218:67–79

Cirilli S, Panfili G, Buratti N, Frixa A (2018) Paleoenvironmental reconstruction by means of palynofacies and lithofacies analyses: an example from the Upper Triassic subsurface succession of the Hyblean Plateau Petroleum System (SE Sicily, Italy). Rev Palaeobot Palynol 253:70–87

CNR – P.F. Geodinamica (1987) Neotectonic Map of Italy, scale 500.000, 1990, Sheet n. 1–6. Quad Ric Sci, 114, S.EL.CA., Firenze

CNR – P.F. Geodinamica (1990) Structural Model of Italy 1:500.000 and Gravity Map, Sheet n. 1–6. Quad Ric Sci, 114, S.EL.CA., Firenze

Cohen KM, Gibbard P (2011) Global chronostratigraphical correlation table for the last 2.7 million years. Subcommission on Quaternary Stratigraphy (International Commission on Stratigraphy), Cambridge

Cosentino D, Cipollari P, Marsili P, Scrocca D (2010) Geology of the central Apennines: a regional review. In: Beltrando M, Peccerillo A, Mattei M, Conticelli S, Doglioni C (eds) Journal of the Virtual Explorer 36(11)

Cuffaro M, Riguzzi F, Scrocca D, Doglioni C (2011) Coexisting tectonic settings: the example of the southern Tyrrhenian Sea. Int J Earth Sci 100(8):1915–1924

Dahm T, Cesca S, Hainzl S, Braun T, Krüger F (2015) Discrimination between induced, triggered, and natural earthquakes close to hydrocarbon reservoirs: a probabilistic approach based on the modeling of depletion-induced stress changes and seismological source. https://doi.org/10.1002/2014JB011778

Davis SD, Frohlich C (1993) Did (or will) fluid injection cause earthquakes? – criteria for a rational assessment. Seismol Res Lett 64(3–4):207–224 de Nardis R, Filippi L, Costa G, Suhadolc P, Nicoletti M, Lavecchia G (2014) Strong motion recorded during the Emilia 2012 thrust earthquakes (Northern Italy): a comprehensive analysis. Bull Earthq Eng 12(5):2117–2145

Devoti R, D'Agostino N, Serpelloni E, Pietrantonio G, Riguzzi F, Avallone A, Cavaliere A, Cheloni D, Cecere G, D'Ambrosio C, Falco L, Selvaggi G, Métois M, Esposito A, Sepe V, Galvani A, Anzidei M (2017) A combined velocity field of the Mediterranean Region. Ann Geophys 60(2):S0215

Di Bucci D, Burrato P, Vannoli P, Valensise G (2010) Tectonic evidence for the ongoing Africa-Eurasia convergence in central Mediterranean foreland areas: a journey among long-lived shear zones, large earthquakes, and elusive fault motions. J Geophys Res. https://doi.org/10.1029/2009JB006480

Di Giulio G, de Nardis R, Boncio P, Milana G, Rosatelli G, Stoppa F, Lavecchia G (2016) Seismic response of a deep continental basin including velocity inversion: the Sulmona intramontane basin (Central Apennines, Italy). Geophys J Int 204(1):418–439

Di Naccio D, Boncio P, Brozzetti F, Pazzaglia J, Lavecchia G (2013) Morphotectonic analysis of the Lunigiana and Garfagnana grabens (northern Apennines, Italy): implications for active normal faulting. Geomorphology 201:293–311

DISS Working Group (2018) Database of individual seismogenic sources (DISS), version 3.2.1: a compilation of potential sources for earthquakes larger than M 5.5 in Italy and surrounding areas. http://diss.rm.ingv.it/diss/, Istituto Nazionale di Geofisica e Vulcanologia; https://doi.org/10.6092/INGV.IT-DISS3.2.1

Ellsworth WL (2013) Injection-induced earthquakes. Science 341(6142):1225942

Fantoni R, Bertello F, Franciosi R (2008) Reservoirs and source rocks in Mesozoic carbonate units of Italy. Rend online Soc Geol It 3:365–366

Ferrarini F, Lavecchia G, de Nardis R, Brozzetti F (2014) Fault geometry and active stress from earthquakes and field geology data analysis: the Colfiorito 1997 and L'Aquila 2009 cases (central Italy). Pure Appl Geophys 172:1079–1103. https://doi.org/10.1007/s00024-014-0931-7

Ferrarini F, Boncio P, de Nardis R, Pappone G, Cesarano M, Aucelli PPC, Lavecchia G (2017) Segmentation pattern and structural complexities in seismogenic extensional settings: the North Matese Fault System (Central Italy). J Struct Geol 95:93–112

Finetti IR, Boccaletti M, Bonini M, Del Ben A, Geletti R, Pipan M, Sani F (2001) Crustal section based on CROP seismic data across the North Tyrrhenian-Northern Apennines-Adriatic Sea. Tectonophysics 343:135–163

Finetti IR, Boccaletti M, Bonini M, Del Ben A, Pipan M, Prizzon A, Sani F (2005) Lithospheric tectono-stratigraphic setting of the Ligurian Sea-Northern Apennines-Adriatic Foreland from Integrated CROP Seismic Data. In: Finetti I (ed) CROP project, deep seismic exploration of the Central Mediterranean Region. Elsevier, Amsterdam, pp 119–158

Frepoli A, Amato A (1997) Contemporaneous extension and compression in the Northern Apennines from earthquake fault plane solutions. Geophys J Int 129:368–388

Frohlich C (1992) Triangle diagrams: ternary graphs to display similarity and diversity of earthquake focal mechanisms. Phys Earth Planet Int 75:193–198

Gasparini C, Iannaccone G, Scarpa R (1985) Fault-plane solutions and seismicity of the italian peninsula. Tectonophysics 117:59–78

Ghielmi M, Minervini M, Nini C, Rogledi S, Rossi M, Vignolo A (2010) Sedimentary and tectonic evolution in the eastern Po-Plain and northern Adriatic Sea area from Messinian to Middle Pleistocene (Italy). Rendiconti Lincei-Scienze Fisiche e Naturali 21:131–166

Gibson G, Sandiford M (2013) Seismicity & induced earthquakes. Available at https://www.chiefscientist.nsw.gov.au/__data/assets/pdf_file/0017/31616/Seismicity-and-induced-earthquakes_Gibson-and-Sandiford.pdf. Accessed 30 Apr 2014

Grasso JR (1992) Mechanism of seismic instability induced by the recovery of hydrocarbons. Pure Appl Geophys 139:507–534

Grigoli F, Cesca S, Priolo E, Rinaldi AP, Clinton JF, Stabile TA, Dost B, Fernandez MG, Wiemer S, Dahm T (2017) Current challenges in monitoring, discrimination, and management of induced

seismicity related to underground industrial activities: a European perspective. Rev Geophys 55:310–340. https://doi.org/10.1002/2016RG000542
Grigoli F, Cesca S, Rinaldi AP, Manconi A, López-Comino JA, Clinton JF, Westaway R, Cauzzi C, Dahm T, Wiemer S (2018) The November 2017 Mw 5.5 Pohang earthquake: a possible case of induced seismicity in South Korea. F Science 360:1003–1006
Guidoboni E, Ferrari G, Mariotti D, Comastri A, Tarabusi G, Sgattoni G, Valensise G (2018) CFTI5Med, Catalogo dei Forti Terremoti in Italia (461 aC-1997) e nell'area Mediterranea (760 a C-1500). https://doi.org/10.6092/ingv.it-cfti5
Gulia L, Weimer S (2010) The influence of tectonic regimes on the earthquake size distribution: a case study for Italy. Geophys Res Lett 37:L10305. https://doi.org/10.1029/2010GL043066
Holton J (1999) Four geologic settings dominate oil, gas fields of Italy, Sicily. Oil Gas J 97(49):81–84
ICHESE (International Commission on Hydrocarbon Exploration and Seismicity in the Emilia region) (2014) Report on the hydrocarbon exploration and seismicity in Emilia Region. Available at http://mappegis.regione.emilia-romagna.it/gstatico/documenti/ICHESE/ICHESE_Report.pdf
Improta L, Valoroso L, Piccinini D, Chiarabba C (2015) A detailed analysis of wastewater-induced seismicity in the Val d'Agri oil field (Italy). Geophys Res Lett 42:2682–2690. https://doi.org/10.1002/2015GL063369
ISIDe Working Group (2016) Version 1.0, (Italian Seismological Instrumental and Parametric Database, 2005-30 Sept. 2017)
Juanes R, Jha B, Hager BH, Shaw JH, Plesch A, Astiz L, Dieterich JH, Frohlich C (2016) Were the May 2012 Emilia-Romagna earthquakes induced? A coupled flow-geomechanics modeling assessment. Geophys Res Lett 43:6891–6897. https://doi.org/10.1002/2016GL069284
Keranen KM, Savage HM, Abers GA, Cochran ES (2013) Potentially induced earthquakes in Oklahoma, USA: links between wastewater injection and the 2011 M w 5.7 earthquake sequence. Geology 41(6):699–702
Lavecchia G (1988) The Tyrrhenian-Apennines system: structural setting and seismotectogenesis. Tectonophysics 147:263–296
Lavecchia G, Bell K (2012) Magmatectonic Zonation of Italy: A tool to understanding mediterranean geodynamics. In: Stoppa F (ed) Updates in Volcanology – a comprehensive approach to the volcanological problems. InTech, Croatia, ISBN 978-953-307-434-4, pp 153–178
Lavecchia G, Brozzetti F, Barchi M, Menichetti M, Keller JVA (1994) Seismotectonic zoning in east-central Italy deduced from an analysis of the Neogene to present deformations and related stress-fields. Geol Soc Am Bull 106:1107–1120
Lavecchia G, Boncio P, Brozzetti F, Stucchi M, Leschiutta I (2002) New criteria for seismotectonic zoning in Central Italy: insights from the Umbria-Marche Apennines. Boll Soc Geol It 1:881–890
Lavecchia G, Boncio P, Creati N (2003) A lithospheric-scale seismogenic thrust in central Italy. J Geodyn 36:79–94
Lavecchia G, de Nardis R, Visini F, Ferrarini F, Barbano MS (2007a) Seismogenic evidence of ongoing compression in eastern-central Italy and mainland Sicily: a comparison. Boll Soc Geol It 126:209–222
Lavecchia G, Ferrarini F, de Nardis R, Visini F, Barbano MS (2007b) Active thrusting as a possible seismogenic source in Sicily (Southern Italy): some insights from integrated structural-kinematic and seismological data. Tectonophysics 445:145–167
Lavecchia G, Boncio P, Brozzetti F, de Nardis R, Visini F (2011a) The contribution of structural geology and regional tectonics to the definition of large-scale seismotectonic provinces and individual seismogenic sources: application to the extensional belt of central Italy. In: Guarnieri P (ed) Recent progress on earthquake geology. Nova Science Publishers, London, pp 165–187
Lavecchia G, Boncio P, Brozzetti F, de Nardis R, Di Naccio D, Ferrarini F, Pizzi A, Pomposo G (2011b) The April 2009 L'Aquila centrseismic sequence (Mw 6.3): a preliminary seismotectonic picture. In: Guarnieri P (ed) Recent progress on earthquake geology. Nova Science Publishers, London, pp 1–17

Lavecchia G, Ferrarini F, Brozzetti F, de Nardis R, Boncio P, Chiaraluce I (2012a) From surface geology to aftershock analysis: constraints on the geometry of the L'Aquila 2009 seismogenic fault system. Ital J Geosci 131:330–347. https://doi.org/10.3301/IJG.2012.24

Lavecchia G, de Nardis R, Cirillo D, Brozzetti F, Boncio P (2012b) The May-June 2012 Ferrara Arc earthquakes (northern Italy): structural control of the spatial evolution of the seismic sequence and of the surface pattern of coseismic fractures. Ann Geophys 55(4):533–540. https://doi.org/10.4401/ag-6173

Lavecchia G, de Nardis R, Costa G, Tiberi L, Ferrarini F, Cirillo D, Brozzetti F, Suhadolc P (2015) Was the Mirandola thrust really involved in the Emilia 2012 seismic sequence (northern Italy)? Implications on the likelihood of triggered seismicity effects. Boll Geofis Teor Appl 56(4):461–488

Lavecchia G, Castaldo R, de Nardis R, De Novellis V, Ferrarini F, Pepe S, Brozzetti F, Solaro G, Cirillo D, Bonano M, Boncio P, Casu F, De Luca C, Lanari R, Manunta M, Manzo M, Zinno I, Tizzani P (2016) Ground deformation and source geometry of the August 24, 2016 Amatrice earthquake (Central Italy) investigated through analytical and numerical modeling of DInSAR measurements and structural-geological data. Geophys Res Lett 43(12):389–398. https://doi.org/10.1002/2016GL071723

Lavecchia G, Adinolfi GM, de Nardis R, Ferrarini F, Cirillo D, Brozzetti F, De Matteis R, Festa G, Zollo A (2017) Multidisciplinary inferences on a newly recognized active east-dipping extensional system in Central Italy. Terra Nova 29:77–89. https://doi.org/10.1111/ter.12251

Macchiavelli C, Mazzoli S, Megna A, Saggese F, Santini S, Vitale S (2012) Applying the Multiple Inverse Method to the analysis of earthquake focal mechanism data: new insights into the active stress field of Italy and surrounding regions. Tectonophysics 580:124–149

Malagnini L, Herrmann RB, Munafo I, Buttinelli M, Anselmi M, Akinci A, Boschi E (2012) The 2012 Ferrara seismic sequence: regional crustal structure, earthquake sources, and seismic hazard. Geophys Res Lett 9:L19302. https://doi.org/10.1029/2012GL053214

Malinverno A, Ryan WBF (1986) Extension in the Tyrrhenian Sea and shortening in the Apennines as a result of the arc migration driven by sinking of the lithosphere. Tectonics 5:227–245

Maschio L, Ferranti L, Burrato P (2005) Active extension in Val d'Agri area, Southern Apennines, Italy: implications for the geometry of the seismogenic belt. Geophys J Int 162:591–609

Mattavelli L, Pieri M, Groppi G (1993) Petroleum-exploration in Italy – a review. Mar Petrol Geol 10:410–425

McGarr A, Simpson D, Seeber L (2002) Case histories of induced and triggered seismicity. In: Lee WW (ed) International handbook of earthquake and engineering seismology, vol 81A. Academic Press, San Diego, pp 647–661

McNamara DE, Hayes GP, Benz HM, Williams RA, McMahon ND, Aster RC, Holland A, Sickbert T, Herrmann R, Briggs R, Smoczyk G, Bergman E, Earle P (2015) Reactivated faulting near Cushing, Oklahoma: increased potential for a triggered earthquake in an area of United States strategic infrastructure. Geophys Res Lett 42:8328–8332. https://doi.org/10.1002/2015GL064669

Meletti C, Patacca E, Scandone P (2000) Construction of a seismotectonic model: the case of Italy. Pure Appl Geophys 157:11–35

Meletti C, Galadini F, Valensise G, Stucchi M, Basili R, Barba S, Vannucci G, Boschi E (2008) A seismic source zone model for the seismic hazard assessment of the Italian territory. Tectonophysics 450:85–108

Michetti AM, Serva L, Vittori E (2000) ITHACA Italy Hazard from Capable Faults: a database of active faults of the Italian onshore territory, CD-Rom Explan notes, 1–150

Monaco C, Mazzoli S, Tortorici L (1996) Active thrust tectonics in western Sicily (southern Italy): the 1968 Belice earthquake sequence. Terra Nova 8:372–381

Montone P, Mariucci MT (2016) The new release of the Italian contemporary stress map. Geophys J Int 205(1):525–531. https://doi.org/10.1093/gji/ggw100

Mucciarelli M, Priolo E, Santulin M, Tamaro A (2015) Seismic hazard from natural and induced seismicity: a comparison for Italy. BGTA 56:519–526

Mulargia F, Bizzarri A (2014) Anthropogenic triggering of large earthquakes. Sci Rep 4:1–5. https://doi.org/10.1038/srep06100

Neri G, Barberi G, Oliva G, Orecchio B (2005) Spatial variations of seismogenic stress orientations in Sicily, south Italy. Phys Earth Planet Inter 148:175–191

Nicholson C, Wesson RL (1992) Triggered earthquakes and deep well activities. Pure Appl Geophys 139:561–578

Olea RA (2017) Carbon dioxide enhanced oil recovery performance according to the literature. In: Verma MK (ed) Three approaches for estimating recovery factors in carbon dioxide enhanced oil recovery, US geological survey scientific investigations report 2017–5062, pp D1–D21. https://doi.org/10.3133/sir20175062D

Pantosti D, Valensise G (1990) Faulting mechanism and complexity of the November 23, 1980, Campania-Lucania earthquake, inferred from surface observations. J Geophys Res 95:15319–15341

Parolai S, Trojani L, Monachesi G, Frapiccini M, Cattaneo M, Augliera P (2001) Hypocenter location accuracy and seismicity distribution in the Central Apennines (Italy). J Seismol 5:243–261

Patacca E, Scandone P (2004) Special Volume of the Italian Geological Society for the IGC 32 Florence 2004. IGC, Florence, pp 93–129

Patacca E, Sartori R, Scandone P (1990) Tyrrhenian basin and Apenninic arcs: kinematic relations since late Tortonian times. Mem Soc Geol It 45:425–445

Piccinini D, Chiarabba C, Augliera P, Monghidoro Earthquake Group (2006) Compression along the northern apennines? Evidence from the Mw 5.3 Monghidoro earthquake. Terra Nova 18:89–94. https://doi.org/10.1111/j.1365-3121.2005.00667.x

Picotti V, Pazzaglia FJ (2008) A new active tectonic model for the construction of the Northern Apennines mountain front near Bologna (Italy). J Geophys Res 13:B08412. https://doi.org/10.1029/2007JB005307

Pierdominici S, Heidbach O (2012) Stress field of Italy – mean stress orientation at different depths and wave-length of the stress pattern. Tectonophysics 532:301–311

Pieri M (2001) In: Vai GN, Martini PI (eds) Italian petroleum geology in anatomy of an Orogen: the Apennines and adjacent Mediterranean Basins. Kluwer Academic Publisher, Dordrecht, pp 533–549

Pieri M, Groppi G (1981) Surface geological structure of the Po Plain, Italy. In: CNR Publ, 414, Progetto Finalizzato Geodinamica

Pondrelli S, Salimbeni S (2006) Italian CMT Dataset. Istituto Nazionale di Geofisica e Vulcanologia (INGV) https://doi.org/10.13127/rcmt/italy

Presti D, Billi A, Orecchio B, Totaro C, Faccenna C, Neri G (2013) Earthquake focal mechanisms, seismogenic stress, and seismotectonics of the Calabrian Arc, Italy. Tectonophysics 602:153–175

Reiter L (1991) Earthquake hazard analysis: issues and insights. Colombia University Press, New York. https://doi.org/10.1002/eqe.4290201211

Romano MA, de Nardis R, Garbin M, Peruzza L, Priolo E, Lavecchia G, Romanelli M (2013) Temporary seismic monitoring of the Sulmona area (Abruzzo, Italy): a quality study of microearthquake locations. Nat Hazards Earth Syst Sci 13(11):2727–2744

Romeo R, Paciello A, Rinaldis R (2000) Seismic hazard maps of Italy including site effects. Soil Dyn Earthq Eng 20:85–92

Rovida A, Locati M, Camassi R, Lolli B, Gasperini P (eds) (2019) Italian Parametric Earthquake Catalogue (CPTI15), version 2.0, Istituto Nazionale di Geofisica e Vulcanologia (INGV). https://doi.org/10.13127/CPTI/CPTI15.2

Santilano A, Trumpy E, Gola G, Donato A, Scrocca D, Ferrarini F, Brozzetti F, de Nardis R, Lavecchia G, Manzella A (2019) A methodology for assessing the favourability of geopressured-geothermal systems in sedimentarybasin plays: a case study in Abruzzo (Italy). Geofluids, ges. https://doi.org/10.1155/2019/4503943

Santulin M, Tamaro A, Rebez A, Slejko D, Sani F, Martelli L, Bonini M, Corti G, Poli ME, Zanferrari A, Marchesini A, Busetti M, Dal Cin M, Spallarossa D, Barani S, Scafidi D, Barreca

G, Monaco C (2017) Seismogenic zonation as a branch of the logic tree for the new Italian seismic hazard map – MPS16: a preliminary outline. Boll Geof Teor Appl 58:313–334

Scafidi D, Barani S, De Ferrari R, Ferretti G, Pasta M, Pavan M, Spallarossa D, Turino C (2015) Seismicity of Northwestern Italy during the last 30 years. J Seismol 19:201–218. https://doi.org/10.1007/s10950-014-9461-0

Scognamiglio L, Tinti E, Quintiliani M (2006) Time domain moment tensor [Data set]. https://doi.org/10.13127/TDMT

Sgroi T, de Nardis R, Lavecchia G (2012) Crustal structure and seismotectonics of central Sicily (southern Italy): new constraints from instrumental seismicity. Geophys J Int 189:1237–1252

Speranza F, Chiappini M (2002) Thick-skinned tectonics in the external Apennines, Italy: new evidence from magnetic anomaly analysis. J Geophys Res. https://doi.org/10.1029/2000JB000027

Stabile TA, Giocoli A, Perrone A, Piscitelli S, Lapenna V (2014) Fluid injection induced seismicity reveals a NE dipping fault in the southeastern sector of the High Agri Valley (southern Italy). Geophys Res Lett 41:5847–5854. https://doi.org/10.1002/2014GL060948

Stein S, Geller RJ, Liu M (2012) Why earthquake hazard maps often fail and what to do about it? Tectonophysics 562:1–25

Trippetta F, Petricca P, Billi A, Collettini C, Cuffaro M, Lombardi AM, Scrocca D, Ventura G, Morgante A, Doglioni C (2019) From mapped faults to fault-length earthquake magnitude (FLEM): a test on Italy with methodological implications. Solid Earth 10:1555–1579. https://doi.org/10.5194/se-10-1555-2019

Vai GB, Martini PI (2001) Anatomy of an Orogen: the Apennines and adjacent Mediterranean Basins. Kluwer Academic Publisher, Dordrecht, pp 533–549

Valoroso L, Improta L, Chiaraluce L, Di Stefano R, Ferranti L, Govoni A, Chiarabba C (2009) Active faults and induced seismicity in the Val d'Agri area (Southern Apennines, Italy). Geophys J Int 178:488–502

VIDEPI Project (2014) Visibility of petroleum exploration data in Italy. Ministry for Economic Development UNMIG. http://unmig.sviluppoeconomico.gov.it/videpi/en/pozzi/pozzi.asp

Viganò A, Della Vedova B, Ranalli G, Martin S, Scafidi DE (2012) Geothermal and rheological regime in the Po plain sector of Adria (Northern Italy). Ital J Geosci 131:228–240. https://doi.org/10.3301/IJG.2012.09

Visini F, de Nardis R, Lavecchia G (2010) Rates of active compressional deformation in central Italy and Sicily: evaluation of the seismic budget. Int J Earth Sci 99:243–264

Waschbusch PJ, Royden LH (1992) Spatial and temporal evolution of foredeep basins: lateral strength variations and inelastic yielding in continental lithosphere. Basin Res 4:179–196

Westaway R (1993) Fault rupture geometry for the 1980 Irpinia earthquake. Ann Geofis 36(1):51–70. https://doi.org/10.4401/ag-4301

Zappaterra E (1994) Source-rock distribution model of the Periadriatic Region. AAPG Bull 78:333–354

Zappaterra E (2013) Italy Mesozoic oils – implications for exploration in the Adriatic Sea. AAPG European Regional Conference & Exhibition, Barcelona

Part III
Key Studies for Seismic Hazard Assessment

Chapter 8
The 2020 National Seismic Hazard Model for Georgia (Sakartvelo)

Nino Tsereteli, Laurentiu Danciu, Otar Varazanashvili, Karin Sesetyan, Lali Qajaia, Tatia Sharia, David Svanadze, and Irine Khvedelidze

Abstract The Earthquake Model of the Middle East (EMME) project paved the road to a state-of-the-art seismic hazard assessment in the region, including Georgia as one of the key participants. After the EMME project, the Institute of Geophysics initiated the revision of the national seismic hazard model of Georgia, in light of the new findings of the EMME project, new harmonized datasets, fully aligned with the probabilistic framework promoted by the Global Earthquake Model (GEM). In this contribution, we present the main elements of a newly developed seismic hazard model for Georgia. We started with the updating of the regionally harmonized datasets (i.e. earthquake catalogues, area seismic sources) with focus on data collected in the recent years. The seismogenic source model consists of two key components: area sources and active faults combined with background seismicity. The main features of the seismo-tectonic domains in the Caucasus region were summarized into three seismogenic classes (i.e. shallow crust, volcanic sources and deep seismicity). Given this classification, a set of ground motion prediction models was selected for each tectonic region and used to quantify the inherent uncertainties of ground motion. The probabilistic seismic hazard was computed for the entire region using the OpenQuake engine; the results including mean and quantile hazard maps, hazard curves and hazard spectra. Comparisons with previous probabilistic seismic hazard maps, followed by a discussion and outlook, conclude the paper.

Keywords Georgia · Earthquake · Seismogenic sources · Seismic hazard

---

N. Tsereteli (✉) · O. Varazanashvili · L. Qajaia · T. Sharia · D. Svanadze · I. Khvedelidze
M. Nodia Institute of Geophysics, I. Javakhisvili Tbilisi State University, Tbilisi, Georgia
e-mail: nino_tsereteli@tsu.ge

L. Danciu
Swiss Seismological Service (SED), ETH Zürich, Zürich, Switzerland

K. Sesetyan
Kandilli Observatory and Earthquake Research Institute, Boğaziçi University, Istanbul, Turkey

© Springer Nature B.V. 2021

F. L. Bonali et al. (eds.), *Building Knowledge for Geohazard Assessment and Management in the Caucasus and other Orogenic Regions*, NATO Science for Peace and Security Series C: Environmental Security,
https://doi.org/10.1007/978-94-024-2046-3_8

8.1 Introduction

The Republic of Georgia is located in the South Caucasus between the Russian Federation to the north and the Republic of Turkey as well as the Republic of Armenia to the south. To the east, it is bounded by Azerbaijan and to the west by the Black Sea. Georgia is highly vulnerable to various geohazards (e.g. earthquakes, floods, landslides and rock-falls), of which the most dangerous and devastating are associated with earthquake activity (Varazanashvili et al. 2012). The latest, major earthquake occurred in 1991 in Racha, with magnitude Mw = 6.9, and tragic consequences: more than 100 casualties and damage to residential houses and infrastructures amounting to 10 billion rubles- as of 1991, approximately 5.5 billion USD (Varazanashvili et al. 2012).

In this contribution, we describe a newly developed probabilistic seismic hazard model for Georgia, with a focus on the most recent earthquake data compilation and its statistical analysis. Intrinsic uncertainties of data and model are accounted for and handled with a logic tree approach. The spatial and temporal variability of earthquake rates forecast are built upon two key hypotheses: (i) past seismicity provides the proxy for future location of seismicity; (ii) future seismicity will take place, cluster and migrate along identified active faults. A fully probabilistic framework has been adopted, and Fig. 8.1 summarises the main steps of the seismic hazard model development considered in the present study.

In the next section, we first provide a summary of the evolution of the national seismic hazard models in Georgia, while in the subsequent sections we elaborate on individual steps of the model development mentioned in Fig. 8.1. In the final two sections, we provide new probabilistic ground shaking estimates for Georgia, and more specifically for the capital city, Tbilisi, with comparisons with previous hazard models. We conclude with a discussion and an outlook on the proposed seismic hazard model.

8.2 Historical Review of the Development of Seismic Hazard Maps for Georgia

Seismic hazard assessment in Georgia before 1991 was conducted within initiatives promoted by the former Soviet Union. (Bune et al. 1974; Seismicheskoe 1980; Mokrushina and Shebalin 1982, 1991; Tsipenyuk 1988; Gusev 1991; Gusev and Shumilina 1995; Ulomov and Shumilina 1999). The first design map of the former Soviet Union including the territory of Georgia was published in 1937 (Gorshkov 1938). In 1947, an update of the seismic zonation was proposed (Gorshkov 1947) and approved in 1949 as a normative map (Fig. 8.2a). Both the 1937 and the 1949 maps were deterministically derived and expressed in terms of macroseismic intensity. The key assumption behind these maps was that past seismicity i.e. (strong earthquakes) would forecast future seismicity. The same approach was

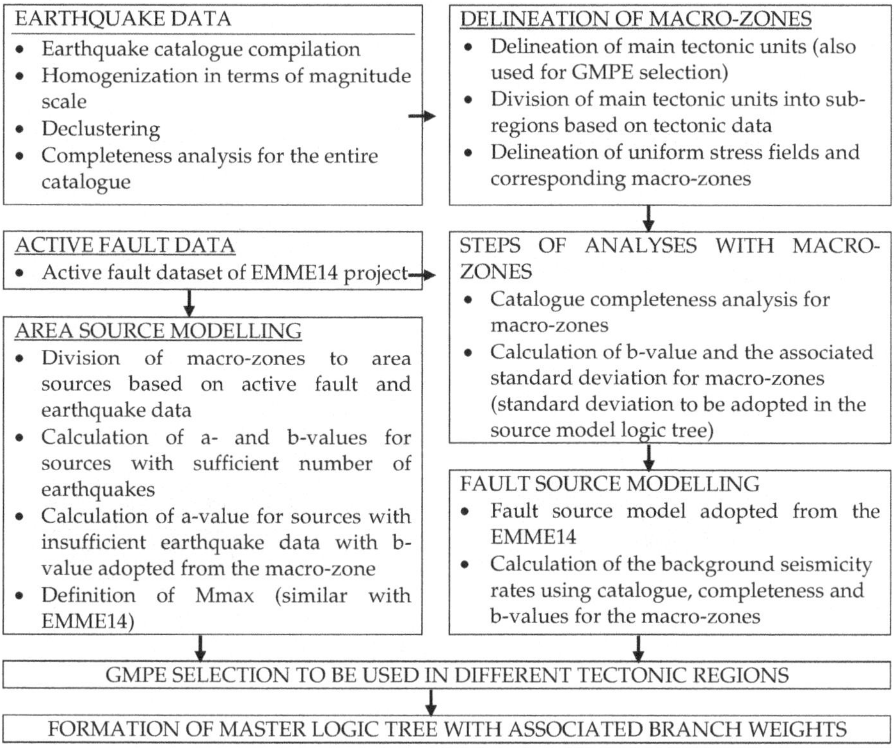

Fig. 8.1 The main steps of the seismic hazard analysis conducted in the present study

also considered in the 1957 update of the normative map (eds. S. Medvedev, B. Petrushevsky, Fig. 8.2b) (Medvedev 1958).

A two-step method for estimating seismic hazard as proposed by Gorshkov (1947), Medvedev (1947), Gubin (1950) and Riznichenko (1965) was considered in the seismic hazard/zoning (SZ) map adopted in 1968 (Seismic zoning, 1968, ed. C. Medvedev, Fig. 8.2c). In 1978, for the first time, the concept of probabilistic characterisation of ground shaking was introduced (ed. M. Sadovsky, Fig. 8.2d). In particular, indices 1, 2 and 3 of intensity on this map (e.g., 7_1, 7_2, 7_3) reflect the occurrence of seismic shaking with intensity 7 once with average return periods of 100, 1000 and 10,000 years, respectively (Seismicheskoe 1980).

With the collection of new information about earthquakes and new knowledge, probabilistic seismic hazard maps have improved steadily. Interestingly, such estimates were constantly changed almost after every destructive earthquake that struck the region. For example, the epicentral regions of the destructive earthquakes of Spitak in 1988 (Armenia) with I_o = X and Racha in 1991 (in Georgia) with I_o = IX (both in MSK scale), were associated with estimated intensities of VII on the SZ-78 map. This initiated the preparation of a new SZ-91 map for Georgia

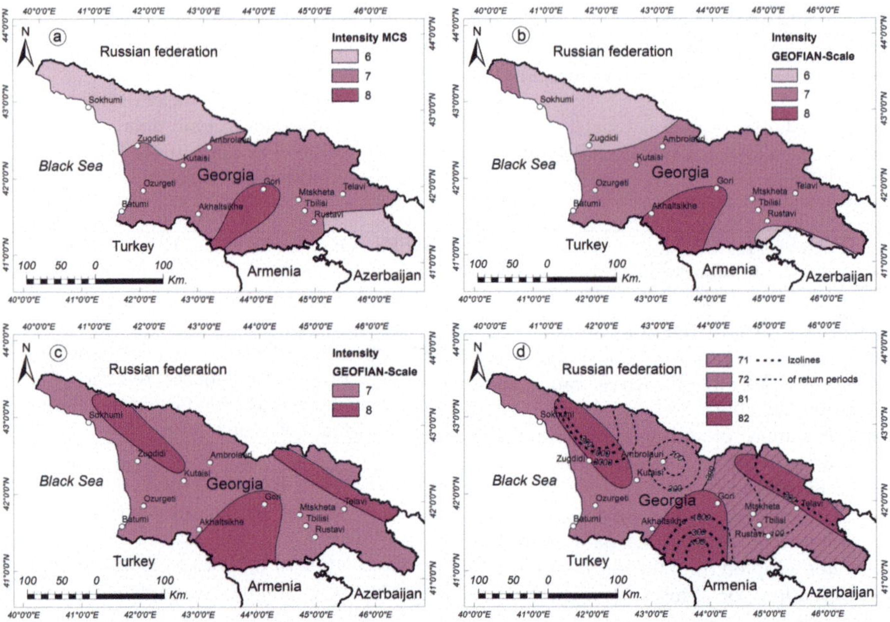

Fig. 8.2 Seismic zonation maps for Georgia, (**a**) SZ-49 (editor: G. Gorshkov); (**b**) SZ-57 (editors: S. Medvedev, B. Petrushevsky); (**c**) SZ-68 map for Georgia (authors: Y. Bius, A. Tskhakaya, M. Rubinstein); (**d**) SZ-78 (authors: I. Aivazishvili, E. Jibladze, V. Papalashvili, M. Rubinstein)

(ed. M. Aleksidze, Gotsadze et al. 1996, Fig. 8.3), which was rather based on a deterministic approach.

In 1995, the first probabilistic seismic hazard maps were computed showing MSK intensities with 10 and 2% probabilities of exceedance in 100 years (Jibladze et al. 1995) using the methodology developed by Riznichenko (1979).

The first Probabilistic Seismic Hazard Analysis (PSHA) for the Caucasus was undertaken in the framework of the GSHAP project (Giardini 1999; Balassanian et al. 1999). Based on the new methodology developed in the GSHAP project, a set of probability maps were later obtained by several authors in individual studies (Ulomov and Shumilina 1999; Varazanashvili 1999; Chelidze et al. 2012; Slejko et al. 2008; Tsereteli et al. 2011, 2012, 2014).

More recently, within several regional and global initiatives, new investigations have been conducted. Namely, a seismic hazard model of Caucasus has been developed in the framework of the regional Earthquake Model of Middle East (i.e. Sesetyan et al. 2018a) and global initiative of the Global Earthquake Model (GEM, Pagani et al. 2018).

The 2014 Earthquake Model of the Middle East (EMME14) is aimed at the assessment of seismic hazard, the associated risk in terms of structural damages, casualties and economic losses and also at the evaluation of the effects of relevant mitigation measures in the Middle East region. Some of the main highlights

Fig. 8.3 SZ-91 map for Georgia (authors: O. Varazanashvili, O. Gotsadze, E. Jibladze)

of EMME14 include the cross-border harmonization of the main datasets (i.e. earthquake catalogue, active faults, ground motion models). Seismogenic sources combine active faults and seismicity for the first time in the region, in-line with other regional models (ESHM13, Woessner et al. 2015, Pagani et al. 2018). EMM14 provides the basis for the present model development as some of the main principles of handling the input data, model uncertainty, logic tree template and computational framework are adopted hereinafter.

Nowadays, in Georgia, the seismic design code is based on a seismic hazard map expressing the severity of the impact in terms of two ground motion intensity measures, i.e. peak ground acceleration (PGA) and MSK intensity with 2% probability of exceedance in 50 years (Chelidze et al. 2012). The comparison of the Georgian normative seismic design map with recent seismic hazard maps for the region proposed by Slejko et al. (2008), Tsereteli et al. (2012, 2014), Sesetyan et al. (2018a) show significant differences for the same probability of exceedance. These discrepancies are due to various factors (e.g. earthquake catalogue, active faults, ground motion models) and advance on probabilistic framework of hazard assessment, which might suggest the need to update the seismic hazard model for Georgia. Our work describes a newly developed probabilistic seismic hazard model for Georgia using updated data of earthquake catalogues, new conversion equations for magnitude homogenization, new area seismic sources developed for national level and recently developed regional and global ground motion characteristic models.

8.3 Developing the Seismic Hazard Model of Georgia

The framework for developing the seismic hazard model for Georgia is illustrated in Fig. 8.1. A reliable earthquake catalogue and an active fault database are essential in the specification of main steps involved in developing the seismogenic source model. The development of alternative seismogenic source models is aimed at capturing the spatial variation of seismicity and the intrinsic uncertainties of source zonation, while the use of a set of ground motion prediction models for different tectonic regimes is aimed at capturing ground motion variability. Hereunder we provide details of the steps of the seismic hazard model development framework presented in Fig. 8.1.

8.3.1 Earthquake Catalogue

In the present section, the earthquake catalogue and the main steps to obtain a fully homogeneous earthquake catalogue are summarized. The main steps comprise the compilation of earthquake catalogue, magnitude homogenisation and finally the statistical analysis of the catalogues to characterize the earthquake recurrence rates.

8.3.1.1 Catalogue Compilation

The main source used for the compilation of the earthquake catalogue of the study region is the database of the Institute of Geophysics of Javakishvili Tbilisi State University (IG TSU) (http://ig-geophysics.ge/IG/language/en/archives/884).

The earthquake catalogue of IG TSU features data on more than 80,216 earthquakes and covers the period from ancient times until 2018, with a minimum magnitude of Ml $\geq$ 0.04. The catalogue covers the area between 38.5°E–47°E and 41°N–44.5°N, including the entire territory of Georgia and surrounding areas of Turkey, Armenia, Iran, Azerbaijan and the Northern Caucasus. From this database, 45,742 events belong to the territory of Georgia. The earthquake data for surrounding areas were obtained by the local sources, mainly the seismic networks of various countries in the region, i.e. the Kandilli Observatory and Earthquake Research Institute for the eastern part of Turkey near the border with Georgia, the International Institute of Earthquake Engineering and Seismology (IIEES) for Iran, Sargsyan et al. (2017) and the Institute of Geology of Azerbaijan for Azerbaijan. The catalogue for the Northern Caucasus after 2008 was updated by IG TSU using international data. For the post-2014 period, the earthquake catalogue of Georgia was updated by IG TSU using international data, with Ml larger than 3.0, as unfortunately, initial national waveform data by Ilia State University (ILIAUNI) are not accessible for Georgian scientists from other universities and calculation or relocation of small events is not possible.

Historical earthquake data for Georgia are based on written sources, epicentral intensities are estimated using the MSK64 scale and estimated Ms are based on correlation equations between Ms and Intensity (Varazanashvili et al. 2018). Complete databases of historical earthquakes and correct estimations of associated parameters are crucial for seismic hazard assessment. They also form the unique information to test the reliability of seismic hazard estimates by various methods. The historical catalogue of Georgia covers the period from 1250 BC up to 1900 and includes moderate to strong earthquakes with magnitude Ms $\geq$ 3.8.

Uncertainties associated with earthquake location and magnitude have a significant influence on seismic hazard assessment. Estimation errors of instrumental location relate to seismic station distributions and velocity models. The instrumental monitoring of earthquakes in Georgia started at the beginning of twentieth century. The instrumental observation period can be subdivided into five periods, with respect to the development of the seismic network in Georgia:

	Period (years)	Number of Seismic Stations in operation during the last decades
I	1900–1935	6
II	1936–1959	10
III	1960–1995	40
IV	1996–2005	10
V	2006–2018	25

8.3.1.2 Magnitude Homogenization

Another important issue is the homogenization of the earthquake catalogues in terms of moment magnitude (*Mw*), as most ground motion prediction models are derived for Mw. Until 2003, in Georgia and in the entire Caucasus region, the energetic parameters of earthquakes were estimated in terms of energy class (K) for small and moderate earthquakes and in terms of surface wave magnitude (M_S) mostly for moderate to strong earthquakes with Ms $\geq$ 4.5. Whenever the direct determination of $\mathrm{M_S}$ was impossible, the size of the events was estimated according to the equation Ms = 0.56 K – 2.2, based on Rautian (1964) and Rautian and Khalturin (1978). Often, $M_b \cong M_{PV}$ (magnitude estimated from the vertical component of p waves) was also assumed. There is an empirical correlation formula between M_{PV} and M_S for the study region: $M_{PV} = 2.5 + 0.63M_S$. Therefore, until 2003, the earthquake catalogues for Georgia and the Caucasus were unified to *Ms*.

Since 2003, because of the reorganization of the Georgian seismic network, only local magnitude values *(Ml)* have been calculated for recorded earthquakes (and hardly ever M_d and M_b values). Mw have been calculated just for earthquakes with *Ml* greater than 4.5 (Adamia et al. 2017; Tsereteli et al. 2016a).

The catalogue of Georgia was unified in terms of *Mw* for the first time in the framework of the EMME (Earthquake Model of Middle East) project, based on the regional correlation equations between different magnitude scales by Zare et al. (2014).

Recently, a relationship between *Mw* and *K* based on Azerbaijan data was derived for 85 earthquakes within the ranges of *Mw* 3.6–6.5 and *K* 9.6–14.2 (Onur et al. 2019):

$$\mathrm{Mw} = 0.5673\ \mathrm{K}{-}1.8244 \tag{8.1}$$

Based on national data, a relationship between Ml and K was derived for 567 earthquakes in Georgia for the period 2003–2005, when both old and new seismic network were operating (Tibaldi et al. 2020):

$$\mathrm{Ml} = 0.5494\ \mathrm{K} - 1.9308 \qquad \sigma = 0.184 \tag{8.2}$$

Equations (8.1) and (8.2) allowed for the development of a relationship between Ml and Mw:

$$\mathrm{Mw} = 0.1694 + 1.0322\mathrm{Ml} \qquad \sigma = 0.228 \tag{8.3}$$

We investigated various relationships between Ms ad Mw, that were published after the EMME project (e.g. Lolli et al. 2014; Storchak et al. 2015). Conversion by Lolli et al. (2014) for Italian data gave lower Mw values towards small Ms:

$$\mathrm{Mw} = \exp\ (1.421 + 0.108\mathrm{Ms}) - 1.863 \qquad \sigma = 0.15 \tag{8.4}$$

Based on the Eqs. (8.1), (8.3) and (8.4), we harmonized our catalogue in terms of Mw. In particular, up to 2003 earthquake magnitudes estimated by *K* were converted using (8.1), earthquakes that were estimated directly by *Ms* were converted according to (8.4). After 2003, earthquakes estimated by *Ml* were converted according to (8.3) and some of them were estimated directly by *Mw* using the moment tensor inversion technique by (Barth et al. 2007). It should be noted that these conversion methods were applied to the catalogue covering the Caucasus region (Georgia, Armenia, Azerbaijan and northern Caucasus) i.e. former USSR territories, where earthquake magnitudes were calculated by similar methods. Data from IIEES (Iran) and KOERI (Turkey) were already expressed in the moment magnitude scale, and were directly adopted. The final processed catalogue consists of 14,978 events with Mw larger than 3.3 (Fig. 8.4).

8.3.1.3 Declustering of the Earthquake Catalogue

For any seismic hazard assessment study, following the classical probabilistic approach established by Cornell (1968), the declustering of the earthquake catalogue compiled for the region has become a standard practice. As the earthquake occurrences in each seismic source are considered to follow a Poisson process, the

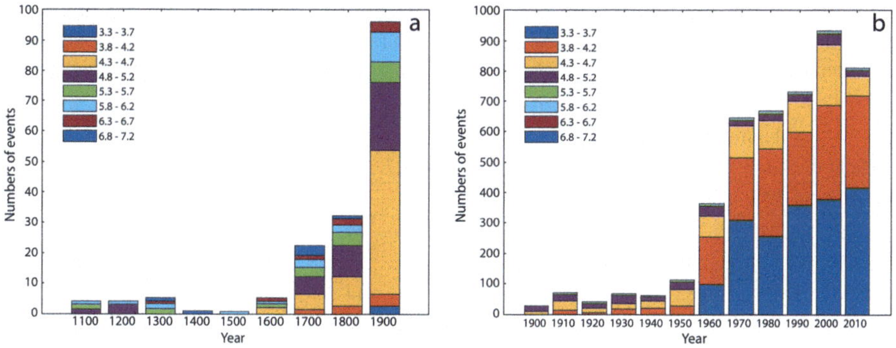

Fig. 8.4 Catalogue magnitude timeline, (**a**): for the historical period between 1100 and 1900; (**b**): for the instrumental period between 1900 and 2018

catalogue that is used to estimate future earthquake occurrences must be devoid of dependent events, i.e. fore and aftershocks that are triggered by a main event have to be removed. The identification of the events that belong to the same family generated by a parent earthquake, however, is not a standard process. Due to the complex nature of earthquake generation, decisions about events that are more distant in time or space can be subject to discussion and these may vary from region to region and even from earthquake to earthquake. To overcome the difficulty of the process, several identification algorithms have been proposed. Among these, the windowing method proposed by Gardner and Knopoff (1974) is one of the most widely used techniques. In this method, magnitude dependent distance and time windows are suggested for the identification of aftershocks. The distance and time windows increase with increasing magnitude of the main shock, suggesting that larger magnitude events will generate aftershocks across a wider area and over a longer time period. The window sizes are determined based on the analysis of earthquake data and different region-dependent windows were later proposed by other researchers (e.g. Grunthal 1985). Another method used for the identification of main and dependent events is the cluster-link method that was proposed by Raesenberg (1985), in which each new event is considered as a main shock and the catalogue is searched for the events triggered by the previous main event, enabling the identification of events triggered not only the major main shock, but by the aftershocks as well. Due to the chained structure of the Raesenberg (1985) method, the completeness of the catalogue becomes an important factor, as a missing event tends to hinder the identification of the next dependent event. On the other hand, the parameters used to define the clusters can be modified to adapt the method to the region under study. Eroğlu-Azak et al. (2018) analysed the effect of using various declustering algorithms and their success in the identification of the aftershocks for the updated post-1900 period earthquake catalogue of Turkey (Kadirioğlu et al. 2018). Their conclusion was that among the employed methods, each algorithm had advantages or disadvantages over the other ones. However, the authors suggested the method by Gardner and Knopoff (1974) as the most suitable

Table 8.1 Number of total events with Mw ≥ 3.3 and number of events per magnitude bin for different methods. GK74 – Gardner and Knopoff (1974); GR85 – Grunthal (1985); UR86 – Uhrhammer (1986); R85 – Raesenberg (1985); R85(1) – Raesenberg with parameters suggested by Eroğlu-Azak et al. (2018)

	Magnitude range (Mw)								
Method	3.3–3.7	3.8–4.2	4.3–4.7	4.8–5.2	5.3–5.7	5.8–6.2	6.3–6.7	6.8–7.2	Mw ≥ 3.3
GK74	1936	1613	871	297	93	43	20	14	6837
GR85	1245	1125	691	274	90	41	20	14	4781
UR86	3134	2391	1153	349	98	44	20	14	10,098
R85	3770	2877	1329	395	107	44	20	14	11,912
R85 (1)	2760	2252	1131	353	100	44	20	14	9301

one for the recently compiled Turkish earthquake catalogue as the post-declustering analysis of the main and aftershocks revealed it as the most successful one in the identification of the dependent events. The original catalogue compiled for Georgia was declustered using the time-space windows of Gardner and Knopoff (1974), Grunthal (1985), Uhrhammer (1986) as implemented in the ZMAP (Wiemer 2001), Raesenberg (1985) with original parameters and Raesenberg (1985) and also with parameters suggested by Eroğlu-Azak et al. (2018) for Turkey. The declustering was performed for the period 1900–2018, as in the historical part we assume to have only independent events. Declustering results indicate differences in the numbers of events identified for each method, as reported in Table 8.1 for magnitude bins of 0.5. It is worth noting that significant differences exist in the range of Mw 3.3–4.7, as it is expected given that the methods use different assumptions and algorithms to identify the aftershocks or the clustered events. Furthermore, to validate the model selection, we decided to verify whether the obtained catalogues followed a Poisson process:

$$y1 = poisson\,(x, l) = \frac{l^x}{x!}e^l \tag{8.5}$$

were x takes values 0, 1, 2, …

$$\lambda = \text{n/tk} \tag{8.6}$$

n – is the total number of events during the period of investigation and tk is the interval according to Luen and Stark (2012). For each declustered catalogue, we tried to find the value of tk, for which the condition of function χ^2(h, p, st) is satisfied, namely when h = 0 and $p \geq 0.05$, the Poisson hypothesis is true, otherwise the hypothesis is false. Table 8.2 presents the estimated value of h and p for all declustered catalogues. As can be observed in Table 8.2, the Poisson hypothesis is true for both Gardner and Knopoff (1974) algorithms and for the Grunthal (1985) algorithm. Different numbers of small events that come from these three methods of declustering can influence the slope of the magnitude frequency relationship

Table 8.2 Value of h and p of function χ^2 for all declustered catalogues

catalogues	p	h
GK74	0.5279	0
GR85	0.4096	0
UR86	0.00001	0
R85	0.00001	0
R85 (1)	0.00001	0

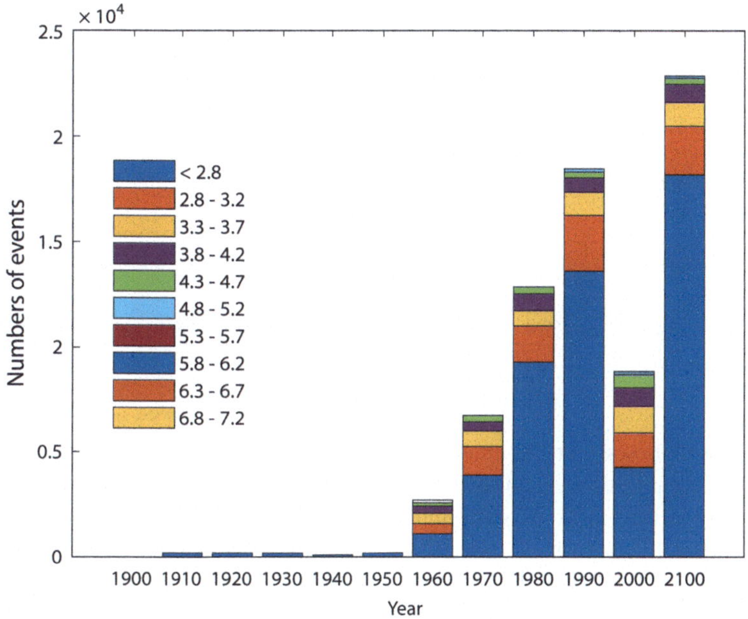

Fig. 8.5 Magnitude timeline of the catalogue declustered by the Gardner and Knopoff (1974) method for the period 1900–2018

(b-value), which should be considered in probabilistic seismic hazard assessment (PSHA). This issue is described in the section "Seismogenic Sources". Thus, as the final and reference catalogue, we considered the catalogue declustered by Gardner and Knopoff (1974) with the best value of h and p of function χ^2 (Fig. 8.5).

8.3.1.4 Magnitude and Time Completeness

Before estimating the earthquake rates or seismic productivity, the completeness of the earthquake catalogue for different magnitude thresholds versus time should be considered. For this purpose, we first performed a simple inspection of the catalogue to constrain cumulative number of events as a function of time. Figure 8.6 presents the cumulative plot for a threshold magnitude of Mw 3.3 as an example.

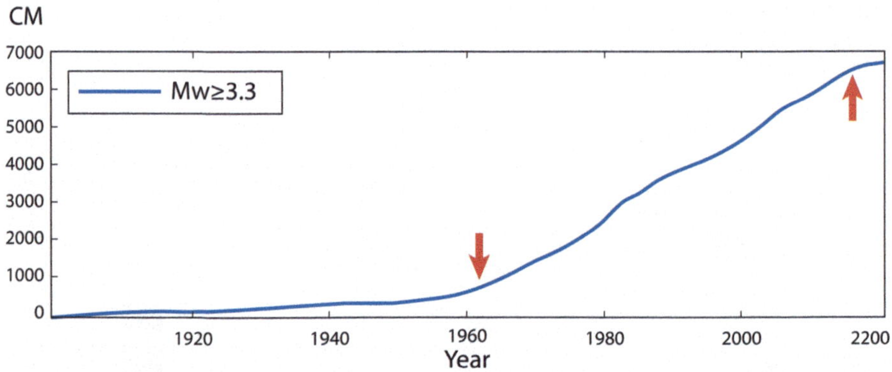

Fig. 8.6 Cumulative plots of magnitudes (CM) as a function of time for the threshold magnitude of Mw = 3.3

Table 8.3 Completeness assessment for the catalogued eclustered by the Gardner and Knopoff (1974) method

Time	2000	1970	1960	1930	1900	1860	1860	1600
Magnitude bin	3.3–3.7	3.8–4.2	4.3–4.7	4.8–5.2	5.3–5.7	5.8–6.2	6.3–6.7	6.8–7.2

Figure 8.6 shows that the change in slope occurs for the years 1962 and 2016. The last 2 years seem lacking earthquakes. As mentioned above, as of today, national waveform data of earthquakes are not accessible in Georgia and we collected initial data (arrival time of waves) from international sites, where small earthquakes with Ml less than 4 can be missing. Earthquake data with Ml ≥ 3 are officially open and provided by the Institute of Earth Sciences and National Seismic Monitoring Center of ILIAUNI (https://ies.iliauni.edu.ge/?page_id=183&lang=en) but as our investigation shows, the list of events is not complete for the last 2 years. Therefore, we consider the earthquake catalogue until 2016 for small earthquakes. The completeness of the catalogue of the post-1900 period declustered by Gardner and Knopoff was investigated for different magnitude bins and time windows using the methodology proposed by Stepp (1972), while the completeness of the pre-1900 period was evaluated based on historical considerations (Table 8.3). The completeness periods obtained from the statistical analysis reflect the various periods of network quality and density, not only in the Georgian territory, but also in the neighbouring regions, and as such they are, in general, more conservative than the individual periods described in the catalogue compilation section above.

8.3.2 *Regional Geology and Tectonics*

The main geology of Georgia is well described in the recent literature (Adamia et al. 2008, 2010, 2011; Westen et al. 2012; Bolashvili et al. 2018). Some of the main points from these studies can be summarized as follows: the territory of

Georgia, as a part of the Caucasus region, is located within the collision zone of the Eurasian and Afro-Arabian plates, in the Mediterranean (Alpine-Himalayan) belt, at the interface of the European and Asian segments. The metamorphic grade, the lithological characteristics of the sedimentary successions and the composition of submarine volcanites are quite different within the different tectonic zones of Georgia, mainly built of Mesozoic and Cenozoic sediments. Early (Precambrian and Paleozoic) formations occupy smaller areas. The following tectonic units can be distinguished in Georgia: (1) The folded system of the Greater Caucasus; (2) The Transcaucasian intermountain zone; and (3) The folded system of the Lesser Caucasus. Each tectonic unit is subdivided into secondary and tertiary ones. The oldest rocks (Precambrian and lower-middle Paleozoic) are found in all tectonic units and the youngest rocks, represented by Quaternary deposits, are found almost throughout the territory of Georgia.

The tectonics and geodynamics of Georgia are well described in numerous publications (Adamia et al. 1981, 2008, 2010; Jackson and Ambraseys 1997; Banks et al. 1997; Allen et al. 2004; Reilinger et al. 2006; Tan and Taymaz 2006; Kadirov et al. 2008, 2012; Mosar et al. 2010; Sosson et al. 2010; Pasquaré et al. 2011; Tibaldi et al. 2017a, b, 2018, 2020), of which, in the following paragraphs, we highlight some main aspects.

The geodynamics of Georgia and the adjacent territories of the Black Sea-Caspian Sea region indicate ongoing convergence process between the Eurasian and Africa-Arabian plates. This process determined the main morphological units of Georgia represented by the mountain ranges of the Greater and Lesser Caucasus and inter-mountain depressions. The major geodynamic activity of the region results in frequent moderate to large earthquakes. According to GPS data, the rate of this convergence is up to 30 mm/year (DeMets et al. 1990, 1994; Triep et al. 1995; Reilinger et al. 2006). Tectonic stresses induced by the northward motion of the Arabian plate are absorbed to a considerable degree in the Periarabian Ophiolitic Suture Zone and in the Zagros fold-thrust belts. North of these structures, the stresses are propagated towards the Central Caucasus and cause crustal shortening. The convergence rate across the lesser Caucasus is up to 10 mm/year, while in the Greater Caucasus it reaches values up to 4 mm/year (McClusky et al. 2000; Reilinger et al. 2006; Sokhadze et al. 2018).

The orientation of the active faults in the region follows three main directions: a "Caucasian" trend, from WNW/W to ESE/E and two transversal directions: NE-SW and NW-SE (Adamia et al. 2008; Tsereteli et al. 2016a, b). Focal mechanism solutions (FMS) along active faults indicate three stress regimes (strike slip, thrust, and transpressional) for moderate and strong earthquakes. "Caucasian" strike active faults are represented by thrust and transpressional faults, while in the transversal active faults, the strike-slip mechanism is dominant (Tsereteli et al. 2016a, b; Adamia et al. 2017). Most transversal active faults are located in the Javakheti volcanic highland of the South Georgia. The highland represents a young tectonic unit formed during Neogene-Quaternary era and active faults here are represented mainly by sinistral and dextral nodal planes of FMS. Here, a huge number of small earthquakes with local magnitude ($Ml < 2.5$) extend up to 5 km depth. A recent investigation by Mumladze et al. (2015) for the central and eastern Greater Caucasus reveal that earthquake depths extend as deep as 158 km with a dip of 45°NE. Dip seismicity was explained by a subducted slab that could cause strong earthquakes with moment magnitudes M > 8, hence indicating a potentially high seismic hazard for Georgia.

8.3.3 Seismogenic Source Models

The development of the seismogenic source model for the region of interest consists of two major steps (a) identifying potentially active earthquake sources in the region and (b) estimating the size and rate of activity associated with these sources. We consider two types of source models: area sources and active faults combined with background seismicity. The joint use of these two independent seismogenic source models effectively captures the inherent spatial variability of the seismicity and the epistemic uncertainty of the temporal characteristics of the future recurrence rates and size of earthquakes.

8.3.3.1 Area Source Model

Increasing seismic data and numerous investigations on seismicity distribution have shown a complex pattern of seismicity indicating a heterogeneous seismicity characterized by clusters, chains and gaps. This brought forward the development of an area seismic source model, which we have defined following tectonic evidence (Danciu and Giardini 2015) as part of the GSHAP compilation (Giardini 1999). A most recent area source model for Georgia was developed in EMME14 (Sesetyan et al. 2018a; Giardini et al. 2018; Danciu et al. 2018a) at regional level, which, in the present study, was updated at the national level. We divided the main tectonic units into sub-regions according to the recent tectonic map of Adamia et al. (2011) and topography of the region (Fig. 8.7). Further division into sub-regions was performed by delineation of zones of uniform stress field. For this purpose, the availability of reliable fault plane solutions (FPS) is essential. The homogeneity of the stress field for selected areas was estimated by seismic consistency (Sc) as defined by Apperson (1991). Sc-factor is a statistic measure showing how similar earthquakes are in a pre-defined region, Cs = 1 meaning that all earthquakes have identically orientated moment tensors, while Cs = 0 meaning that earthquake sources cancel each other. The database of FPSs in the present study includes information about 950 earthquakes in the Caucasus (Tsereteli et al. 2016a, b; Adamia et al. 2017; Tibaldi et al. 2020). Using this procedure, sub-regions were divided into macro-zones (MZ) with Cs close to one. Stress inversion by Michael (1984) was applied for each MZ. Stress inversion including both nodal planes of each FPS enables finding the best fitting planes during the procedure (Barth and Wenzel 2010) with orientation of the maximum horizontal stress (SH max). Figure 8.8 presents some FPSs from the

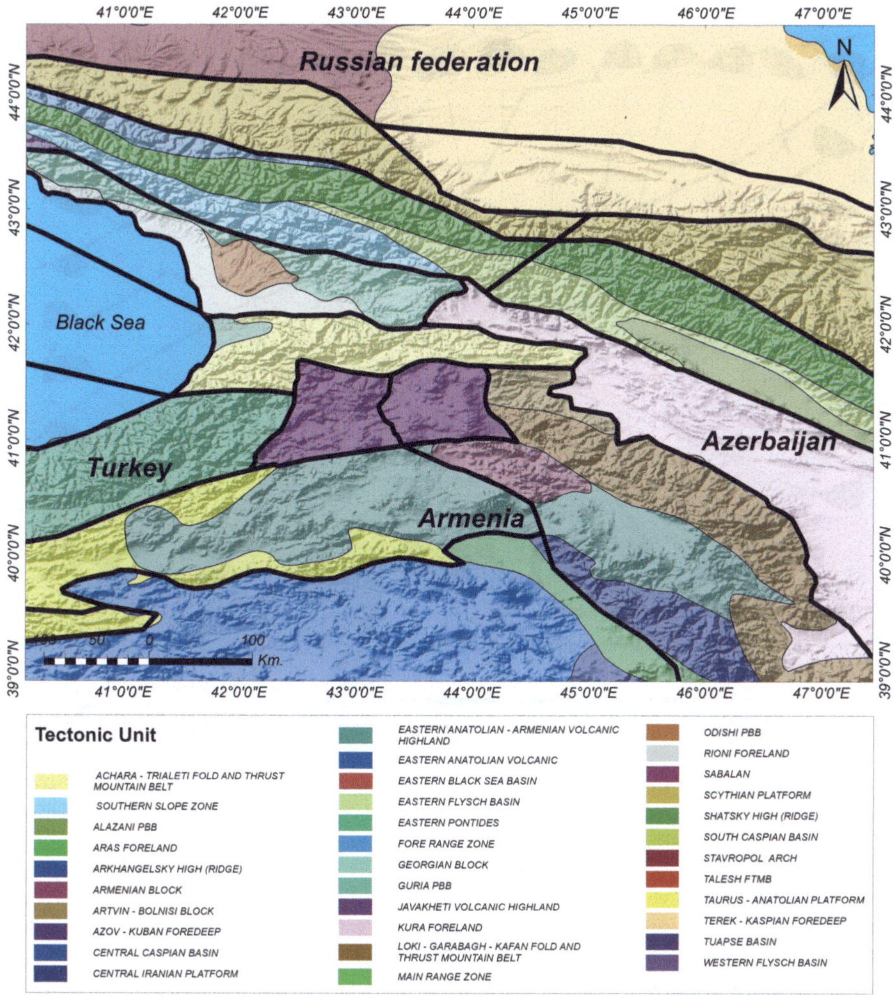

Fig. 8.7 Main tectonic units of the Caucasus (after Adamia et al. 2011) and selected sub-regions (black lines)

database, which are most frequently observed. Black arrows show orientations of the SH max, the changes of which indicate changes in the stress field and therefore justify the definition of different macro-zones. Finally, sub-regions were subdivided into 16 macro-zones. Active faulting data and seismicity patterns were further used to define the areas seismic source model, presented in Fig. 8.9.

Individual area sources are characterized by the activity rate *a* and *b* value of the Gutenberg–Richter (GR) law (Gutenberg and Richter 1944), that for a given source, describes the magnitude – frequency distribution of earthquake occurrence

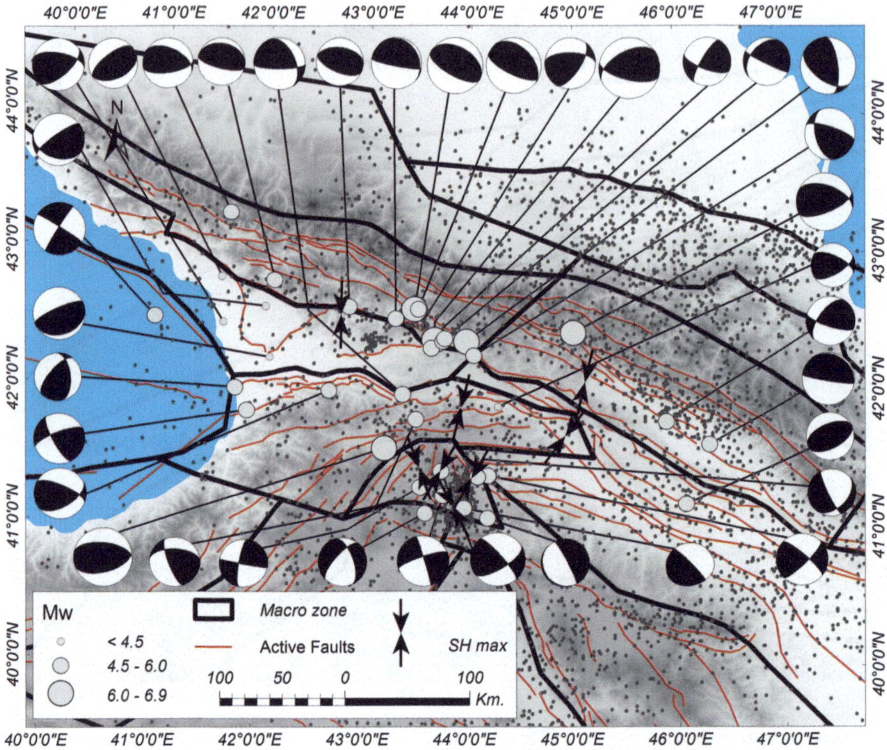

Fig. 8.8 Prevalent FPS for Caucasus. SH max – maximum horizontal stress obtained by stress inversion from macro-zones with Cs close to unity. The black dots indicate the epicentres derived from the earthquake catalogue

by the relationship $\text{Log}_{10}N(\text{M}) = a - b{*}M$, where N(M) is the cumulative number of earthquakes per unit time with magnitude larger or equal to M.

The seismic activity rate *a-value* represents log of the number of events with magnitude $\text{M} \geq 0$ and defines the total seismic productivity of a given area source. The *b-value* is a scaling parameter – the negative value of magnitude-frequency occurrence curve – and is vital for estimation of earthquake occurrence probability. The parameters *a* and *b* of earthquake recurrence rates are estimated for independent events (earthquake catalogue declustered by the Gardner and Knopoff 1974 method) and the completeness intervals of different magnitude bins proposed for individual macro-zones. As mentioned above, the completeness intervals were estimated for the macrozones by means of the Stepp (1972) procedure (Table 8.4) and were extended to the area sources (ASSs) that were included in corresponding macrozones. The slope of the recurrence curve, *b,* was first calculated for the 16 macrozones shown in Fig. 8.9 through the maximum likelihood (MLE) procedure of Weichert (1980). The *b*-value obtained for each MZ is generalized for all ASS falling into given MZ. Then, the recurrence parameters *a* and *b* were independently

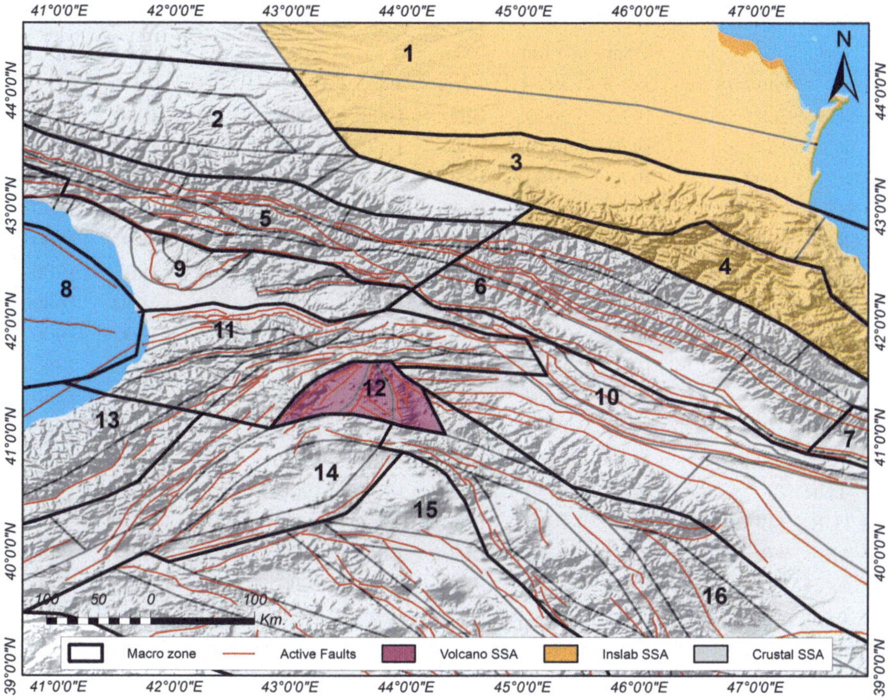

Fig. 8.9 Macrozones and area seismic sources for Georgia and adjacent areas

estimated for ASSs with number of earthquakes more than 20 with the same MLE procedure. For ASSs with a smaller number of earthquakes (less than 20 events), the *b*-value was adopted from corresponding MZ and the corresponding activity rate, *a,* was estimated by the distribution of the total number of events in the MZ to the smaller ASS. The same procedure was repeated using the earthquake catalogues declustered by way of the other methods described in the previous chapter, and standard deviation of the *b*-value for all 16 MZs were calculated. The average standard deviation of the *b*-value was less than 0.02 for all investigated catalogues, which was adopted as the standard deviation to be used in the logic tree structure. For each source, the minimum magnitude was set to Mw = 4.5 and the upper magnitude was defined as the maximum possible earthquake within each source. The area sources are defined as regions of equal probability, memory less of earthquake occurrence hence appropriate for time-independent hazard estimates.

Table 8.4 Completeness of macrozones

		Magnitude bins							
Zone	Num. of Eqs.	3.3–3.7	3.8–4.2	4.3–4.7	4.8–5.2	5.3–5.7	5.8–6.2	6.3–6.7	6.8–7.2
1_NC	209	1970	2000	1960	1900	1900	1900	1880	1880
2_NC	153	1970	1970	1940	1900	1900	1900	1900	1900
3_NC	1009	1970	1960	1950	1950	1950	1900	1900	1900
4_NC	273	1970	1970	1950	1890	1890	1870	1870	1870
5_GCW	287	1980	1970	1930	1900	1900	1860	1860	1100
6_GCE	773	1960	1950	1900	1900	1900	1900	1900	1700
7_SHE	233	1990	1970	1950	1850	1850	1850	1850	1600
8_BS	47	2000	1950	1950	1930	1930	1930	1900	1900
9_FL	184	1970	1960	1910	1910	1900	1860	1860	1860
10_FL	975	1970	1970	1950	1900	1850	1850	1850	1850
11_LC	239	1980	1950	1950	1880	1860	1860	1200	1200
12 JAV	332	1970	1950	1920	1920	1900	1880	1880	1880
13_TUR	73	1960	1950	1950	1920	1900	1900	1900	1900
14 TUR	349	1980	1960	1960	1900	1900	1900	1900	1900
15 AR	978	1980	1980	1960	1960	1900	1800	1800	1800
16 AZ	723	1980	1980	1980	1900	1900	1850	1850	1850

NC North Caucasus, *GCW* West Greater Caucasus, *GCE* East Greater Caucasus, *SHE* Shemakh region, *BS* Black sea, *FL* foreland, *LC* Lesser Caucasus, *Jav* Javakhety upland, *Tur* Turkey, *AR* Armenia, *AZ* Azerbaijan

8.3.3.2 Active Faults and Background Seismicity Model

The first active fault source (AF) model for Georgia was developed during the EMME project (Giardini et al. 2018; Sesetyan et al. 2018a) and was also adopted in the present study (Fig. 8.8).

The quality level of active fault data is Class A (see Danciu et al. 2018a), i.e. including full fault geometry and long-term slip rates, information supported by historical and instrumental observations and geological field investigations. The AF database (as compiled by Gülen et al. 2014) contains information about more than 30 parameters, such as fault section name, fault section identity, fault trace, fault type, average strike, dip, and rake values, fault section length, lower and upper seismogenic depth, horizontal and vertical slip rates, and average aseismic slip factor.

Slip rates of active faults are estimated from geological observation by the offset of stratigraphic markers. The seismic productivity of active faults is defined in agreement with Model 2 of Anderson and Luco (1983) based on the long-term fault slip rates. This was described extensively in Danciu et al. (2018a) and will not be repeated here. To avoid double counting of the earthquake recurrence rates in the vicinity of faults, asymmetric buffer zones around active faults are defined. In such buffer zones, the lower bound of the magnitude frequency distribution of active faults is set to a magnitude Mw = 5.5, whereas the background seismicity accounts for smoothing-specific earthquake rates up to 5.5. (Danciu et al. 2018a). The background seismicity rates were estimated using the declustered earthquake catalogue and the completeness intervals for the macrozones presented in Fig. 8.9 and Table 8.4. The smoothing algorithm of Frankel (1995) was used with a kernel bandwidth of 20 km.

8.3.3.3 Maximum Magnitude

The estimation of the maximum magnitude is one of the significant steps in PSH, as it determines the upper bounds of the earthquake size thought to be possible within a specific active fault or seismo-tectonic domain. An extensive literature provides various options and procedures to assess this parameter i.e. empirical correlations between fault geometries and earthquake size (Wells and Coppersmith 1994; Leonard 2010), dynamic links of fault sections (Mignan et al. 2015), statistical estimates (Kijko and Singh 2011), historical record and analogies to similar tectonic domains (Wheeler 2009), tectonic proxies and statistical estimates (EPRI, Johnston et al. 1994). In this investigation, the Mmax can be estimated by several approaches, particularly:

- by the observed maximum earthquake over historical seismicity record considering uncertainty in location;
- by increments corresponding to magnitude estimation errors;
- by fault geometry and scaling relations (e.g. Wells and Coppersmith 1994);
- by the length (L_{max}) of an entire active fault which falls into a given ASS when 10% of the entire fault length falls into the SSZ (Shebalin et al. 2000).

The final values of the M_{max} were estimated in the investigated region by averaging individual values obtained through the above described methods. The epistemic uncertainties around this parameter are represented in a logic tree structure described in *Section: Logic Tree and Model Uncertainties*. Hereinafter, two values of increment (+0.3 and 0.6 in moment magnitude units) are used to augment the maximum magnitude, and they are meant to depict the uncertainties of the regional catalogue magnitude (Zare et al. 2014). According to Wheeler (2009), the two added increments indicate an extension of the historical record by a factor of 1.35–1.7, assuming a regional b-value of 0.9–1.00. With the two factors, the historical record is extended to 4000 and 5000 years to provide conservatism against earthquakes larger than the measured maximum magnitude (Danciu et al. 2018a).

8.4 Ground Motion Modelling

The modelling of the ground motion, resulting from individual rupture scenarios generated by the source model, represents the second major part of seismic hazard assessment effort.

In recent years, ground motion modelling in seismic hazard assessment has constantly progressed from the use of a single preferred (or regionally available) ground motion prediction model (Danciu and Giardini 2015) to multiple models (Delavaud et al. 2012; Danciu et al. 2018b) and/or backbone models (Douglas 2018). The selection of ground shaking models is a complex task, given the increased number of ground motion predictive models (see the online compendium of ground motion models at http://www.gmpe.org.uk/; Douglas and Edwards 2016) and the relatively low number of ground shaking recordings in Georgia that can be used for testing. Similarly, the lack of a sufficient number of strong motion data so far did not allow for the development of local ground motion prediction models for Georgia. To overcome this problem, Tsereteli et al. (2016a, b) developed a hybrid-empirical model using data from two neighbouring countries, Turkey and Iran, providing a set of estimations for PGA and spectral accelerations at selected periods. Comparison of the estimated values with the available recordings of the 15 June 1991 Mw 6.1 earthquake in Georgia yielded favourable results; however, due to the fact that the model provides only median estimations of ground motions, without the associated uncertainties, we were not able to use it in our logic tree structure.

As such, the selection of ground motion models to be used in the hazard analysis has been based on a consideration of recent developments in regional (Danciu et al. 2018b) and global models (Pagani et al. 2018). Several ground motion prediction equations have been selected, given their validity for similar tectonic and seismogenic context (e.g. active shallow crustal, deep seismicity, volcanic). As in past studies the hazard was quantified and estimated in terms of macroseismic intensity – the starting point of the present study was the ground motion logic tree of EMME14 (Danciu et al. 2018b). The ground motion models used to describe the inherent variability of ground shaking for shallow crust seismicity were derived from Akkar et al. (2014), Chiou and Youngs (2008), Akkar and Cagnan (2010) and Zhao et al. (2006). The models selected for the deep seismicity were those by Lin and Lee (2008) and Youngs et al. (1997). The weights of these models were assigned following an extensive sensitivity analysis, with focus on removing the cognitive bias of expert elicitation. It needs to be pointed out that the logic tree of the EMME14 is outdated – given the rapidly increasing number of predictive models worldwide, made possible by the availability of new strong motion datasets. The most recent versions of the NGA-West2 (Bozorgnia et al. 2014) include models developed by Abrahamson et al. (2014), Boore et al. (2014), Campbell and Bozorgnia (2014), Chiou and Youngs (2014), Idriss (2014). The NGA models are evolutionary, as they were developed to replace the previous versions i.e., NGA-West1. These equations were used to compute the 2014 U.S. Geological Survey (USGS) National Seismic Hazard Maps (Petersen et al. 2015), particularly for calculating crustal earthquakes in the Western US. Bindi et al. (2014) proposed a new GMPE derived on a dataset consisting of predominantly Italian strong-motion data. Cauzzi et al. (2014) also provided an update of the 2008 GMPE, with an improved functional form that accounts for large magnitudes saturation. In Kotha et al. (2016) can be seen the European update of the Bindi et al. (2014). We used

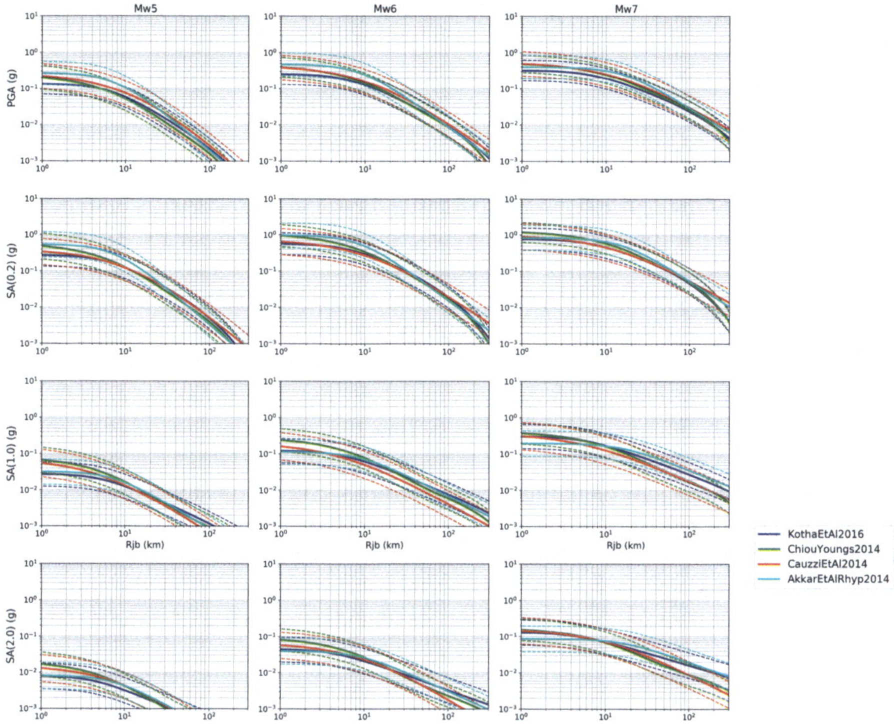

Fig. 8.10 Trellis plots depicting median and one standard deviation for the selected GMPEs for different magnitudes (Mw = 5, Mw = 6, Mw = 7) and a range of distances from 1 to 300 km

trellis plots to compare the ground motion models selected in EMME14 and the new models.

Figure 8.10 shows trellis plots for different magnitude and distance bins, selected to represent the dominant scenarios of shallow crustal earthquakes in Georgia. Median values of spectral accelerations are estimated by each ground motion model for a reference rock site class i.e. Vs30 = 800 m/s. Every earthquake scenario is described by a vertical fault, strike-slip fault mechanism, located at a depth of 10 km and size scaled to the given magnitude. Additional parameters required by the NGA predictive models are estimated with the default parameters provided by their reference papers. Danciu et al. (2018b) recommend as a good practice "*to select at least one predictive equation from (1) local models describing the local characteristics of the strong-motion data, (2) pan- regional GMPEs to account for completeness of the dataset at large tectonic scale, (3) NGA models due to their overall performance, complex functional forms and aleatory uncertainty models, and (4) global predictive models to enhance the data of large magnitude events.*" Grunthal et al. (2018) recommend also the use of different ground motion datasets (i.e. European, RESORCE-DB, NGA-West2 and Japanese DB) to capture the epistemic uncertainties of the ground motion models. Thus, the following models have been selected: Akkar et al. (2014) and Kotha et al. (2016) – pan-regional models; Chiou and Youngs (2014) an NGA West2 model; Cauzzi et al. (2014) – a global model (includes Japanese data too). It is worth noting that, as there is no local model available – the Kotha et al. (2016) is the lower bound of the ground motion, hence it might be reasonable to describe the local variability of the ground motion. The weights are symmetrical: Kotha et al. (2016) and Chiou and Youngs (2014) are the lower and the upper bound respectively, hence a 0.15 weight is assigned to them; the central part of the ground motion distribution is captured by Akkar et al. (2014) and Cauzzi et al. (2014), with equal weights of 0.35. Two ground motion models, i.e. Bradley (2013) and Faccioli et al. (2010) are selected and used for volcanic sources. For deep seismicity sources the models of Lin and Lee (2008), Abrahamson et al. (2016), Montalva et al. (2017) are used. The weights of the logic tree branches are presented in the next section. Note that the current trend in PSHA is to use a backbone approach when modelling the ground motion characteristic model (Atkinson et al. 2014). In such approach, one model (GMPE) is selected to represent the central part of the model and various criteria (residuals trends, site-specific terms, regional variability factors) are used to scale the upper and the lower range, in order to represent the epistemic range of ground motion expected at a given site. The backbone approach might be a viable solution for modelling the epistemic uncertainties, but their calibration requires an abundant database of ground motion recordings in the region of interest (Douglas 2018).

8.5 Logic Tree and Model Uncertainties

Uncertainties are inherent during the entire model building process, from collection and harmonization of the earthquake catalogue, compilation and curations of active faults, statistical analysis of the earthquake catalogue, fitting techniques, source delineation and subjectivism of model's selection. A logic tree framework is adopted herein to capture such uncertainties. Generally, the uncertainties are separated as aleatory variability and epistemic uncertainty. Aleatory variability is intrinsic to the process, whereas the epistemic one can be reduced and depends on lack of data or knowledge (Bommer and Scherbaum 2008). The present hazard model treats epistemic uncertainties in all hazard components through incorporation of alternative models in a logic tree structure and associated weighting. The structure of the logic tree is given in Fig. 8.11, and consists of two sections: a source model logic tree and a ground motion logic tree. The seismogenic source logic tree consists of three branching levels: source model, *b*-value and maximum magnitude (i.e. Mmax). The first branch describes the spatial variability of the seismogenic sources (i.e. area source vs. active faults + background seismicity). The temporal variability of the source models is described by the *b*-value and Mmax branches. Ideally, each branch of the logic tree should represent an independent and alternative model, however this is not possible as the earthquake catalogues is represented in both branches of the seismogenic sources.

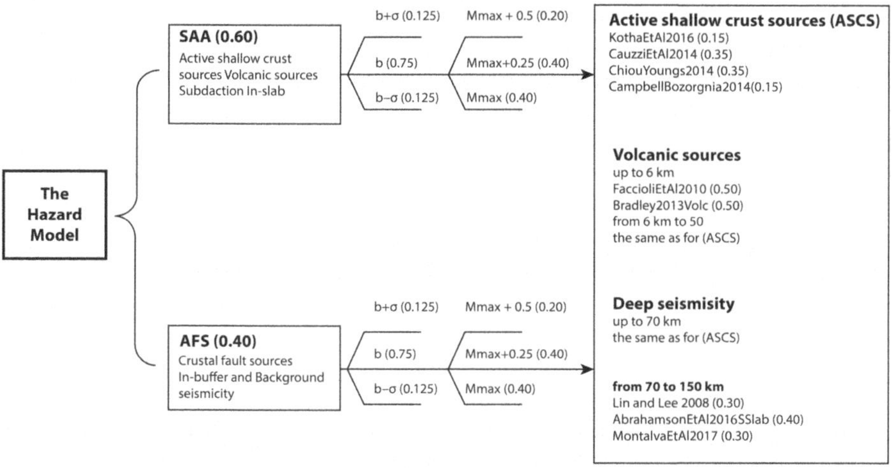

Fig. 8.11 PSHA logic tree for the epistemic uncertainty in the seismic sources, recurrence parameters, Mmax and GMPEs (note that the depth range represents an arbitrarily defined range to capture the data and the seismogenic domain, not the actual depth range, which ranges from 0 to 30 or 50 km)

PSH was obtained from a set of realizations, in accordance with the four main levels of logic tree branches: seismic sources models, recurrence parameter *b*-value, Mmax and GMPEs. The final set of GMPEs to be associated with individual seismic sources were representative for three different tectonics: active shallow crustal sources (ASCS), volcanic sources and subduction in-slab sources. Uncertainty in b-value was estimated for all macrozones for selected declustered catalogues. The lower seismogenic depth of the model for ASCS is up to 70 km depth for the region overlapping the in-slab sources, otherwise the seismogenic layer is confined within the 0–30 km (and 50 km in some sources). The volcanic sources are confined within the first 10 km of the seismogenic layer, i.e. the volcanic sources were considered to be within a depth range of 0–6 km, and subduction in-slab sources were considered to be confined within a depth of 70–150 km. The depth distribution of shallow sources is described by a range from 10 to 30 km, while the depth distribution of the sources within the subduction in-slab is described by a range of 70–150 km.

8.6 Ground Motion Hazard Assessment: Results

The probabilistic seismic hazard analysis identifies, for each site, the annual probability of exceedance of various ground-motion levels defined in terms of selected ground motion intensity measures, such as PGA or spectral accelerations corresponding to various structural periods, due to all possible future earthquake scenarios.

The OpenQuake software (Pagani et al. 2014) has been used to implement the hazard model of Georgia. Seismic hazard estimates can be affected by model parameterization and implementation along the course of modelling.

OpenQuake is used in all recent regional hazard computation projects (e.g. EMME14, ESHM13 (Woessner et al. 2015) and GEM19 (Pagani et al. 2020) but also in national models (Italy – Meletti et al. 2019; Turkey – Sesetyan et al. 2018b; Canada- Adams et al. 2019; Switzerland – Wiemer et al. 2016; Australia – Allen et al. 2018). A key feature of OpenQuake is the use of state-of-the art seismic source representation combined with logic tree options for accounting the input uncertainties. Monelli et al. (2014) have shown the significant impact of the use of extended ruptures on the hazard estimates with respect to point-rupture approximation, which leads to a significant increase in the probabilities of exceedance for specific levels of ground motion. At this point, it is not clear whether any of the previously computed models for Georgia were considering any proxy for extended ruptures. Using the seismic hazard model developed and presented in this manuscript, we have calculated, with OpenQuake, the exceedance probabilities (POEs) of specific ground motion levels for PGA and pseudo-accelerations (SA) spectra with 5% critical damping within 50 years at all sites within the Georgian territory. Results are presented for the reference rock only (Vs30 = 800 m/s), in the form of hazard curves, for the 5th, 16th,15th, 50th(median), 85th, 95thpercentiles and mean estimates of the geometric mean of the two horizontal components of the ground motion. The quantiles are an indication of the degree of the epistemic uncertainty in the seismic hazard assessment and often a ratio between the lower and upper quantiles is used to measure such dispersion (Sesetyan et al. 2018a). Hazard curves are presented at equally spaced grid points of 5 by 5 km. The minimum magnitude for hazard integration is 4.5 Mw.

Furthermore, uniform hazard spectra (UHS) are also available depicting spectral acceleration values of uniform probability levels for a range of spectral periods. UHS are the basis for defining the seismic design level applicable to new buildings and median (50th) or different percentiles are used to shape these design spectra according to the importance of the structure under design. Seismic hazard maps representing the spatial distribution of the ground shaking for a fixed return periods of 72, 95, 475, 975, 2475, 4975 and 9975 years, corresponding to 50, 40, 10, 5, 2, 1 and 0.5% exceedance in 50-year time-span. Note that the return periods and the POEs in 50 years terms are to be used only in a statistical sense. A return period of 475 years does not imply an expected hazard in the next 475 years; it rather indicates that the hazard level in the next 1 year has a very low probability (i.e. 1/475 years ~ 0.002107 per annum) of occurring (Musson and Winter 1996).

In Figs. 8.12, 8.13 and Fig. 8.14 we present the spatial distributions of PGA and 5% damped SA (0.2 sec and 1.0 sec) for 10 and 2% probabilities of exceedance in 50 years (corresponding to mean return periods of 475 and 2475 years respectively).

Figure 8.12 also shows the epicentral locations of moderate to large magnitude earthquakes from our catalogue. Comparison of obtained probabilistic seismic hazard map with observed seismicity shows good correlation. Zones with strong earthquakes Mw > 6.1 are characterized by mean PGA in the range of 0.32–0.53 g

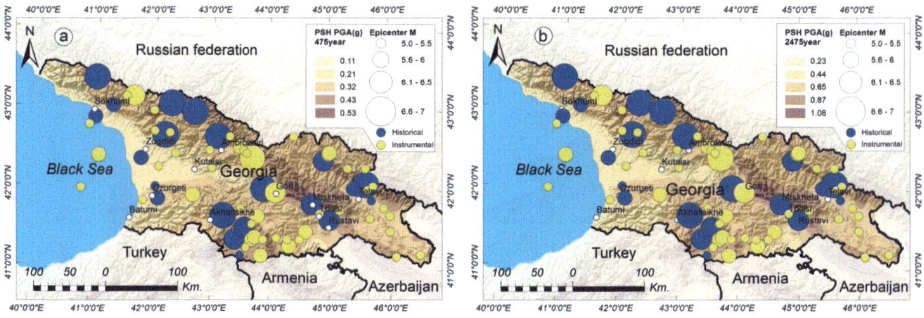

Fig. 8.12 Hazard map showing the mean PGA distribution with 10% (**a**) and 2% (**b**) probabilities of exceedance in 50 years for a reference rock (Vs30 [800 m/s), overlain with epicentral distribution of medium to large magnitude earthquakes

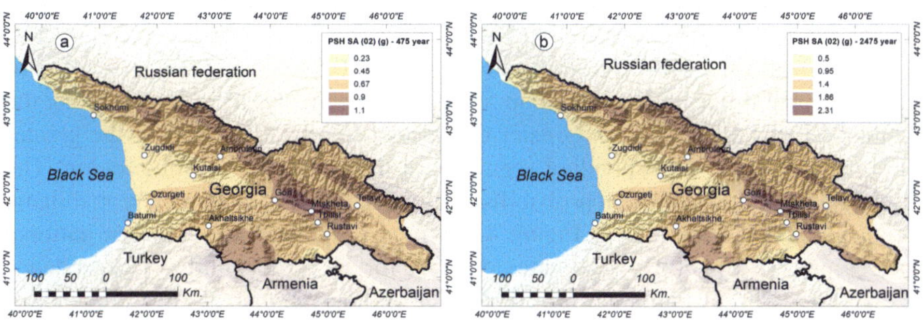

Fig. 8.13 Hazard map showing the mean SA distribution at T = 0.2 sec with 10% (**a**) and 2% (**b**) probabilities of exceedance in 50 years for a reference rock (Vs30 [800 m/s)

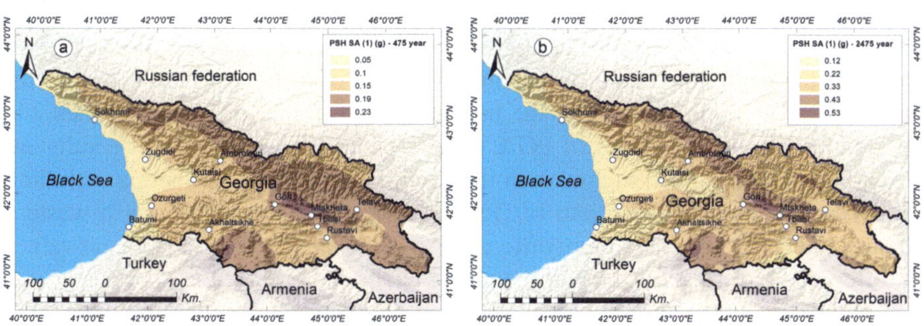

Fig. 8.14 Hazard map showing the mean SA distribution at T = 1.0 sec with 10% (**a**) and 2% (**b**) probabilities of exceedance in 50 years for a reference rock (Vs30 [800 m/s)

for 10% probability of exceedance in 50 years (475 years return period) and in the range of 0.66–1.08 g for 2% probability of exceedance in 50 years (2475 years return period).

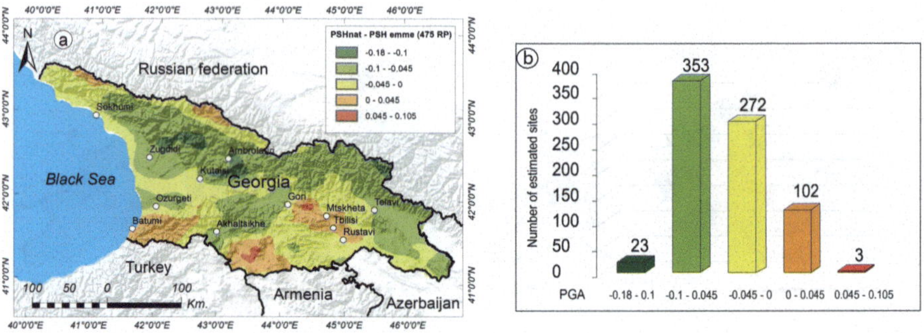

Fig. 8.15 (**a**) Differences in PGA for 10% probability of exceedance in 50 years estimated at national (this study) and regional (EMME 2014) levels; (**b**) number of sites per bins of PGA differences

The highest values of PGA, in the order of 0.43–0.530 g and 0.87–1.08 g for 10 and 2% probabilities of exceedance in 50 years respectively, are observed along the southern slopes of the eastern Greater Caucasus (Fig. 8.12), caused by the greater slip rates along the corresponding segments of the active faults in the region. The results obtained in the present probabilistic model are comparable with probabilistic estimates of the EMME2014. Figure 8.15a presents differences between national and regional (EMME2014) estimates of PGA for 10% probability of exceedance in 50 years. The histogram in Fig. 8.15b shows the number of sites per bins of PGA differences according to the legend presented in Fig. 8.15a. We observe lower estimated values of PGA up to 86% of the whole territory of Georgia. Lower values of PGA with respect to the EMME2014 results in the range of −0.18 −0.045 (g) correspond to about 50% of the whole territory of Georgia and are observed in most part of Greater Caucasus, in the central and western parts of Georgia. Slightly lower values in the negligible range (less than −0.0045 g of difference) are observed for 36% of the Georgian territory in the background of higher differences zones. On the other hand, higher values of PGA with respect to the EMME2014 results are observed in the Javakheti highland, in the central eastern part of Georgia including Tbilisi, in the south-western part of Georgia and some parts of the north-western Greater Caucasus. The maximum difference in PGA value is 0.105 g. Greater values of difference govern about 14% of the whole territory.

Seismic hazard estimates are exemplified for two sites in Tbilisi, where the computed seismic hazard is significantly higher in the northern than in the southern sector of the city (Fig. 8.16). We need to highlight the minimum and maximum values of mean PGA across the city, which are 0.28 g and 0.36 g for 10% of exceedance in 50 years. Such differences can be explained by the presence of an active fault with relatively high slip rates to the northern part of the city. Figures 8.17a, b present the mean 5% damped uniform hazard spectra for six return periods (i.e. 72, 475, 2475, 4975 and 9975 years) for southern and northern

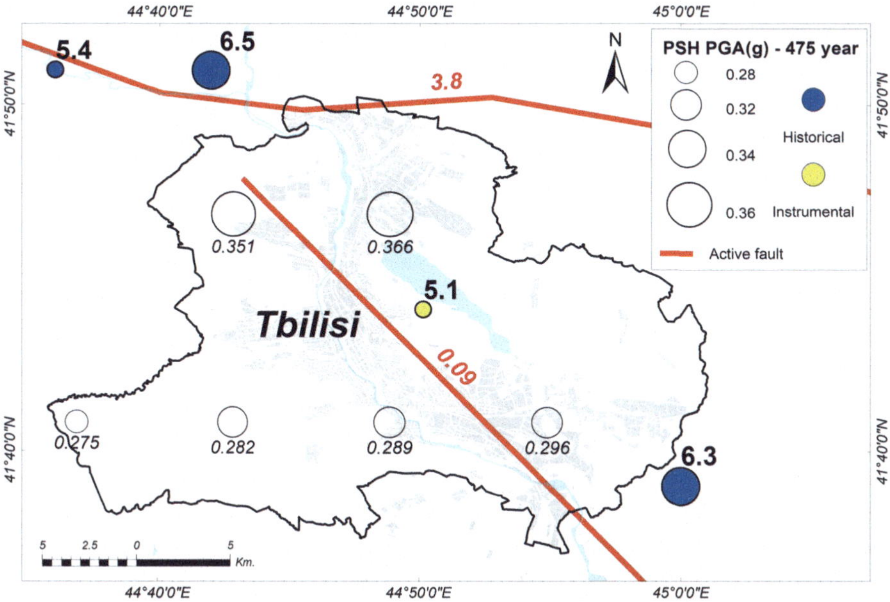

Fig. 8.16 Hazard map for Tbilisi, showing the mean PGA distribution with 10% with active faults and minimum slip rates along them

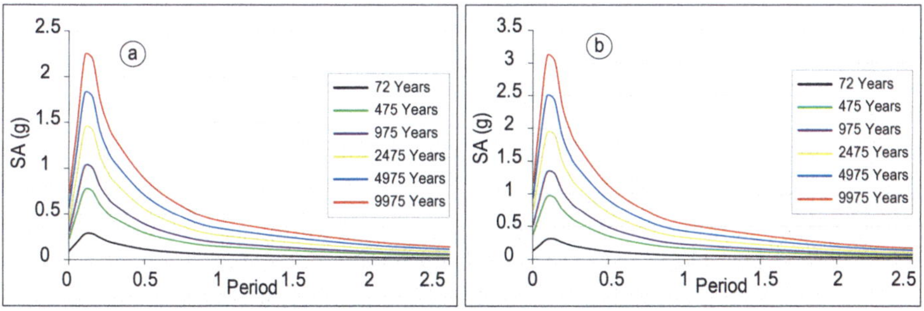

Fig. 8.17 Mean 5% damped uniform hazard spectra for six return periods (i.e. 72, 475, 2475, 4975 and 9975 years) for the southern (**a**) and for the northern part of Tbilisi (**b**)

Tbilisi, respectively, while Figs. 8.18a, b present the hazard curves in terms of PGA corresponding to mean and 5, 16, 50, 84 and 95% quantile computations for southern and northern Tbilisi, respectively. Such estimations provide information about the uncertainty ranges associated with the ground motion estimations at all return periods.

We have also investigated the differences between the new seismic hazard map and the normative map of Georgia. For this purpose, we compare, in Fig. 8.19a, the distribution of hazard values computed for the same intensity measure type and the

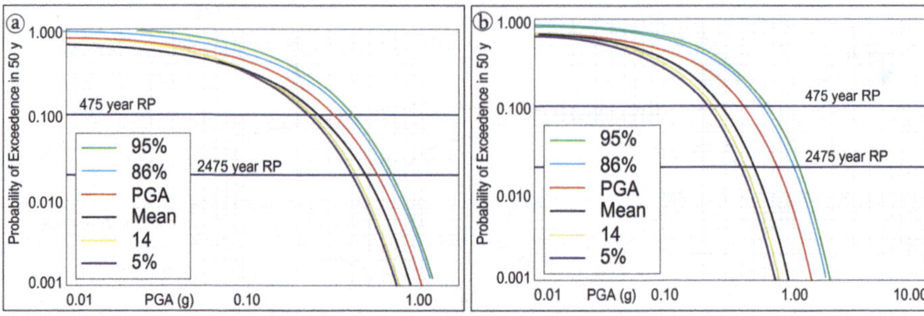

Fig. 8.18 Mean and 5, 16, 50, 84 and 95% quantile PGA hazard curves for southern (**a**) and northern Tbilisi (**b**)

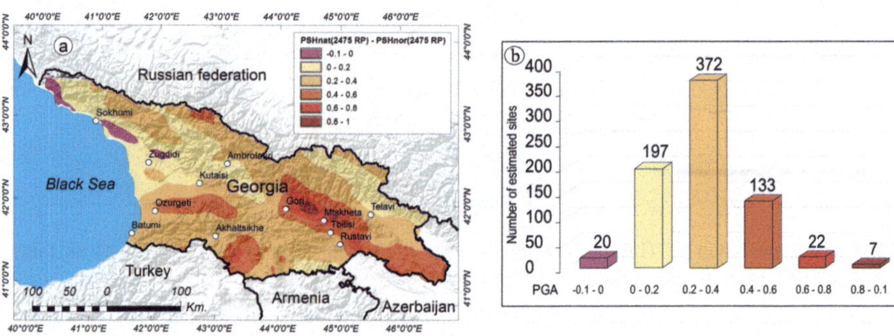

Fig. 8.19 (**a**) Differences in PGA for 2% probability of exceedance in 50 years estimated at new national (this study) and old national (normative map of Georgia) levels; (**b**) number of sites per bins of PGA differences

same probability of exceedance in a given time interval (PGA with 2% probability of exceedance in 50 years). We point out significant differences between the two maps, with values larger than the overall epistemic uncertainty computed in this analysis. The histogram in Fig. 8.19b shows the number of sites per bins of PGA differences with 0.2 g increment, where we observe higher estimated values of PGA for 92% of the whole territory of Georgia.

8.7 Hazard Disaggregation for the Selected Site

For intrinsically probabilistic applications, the selection of a scenario earthquake is based on the disaggregation of the hazard to show which events contribute the most (or for performance-based applications, to damage, even loss). The probabilistic seismic-hazard disaggregation involves the determination of earthquake variables, principally magnitude, distance and attenuation equation epsilon for each end brunch of the logic tree structure defining seismic events that contribute to a selected

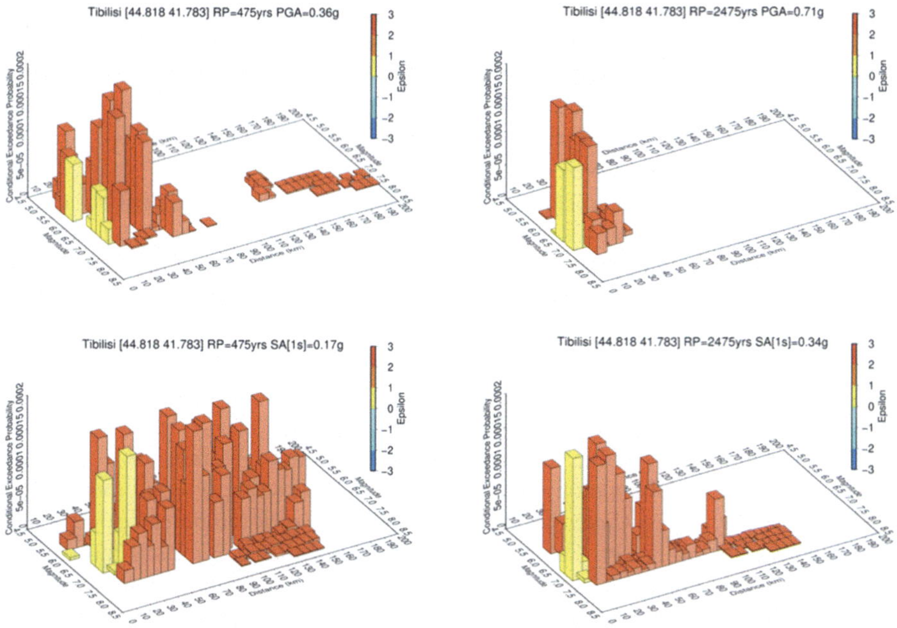

Fig. 8.20 Disaggregation plots for Tbilisi for two parameters (PGA and SA(1 s) for the 475-year and 2475-year return periods

seismic-hazard level (McGuire 1995; Bazzurro and Cornell 1999). Figure 8.20 displays disaggregation plots for Tbilisi as an example for two intensity measure types (PGA and SA (1 s) for 475 and 2475 years return periods). We observed that sources at relatively close distances dominate the disaggregated level of PGA (i.e. 0.36 g), though for 475 years return period a minor contribution from deep/in-slab seismicity sources is visible. A significantly high contribution from large magnitude bins across large distance sources for the medium frequency ground motions (SA at T = 1.0 sec) is observed. The same pattern is observed for the 2475-year return periods and both intensity measures.

8.8 Conclusions and Future Perspectives

In this work, we presented an updated probabilistic seismic hazard model for Georgia obtained leveraging from the experience in the EMME project. Input seismic data were homogenised using national data and seismic source models were specified at national level.

Statistical investigation of catalogues has shown that conversion equations between different types of magnitude have a significant influence on the magnitude time distribution of earthquakes especially for the small to medium magnitude ranges (up Mw 5.5), governing the lower boundary of the magnitude frequency relationship. On the other hand, the non-accessibility of national data has limited our investigation in terms of small magnitude earthquakes recorded during the most recent period. The effects of the use of different magnitude conversion formula remains to be studied in future efforts.

The study region is seismically active and has a complex geological structure. Although the present model has attempted to incorporate all the available geometric and kinematic information about the active faults in the region, there are still several ongoing efforts for the characterisation of already known and newly discovered sources (e.g. Tibaldi et al. 2017a, b). The active fault structures in the Black Sea area are not fully characterised yet. Therefore, future efforts should be in the direction of improving the faults source models of the region, by incorporating most recent studies.

The availability of reliable fault plane solutions is a key parameter in seismic source delineation at the national level. Limited data of strong ground motion for Georgia and the whole Caucasus do not allow for the development of regional ground motion models yet. Fault slip rates were estimated for each fault from offset stratigraphic markers. Seismic hazard estimates are sensitive to variations in fault slip rates and more reliable estimations are highly recommended for future efforts. The spatial variability of earthquake processes was captured by mixing two seismogenic source models, area sources and active faults, whereas the temporal variability was captured by alternative magnitude frequency distributions (e.g. based on seismicity for area sources and slip-rate based for active faults). For each seismogenic source, three alternative models were used as defined by pairs of a- and b-values and variable maximum magnitudes. Such alternative models provide discrete formulation of the continuous probability distribution, in which the range of each parameter is represented by a lower, a central and an upper value. The propagation of data variability and model uncertainties might explain the width of the percentiles around the median hazard (see Figs. 8.17 and 8.18). The percentile range around the median or mean estimates might also describe the natural variability of the model building process. Also, the percentile range might provide a meaning for unforeseen and/or unlikely effect not captured in the current hazard model i.e. infrequent occurrence of earthquake in the next observation period, inconsistent or biased earthquake rate forecast within the region, significant different ground motion levels (lower or higher) produced by future earthquakes in the region, unexpected large magnitude earthquakes, etc. Hence, the range around the hazard estimates are of critical importance and need to be properly communicated.

The ground motion variability has been represented by a set of regional or global ground motion prediction models, selected for each tectonic region. Although we were not able to include in the ground motion logic tree any local model for Georgia, incorporation of the hybrid-empirical model by (Tsereteli et al. 2016b) and evaluation of the hazard estimates with respect to this local model will be one of the subjects of future updates.

A comparison of national model to regional model (Sesetyan et al. 2018a) shows differences more than 0.06 g in some places, particularly the present model shows higher values in the south-western part of Georgia, Javakheti highland and in the central eastern part of Georgia, including Tbilisi. On the other hand, in some parts of the Greater Caucasus, the regional model of higher values in ground shaking estimations, the observed differences in the results are within the range of epistemic uncertainty of the ground motion models.

The 2020 national hazard model for Georgia replaces the prior hazard models and hazard results derived in 2010. The seismic hazard 2020 reflects the latest scientific findings (principle of stationary seismicity, active faults and uncertainties quantification) and updated datasets i.e. updated earthquake catalogue, revised estimates of the historical catalogue, a homogeneous magnitude scale, improved seismogenic source models and rupture forecast model (mainly area sources).

The hazard model building is a complex integration process that results in alternatives and discrepancies, hence a unique and universal solution does not exist.

Rigorous statistical validation of ground motion probabilistically estimated is merely possible to date (e.g. Beauval et al. 2008) due to short observational time of the main input datasets (i.e. although the earthquake catalogue spans over 100 years, ground motion data have been recorded only in the last decades). Sensitivity analysis as well as model transparency, documentation, input datasets and hazard computational files are mandatory to secure reproducibility of the results (http://ig-geophysics.ge/IG/language/en/seismology-seismic-hazard-and-disaster-risk-sector/multy-hazard).

Finally, the ultimate goal of every seismic hazard assessment is to determine ground shaking at a site, as a basis for implementing adequate earthquake mitigation and protection measures. Proper interpretation of the hazard estimates requires the application of professional judgment and expertise. It is highly recommended that the engineering community and the decision-makers use the most updated, robust, technically and scientifically sound version of the seismic hazard model, endorsed by local and international experts in the field of seismology, geology, geophysics and engineering.

8.9 Data and Resources

The local catalogue is taken from the internal bulletin of M. Nodia Institute of geophysics of Tbilisi State University; after 2014, earthquake data for events with $M_L \geq 3$ are taken from the official webpage of the Institute of Earth Sciences and National Seismic monitoring center of ILIAUNI (https://ies.iliauni.edu.ge/?page_id=183&lang=en). The earthquake catalogue for the eastern part of Turkey, near the border of Georgia, was provided by Kandilli Observatory and Earthquake Research Institute. For Iran, the earthquake catalogue was provided by International Institute of Earthquake Engineering and Seismology (IIEES). The earthquake catalogue for Azerbaijan was provided by the Institute of Geology of Azerbaijan. The catalogue for the northern Caucasus, after 2008, was updated by IG TSU using international data:

IRIS (http://www.iris.washington.edu/ieb/index.html);
GSRAS (http://www.ceme.gsras.ru/eng/index.htm);
NEIC USGS (http://earthquake.usgs.gov/earthquakes/search/);
OBN (http://www.wdcb.ru/sep/seismology/cat_OBN.html);
EMSC (http://www.emsc-csem.org/#2).

The regional catalogue used in this study is the EMME14 (Zare et al. 2014).

The hazard computation has been performed by using OpenQuake, developed by the Global Earthquake Model (GEM) for seismic hazard and risk analysis (https://github.com/gem/oq-engine).

Acknowledgements We would like to express our special thanks to Dr. Marco Pagani for his support, to the Editors of this Special Issue as well as the reviewers for their insightful comments, which significantly helped us to improve the quality of our work.

References

Abrahamson NA, Silva WJ, Kamai R (2014) Summary of the ASK14 ground motion relation for active crustal regions. Earthquake Spectra 30(3):1025–1055

Abrahamson NA, Nicholas G, Kofi A (2016) BC hydro ground motion prediction equations for subduction earthquakes. Earthquake Spectra 32(1):23–44

Adamia S, Chkhotua T, Kekelia M et al (1981) Tectonics of the Caucasus and adjoining regions: implications for the evolution of the Tethys Ocean. J Struct Geol 3(4):437–447. https://doi.org/10.1016/0191-8141(81)90043-2

Adamia S, Mumladze T, Sadradze N et al (2008) Late Cenozoic tectonics and geodynamics of Georgia (SW Caucasus). Georgian Int J Sci Tech (Nova Science Publisher) 1(1):77–107

Adamia S, Alania V, Chabukiani A et al (2010) Evolution of the late Cenozoic basins of Georgia (SW Caucasus): a review. In: Sosson M, Kaymakçı N, Stephenson R, Bergerat F (eds) Sedimentary basin tectonics from the Black Sea and Caucasus to the Arabian platform, Special Publication, vol 340. Geological Society, London, pp 239–259

Adamia S, Zakariadze G, Chkhotua T et al (2011) Geology of the Caucasus: a review. Turk J Earth Sci 20:489–544

Adamia S, Alania V, Tsereteli N et al (2017) Post-collisional tectonics and seismicity of Georgia. In: tectonic evolution, collision, and seismicity of Southwest Asia: in honor of manuel Berberian's forty-five years of research contributions. Geol Soc Am Spec Pap 525:535–572

Adams J, Trevor A, Stephen H et al (2019) Canada's 6th generation seismic hazard model, as prepared for the 2020 national building code of Canada. Paper presented at the 12th Canadian Conference on Earthquake Engineering, Quebec City, 17–20 June 2019. http://earthquakescanada.nrcan.gc.ca/hazard-alea/recpubs-en.php

Akkar S, Çağnan Z (2010) A local ground-motion predictive model for Turkey, and its comparison with other regional and global ground-motion. Bull Seismol Soc Am 100:2978–2995

Akkar S, Sandikkaya MA, Bommer JJ (2014) Empirical ground-motion models for point- and extended-source crustal earthquake scenarios in Europe and the Middle East. Bull Earthq Eng 12(1):359–387

Allen M, Jackson J, Walker R (2004) Late Cenozoic reorganization of the Arabia-Eurasia collision and the comparison of short-term and long-term deformation rates. Tectonics 23:1–16

Allen T, Griffin J, Leonard M et al (2018) The 2018 national seismic hazard assessment for Australia: model overview. Geoscience Australia record 2018/27. https://d28rz98at9flks.cloudfront.net/123020/Rec2018_027.pdf

Anderson JG, Luco JE (1983) Consequences of slip rate constraints on earthquake occurrence relations. Bull Seismol Soc Am 73:471–496

Apperson KD (1991) Stress fields of the overriding plate at convergent margins and beneath active volcanic arcs. Science 254(5032):670–678

Atkinson GM, Bommer JJ, Abrahamson NA (2014) Alternative approaches to modeling epistemic uncertainty in ground motions in probabilistic seismic-hazard analysis. Seismol Res Lett 85(6):1141–1144

Balassanian S, Ashirov T, Chelidze T et al (1999) Seismic hazard assessment for the Caucasus test area. Ann di Geofis 42(6):1139–1164

Banks C, Robinson A, Williams M (1997) Structure and regional tectonics of the Achara-Trialeti fold belt and the adjacent Rioni and Kartli foreland basins, Republic of Georgia. In: Robinson AG (ed) Regional and petroleum geology of the Black Sea and surrounding region, vol 68. American Association of Petroleum Geologists Memoir, Tulsa, pp 331–336

Barth A, Wenzel F, Giardini D (2007) Frequency sensitive moment tensor inversion for light to moderate magnitude earthquakes in eastern Africa. Geophys Res Lett 34:L15302. https://doi.org/10.1029/2007GL030359

Barth A, Wenzel F (2010) New constraints on the intraplate stress field of the Amurianplate deduced from light earthquake focal mechanisms. Tectonophysics 482:160–169. https://doi.org/10.1016/j.tecto.2009.01.029

Bazzurro P, Cornell CA (1999) Deaggregation of seismic hazard. Bull Seismol Soc Am 89:501–520

Beauval C, Bard PY, Hainzl S et al (2008) Can strong-motion observations be used to constrain probabilistic seismic-hazard estimates? Bull Seismol Soc Am 98:509–520. https://doi.org/10.1785/0120070006

Bindi D, Massa M, Luzi L et al (2014) Pan-European ground-motion prediction equations for the average horizontal component of PGA, PGV and 5%-damped PSA at spectral periods up to 30 s using the RESORCE dataset. Bull Earthq Eng 12:391–430

Bolashvili N, Dittmann A, King L et al (eds) (2018) National Atlas of Georgia. Ivane Javakhishvili Tbilisi State University, Vakhushti Bagrationi Institute of Geography, Justus Liebig University of Giesses, Insitute for Geography

Bommer JJ, Scherbaum F (2008) The use and misuse of logic trees in probabilistic seismic hazard analysis. Earthquake Spectra 24:997–1009. https://doi.org/10.1193/1.2977755

Boore DM, Stewart JP, Seyhan E et al (2014) NGA-West2 equations for predicting PGA, PGV, and 5% damped PSA for shallow crustal earthquakes. Earthquake Spectra 30(3):1057–1085

Bozorgnia Y, Abrahamson NA, Atik LA et al (2014) NGA-West2 research project. Earthquake Spectra 30(3):973–987

Bradley BA (2013) A summary of the observed ground motions from the 21 July 2013 Mw 6.6 earthquake in Cook Strait, off the coast of Seddon. https://sites.google.com/site/brendonabradley/

Bune VI, Medvedev SV, Risnichenko YV et al (1974) Successes and hopes seismic zoning of the USSR. Izv Acad Sci USSR Phys Solid Earth 10:95–102

Campbell KW, Bozorgnia Y (2014) NGA-West2 ground motion model for the average horizontal components of PGA, PGV, and 5% damped linear acceleration response spectra. Earthquake Spectra 30(3):1087–1115

Cauzzi C, Faccioli E, Vanini M et al (2014) Updated predictive equations for broadband (0.01-10 s) horizontal response spectra and peak ground motions, based on a global dataset of digital acceleration records. Bull Earthq Eng 13(6):1587–1612

Chelidze T, Javakhishvili Z, Varazanashvili O et al (2012) Seismic hazard assessment of Georgia (probabilistic approach). Institute of Geophysics, Academy of Sciences of Georgia

Chiou BSJ, Youngs RR (2008) An NGA model for the average horizontal component of peak ground motion and response spectra. Earthquake Spectra 24:173–215

Chiou BSJ, Youngs RR (2014) Update of the Chiou and Youngs NGA model for the average horizontal component of peak ground motion and response spectra. Earthquake Spectra 30(3):1117–1153

Cornell CA (1968) Engineering seismic risk analysis. Bull Seismol Soc Am 58:1583–1606
Danciu L, Giardini D (2015) Global seismic hazard assessment program-GSHAP legacy. Ann Geophys 58:1. https://doi.org/10.4401/ag-6734
Danciu L, Şeşetyan K, Demircioglu M et al (2018a) The 2014 earthquake model of the Middle East: seismogenic sources. Bull Earthq Eng 16(8):3465–3496
Danciu L, Kale Ö, Akkar S (2018b) The 2014 earthquake model of the Middle East: ground motion model and uncertainties. Bull Earthq Eng 16(8):3497–3533. https://doi.org/10.1007/s10518-016-9989-1
Delavaud E, Cotton F, Akkar S et al (2012) Toward a ground-motion logic tree for probabilistic seismic hazard assessment in Europe. J Seismol 16:451–473
DeMets C, Gordon R, Argus D et al (1990) Current plate motions. Geophys J Int 101:425–478. https://doi.org/10.1111/j.1365-246X.1990.tb06579.x
Demets C, Gordon R, Argus D et al (1994) Effect of recent revisions to the geomagnetic reversal time scale on estimates of current plate motions. Geophys Res Lett 21(20):2191–2194. https://doi.org/10.1029/94GL02118
Douglas J (2018) Calibrating the backbone approach for the development of earthquake ground motion models. Paper presented at Best Practice in Physics-based Fault Rupture Models for Seismic Hazard Assessment of Nuclear Installations: Issues and Challenges Towards Full Seismic Risk Analysis, Cadarache, France, 14–17 May 2018
Douglas J, Edwards B (2016) Recent and future developments in earthquake ground motion estimation. Earth Sci Rev 160:203–219
Eroğlu Azak T, Kalafat D, Şeşetyan K et al (2018) Effects of seismic declustering on seismic hazard assessment: a sensitivity study using the Turkish earthquake catalogue. Bull Earthq Eng 16(8):3339–3366. https://doi.org/10.1007/s10518-017-0174-y
Faccioli E, Bianchini A, Villani M (2010) New ground motion prediction equations for t>1 s and their influence on seismic hazard assessment. In: Proceedings of the University of Tokyo Symposium on Long- Period Ground Motion and Urban Disaster Mitigation, Tokyo, Japan, 17-18 March 2010
Frankel A (1995) Mapping seismic hazard in the central and eastern United States. Seismol Res Lett 66:8
Gardner JK, Knopoff L (1974) Is the sequence of earthquakes in southern California, with aftershocks removed, Poissonian? Bull Seismol Soc Am 64(5):1363–1367
Giardini D (1999) The global seismic Hazard assessment program (GSHAP) -1992/1999. Ann di Geofis 42(6):957–974
Giardini D, Danciu L, Erdik M et al (2018) Seismic hazard map of the Middle East. Bull Earthq Eng 16(8):3567–3570. https://doi.org/10.1007/s10518-018-0347-3
Gorshkov GP (1938) On the seismic zoning of Middle Asia. Proc Inst Seismol AS USSR 79(6):67–71
Gorshkov GP (1947) Scheme of seismic zoning of the USSR. In: Jubilee Collection, part I. AS USSR, p 454
Gotsadze O, Varazanashvili O, Jibladze E (1996) Engineering analysis of the implications of the 1991 Racha earthquake in Georgia. Metsniereba Publ House, Tbilisi
Grunthal G (1985) The updated earthquake catalogue for the German Democratic Republic and adjacent areas statistical data characteristics and conclusions for hazard assessment. In: Proceedings of the 3rd international symposium on the analysis of seismicity and on seismic risk, Prague, 17–22 June 1985, vol I
Grunthal G, Stromeyer D, Bosse C et al (2018) The probabilistic seismic hazard assessment of Germany-version 2016, considering the range of epistemic uncertainties and aleatory variability. Bull Earthq Eng 16(10):4339–4395
Gubin IE (1950) Seismotectonic method of seismic zoning. Proc Geophys Inst AS USSR 13:140
Gülen L, Sesetyan K, Adamia S et al (2014) Earthquake model of the Middle East (EMME) project: active faults and seismic sources. In: Abstract of the second European Conference on Earthquake Engineering and Seismology (2ECEES), Istanbul, Turkey, 24–29 August 2014

Gusev AA (1991) About the need to adjust the documents regulating anti-seismic measures in the USSR. Integrated assessment of seismic hazard. Probl Earthq Eng 32:147–160

Gusev AA, Shumilina LS (1995) Some problems concerning the methods of general seismic zoning. In: Seismicity and seismic zoning of North Eurasia, Moscow, vol 2–3. OIFZ RAN, pp 289–300

Gutenberg B, Richter CF (1944) Frequency of earthquakes in California. Bull Seismol Soc Am 34:184–188

Idriss IM (2014) An NGA-West2 empirical model for estimating the horizontal spectral values generated by shallow crustal earthquakes. Earthquake Spectra 30(3):1155–1177

Jackson J, Ambraseys N (1997) Convergence between Eurasia and Arabia in eastern Turkey and the Caucasus. In: Giardini D, Balassanian S (eds) Historical and Prehistorical earthquakes in the Caucasus: North Atlantic Treaty Organization (NATO) Asian series 2, environment, vol 28. Kluwer Academic Publishers, Dordrecht, pp 79–90

Jibladze E, Butikashvili N, Tsereteli N (1995) Seismic regime, seismic hazard and seismotectonic deformation in the Caucasus. Institute of Geophysics, Georgian Academy of Sciences Press, Tbilisi

Johnston AC, Coppersmith KJ, Kanter LR et al (1994) The earthquakes of stable continental regions, vol 1: assessment of large earthquake potential, Technical report TR-102261-V1. Electric Power Research Institute, pp 1–368

Kadirioğlu FT, Kartal RF, Kılıç T et al (2018) An improved earthquake catalogue (M≥4.0) for Turkey and its near vicinity (1900–2012). Bull Earthq Eng 16:3317–3338

Kadirov F, Mammadov S, Reilinger R et al (2008) Some new data on modern tectonic deformation and active faulting in Azerbaijan (according to Global Positioning System measurements). Proc Azerbaijan Nat Acad Sci Earth Sci 1:82–88

Kadirov F, Floyd M, Alizadeh A et al (2012) Kinematics of the Caucasus near Baku, Azerbaijan. J Nat Hazards 63:997–1006. https://doi.org/10.1007/s11069-012-0199-0

Kijko A, Singh M (2011) Statistical tools for maximum possible earthquake magnitude estimation. Acta Geophys 59(4):674–700. https://doi.org/10.2478/s11600-011-0012-6

Kotha SR, Bindi D, Cotton F (2016) Partially non-ergodic region specific GMPE for Europe and Middle-East. Bull Earthq Eng 4(4):1245–1263

Leonard M (2010) Earthquake fault scaling: self-consistent relating of rupture length, width, average displacement, and moment release. Bull Seismol Soc Am 100:1971–1988. https://doi.org/10.1785/0120090189

Lin PS, Lee CT (2008) Ground-motion attenuation relationships for subduction-zone earthquakes in northeastern Taiwan. Bull Seismol Soc Am 98(1):220–240. https://doi.org/10.1785/0120060002

Lolli B, Gasperini P, Vannucci G (2014) Empirical conversion between teleseismic magnitudes (mb and Ms) and moment magnitude (Mw) at the Global, Euro-Mediterranean and Italian scale. Geophys J Int 199:805–828

Luen B, Stark PB (2012) Poisson tests of declustered catalogs. Geophys J Int 189(1):691–700. https://doi.org/10.1111/j.1365-246X.2012.05400.x

McClusky S, Balassanian S, Barka A et al (2000) Global positioning system constraints on plate kinematics and dynamics in the Eastern Mediterranean and Caucasus. J Geophys Res 105(3):5695–5719. https://doi.org/10.1029/1999JB900351

McGuire RK (1995) Computations of seismic Hazard. Ann Geofis 36:181–200

Medvedev SV (1947) On the problem of accounting of seismic activity of area at construction. In: Proceedings of Inst Seismol AS USSR, 119

Medvedev SV (1958) Seismic zoning map of the USSR territory 1957. Proc Inst Phys Solid Earth Acad Sci USSR 1(158):3–28

Meletti C, Marzocchi W, D'Amico V et al (2019) MPS19: the updated Italian Seismic Hazard model. In: Abstracts of the 2019 EGU General Assembly, Vienna, Austria, 7–12 April 2019. https://meetingorganizer.copernicus.org/EGU2019/EGU2019-7372.pdf

Michael AJ (1984) Determination of stress from slip data: faults and folds. J Geophys Res 89:11517–11526

Mignan A, Danciu L, Giardini D (2015) Reassessment of the maximum fault rupture length of strike-slip earthquakes and inference on M_{max} in the Anatolian Peninsula, Turkey. Seismol Res Lett 86(3):890–900

Mokrushina NG, Shebalin NV (1982) Assessment of the quality of seats strong shakings on seismic zoning map of the USSR. Probl Earthq Eng 23:97–113

Mokrushina NG, Shebalin NV (1991) Assessment of the quality of seats strong shakings on seismic zoning map of the USSR. Part II. Map GSZ-78. Integrated assessment of seismic hazard. Probl Earthq Eng 31:122–125

Monelli D, Pagani M, Weatherill G et al (2014) Modeling distributed seismicity for probabilistic seismic-hazard analysis: implementation and insights with the OpenQuake engine. Bull Seismol Soc Am 104(4):1636–1649. https://doi.org/10.1785/0120130309

Montalva GA, Bastías N, Rodriguez-Marek A (2017) Ground-motion prediction equation for the Chilean subduction zone. Bull Seismol Soc Am 107(2):901–911. https://doi.org/10.1785/0120160221

Mosar J, Kangarli T, Bochud M et al (2010) Cenozoic–recent tectonics and uplift in the Greater Caucasus: a perspective from Azerbaijan. In: Sosson M, Kaymakçı N, Stephenson R, Bergerat F (eds) Sedimentary basin tectonics from the Black Sea and Caucasus to the Arabian Platform, Special Publication, vol 340. Geological Society, London, pp 261–279

Mumladze T, Forte AM, Cowgill ES et al (2015) Subducted, detached, and torn slabs beneath the Greater Caucasus. Geo Res J 5:36–46

Musson RMW, Winter PW (1996) Seismic hazard maps for the U.K. Nat Hazards 14(2):141–154. https://doi.org/10.1007/BF00128262

Onur T, Gok R, Godoladze T et al (2019) Probabilistic seismic hazard assessment for Georgia. Technical report LLNL-TR-771451 963125. https://www.osti.gov/biblio/1511856-probabilistic-seismic-hazard-assessment-georgia

Pagani M, Monelli D, Weatherill G et al (2014) OpenQuake-engine: an openhazard (and risk) software for the Global Earthquake Model. Seismol Res Lett 85:692–702. https://doi.org/10.1785/0220130087

Pagani M, Garcia-Peleaz J, Gee R, Johnson K, Poggi V, Silva V, Simionato M, Styron R, Vigano D, Danciu L, Monelli D, Weatherill G (2020) The 2018 version of the global earthquake model: Hazard component. Earthquake Spectra. https://doi.org/10.1177/8755293020931866

Pagani M, Garcia-Pelaez J, Gee R et al (2018) Global Earthquake Model (GEM) Seismic Hazard Map (version 2018.1 – December 2018). https://doi.org/10.13117/GEM-GLOBAL-SEISMIC-HAZARD-MAP-2018.1

Pasquaré F, Tormey D, Vezzoli L et al (2011) Mitigating the consequences of extreme events on strategic facilities: evaluation of volcanic and seismic risk affecting the Caspian oil and gas pipelines in the Republic of Georgia. J Env Manag 92:1774–1782

Petersen MD, Moschetti MP, Powers PM et al (2015) The 2014 United States national seismic hazard model. Earthquake Spectra 31:S1):S1–S1)S30. https://doi.org/10.1193/120814EQS210M

Rautian T (1964) About the determination of the earthquake energy in the distance to 3000 km. In: Riznichenko Y (ed) Experimental seismicity. Academy of Sciences of USSR, Moscow, pp 88–89

Rautian T, Khalturin VI (1978) The use of coda for determination of the earthquake source spectrum. Bull Seismol Soc Am:68904–68922

Reasenberg P (1985) Second-order moment of Central California seismicity, 1969–1982. J Geophys Res 90(7):5479–5495

Reilinger R, McClusky S, Vernant P et al (2006) GPS constraints on continental deformation in the Africa-Arabia-Eurasia continental collision zone and implications for the dynamics of plate interactions. J Geophys Res 111(B5):B05411. https://doi.org/10.1029/2005JB004051

Risnichenko YV (1979) Seismic shaking of the USSR territory. Nauka Publ House, Moscow

Riznichenko YV (1965) From the activity of earthquake sources to the shakeability of earth's surface. Izv Acad Sci USSR Phys Solid Earth 11:1–12

Sargsyan GV, Abgaryan GR, Mkhitaryan KA et al (2017) Seismic conditions of the territory of Armenia and adjacent areas after the 1988 Spitak earthquake. Gyumri, Pub House of Eldorado Seismic Zoning of the USSR (1968). Moscow, Nauka Publ House

Seismicheskoe raionirovanie territorii SSSR (1980) Metodicheskieosnovyi regionalnoe opisanie karty 1978 goda (seismic zoning of the USSR territory: underlying methods and regional legend of the map of 1978). Nauka Publ House, Moscow

Sesetyan K, Danciu L, Demircioglu MB et al (2018a) The 2014 Seismic hazard model of the Middle East: overview and results. Bull Earthq Eng 16(8):3535–3566. https://doi.org/10.1007/s10518-018-0346-4

Sesetyan K, Demircioğlu MB, Duman TY et al (2018b) A probabilistic seismic hazard assessment for the Turkish territory-part I: the area source model. Bull Earthq Eng 16(8):3367–3397. https://doi.org/10.1007/s10518-016-0005-6

Shebalin NV, Trifonov VG, Kozhurin AI et al (2000) A unified seismotectonic zonation of Northern Eurasia. J Earthq Predict Res 8:8–31

Slejko D, Javakhishvili Z, Rebez A et al (2008) Seismic hazard assessment for the Tbilisi test area (eastern Georgia). Istituto Nazionale di Oceanografia e di Geofisica Sperimentale, Trieste (Italy) 49:37–58

Sokhadze G, Floyd M, Godoladze T et al (2018) Active convergence between the Lesser and Greater Caucasus in Georgia: constraints on the tectonic evolution of the Lesser–Greater Caucasus continental collision. Earth Planet Sci Lett 481:154–161. https://doi.org/10.1016/j.epsl.2017.10.007

Sosson M, Rolland Y, Danelian T et al (2010) Subductions, obduction and collision in the Lesser Caucasus (Armenia Azerbaijan, Georgia), new insights. In: Sosson M, Kaymakci N, Stephanson R, Bergarat F, Storatchenoko V (eds) Sedimentary basin tectonics from the Black Sea and Caucasus to the Arabian Platform, Special Publication, vol 340. Geological Society of London, London, pp 329–352

Stepp JC (1972) Analysis of the completeness of the earthquake sample in the Puget Sound area and its effect on statistical estimates of earthquake hazard. In: Proceedings of the international Conf on microzonation for safer construction: research and application, Seattle, Washington, 30 October-3 November 1972, vol 64, pp 1189–1207

Storchak DA, Di Giacomo D, Bondár I et al (2015) ISC-GEM global instrumental earthquake catalogue (1900–2009). Phys Earth Planet Inter 239:48–63. https://doi.org/10.1016/j.pepi.2014.06.009

Tan O, Taymaz T (2006) Active tectonics of the Caucasus: earthquake source mechanisms and rupture histories obtained from inversion of teleseismic body waveforms. In: Dilek Y, Pavlides S (eds) Postcollisional tectonics and magmatism in the Mediterranean region and Asia, Special Paper, vol 409. Geological Society of America, Boulder, pp 531–578

Tibaldi A, Alania V, Bonali FL et al (2017a) Active inversion tectonics, simple shear folding and back-thrusting at Rioni Basin, Georgia. J Struct Geol 96:35–53

Tibaldi A, Russo E, Bonali FL et al (2017b) 3-D anatomy of an active fault-propagation fold: a multidisciplinary case study from Tsaishi, western Caucasus (Georgia). Tectonophysics 717:253–269

Tibaldi A, Bonali FL, Russo E et al (2018) Structural development and stress evolution of an arcuate fold-and-thrust system, southwestern Greater Caucasus, Republic of Georgia. J Asian Earth Sci 156:226–245

Tibaldi A, Tsereteli N, Varazanashvili O et al (2020) Active stress field and fault kinematics of the Greater Caucasus. J Asian Earth Sci 188:104108. https://doi.org/10.1016/j.jseaes.2019.104108

Triep EG, Abers GA, Lerner-Lam AL et al (1995) Active thrust front of the Greater Caucasus: the April 29, 1991, Racha earthquake sequence and its tectonic implications. J Geophys Res 100(B3):4011–4033. https://doi.org/10.1029/94JB02597

Tsereteli N, Varazanashvili O, Arabidze V et al (2011) Seismic risk assessment for Tbilisi – new approaches. In: Avagyan A, Barry DL, Coldewey WG, Reimer DWG (eds) Stimulus for human and societal dynamics in the prevention of catastrophes, vol 80. IOS Press, Amsterdam, pp 109–130

Tsereteli N, Tanircan G, Safak E et al (2012) Seismic hazard assessment for southern Caucasus – Eastern Turkey energy corridors: the example of Georgia. In: Barry DL, Coldewey WG, Reimer DWG, Rudakov DV (eds) Correlation between human factors and the prevention of disasters, vol 94. IOS Press, Amsterdam, pp 96–111

Tsereteli N, Arabidze V, Varazanashvili O et al (2014) Vulnerability analysis and GIS based seismic risk assessment Georgia case. Improving disaster resilience and mitigation-IT Meansand tools. In: NATO science for peace and security, series C, environmental security, vol 143. Springer, Dordrecht, pp 307–321. http://link.springer.com/chapter/10.1007/978-94-017-9136-6_20#page-1

Tsereteli N, Tibaldi A, Alania V et al (2016a) Active tectonics of central-western Caucasus, Georgia. Tectonophysics 691:328–344. https://doi.org/10.1016/j.tecto.2016.10.025

Tsereteli N, Askan A, Hamzehloo H (2016b) Hybrid-empirical ground motion estimations for Georgia. Acta Geophys 64(5):1225–1256. https://doi.org/10.1515/acgeo-2016-0048

Tsipenyuk IF (1988) Damageability and reliability of large-panel buildings under seismic actions. Investigations on seismic hazard. Probl Earthq Eng 29:141–153

Uhrhammer R (1986) Characteristics of Northern and Central California Seismicity. Earthq Notes 57(1):21

Ulomov VI, Shumilina LS (1999) Komplekt kart obshchego seismicheskogo raionirovaniya territorii Rossiiskoi Federatsii OSR-97. Masshtab 1: 8000000. Obyasnitelnaya zapiska i spisok gorodovi naselennykh punktov, raspolozhennykh v seismo opasnykhraionakh (A Set of 1:8000000 General Seismic Zoning Maps of the Russian Federation (GSZ-97): Explanatory Note and the List of Towns and Settlements Located in Regions of Seismic Hazard), Moscow, OIFZ

Varazanashvili O (1999) Seismic hazard assessment of Georgia by deterministic and probabilistic methods. J Georgian Geophys Soc (A) 4:35–45

Varazanashvili O, Tsereteli N, Amiranashvili A et al (2012) Vulnerability, hazards and multiple risk assessment for Georgia. J Nat Hazards 64(3):2021–2056. https://doi.org/10.1007/s11069-012-0374-3

Varazanashvili O, Tsereteli N, Bonali FL et al (2018) GeoInt: the first macroseismic intensity database for the Republic of Georgia. J Seismol 22(3):625–667

Weichert DH (1980) Estimation of the earthquake recurrence parameters for unequal observation periods for different magnitudes. Bull Seismol Soc Am 70(4):1337–1346

Wells DL, Coppersmith KJ (1994) New empirical relationships among magnitude, rupture length, rupture width, rupture area, and surface displacement. Bull Seismol Soc Am 84:974–1002

Westen CJ, Straatsma M, Turdukulov U et al (2012) Atlas of natural hazards and risks of Georgia: e-book. Caucasus Environmental NGO Network (CENN), Tbilisi

Wheeler RL (2009) Methods of Mmax estimation east of the Rocky mountains. Open-File Report 2009-1018, prepared with funding from the U.S. Nuclear Regulatory Commission. https://doi.org/10.3133/ofr20091018

Wiemer S (2001) A software package to analyse seismicity: ZMAP. Seismol Res Lett 72(3):373–382

Wiemer S, Danciu L, Edwards B et al (2016) Seismic Hazard Model 2015 for Switzerland (SUIhaz2015). Report, Swiss Seismological Service (SED) at ETH Zurich. https://doi.org/10.12686/a2

Woessner J, Danciu L, Giardini D et al (2015) The 2013 European seismic hazard model: key components and results. Bull Earthq Eng 13(12):3553–3596

Youngs RR, Chiou BSJ, Silva WJ et al (1997) Strong ground motion attenuation relationships for subduction zone earthquakes. Seismol Res Lett 68:58–73

Zare M, Amini H, Yazdi P et al (2014) Recent developments of the Middle East catalog. J Seismol 18(4):749–772. https://doi.org/10.1007/s10950-014-9444-1

Zhao JX, Zhang J, Asano A et al (2006) Attenuation relations of strong ground motion in Japan using site classifications based on predominant period. Bull Seismol Soc Am 96:898–913

Chapter 9
Non-ergodic Ground-Motion Models for Crustal Earthquakes in Georgia

Norman Abrahamson and Nicolas Kuehn

Abstract A sparse data set consisting of 33 local strong-motion recordings from earthquakes with magnitudes between 4.0 and 7.1 and distances between 8 and 190 km is used to develop an initial non-ergodic ground-motion model for the Georgia Republic for spectral acceleration at T = 0.2 s. The non-ergodic adjustments capture the local differences in the linear site effects and linear path effects. The non-ergodic GMMs have reduced aleatory variability along with shifts (increases or decreases) to the median ground motion for a given site/source pair. For sites in central Georgia, there is some local data to constrain the shift in the median ground motion from the base ergodic GMM. For sites far from the available local ground-motion data, such as the Enguri dam site, there is larger epistemic uncertainty in the shifts to median. In this case, the reduced aleatory variability will be offset by the larger epistemic uncertainty. Although there is large uncertainty, the advantage of using the non-ergodic approach in regions with sparse local ground-motion data is that the large epistemic uncertainty range is properly quantified and it shows the potential change to the hazard that can be expected as local ground-motion data are collected.

9.1 Introduction

Ground-motion models (GMMs) are an important ingredient of probabilistic seismic hazard analysis (PSHA). They model the conditional distribution of some

N. Abrahamson (✉)
University of California, Berkeley, CA, USA
e-mail: abrahamson@berkeley.edu

N. Kuehn
Civil and Environmental Engineering Department, University of California, Los Angeles, CA, USA
e-mail: kuehn@ucla.edu

© Springer Nature B.V. 2021

F. L. Bonali et al. (eds.), *Building Knowledge for Geohazard Assessment and Management in the Caucasus and other Orogenic Regions*, NATO Science for Peace and Security Series C: Environmental Security,
https://doi.org/10.1007/978-94-024-2046-3_9

ground-motion parameter of interest (e.g., peak ground acceleration (PGA) or response spectral ordinates (PSA)), given a source/site scenario (i.e., given magnitude, distance, site conditions, and other parameters). Typically, empirical GMMs are estimated by gathering observed strong-motion data from different events and at different stations, and then estimating a model that describes the average scaling of the ground-motion parameter of interest with the source, path, and site related parameters such as magnitude, distance, and V_{S30}. Traditionally, empirical GMMs have been based on the ergodic assumption (Anderson and Brune, 1999). The ergodic assumption relates to trading space for time: we assumes that the ground-motion distribution at a specific site over time (what we are interested in PSHA) is the same as the ground-motion distribution observed at different sites (what is easily observable). The ergodic assumption allows the aggregation of ground-motion data from different regions and different stations to estimate an average GMM for the combined regions and sites which is advantageous because the empirical ground-motion data for a specific site is sparse, particularly at large magnitudes and short distances.

While the ergodic assumption allows us to increase the size of the data set used to develop GMMs, the large increases in ground-motion data sets around the world over the last decade have shown just how poor the ergodic assumption is. Ground-motion studies using data from sites that have recorded multiple earthquakes have shown that 60–70% of the variance in an ergodic GMM is actually due to systematic source/path/site effects that are not applicable to all sites and all earthquakes. By ignoring these systematic effects, we group together ground-motion data that does not have the same median. As a consequence, the aleatory variability is too large because it includes both the aleatory variability in the ground motion for a single source/site pair and the differences in the median ground motion for different source/site pairs. Because the aleatory variability controls the shape of the hazard curve, this overestimation of the aleatory variability can have a large effect on the hazard for long return periods (Bommer and Abrahamson, 2006).

While the ergodic assumption is convenient for constraining the average magnitude and distance scaling in GMMs, the ergodic assumption is not required for PSHA. The ergodic assumption can be relaxed by taking into account the systematic source, path, and site effects. Relaxing the ergodic assumption comes in many flavors: accounting for differences in the site-specific effects by using single-station sigma (Rodriguez-Marek, 2013), regionalizing GMMs in which some of the GMM coefficients differ between broad region (e.g., Abrahamson et al. 2014; Boore et al. 2014; Campbell and Bozorgnia 2014; Chiou and Youngs 2014; Kotha et al. 2016), direct modeling of source/site-specific path effects for the anelastic attenuation term (e.g., Anderson and Uchiyama 2011; Dawood and Rodriguez-Marek 2013; Kuehn et al. 2019), or modeling the spatial variations of the source, path, and site terms within a broad region (e.g., Landwehr et al. 2016; Kuehn and Abrahamson 2020). Fully non-ergodic empirically-based GMMs are rare. To date, only the model of Landwehr et al. (2016) claims to be fully non-ergodic, but many more fully non-ergodic GMMs are expected over the next few years.

In general, GMMs that model non-ergodic effects (systematic source, path, and site effects) are developed in data-rich regions because we need the site-specific data to be able to estimate the systematic deviations from an ergodic base model. This has lead to the perception that non-ergodic PSHA can only be performed in regions with a lot of data. As a result, regions with sparse data continue to use ergodic GMMs. The problem with this approach is that lack of data does not imply that the ground motion in the region will be consistent with an ergodic GMM. Instead, lack of data implies large epistemic uncertainties in the GMM. Nonergodic GMMs should also be used in regions where data are sparse; however, the systematic effects will be associated with large epistemic uncertainties.

In this paper, we go through the basic concepts of non-ergodic GMMs, how they are developed, and what is needed for prediction of the systematic effects. Finally, we explain how they can be used in a data-sparse region such as the Georgia Republic.

9.2 Nonergodic Seismic Hazard Analysis

The goal of PSHA is to calculate the expected rate of exceedance of a ground-motion level A at a site of interest, for a ground-motion parameter of interest Y (e.g., PGA); we call this rate of exceedance $\nu(Y > A)$. To calculate $\nu(Y > A)$, we need a model for earthquake occurrence, which gives the joint occurrence probability of an event of a certain magnitude in a certain distance, $p(M, R)$; and a GMM, which describes the probability that $Y > A$ for a given earthquake scenario (given magnitude and distance), $p(Y > A)$. Together with a base rate for earthquake occurrence ν_{base}, one can then calculate the rate of exceedance by integrating over possible magnitudes and distances:

$$\nu(Y > A) = \nu_{base} \int_M \int_R p(m, r) p(Y > A|m, r)\, dr\, dm \tag{9.1}$$

The distribution of ground-motion is typically modeled as a lognormal distribution, where the median and standard deviation are functions of magnitude, distance, and other source and site related parameters:

$$\mu = f(M, R, \ldots; \boldsymbol{\theta}) \tag{9.2}$$

$$\sigma = g(M, R, \ldots; \boldsymbol{\beta}) \tag{9.3}$$

where $\boldsymbol{\theta}$ and $\boldsymbol{\beta}$ are coefficients of the functions f and g, respectively. In Equation (9.2), the median prediction is ergodic, because it does not depend on the location of the source or the site. Hence, for the same magnitude, distance, and site condition, it will predict the same median ground motion for all site/source pairs, ignoring local, systematic differences in the source, path, and site effects. To go from an ergodic to a non-ergodic ground-motion model, the median prediction

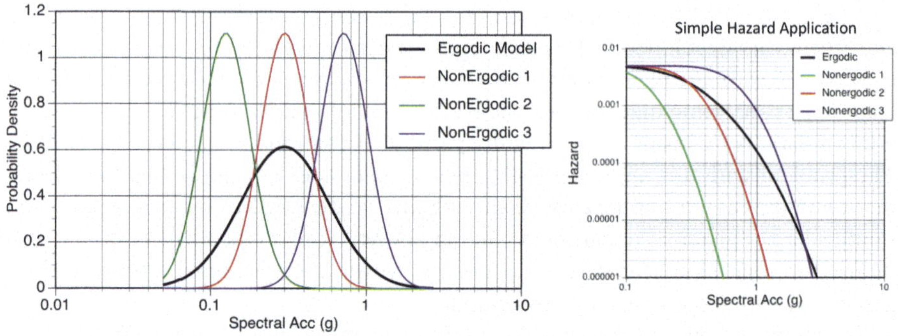

Fig. 9.1 Principle of non-ergodic PSHA. On the left, the density of an ergodic model is compared to the density of three non-ergodic models, which have a smaller standard deviation, but different medians. On the right, the hazard curves are shown for a simplified scenario–the non-ergodic models lead to steeper hazard curves, but a wide spread

needs to also become a function of source and site location. This can be done in different ways: one can make the coefficients $\boldsymbol{\theta}$ location-dependent (Landwehr et al., 2016), or one can take the ergodic prediction, and add location-dependent adjustments. Both of these models can be modeled mathematically in the same way. We will describe non-ergodic models using an ergodic base model plus adjustment terms. For the adjustment to an ergodic base model approach, the median prediction becomes

$$\mu = f_{erg}(M, R, \ldots) + \delta L(\mathbf{x}_{source}) + \delta S(\mathbf{x}_{site}) + \delta P(\mathbf{x}_{source}, \mathbf{x}_{site}) \tag{9.4}$$

where δL is a systematic source effect, δS is a systematic site effect, and δP is a systematic path effect; $\mathbf{x}_{source}$ and $\mathbf{x}_{site}$ are the source and site coordinates, respectively. Because we are now incorporating previously unmodelled effects into the median prediction, the value of the aleatory standard deviation decreases. For an ergodic model, the systematic source, path, and site effects are mapped into the aleatory variability.

The systematic effects in Equation (9.4) can be estimated from data. If data is sparse, there are still systematic effects, but they are very uncertain. This epistemic uncertainty in the systematic effects needs to be incorporated into PSHA calculations to reflect the range of possible adjustment factors for each source/site pair. If the epistemic uncertainty is modeled properly, then non-ergodic GMM can be used in regions with sparse data.

Figure 9.1 shows the principle of non-ergodic PSHA. The ground-motion distribution for a single source-site scenario is plotted for an ergodic model and three possible adjustment terms. If we do not know the value of the adjustment terms, then they can be either positive or negative, and we have to take that uncertainty into account in the hazard calculation. This is typically done by putting different values of the adjustment terms on different branches of a logic tree. If there is no local data to constraint the adjustment terms, then the mean hazard from the different non-ergodic branches will be the same as the ergodic mean hazard, but there are much wider fractiles. Once we collect more data and can better constrain the estimate of the adjustment terms, the mean non-ergodic hazard will become different from the ergodic hazard and the non-ergodic hazard fractiles will decrease. This effect is shown for three sites in California sites with different number of data to constrain the non-ergodic adjustment terms in Abrahamson et al. (2019).

9.2.1 Non-ergodic GMMs

In Equation (9.4), we described how a nonergodic GMM can be derived as a location-dependent adjustment to an ergodic GMM. The non-ergodic adjustment terms are specific for each source and/or site location. Given all of the possible source and site locations, there is a large number of non-ergodic terms to estimate. For example, using 5×5 km spacing in California, there would be about 20,000 possible site and source locations. If there are five non-ergodic terms, this leads to about 100,000 adjustment coefficients. Rather than directly estimating the individual adjustment terms, we can instead use a Gaussian Process (GP) model (Rasmussen and Williams, 2006) in which the likelihood only depends on the mean, variance and correlation lengths of the coefficients, which reduces the number of terms that are estimated from the data to 10–15 terms depending if the mean is constrained to zero or not. To use the GP method, we make the assumption that the adjustment terms are a function of source and/or site location. Because we do not know the functional form of the spatial dependency, we model it as a non-parametric, flexible function. In Bayesian non-parametrics, GP is used as a prior for an unknown function. Thus, we model the dependency of the adjustment terms on location as a GP, which can be written as follows for the source term:

$$\delta L(\mathbf{x}_{source}) \sim GP(m(\mathbf{x}_{source}), k(\mathbf{x}_{source}, \mathbf{x}'_{source})) \tag{9.5}$$

where $m(\mathbf{x}_{source})$ is a mean function, and $k(\mathbf{x}_{source}, \mathbf{x}'_{source})$ is a covariance function. Together, the mean and covariance function fully specify the GP. Since δL is an adjustment term, we set $m(\mathbf{x}) = 0$.

Modelling the adjustment terms as a function with a GP prior is mathematically equivalent to treating them as spatially correlated. Thus, we implicitly assume that systematic effects for sources/sites that are close to each other are similar. The amount of similarity is determined by the covariance function, which is typically modeled by an exponential covariance function

$$k(\mathbf{x}, \mathbf{x}') = \theta^2 \exp \frac{-|\mathbf{x} - \mathbf{x}'|}{\rho} \tag{9.6}$$

where ρ is a length-scale which describes how fast the systematic effects can change over space, and θ^2 is a variance that determines how much the systematic effect can vary.

Since the GP basically specifies a multivariate normal distribution for the systematic effects, one can easily calculate the distribution of a systematic effect at a new location, conditional on the observed data. The mean μ_* and epistemic standard deviation ψ_* of this distribution can be calculated from (see also Equation (10) and (11) in Landwehr et al. 2016):

$$\mu_* = \mathbf{k}^T(\mathbf{K} + \sigma^2 I)^{-1}\mathbf{y}$$
$$\psi_* = k_* - \mathbf{k}^T(\mathbf{K} + \sigma^2 I)^{-1}\mathbf{k} \tag{9.7}$$

where $\mathbf{K}$ is the covariance matrix of the data, as specified by the GP; $\mathbf{k}$ is a vector describing the covariance between the data and the new location; and σ^2 is the aleatory standard variance. For details, see Landwehr et al. (2016) and Rasmussen and Williams (2006).

In Equation (9.5), we used the source effects as an example to define the GP, but the model can also be used for site and path effects, using the appropriate locations as input. For path effects $\delta P(\mathbf{x}_{source}, \mathbf{x}_{site})$, there is an additional complexity in that the path effects depend on both source and site coordinates. Theoretically, this can be included in a GP by calculating the absolute distance $|\mathbf{x} - \mathbf{x}'|$ in Equation (9.6) for a four-dimensional vector. Alternatively, the model can be simplified by using the midpoint between source and site coordinates as the coordinate for path effects. In Landwehr et al. (2016), the correlation for path effects was modeled using the source coordinate. All of these approaches are isotropic models for the path effects. Non-isotropic path effects can also be included in the non-ergodic GMM. For example, in Dawood and Rodriguez-Marek (2013) and Kuehn et al. (2019), the path effects are modeled as the sum of anelastic attenuation effects over different cells, which allows the anelastic attenuation to vary across space and the path effects reflect the specific ray path from the source to the site.

9.3 Non-ergodic GMMs for California

Due to the large number of strong-motion data in California, it is possible to develop non-ergodic GMMs that show significant differences in the adjustment terms for different locations. In Landwehr et al. (2016), a non-ergodic GMM was developed, where adjustment terms were modeled in a GP framework similar to Equation (9.5), but applied to the coefficients of the model. This is called a varying coefficient model (VCM). Their model can be written as

$$\mu = f_{base}(M, R, V_{S30}, FM)$$
$$+ \delta\theta_{-1}(\mathbf{x}_{source}) + \delta\theta_{gs}(\mathbf{x}_{source}) \ln\sqrt{R^2 + 6^2}$$

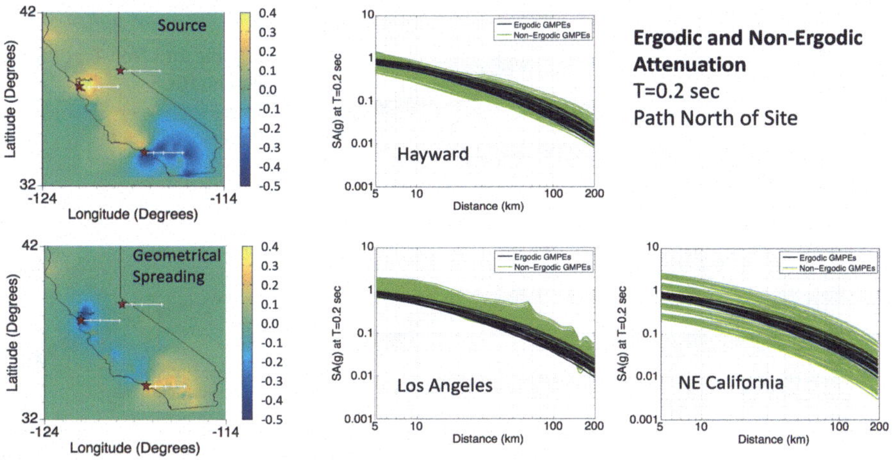

Fig. 9.2 Non-ergodic GMMs for three sites in California for sources located to the east of each site. The left-hand panels show the map of the non-ergodic source term (upper figure) and geometrical-spreading term (lower figure). The right-hand panels compare the distance attenuation for M6.5 earthquakes for the range of ergodic models (black curves) and the range of non-erdodic models (green curves). Reference system: NAD83/geographic coordinates

$$
\begin{aligned}
&+ \delta\theta_0(\mathbf{x}_{site}) + \delta\theta_{vs}(\mathbf{x}_{site}) \ln V_{S30} \\
&+ \delta\theta_7(\mathbf{x}_{source}) R
\end{aligned} \tag{9.8}
$$

where each $\delta\theta$ terms are modeled as an individual GP. Equation (9.8) looks more complicated than Equation (9.4); the main addition is to model the source and site adjustments as a sum of a constant and an adjustment to the geometrical spreading and linear site scaling.

Figures 9.2 and 9.3 show maps of the mean adjustment terms for geometrical spreading and the source constant. The mean is zero in regions where there is no data, but is adjusted in the vicinity of data. These figures also show examples of the impact of the non-ergodic adjustment terms on the distance attenuation for different sites in California. The green curves show the non-ergodic attenuation curves to the East (Fig. 9.2) and North (Fig. 9.3) for each site. The suite of the green curves is developed by sampling the uncertainty in the adjustment terms. For comparison, the black curves show the uncertainty range from ergodic GMMs commonly used for California. The ergodic curves are the same for all sites and all directions, but the non-ergodic curves vary by the site location and the direction to the earthquake source. For each site, the epistemic uncertainty for the non-ergodic GMM depends on the number of observed data available in the vicinity of the site. For sites that have data in their vicinity (Hayward and Los Angeles), the range of the non-ergodic adjustments is relatively narrow, and the mean of the adjustment terms is not zero. For the site without data in the vicinity (North-East California), the range is wide, and the mean is zero. In contrast, the ergodic model shows the same narrow

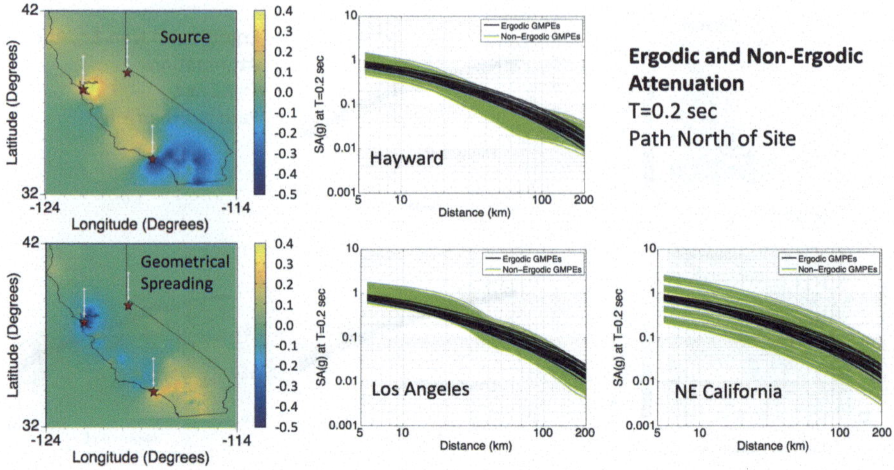

Fig. 9.3 Non-ergodic GMMs for three sites in California for sources located to the north of each site. The left-hand panels show the map of the non-ergodic source term (upper figure) and geometrical-spreading term (lower figure). The right-hand panels compare the distance attenuation for M6.5 earthquakes for the range of ergodic models (black curves) and the range of non-erdodic models (green curves). Reference system: NAD83/geographic coordinates

epistemic uncertainty band for all sites and does not reflect that there is better control on the GMM in regions with data.

9.4 Ground-Motion Data Set for Georgia

A sparse ground-motion data set for the Georgia Republic was provided by Nino Tsereteli (personal communication). This data set contains 88 recordings from 1990 to 1991. The magnitude-distance distribution of the Georgian data is shown in Fig. 9.4. For earthquakes with Ms greater than 5, the Ms magnitude is used; otherwise, if the mb is greater than 4.8, then the mb is used. For smaller earthquakes M_L is used. For this paper, we selected the (Abrahamson et al., 2014) GMM (ASK14) for crustal earthquakes in active regions as the ergodic base GMM. To keep this example application simple, we did not include epistemic uncertainty in the ergodic base GMM.

The residuals of the Georgian data are computed with respect to the ASK14 GMM. Following the standard approach used for developing GMMs, the residuals are separated into between-event residuals and within-event residuals using random effects (Abrahamson and Youngs, 1992). The between-event residuals for spectral acceleration (SA) for T = 0.2 sec are shown in Fig. 9.5. For magnitudes greater than 4, the between-event residuals are centered about 0.25 and do not show a trend with magnitude. In contrast, for magnitudes less than 4, there is a large offset in the

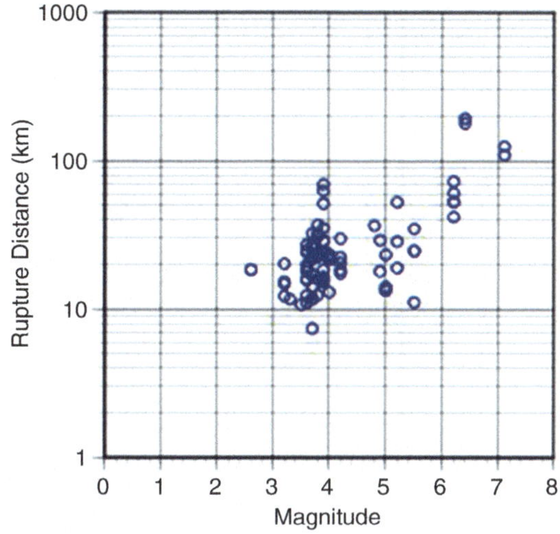

Fig. 9.4 Distributions of magnitudes and distances in the Georgia strong-motion data set

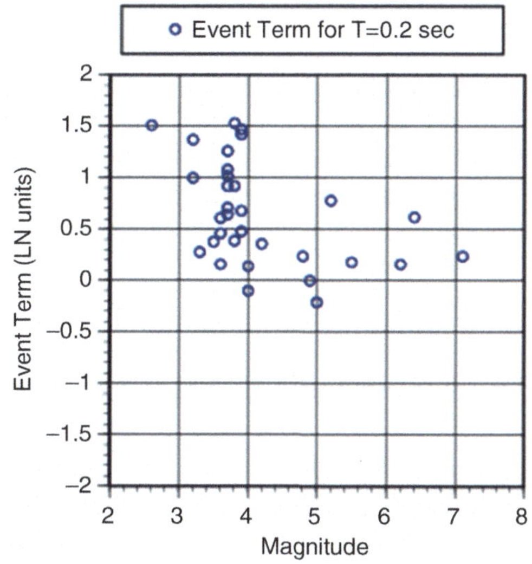

Fig. 9.5 Event terms for the Georgia data set relative to the ASK14 GMM for T = 0.2 s

between-event residuals and the mean is shifted to about 0.7. This type of offset in the between-event residuals can reflect a change in the method used to compute the earthquake magnitude for small and large events.

In developing the non-ergodic GMM for Georgia, we do not want to map magnitude bias for the small magnitudes into systematic source, path, and site terms that are applied to the larger magnitudes used in PSHA. Correcting the estimates of the moment magnitudes for small magnitudes is beyond the scope of this study, but this should be addressed in future ground-motion studies for this region. Therefore, we removed the data for magnitudes less than 4 from our analysis. This reduced the local data set from 88 recordings to 33 recordings. The locations of the earthquakes

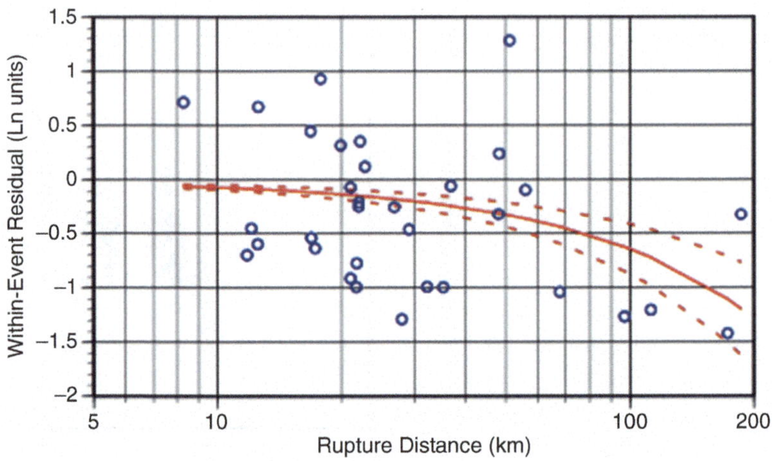

Fig. 9.6 Within-Event residuals for the Georgia data set relative to the ASK14 GMM for T = 0.2 s

and station in the selected subset of data from earthquake with magnitudes greater or equal to 4 are shown in Fig. 9.7. The within-event residuals for T = 0.2 sec relative to the ASK14 GMM is shown in Fig. 9.6. In this data set, the few available recordings at distances greater than 60 km have negative within-event residuals indicating a stronger attenuation with distance (lower Q), but this is not well constrained due to the sparse data set.

9.5 Non-ergodic GMM for Georgia

There is not enough local data to estimate region-specific hyperparameters (variances and correlation lengths of the source and site-specific adjustment terms). Therefore, we assumed that the hyperparameters (correlation lengths and variances) found for California are applicable to the Georgia Republic. By doing so, we are making an ergodic assumption about the hyperparameters, but we use the local data to estimate the spatial distribution of the systematic differences in the median ground motion. The basis for assuming that the California hyperparameters are applicable to other regions is that the hyperparameters are related to the tectonic setting and it is reasonable to assume that the correlation lengths and variances of the ground motions will be similar for regions from the same broad tectonic category. A comparison of spatial correlation lengths and variances estimated for the Anza array region in California and those estimated for Taiwan found that they are similar (Kuehn and Abrahamson, 2020). Currently, when we apply the ergodic assumption for GMMs, we assume that both the median and variability of the ground motion for a given scenario (magnitude, distance, site) is applicable to all regions in the

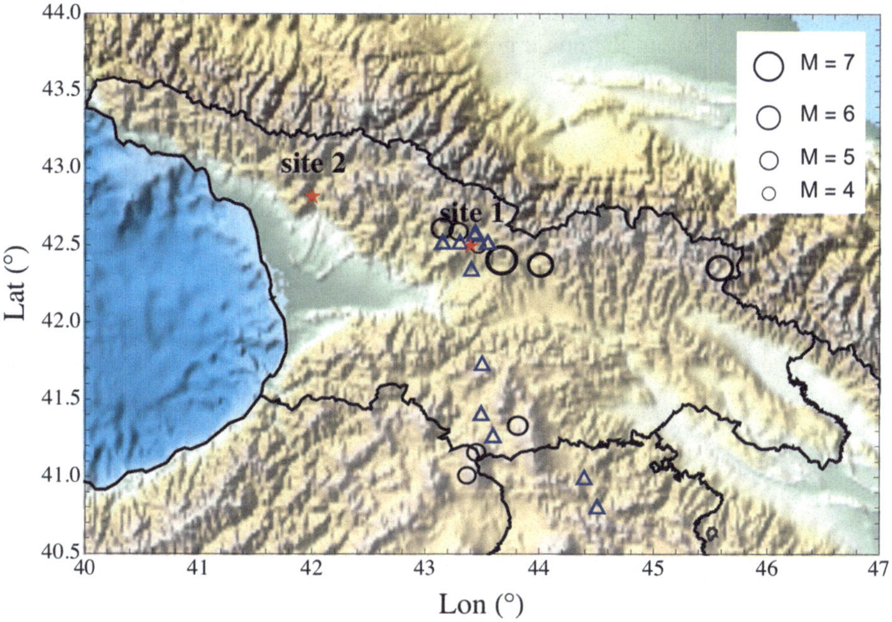

Fig. 9.7 Earthquakes and station locations in the Georgia strong-motion data set. Site 1 and Site 2 show the example locations for the non-ergodic GMM for Georgia. Reference system: ITRF00/geographic coordinates

same broad tectonic category. Although we are assuming that the hyperparameters for California are applicable to other active crustal regions such as the Georgia Republic, it is a much weaker assumption than the standard ergodic assumption because it only involves the variability and not also the median values.

Even with a small amount of local ground-motion data, we can still use the non-ergodic framework for the GMM. Given the limited data, we limited the non-ergodic GMM for Georgia to three terms: site constant, source constant, and linear R term for the path term. The non-ergodic GMM for Georgia is given by:

$$ln(PSA) = GMM_{erg}(M, R_{RUP}, V_{S30}, F, Z_{TOR}) \\ + \delta L(\mathbf{x}_{source}) + \delta S(\mathbf{x}_{site}) + \delta\theta_7(\mathbf{x}_{source}, \mathbf{x}_{site})R \tag{9.9}$$

When using GP models with small data sets, the can be over-fitting of the data if the hyperparameters are also estimated from the data. Here, we constrain the hyperparameters to the values from the larger California data set. The California hyperparameter do not give very short correlation lengths which helps to avoid over-fitting the data. Also, as shown in Fig. 9.1 the aleatory variability of the non-ergodic GMM is much smaller than for the ergodic model, a small number of recordings can provide some constraints on the location-specific median. As will be shown later, in parts of Georgia without data, the epistemic uncertainty in the median ground motion is consistent with the epistemic uncertainty in parts of California with little nearby data to constrain the non-ergodic terms, indicating that we have not overfit the Georgia data and overestimated the epistemic uncertainty.

We used four NGA-West2 GMMs with the additional epistemic uncertainty from Al-Atik and Youngs (2014) to set 12 alternative ergodic GMMs. Using Equation (9.7), we then generated 50 realizations of the spatial vectors of non-ergodic terms ($\delta L(\mathbf{x}_{source})$, $\delta S(\mathbf{x}_{site})$, and $\delta\theta_7(\mathbf{x}_{source}, \mathbf{x}_{site})$). The adjustments for the systematic path term are an adjustment to the anelastic attenuation coefficient. We model this as a function of both the event and site coordinate; we use the midpoint of the two.

Figure 9.8 shows the mean μ_* and epistemic standard deviation ψ_* of the non-ergodic adjustment terms for a grid of points around two sites. One site has some data in the vicinity, the other one does not. For site 2 (without local data), the mean adjustment terms are zero for most points. Only for grid points to the East, where there is some local data, does the mean adjustment change. The standard deviation is large, and increases with larger distances; this is because the anelastic attenuation slopes are uncertain, which has a larger effect at large distances. For site 1 (with some local data in the vicinity), the mean adjustments are not zero anymore: the local data reduces the median from earthquakes to the east by up to 10 percent and reduces the median from earthquakes to the south by up to 15 percent. These median adjustments are not large, partly due to the small data set available in the region. The

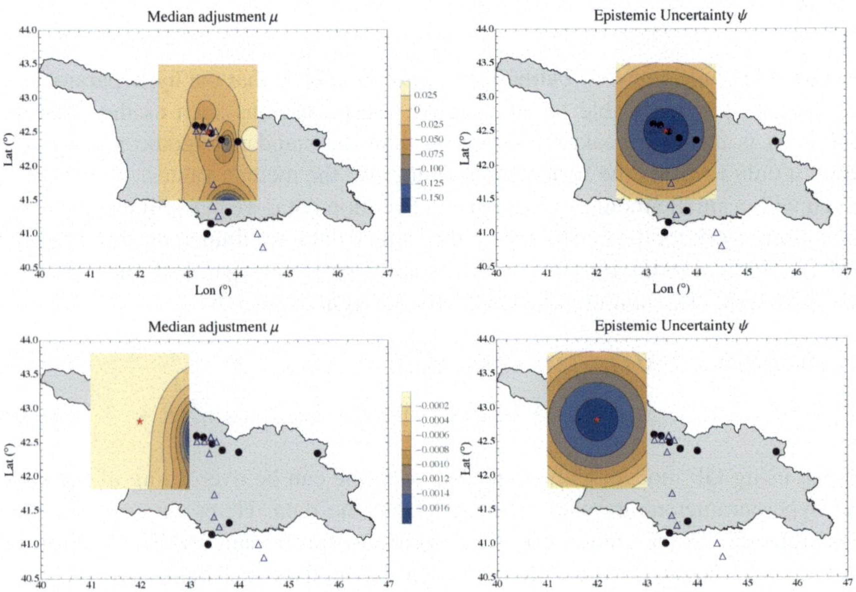

Fig. 9.8 Map of mean (left) and standard deviations (right) for a site with no data in vicinity (bottom), and a site with some data in vicinity (top). Reference system: ITRF00/geographic coordinates

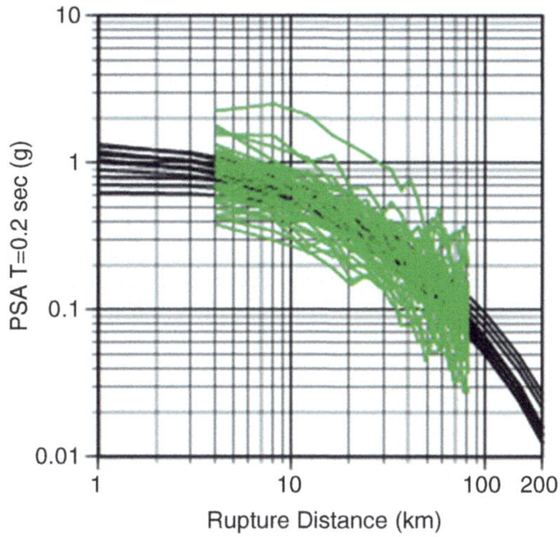

Fig. 9.9 Forty realizations of non-ergodic GMMs for site 1 for earthquakes located east of the site. This site has some local ground-motion data nearby to constrain the non-ergodic terms

epistemic uncertainty is still very large (greater than 0.6 ln units) because there are only 28 observed data points, which is not a large enough number of observations to reduce the epistemic uncertainty by a large amount. To reduce this large epistemic uncertainty, additional data needs to be collected.

The M6.5 distance attenuation for the 12 alternative ergodic GMMs are compared to the suite of 40 non-ergodic GMMs for site 1 (with some local data) in Fig. 9.9 (attenuation to the east) and Fig. 9.10 (attenuation to the south). The epistemic uncertainty range of the non-ergodic GMMs is wider than for the ergodic GMMs. This is because the ergodic GMMs represent average models based on large data sets and these average models are well constrained by the data for M6.5. In contrast, the non-ergdoic GMM is for site 1 captures the range of systematic differences between the source, path, and site effects at site 1 and the average source, path, and site effects in the ergodic GMM. The epistemic uncertainty range is narrow for the attenuation to the south as there are more data in the south direction than in the east direction.

The M6.5 distance attenuation for the ergodic and non-ergodic GMMs for site 2 are compared in Figs. 9.11 and 9.12 for earthquakes located to the east and south of site 2, respectively. For this site, there is no local data to constrain the non-ergodic terms, so the epistemic uncertainty range is wider.

9.6 Conclusions

Ergodic GMMs provide the average base scaling with magnitude, distance, and site condition, but they are not applicable to individual sites because the ergodic GMMs combine data from different regions and sites that have different median ground

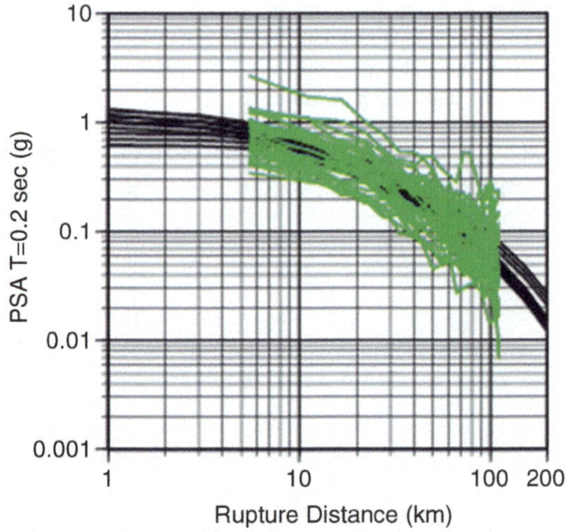

Fig. 9.10 Forty realizations of Non-ergodic GMMs for site 1 for earthquakes located south of the site. In this direction, there is additional local ground-motion data to constrain the non-ergodic terms

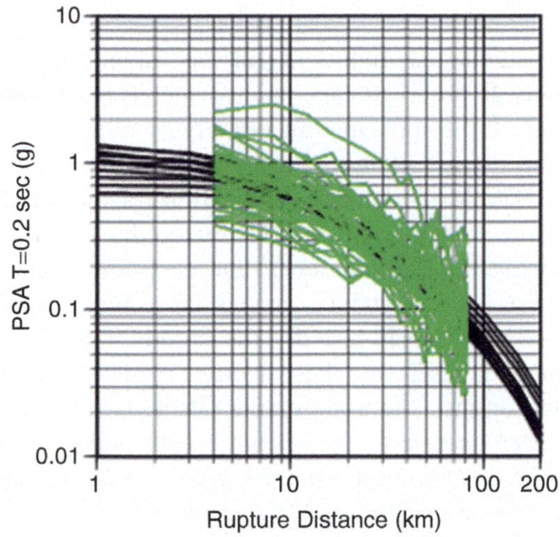

Fig. 9.11 Forty realizations of the non-ergodic GMMs for site2 for earthquakes located east of the site. This site has no local ground-motion data to constrain the non-ergodic terms

motions for a given scenario and site condition. The non-ergodic GMMs address this limitation by modeling the systematic source, path, and site effects using local ground-motion data. There is little local ground-motion data from large earthquakes, but ground-motion data from small magnitudes can be used to constrain the linear path effects and linear site effects. Nonlinear effects need to be constrained by the base ergodic GMM.

The non-ergodic approach provides a framework for using local ground-motion data in a way that can significantly improve the accuracy of the GMM by accounting for systematic source, path, and site effects as the number of recordings in the local region grow. If there are few local recordings, there is much larger epistemic uncertainty in the median ground motion than given by the range from alternative ergodic models, as shown in Figs. 9.9, 9.10, 9.11 and 9.12. By showing the large

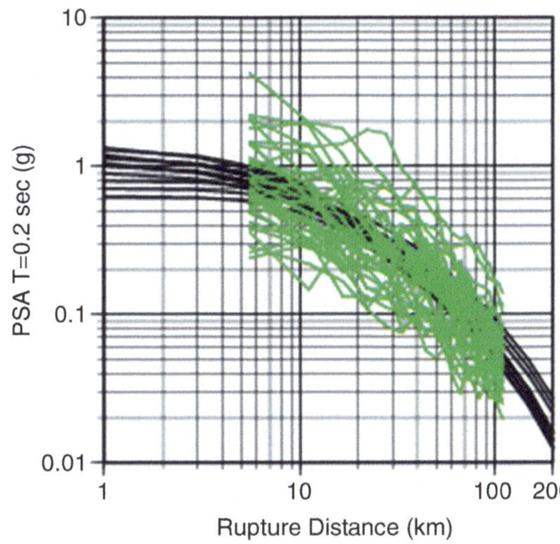

Fig. 9.12 Forty realizations of Non-ergodic GMMs for site2 for earthquakes located south of the site. This site has no local ground-motion data to constrain the non-ergodic terms

epistemic uncertainties in regions with sparse data, this framework can be used to demonstrate the business case for installing additional seismic instrumentation at and near the project site.

The non-ergodic approach can be used for both regions with local data and regions without local data. The key concept is that lack of data does not imply that there are no systematic source, path, and site effects in a region. For regions without local data, the ergodic base GMM as the central estimate but with large epistemic uncertainties about this base GMM. Without local data to constrain the non-ergodic terms, estimates of the seismic hazard and in a specific region and the resulting ground motion at a given return period used for design of structures has much large epistemic uncertainties than indicated using alternative ergodic GMMs. Programs to collect local ground motion data from small and moderate magnitude events or to simulate ground motion data using 3-D crustal models for the region are needed to reduce the epistemic uncertainty in the GMMs and to significantly improve the estimates of seismic hazard. As this data are collected, the non-ergodic methodology provides a smooth transition from regions with data to regions without data.

Acknowledgments The ground motion data from the Georgia Republic were provided by Nino Tsereteli. The non-ergodic attenuation plots for California were generated by Katie Wooddell. This work benefited from discussions with the members of the PEER/UCLA/SCEC Non-Ergodic Working Group.

References

Anderson JG, Brune JN (1999) Probabilistic seismic hazard analysis without the ergodic assumption. Seismol Res Lett 70(1):19–28

Bommer JJ, Abrahamson NA (2006) Why do modern probabilistic seismic-hazard analyses often lead to increased hazard estimates? Bull Seismol Soc Am 96(6):1967–1977

Rodriguez-Marek A (2013) A model for single-station standard deviation using data from various tectonic regions. Bull Seismol Soc Am 103:3149–3163

Abrahamson NA, Silva WJ, Kamai R (2014) Summary of the ASK14 ground motion relation for active crustal regions. Earthquake Spectra 30(3):1025–1055

Boore DM, Stewart JP, Seyhan E, Atkinson GM (2014) NGA-West2 equations for predicting PGA, PGV, and 5% damped PSA for shallow crustal earthquakes. Earthquake Spectra 30(3):1057–1085

Campbell KW, Bozorgnia Y (2014) NGA-West2 ground motion model for the average horizontal components of PGA, PGV, and 5% damped linear acceleration response spectra. Earthquake Spectra 30(3):1087–1115

Chiou BSJ, Youngs RR (2014) Update of the chiou and youngs NGA model for the average horizontal component of peak ground motion and response spectra. Earthquake Spectra 30(3):1117–1153

Kotha SR, Bindi S, Cotton F (2016) Partially non-ergodic region specific GMPE for Europe and Middle-East. Bull Earthq Eng 14(4):1245–1263

Anderson JG, Uchiyama Y (2011) A methodology to improve ground-motion prediction equations by including path corrections. Bull Seismol Soc Am 101(4):1822–1846

Dawood H, Rodriguez-Marek A (2013) A method for including path effects in ground-motion prediction equations: an example using the Mw 9.0 Tohoku earthquake aftershocks. Bull Seismol Soc Am 103(2B):1360–1372

Kuehn NM, Abrahamson NA, Walling MA (2019) Incorporating nonergodic path effects into the NGA-West2 ground-motion prediction equations. Bull Seismol Soc Am 109(2):575–585

Landwehr N, Kuehn NM, Scheffer T, Abrahamson NA (2016) A nonergodic ground-motion model for California with spatially varying coefficients. Bull Seismol Soc Am 106(6):2574–2583

Kuehn NM, Abrahamson NA (2020) Spatial correlations of ground motion for non-ergodic seismic hazard analysis. Earthq Eng Struct Dyn 49(1):4–23

Abrahamson NA, Kuehn NM, Walling MA, Landwehr N (2019) Probabilistic seismic hazard analysis in california using nonergodic ground motion models. Bull Seismol Soc Am 109(4):1235–1249

Rasmussen CE, Williams CKI (2006) Gaussian processes for machine learning. MIT Press, Cambridge

Abrahamson NA, Youngs RR (1992) A stable algorithm for regression analysis using the random effects model. Bull Seismol Soc Am 82(1):505–510

Al-Atik L, Youngs RR (2014) Epistemic uncertainty for NGA-West2 models. Earthquake Spectra 30(3):1301–1318

Chapter 10
Time Series Analysis of Fault Strain Accumulation Around Large Dam: The Case of Enguri Dam, Greater Caucasus

Tamaz Chelidze, Teimuraz Matcharashvili, Vakhtang Abashidze, Nadezhda Dovgal, Ekaterine Mepharidze, and Lia Chelidze

Abstract The 271-m-high Enguri arch dam, still one of the highest of this kind in operation in the world, was built in the canyon of the Enguri river (western Georgia) in the 1970s. It is located in a zone of high seismic hazard and is close to the Ingirishi strike-slip active fault. The strong seismic activity, as well as the great number of people living downstream of the dam, make this facility a potential source of a major catastrophe in Georgia. Thus, the Enguri Dam, with its 1 billion cubic meters water reservoir, should be under permanent monitoring. At the same time, this area is an amazing natural laboratory, where researchers can investigate both tectonic and geotechnical issues and the effects of load-unload processes at the artificial reservoir, i.e. the reaction to a controllable loading of the Earth crust. This is an important scientific topic, connected with the fundamental problem of reservoir-induced earthquakes as well as with environmental geotechnical problems, related to the security of major dams in general. The complexity analysis methods applied to time series of the fault strain monitoring data enables discriminating seismic events derived by the human activities and background seismicity.

Such complexity analysis provides a new approach – Reservoir Synchronized Induced Seismicity, or RSIS – for a quantitative assessment of reservoir water level load on the (synchronized/ordered) temporal pattern of local seismicity. Besides, analysis of complexity in the fault strain monitoring data can provide also some early warning methods for signaling dangerous deviations of fault strains and dam tilts temporal behavior from the stable (background) pattern.

T. Chelidze · T. Matcharashvili · V. Abashidze · N. Dovgal · E. Mepharidze · L. Chelidze
M. Nodia Institute of Geophysics, Tbilisi State University, Tbilisi, Georgia

T. Chelidze (✉) · V. Abashidze
European Centre "Geodynamical Hazards of High Dams" at the Council of Europe, Tbilisi, Georgia

© Springer Nature B.V. 2021

F. L. Bonali et al. (eds.), *Building Knowledge for Geohazard Assessment and Management in the Caucasus and other Orogenic Regions*, NATO Science for Peace and Security Series C: Environmental Security,
https://doi.org/10.1007/978-94-024-2046-3_10

10.1 Introduction

The monitoring of strains and seismic activity in areas occupied by large dams is a key tool for understanding the intimate connections between earthquakes and man-made, regular, quasi-periodic strains in the Earth, created by seasonal water load-unload operations at the reservoir. We can regard the areas where large dams are located, as natural laboratories, providing a chance to study seismic process in controlled (repeated) conditions.

The 271-m-high Enguri arc dam – one of the largest in the world – was built in the canyon of the Enguri river in western Georgia. It is located close to the Ingirishi active fault, in a zone of high seismic hazard capable of triggering earthquakes (Figs. 10.1 and 10.2) with epicentral intensity up to $I_0 = IX$ MMI (Varazanashvili et al. 2011). The volume of the lake at the dam is 10^9 cubic meters and the water level height in the lake varies seasonally by 100 m, which means that Enguri reservoir can activate Reservoir-Triggered Seismicity – RTS (Gupta 1992, 2018). The dominant tectonic feature of the region is the active, ENE-WSW oriented Ingirishi strike-slip fault (Fig. 10.2), located north of the dam: its branch fault crosses the foundation of the Enguri dam (Chelidze et al. 2013).

Taking into account the potential risk, a geophysical monitoring system was organized well before construction works for providing secure exploitation of the large Enguri dam. Due to the high seismic activity of the region, a local seismic

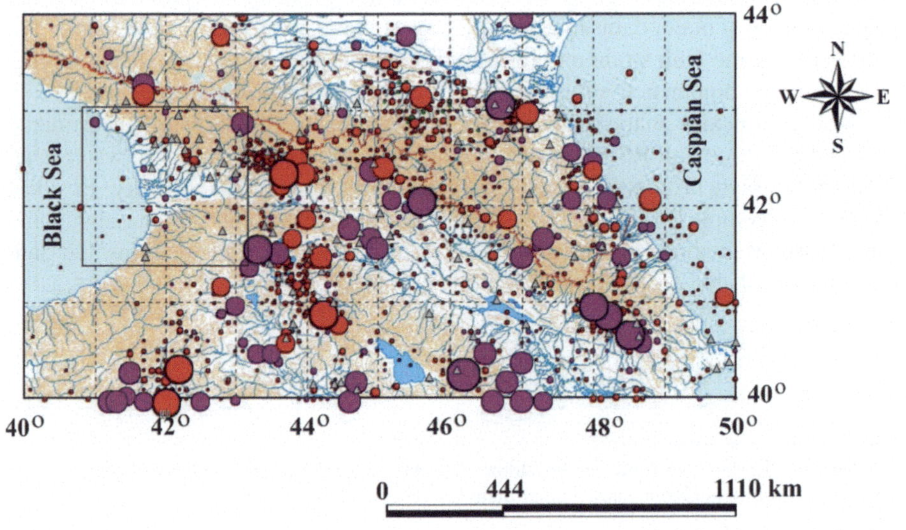

Fig. 10.1 Epicentral distribution for the regional seismicity of the Caucasus region in the 500 BC-2000 AD time window; historical earthquakes (in purple) are until 1909 A.D, after which they belong to the first 80 years of the 1990s; earthquake magnitude is provided in the range (4–7.5); data are from Varazanashvili et al. 2011. The triangles show the location of seismic stations; the black box locates the study area, including the Enguri dam area, shown in Fig. 10.2

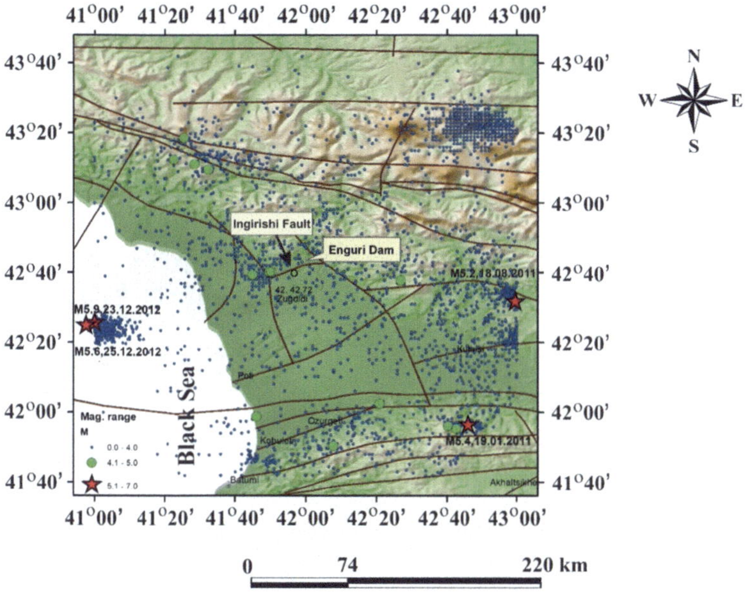

Fig. 10.2 Seismicity for western Georgia, the region where the Enguri dam is located; earthquakes are reported for the 974–2017 AD time window, within a radius of R = 100 km, centered at the dam (earthquakes are from the Catalogue of the Georgian Institute of Geophysics). The scheme for active tectonic faults is from Gamkrelidze et al. (1998); the strike-slip Ingirishi fault, crossing the Enguri dam lake, is also indicated

network was installed in the area also well before its construction, with the aim of studying possible reservoir-triggered activity (Balavadze 1981); the network operated until 1990. The monitoring system of the Enguri dam and its foundation includes a network of tiltmeters, piezometers and reverse plumb-lines into the dam (Chelidze et al. 2013), a weather station, water level gauges for monitoring water level in the lake, as well as complex of strain-meters and tiltmeters, installed in the dam body and at its foundation (Abashidze 2001).

The problem of human-induced earthquakes, including RTS, has become relevant over the last decades (Grigoli et al. 2017; Foulger et al. 2017; Savage et al. 2017). The RTS pattern in the Enguri area should depend on the water level variation regime in the lake (Gupta 1992; Gupta 2018).

The analysis of tilt and branch fault strain monitoring data is presented in several papers (Chelidze et al. 2013, 2019). The present paper is devoted to an application of the complexity theory to the analysis of strain, seismicity and dam tilt time series. We have applied new methods of complexity analysis (Webber and Marwan 2015; Peng et al. 1994, 1995; Hu and Gao 2006) in order to assess, in a quantitative way, a correlation of branch fault strain and local seismicity with water level variations. Our effort has enabled us to define the influence of man-made activity on the seismicity pattern in the area. We show that the complexity

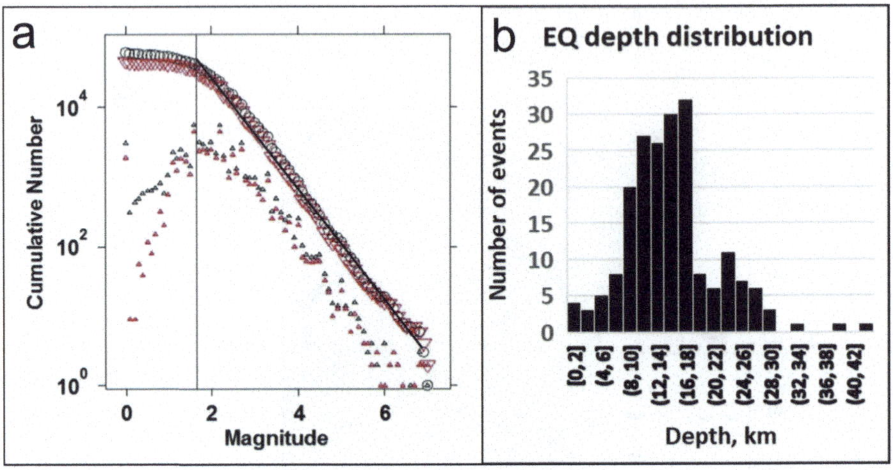

Fig. 10.3 (**a**) Cumulative Gutenberg-Richter plot of the whole (black circles) and aftershock-depleted (upside-down, red triangles) catalogue of Georgia. The plot shows also the binned frequency-magnitude distribution of the whole (upside black triangles) and aftershock-depleted (upside red triangles) catalogues. The cut-off magnitude is M 1.7 (vertical line). (**b**) Earthquake (EQ) depth distribution in a 50-km radius from the dam. The greater frequency for hypocenters is in the 10–20 km range, the highest frequency is for the class 16–18 km

analysis provides a new tool for a quantitative assessment of reservoir water level load on the (synchronized/ordered) temporal pattern of local seismicity, which is hereby referred to as Reservoir Synchronized Induced Seismicity (RSIS). The final section of our work is aimed at showing possible applications of the complexity analysis to dam security problem, using dam tilt monitoring data.

10.2 Data

Two monitoring data sets – seismic and fault zone extension/compaction (Figs. 10.3, 10.5 and 10.6) – are used to reveal the complex dynamics of the area, which represent a combination of natural and man-made processes.

In regard to seismicity, in Fig. 10.1 we present both the general regional seismicity due to active tectonics in the Caucasus region as well as in Georgia, with data from Varazanashvili et al. (2011). Some of the major earthquakes have proven to be devastating, i.e., the Racha earthquake of 29 April 1991, with Ms = 6.9. It occurred 150 km to the West of the Enguri dam and is the strongest event ever recorded in Georgia (Varazanashvili et al. 2018). The fault plane solution data for 130 earthquakes of magnitudes 4–5 show that the territory of Georgia is currently under latitudinal compression and longitudinal extension with an overall crustal thickening (Adamia et al. 2017). According to Tibaldi et al. (2017), the regional stress regime in the Enguri dam area corresponds to a N-S directed contraction, compatible with thrust faulting. Figure 10.2 shows the spatial distribution of seismicity around the Enguri dam region within a R = 100 km from the dam as well as the traces of active faults, according to Gamkrelidze et al. (1998). The most important fault in the dam area is the Ingirishi strike-slip fault. The earthquake time series (ETS) for the Enguri area from 3 January 1974 to 31 December 2016 (the Enguri seismic catalogue) was compiled using the (regional) catalogues from the Institute of Geophysics (Georgia) and from the International Seismological Centre for 1974–2018 A.D., as well as data from Varazanashvili et al. (2018). The locations of earthquakes are shown in Figs. 10.1 and 10.2. The magnitudes until 2003 are Mpv and after 2003 are expressed as Ml. Our study area includes events located at distances either 50 or 100 km from the dam. The completeness magnitude for the whole used catalogue is $Mc = 1.7$ (Fig. 10.3a), but in some cases we confine ourselves to a value of magnitude of 2.2, as in some periods the Mc value increased to 2.2 due to non-stable functioning of the national seismic network. The hypocenters of most of the events are located at a depth from 8 to 18 km, though there is also a second smaller peak in the 24–26 km interval (Fig. 10.3b); a similar bimodal distribution is observed in several regions of Georgia (Tibaldi et al. 2017).

Regarding the monitoring network for strain, the branch fault of the main Ingirishi fault (Figs. 10.4a, b) crosses the foundation of Enguri dam and, thus, poses a significant hazard to the dam. In order to permanently monitor the fault behavior, two years before the first infilling of the reservoir, in December 1974, a quartz strain-meter crossing the fault zone (FZ) was installed 100 m downstream of the dam foundation (Figs. 10.4a, b). The strainmeter's fixed and free parts are located on intact rocks on the opposite sides of the FZ and are separated from this 10 m-wide zone by a 5-m distance (the full length of the quartz tube is 22.5 m). This means that the device records displacement of the intact portions divided by the 10 m wide fault zone, in a direction perpendicular to the fault zone. Therefore, it was possible to document the fault zone's extension/contraction. The free end of the tube initially was equipped with photo-optical recording system (Abashidze 2001). The displacements' sensitivity of this system was of the order of 0.18 μm/mm, which enables also recording a tidal component of the fault zone strain. At present, the laser system (Laser model R-39568, Green HeNe Laser, 633 nm and Laser Position Sensor OBP-A-9L) doubles the photo-optical registration. The laser is attached to the free end of the same quartz tube. Sensitivity of the strain-meter with the laser sensor is one μm/mm.

Regarding the temporal pattern of the strain across the fault in the dam foundation, Fig. 10.5 shows an almost 40-year history of the fault zone extension (FZE), from 1974 to 2017, as well as water level changes in the Enguri reservoir H, from April 1978. According to Fig. 10.5, the dam area experiences stresses of different origin, acting on different time scales, from decades to months and days. Actually, the area under study is a natural large-scale laboratory for the investigation of tectonic, man-made and environmental impacts on the fault zone deformation pattern. The summary contributions of these processes are reflected in the time series of fault zone strain. It is evident that the fault dynamic reflects

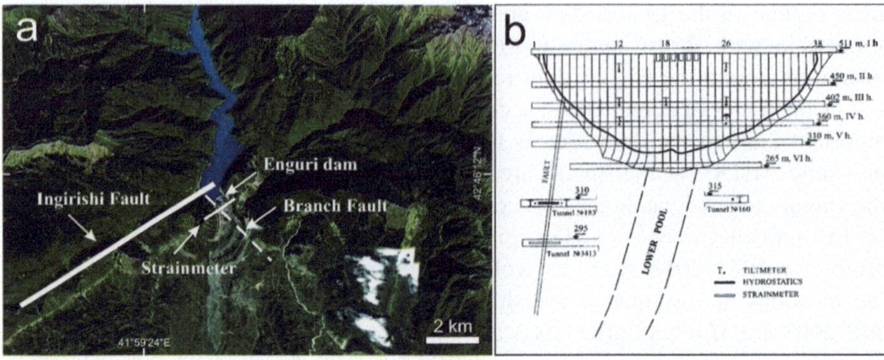

Fig. 10.4 (**a**) Satellite image of the Enguri dam and reservoir area with locations of the main Ingirishi fault and the branch fault, crossing the dam foundation. Google Earth Pro 7.1 Imagery, Enguri dam area, Georgia. Landsat/Copernicus 31/12/2011, http://www.earth.google.com. (**b**) Scheme of the Enguri dam (upstream view. Parallel to the dam) with the tiltmeter network (T) and the strain-meter, installed on the branch fault (in tunnel 3413)

joint influence of two main factors: one (tectonic strain) leads to piecewise linear (in time) displacement, which we define as a trend component and the other one – to quasiperiodic oscillations, decorating the main trend. The well-marked periodic decoration of the main fault extension trend became evident in 1987, 2–3 years after the start of the (transient) load-unload of reservoir in 1984.

The long-term, piecewise-linear trend documents a persistent extension across the fault (Fig. 10.5), amounting to 7000 μm (7 mm) during observation period. FZE (y) relation with time (t) follows a simple linear equation: $y(t) = at + b$. where the coefficient a, the slope of the linear component of the FZE or the strain rate, differs from one period to another (Table 10.1). At the same time, fault zone extension rates change significantly over time, probably reflecting the role played by non-stationary tectonic strain (Fig. 10.5). In Table 10.1, the patterns of both components of strain – the tectonic and anthropogenic ones, are considered. It is worth pointing out that the main, slow-changing component of extension is connected with the general tectonic activity of the fault, as it began even before lake infilling (Fig. 10.5). On the other hand, the most probable source of the transient (arrow n. 2 in Fig. 10.5) and short-term quasiperiodic variations, observed along with the main trend, are due to man-made effects. The first event (the FZE kink in 1978) is due to the first massive recharge of the lake. The regular, quasiperiodic variation in FZE dynamics around the main trend after 1987 (Fig. 10.5) are due to the fluctuation of the water load in the lake. We elucidate six periods with appreciable differences in water level regime and fault-strain-rate behavior (see Table 10.1):

According to the table, the slowly changing component (trend) rate varies in different periods between 250 and 150 μm /year. The sign-changing strain component, marking the beginning of a quasiperiodic (seasonal) WL regime after 1987, became appreciable in the fourth period.

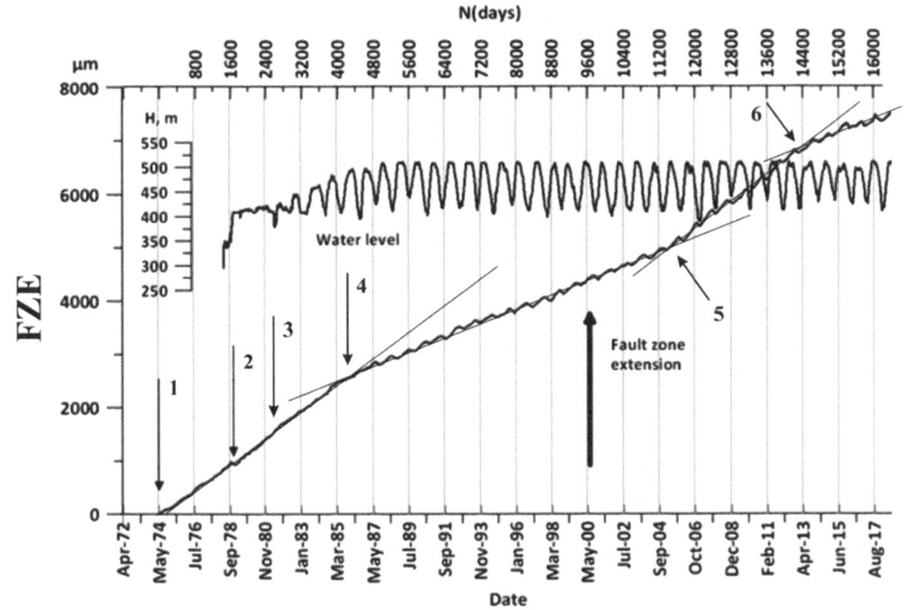

Fig. 10.5 The upper curve shows the variations of water level height at the Enguri lake from 1978 to 2017; the lower curve shows data from the strain-meter, reported for fault zone extension (FZE) along a branch of the major Ingirishi fault, crossing the foundation of the dam. Data are from 1974 to 2017 versus time. Note that there are two scales for the y-axis: water level H in meters (inner scale) and FZE in microns (outer scale). Arrows 1, 2, 3, 4, 5, 6 correspond to the start of 6 periods of fault zone extension, see Table 10.1 for details. Note the small kink marking the episode of fault compaction by approximately 90 microns, as a consequence of a quick, 100-m increase in water level in 1978. The upper horizontal axis shows the number of days after the start of the strain-meter monitoring. The thin, straight lines show six periods of the fault's main trend with different extension rates

From the data (Fig. 10.5), it is possible to observe also that the water load decreased the initial FZE rate recorded before lake recharge, as the total accumulated strain value in 2017 is only 7000: if the initial strain rate were retained, the simple linear interpolation shows that the total FZE would amount to approximately 9500 micron. The accumulated strain decrease can be explained by the favorable for fault compaction strain orientation caused by water load (Gupta 1992; Talwani 1997).

Regarding the temporal behavior of the test area seismicity, Fig. 10.6a displays the earthquake number per month (seismic rate) from 1974 to 2017, within a 100-km epicentral distance Δ from the dam; Fig. 10.6b shows the same data on a shorter time scale, from February 1976 to 1991. It is also highlighted the EQ magnitude M and its epicenter separation from the dam Δ for the strongest events. The epicenters of EQ. M 4.3 (21 Dec 1979) and M 4.3 (27 Dec 1979) are close to the Enguri lake, whereas the EQ. M 5.4 (19 Jan 2011) and M 5.9 (23 Dec 2012), are far from it, on the 80–100 km distance (Fig. 10.6a).

Table 10.1 Subdivision in periods of fault zone extension

Number of periods	Periods	Number of days in the period; in brackets the same from zero day (May 1974) to the end of a given period	Tectonic component of strain rate α μm/year	Pattern of lake impounding (man-made component of strain)
1	May 1974–Apr 1978	1500 (1500)	230	Before lake impounding
2	May 1978–Jan 1981	1300 (2800)	250	WL in the lake raised to 100 m
3	Feb 1981–May 1985	1400 (4200)	250	Irregular load-unload regime
4	Jun 1985–Sep 2004	7000 (11200)	160	Regular quasi-periodic regime
5	Oct 2004–Feb 2013	3200 (14400)	230	Regular quasi-periodic regime
6	Apr 2013–Mar 2018	2000 (16400)	150	Regular quasi-periodic regime

To single out more clearly the man-made impact, we have provided in Fig. 10.6b the detailed seismic rate during the first 200 months after January 1974. The corresponding water level regime is shown in Fig. 10.5: the recharge began in April 1978 (arrow n. 1) and WL was abruptly risen to 100 m in November 1978 (arrow n. 2). Almost simultaneously with the last operation, an abrupt compaction of the fault zone crossing the foundation of the dam, by approximately 90 μ m, was registered by the strainmeter installed on the fault (Figs. 10.4 and 10.5). According to Figs. 10.6a, b, the strong seismic activity very close to the dam ($\Delta = 0$–10 km, hypocenter depth 3–10 km) in December 1979, with 4 events of magnitude M 3.7–M 4.3, followed (with a 14 month-lag) the fast initial recharge of the lake to the critical water level (100 m) in September 1978. According to existing data (Foulger et al. 2017; Gupta 1992, 2018), such water load as a rule generates Reservoir Induced Seismicity (RIS).

In order to better resolve the seismic events related to the filling and exploitation of the reservoir, we have used the cellular approach (Kafka and John 2011). Namely, we have plotted separately the number of all recorded EQs per year (all EQs of magnitudes $M > 1$) in the near-to-dam zone, in the area delimited by the R = 50 km radius from the dam (Figs. 10.7a–c) and in the broader area delimited by the R = 100 km radius (Figs. 10.7b–d) from the dam. In Figs. 10.7c, d, we have shown the same data for the EQs of magnitudes $M \geq 2.2$. We can conclude that

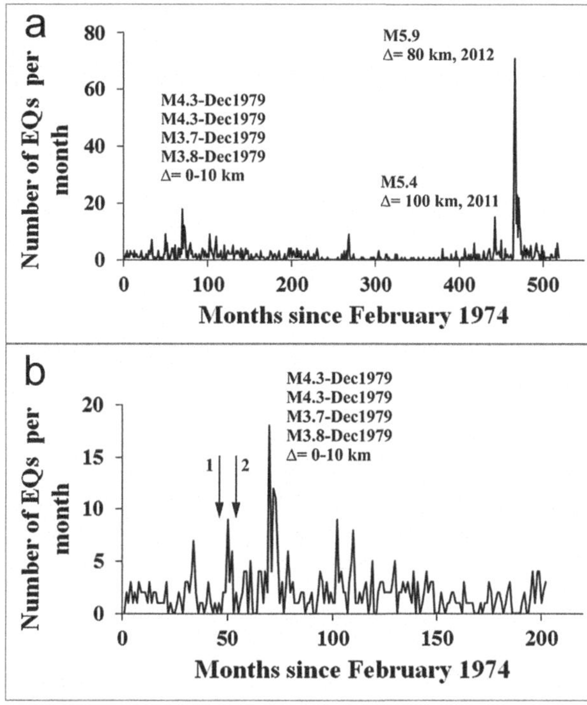

Fig. 10.6 Number of EQs. (M ≥ 2.2) per month, versus month number, within a 100-km radius from the dam; Δ is the epicentral distance from the dam: (**a**) from February 1974 to 2017; (**b**) from February 1974 to 1991. Arrows 1 and 2 in Fig. 10.6b correspond to: (1) beginning of lake filling and (2) 100-m water level increase The two strongest events (M 4.3) occurred on 21.12.1979 and 27.12.1979, 14 months after attaining the critical WL height of 100 m

the EQ statistics within the R = 50 km area (Figs. 10.7a–c) is dominated by events connected with reservoir impoundment, whereas the swarm of EQs with Δ from 0 to 10 km (compare with Fig. 10.6b), i.e. the near zone seismicity, is probably linked to reservoir water level regime in both space and time. On the other hand, most of the 2011–2012 seismic activity in the wider, R = 100 km area (Figs. 10.7b–d) is due to the relatively strong, remote events with M 5.2–5.9 with Δ of the order of 80–100 km. The share of these EQs in the R = 50 km area for the same period is much less (Figs. 10.7a–c). We conclude that the 2011–2012 events are too far away from the Enguri dam and thus belong to the class of regional tectonic events (see also Figs. 10.1 and 10.2).

The above spatial/temporal restriction does not work for the analysis of seismic time series complexity, especially when we analyze EQ recurrence intervals (time periods between successive EQs) because, in this case, the RIS, unlike background seismicity, is marked by quasiperiodic recurrence related to the regularity of reservoir load-unload process. As the application of complexity tools enables to select ordered events, the role of (irregular) background component of seismicity in recurrence interval series is suppressed. This enables to discriminate clearly the recurrence, synchronized with water level change component of ETS even in the far zone (R = 100 km), where the share of irregular component of seismicity is dominant (see Figs. 10.7b–d).

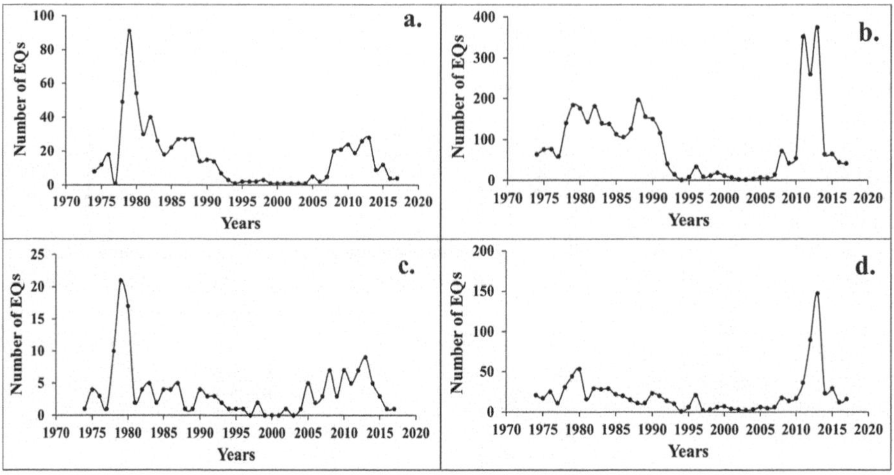

Fig. 10.7 (**a**) EQs number, all events, including those with M < 2.2 in the R = 50 km area; (**b**) EQs number, all events, including those with M < 2.2 in the R = 100 km area; (**c**) EQ number, events including those with M > 2.2 in the R = 50 km area; (**d**) EQ number, events including those with M > 2.2, in the R = 100 km area

10.3 Recurrence Intervals and Magnitudes Time Series Analysis

As we have noted above, the earthquake time series (ETS) presented in Fig. 10.7 are a complex mix of background seismicity, connected with natural, long-term tectonic activity of the test area with a seismic response to the dam construction and lake impoundment, which was irregular until 1985 and later on became quasiperiodic (Fig. 10.5). To single out the dynamic patterns of the seismic datasets connected with water level variations, we have used new, effective methods of complexity analysis (Chelidze et al. 2019) applied to EQ magnitudes and recurrence interval sequences. The complexity analysis allows for the recognition of periods with different level of ordering/determinism, which we connect with transition in water level regime from disordered (1978–1984) to weakly ordered (1985–1986) and finally, to a quasiperiodic pattern (1986-until today). The earthquake time series (ETS) pattern should not be expected to follow exactly that of the relatively simple fault strain variations, described earlier (Fig. 10.5 and Table 10.1). The recurrence in ETS is much more difficult to detect and call for the application of special methods. In our earlier studies devoted to the complexity analysis of the seismic regime in the Enguri dam area, we calculated the variation of the phase diffusion coefficient (Matcharashvili et al. 2008, 2010), performed Visibility Graph Analysis (Telesca and Chelidze 2018) and Singular Spectrum Analysis (Telesca et al. 2012) in order to assess in a quantitative way correlation between local seismic regime and water level variations.

In the present work, we analyze recurrent patterns of local seismicity, using such methods as Recurrence Plots and Recurrent Quantification Analysis (Webber and Marwan 2015), Detrended Fluctuation Analysis and Lempel-Ziv complexity measure in order to confirm the appearance of synchrony between water level variations and local seismic regime.

Recurrence Plots (RP) Recurrence Plots allow visualization of recurrent behavior of dynamical system by plotting the arbitrary close (after some time lag) states in the two-dimensional projection of the high-dimensional phase space trajectory (Eckmann et al. 1987; Webber and Marwan 2015). The RP visualize recurrences in the dynamics of the dynamical system, namely, times when the same state is fixed after some time lag, are plotted on the square matrix. Both axes of the matrix represent time between recurrent points i and j of the trajectory, divided by a time lag less than some threshold time interval (threshold distance) ε_i. If the time lag between i and j is less than ε_i, the situation is considered as a recurrence of a definite state and is marked as one; if not, the state is recognized as not recurrent and is marked by zero. The plot can be constructed not only by zeros and ones, but also by differently colored dots. The RP reveal a structural pattern (diagonal lines, checkerboard structures), which are different for different degrees of determinism in the phase space of the system.

Recurrent Quantification Analysis For a quantitative description of changes in dynamical structure, we used the Recurrence Quantification Analysis approach, RQA (Zbilut and Webber 1992; Webber and Zbilut 1994; Marwan et al. 2007; Webber and Marwan 2015). This is a quantitative extension of the Recurrent Plot (RP) structure (Eckman et al. 1987). If the dynamics of system is more or less deterministic, in recurrence plots diagonal lines appear, which mean that the trajectory of system's states visits the same area of the phase space regularly, but at different times (Zbilut and Webber 1992; Webber and Zbilut 1994; Webber and Zbilut 2005; Marwan et al. 2007; Webber et al. 2009). Diagonal lines are longer for periodic signals, whereas, for stochastic signals, they disappear. Thus, the RQA provides several quantitative measures of complexity based on measurement of the lengths of diagonally and vertically oriented lines in the recurrence plot (Webber and Marwan 2015). We use one of such measures – %DETERMINISM or %DET, which is calculated as a ratio of number of points, forming diagonal lines, to the total number of phase space points. It reveals changes in the extent of regularity (recurrence) in the analyzed data sets.

Detrended Fluctuation Analysis (DFA) Long-range time-correlations in the investigated datasets were assessed by means of the Detrended Fluctuation Analysis (DFA) method (Peng et al. 1994, 1995). This method allows for the detection of long-range correlations embedded in a non-stationary time series through calculation of a quantitative parameter – the DFA scaling exponent. In order to perform DFA at first a given time series of N samples is integrated. Afterwards, the integrated time series is divided into boxes of length n, and in each box, the polynomial local trend is calculated and removed. Then *N/n* mean squared residuals – Detrended Fluctuation Function ($F(n)$), should be calculated for each box of size n:

$$F(n) = \sqrt{\frac{1}{N}\sum_{i=1}^{N}[Y(i) - Y_n(i)]^2}. \quad (10.1)$$

If the analyzed data reveal fractal or self-similar properties, a power-law behavior $F(n)$~n^{α} can be revealed. If a power law scaling exists, the *F(n)* vs. *n* relationship, in double logarithmic fluctuation plot, will be linear or close to be linear and the scaling exponent α can be estimated. If the scaling exponent $\alpha = 0.5$, we deal with the uncorrelated dynamics of random walk type (Peng et al. 1994). If α is different from 0.5, then the time series is regarded as long-range correlated or anticorrelated, with $\alpha > 0.5$ or $\alpha < 0.5$ accordingly (Peng et al. 1994, 1995).

Lempel and Ziv Complexity (LZC) Measure Lempel and Ziv algorithmic complexity (Lempel and Ziv 1976; Aboy et al. 2006; Hu and Gao 2006) is a method for quantification of the complexity level, based on the transformation of a given data sequence into a new symbolic sequence. This original data are converted into (0, 1) sequence by comparing them to a certain threshold value (usually a median of the original data set). Once the symbolic sequence is obtained, it is parsed to obtain distinct words and the words are encoded. Denoting the length of the encoded sequence for those words, the LZ complexity can be defined as $C_{LZ} = L(n)$,where *L(n)* is the length of the encoded sequence and *n* is the total length of sequence (Hu and Gao 2006). Data sequences with a certain regularity are less complex; the LZ complexity increases as the sequence grows in length and irregularity. In our case, the sequence length is constant and LZC depends only on the level of regularity.

10.4 Results

Complexity Analysis of Seismic Time Series In this study, we have analyzed two types of ETS related to the seismicity of the Enguri area: recurrence intervals between sequential EQs and the magnitude sequences. In this way, we can reveal dynamical changes in the RIS characteristics of the investigated seismic area related with the loading regime of the dam. As mentioned above, although the pattern of events in the far zone can be spoiled by the EQs not related to the reservoir-induced strain, the recurrence interval distribution is less sensitive to (background) random events even relatively far from the dam, i.e. for the R = 100 km area, due to recurrence-selective complexity analysis approach.

Results for the Near R = 50 km Area In this case (R = 50 km), we included into the complexity analysis, and namely, in the Recurrence Plot method, the EQs below

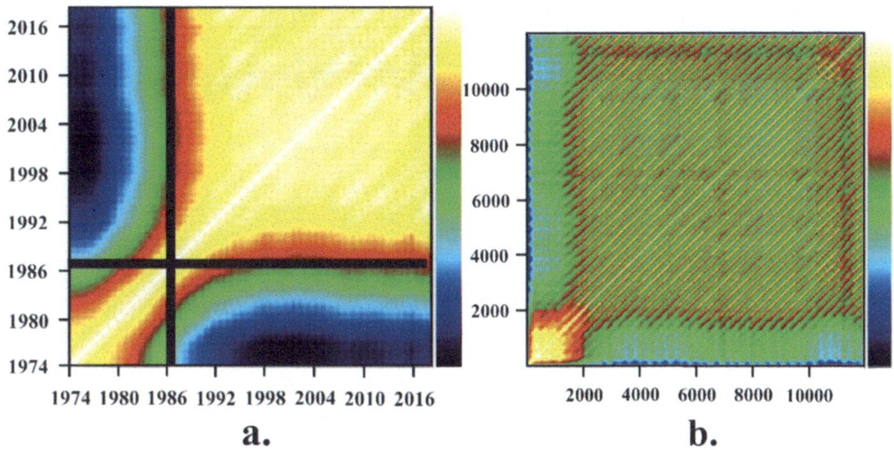

Fig. 10.8 (**a**) Recurrence plots of EQs recurrence intervals, local seismicity from March 1974 to March 2017, area with R = 50 km; (**b**) water level in the Enguri dam reservoir from April 1978 to August 2017; the axes report number of days after the start of recharge. In (**a**) and (**b**), the colored scale bars on the right side of RPs show the decrease of ordering from yellow cells (corresponding to a highly ordered recurrence) to blue ones (low recurrence of events). Black lines in Fig. 10.8a show borders of ordered EQ recurrence intervals. Note: if water level variations became regular after the 3000th day mark (1983) the EQ time series reveals high recurrence level after 1987

the representative magnitude M 1.7, in order to fulfill the conditions of the utilized complexity theory methods – requiring statistics for at least 500 events.

Blue cells in PRs correspond to a less ordered recurrence, whereas yellow ones to better recurrence of events. The distinct transitions from a less regular to a more regular pattern in the seismic regime occur after 1987 (Fig. 10.8a), whereas the water level change in the reservoir became quasiperiodic, at the 3000 day mark in Fig. 10.5, which corresponds to 1983 (Fig. 10.8b). Note that diagonal lines in the RP of seismicity, which mark the beginning of recurrence in ETS, appear after 1987, i.e. approximately with the 3 years' time gap after the setting of quasiperiodic regime in water level variations. The same time gap is observed also between water level variations and fault zone extension (Fig. 10.5).

Results for M > 2.2 Events, R = 100 km Area We used the DFA, RP, RQA and LZC methods for performing a complexity analysis of inter-event and magnitude data sequences of the Enguri area seismic catalogue (1974–2017), involving 913 events above M 2.2, which occurred within 100 km distance from the dam.

In Fig. 10.9, the results of DFA exponent α calculation of recurrence sequences are shown. The calculation was made for 500 data length windows shifted by 1 data. Usually, the DFA analysis is carried out for different polynomial fits. We used different order of the polynomial fitting in order to eliminate trends of certain origin: namely, polynomials of order p ($p = 2, 3, 4, 5$) were used. The similarity of curves at different polynomial fits observed in Fig. 10.9 indicates that the results are almost

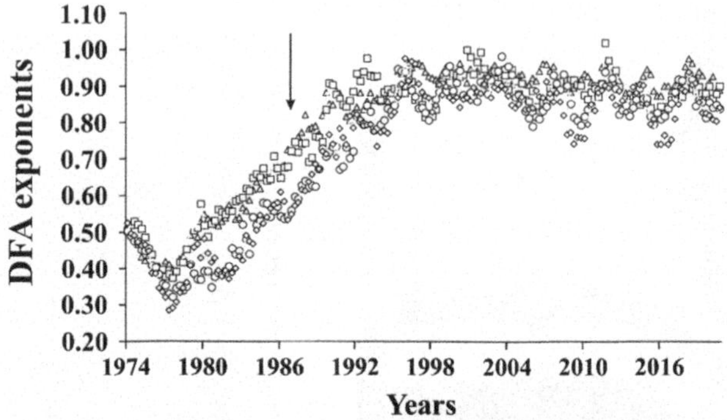

Fig. 10.9 DFA exponents of recurrence sequences (M 2.2 threshold) around the Enguri reservoir, R = 100 km area (1974–2017). Different points show results of fit with a polynomial of different order p: (p = 2, 3, 4, 5). The minimum of DFA corresponds to the beginning of lake recharge in 1978; the arrow marks the year (1987), when the long-range correlation became maximal

Fig. 10.10 %DET of magnitude data sequences (M 2.2 threshold) around the Enguri reservoir (R = 100 km), between 1974 and 2017. Note increased determinism in magnitude distribution after 1987

not influenced by trends. DFA exponents for inter-event times indicate a gradual DFA exponent increase toward the period of reservoir water level quasiperiodic variation. A marked increase up to a DFA exponent value $\alpha = 0.9$ took place in a period starting from 1987 and lasting until 2017. According to these results, under the influence of water level periodic variations in the reservoir, long-range correlation clearly increased in the EQ time distribution. As the DFA for EQ magnitude distribution showed slight or negligible changes in a long-range feature, we chose not to report it here.

We also applied the RQA and LZC methods to recurrence intervals and magnitude data sequences from the Enguri seismic catalogue (1974–2017). The calculation was made for 500 data length windows shifted by 1 data (Figs. 10.10 and 10.11).

It seems that the RQA analysis enables revealing recurrence/synchronization of seismicity with water level variation in both magnitude (Fig. 10.10) and recurrence interval sequences (Fig. 10.11) after 1987. The results of the DFA analysis, as well

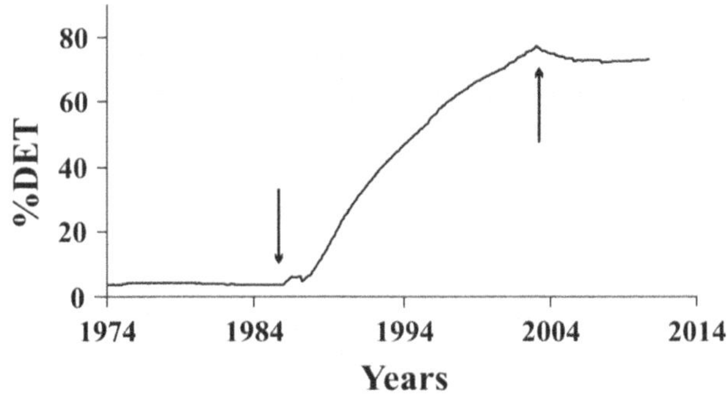

Fig. 10.11 %DET of recurrence interval sequences (M 2.2 threshold) around the Enguri reservoir (R = 100 km), between 1974 and 2017. Note increased determinism in recurrence intervals after 1987

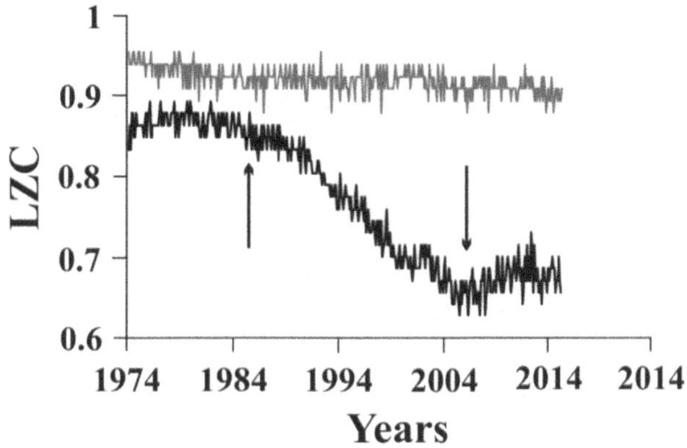

Fig. 10.12 Lempel and Ziv complexity measure calculated for M 2.2 threshold around the Enguri reservoir (R = 100 km), between 1974 and 2017, using 500 data windows shifted by one data. Magnitudes sequence (grey) and inter-event sequence (black) are shown. Note increased determinism (decreased LZC) in recurrence intervals after 1987

as the Lempel and Ziv complexity (Fig. 10.12), indicate that synchronization is not evident in magnitude time series. At the same time, all the used complexity measures confirm the conclusion that the recurrence interval sequence synchronizes with water level change since 1987, i.e. after the 3-year time gap. It is interesting that %DET, DFA and LZC all point to a stabilization of highest stage of determinism in EQ recurrence intervals is attained in 1987 (Figs. 10.11 and 10.12). Afterwards, the determinism value is maximal and almost constant, which points to the establishment of a stable correlation between water level variations and induced seismicity.

10.5 Discussion

Nonlinear Dynamics Patterns in the Strain of the Fault, Crossing Dam Foundation A close correlation between FZE and water level variation is clear, even visually (see Fig. 10.5). The comprehensive complexity analysis of the fault strain time series was presented in a recent paper by Chelidze et al. (2019), where nonlinear dynamics methods, such as RP, RQA, Mutual Information (MI) and Singular Spectrum Analysis (SSA) were first applied in order to reveal hidden structures in the strain data. We have shown here that the intermediate-term effects in FZE are mainly connected with operations at the reservoir, namely with water level quasi-periodic variations, generating one-year cycles as well as higher order harmonics. The FZE cycle during established water level regime (after 1987) lags behind water level changes by approximately 200 days. This lag can be explained in terms of poroelastic effects in the dam foundation rocks, and particularly in terms of the retarded water diffusion from the lake to the strain-meter location (Chelidze et al. 2019). Note that the 200-day FZE lag behind water level changes characterizes the period of established regime of reservoir loading after 1987; the time lag between FZE and water level in the initial (transition) stage 1983–1987 is much longer, 2–3 years, as shown above.

Nonlinear Dynamic Patterns in Seismicity: Synchronization with Water Level Variations in Enguri Lake in the Far Zone ($R = 100$ km) If the correlation of FZ strain with water level variation can be seen even visually (Fig. 10.5), the detection of correlation (ordering, determinism) between water level and local seismicity is much more complicated due to the effect of strong seismic background of tectonic origin, masking the relatively weak, man-induced impacts. In order to reveal the hidden patterns of this weak forcing in the seismic catalogue, it has been necessary to use nonlinear dynamics (complexity theory). Hereunder we summarize the results obtained by means of different complexity analysis approaches: we use the methods' abbreviation with a subscript *m* for magnitude time series and with a subscript *ri* – for the recurrence intervals set, for instance, $(RQA)_m$ and $(RQA)_{ri}$.

(i) We can elucidate the transition in seismic regime by $(DFA)_{ri}$ – the intensive increase of the DFA exponent for the recurrence intervals after 1987 (Fig. 10.9).
(ii) Using the RQA approach, we have revealed drastic changes in ETS dynamics for both magnitudes $(\%DET)_m$ and recurrence intervals $(\%DET)_{ri}$ for 1987 (Figs. 10.10 and 10.11).
(iii) Using LZC we see that there are no changes in $(LZC)_m$, but $(LZC)_{ri}$ undergoes drastic changes in recurrence intervals dynamics around 1987 (Fig. 10.12).

Summing up the above, we can mark the dates of significant changes in the dynamics of recurrence interval series (beginning of significant ordering) after 1987 by all the applied methods, i.e. with a 3-year lag between the start of sign-changing water loading of reservoir. It is important to note that although the number of events in the far zone (100 km) contains a lot of EQs not related to the reservoir-induced quasi-periodic strain (see Figs. 10.7b–d), the complexity analysis enables separating the ordered component in the recurrence intervals' data set even for R = 100 km.

The results of the calculations provided in Figs. 10.8, 10.9, 10.10, 10.11 and 10.12, allow us to conclude that changes occurred in the recurrence intervals data sets are much stronger than the ones occurred in magnitude sequences. Namely, %DET and DFA of recurrence intervals' sequences essentially increase and LZC sequences noticeably decrease: both these effects point to the growth of order (recurrence) in ETS in the period after 1987. According to these results, under the influence of water level periodic variations in the reservoir, long-range correlation clearly increased in earthquakes time distribution, while earthquakes magnitude distribution is characterized by slight or negligible change in long-range features.

Thus, beginning from 1987, the dynamics of local seismicity, especially recurrence intervals, is much more ordered due probably to synchronization of seismic activity with water level variation regular pattern. This conclusion confirms the results from our earlier work, in which we carried out the analysis of seismic activity in the dam area using the Singular Spectrum Analysis (SSA) technique in order to investigate the relationship of local seismicity with reservoir water level variations (Matcharashvili et al. 2008, 2010; Telesca et al. 2012; Telesca and Chelidze 2018). We revealed the dominant one-year period in seismicity, which corresponds to seasonal load-unload of the Enguri lake: this period was absent in ETS of the area, in the reference period, before lake impoundment.

10.6 Detection of an Anomalous Behavior of the Dam Using Complexity Analysis of Dam Tilts

We have also compiled the package (module DAMTOOL) for the complexity analysis of dam dynamics using instrumental monitoring (strains and tilts) time series, which allows for the detection of deviations from the normal pattern of time series of strain/tilt, due to technological or natural forcing (Chelidze et al. 2013) of the data recorded by the tiltmeter (Fig. 10.13), installed into the Enguri dam body.

We hypothesize that the application of complexity theory methods to the analysis of observed strain/tilt/seismicity time series of major engineering facilities (such as

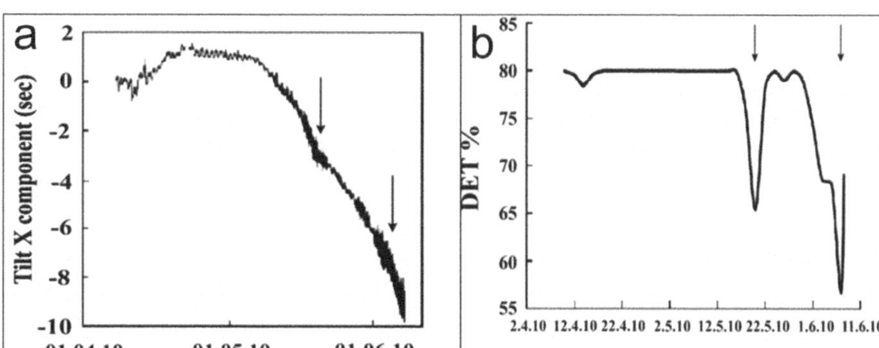

Fig. 10.13 Results of the Recurrence Quantification Analysis (%DET -% of determinism) for the tilt time series from April 2010 to June 2010 at the Enguri dam. (**a**) Original tilt time series (in second vs time); (**b**) RQA %DET vs time. Note high values of DET during regular regime, which point to very regular (normal) behavior and strong deviations from the normal state due to geotechnical impact – addition of high frequency component during intensive discharge of water through the dam outlet for the 12 to 22 May 2010 and 1 to 11 June 2010 time intervals (arrows)

dams), can reveal anomalous nonlinear patterns in the data, which may point to deviations from long-term normal regimes. The comparison of these deviations, obtained during known non-destructive technological and natural events (water discharge, seiches, strong precipitation, etc.) can help to identify the sources of anomalous behaviors of the dam's structure, which are not related to regular technological and natural effects and, thus, can unveil dangerous changes in dam dynamics.

10.7 Conclusions

Based on the Complexity analysis of fault strain and earthquake time series around the major Enguri dam, performed through Recurrent Plots, Recurrent Quantification Analysis, Detrended Fluctuation Analysis and Lempel-Ziv measure, we conclude that the relatively weak influence of water level periodic forcing makes both strain and local earthquake recurrence interval sequences more regular, if compared to periods without such forcing. This means that nonlinear dynamics methods, especially the Recurrent Quantification Analysis, are effective in the detection and quantitative analysis of the Reservoir Induced Seismicity (synchronized with water level variations) near large dams, as they make it possible to divide events, ordered in terms of the impact of reservoir water regular variations, from background seismicity. Thus, the Complexity analysis provides a new tool for quantitative assessment of the effect of water load regular variation on the ordering of temporal pattern of the local seismicity. In our work, this effect is referred to as Reservoir Induced Synchronized Seismicity, or RISS. The further analysis of complexity in monitoring data can provide also some early warning methods for signaling dangerous deviations of fault strains and dam tilts temporal behavior from the stable (background) pattern.

Acknowledgements This work was supported by a grant provided in the framework of the European and Mediterranean Major Hazards Agreement at Council of Europe (EUR-OPA): Ref. No: CA/2017/08 FIMS PO No:537534.

References

Abashidze V (2001) Geophysical monitoring of geodynamical processes at Enguri dam. Publ House Metsniereba, Tbilisi. (in Russian)

Aboy M, Hornero R, Abásolo D et al (2006) Interpretation of the Lempel-Ziv complexity measure in the context of biomedical signal analysis. IEEE Trans Biomed Eng 53(11):2282–2288

Adamia S, Alania V, Tsereteli N et al (2017) Postcollisional tectonics and seismicity of Georgia. In: Sorkhabi R (ed) Tectonic evolution, collision, and seismicity of Southwest Asia, Special Paper, vol 525. Geological Society of America, Washington, DC, pp 1–38

Balavadze B (ed) (1981) Geological-geophysical studies in the region of Enguri Hydro power station. Publ House Metsniereba, Tbilisi. (In Russian)

Chelidze T, Matcharashvili T, Abashidze V et al (2013) Real time monitoring for analysis of dam stability: potential of nonlinear elasticity and nonlinear dynamics approaches. Front Struct Civ Eng 7:188–205. https://doi.org/10.1007/s11709-013-0199-5

Chelidze T, Matcharashvili T, Abashidze V et al (2019) Complex dynamics of fault zone deformation under large dam at various time scales. Geomech Geophys Geo-Energy Geo-Res 5(4):437–455. https://doi.org/10.1007/s40948-019-00122-3

Eckmann JP, Kamphorst S, Ruelle D (1987) Recurrence plots of dynamical systems. EPL 4:973–977

Foulger G, Miles P, Wilson M et al (2017) Global review of human-induced earthquakes. Earth Sci Rev 178:438–514. https://doi.org/10.1016/j.earscirev.2017.07.008

Gamkrelidze I, Giorgobiani T, Kuloshvili S et al (1998) Active deep faults map and catalogue for the territory of Georgia. Bull Georgian Acad Sci 157(1):80–85

Grigoli F, Cesca S, Priolo E et al (2017) Current challenges in monitoring, discrimination, and management of induced seismicity related to underground industrial activities: a European perspective. Rev Geophys 55(2):310–340. https://doi.org/10.1002/2016RG000542

Gupta HK (1992) Reservoir-induced earthquakes. Elsevier, Amsterdam

Gupta HK (2018) Review: reservoir triggered seismicity (RIS) at Koyna, India, over the past 50 Yrs. Bull Seismol Soc Am 108(5B):2907–2918. https://doi.org/10.1785/0120180019

Hu J, Gao J (2006) Analysis of biomedical signals by the Lempel-Ziv complexity: the effect of finite data size. IEEE Trans Biomed Eng 53:2606–2609

Kafka A, John E (2011) Proximity to past earthquakes as a least-astonishing hypothesis for forecasting locations of future earthquakes. Bull Seismol Soc Am 101:1618–1629. https://doi.org/10.1785/0120090164

Lempel A, Ziv J (1976) On the complexity of finite sequences. IEEE Trans Inf Theory 22(1):75–81

Marwan N, Romano MC, Thiel M et al (2007) Recurrence plots for the analysis of complex system. Phys Rep 438:237–329

Matcharashvili T, Chelidze T, Peinke J (2008) Increase of order in seismic processes around reservoir induced by water level periodic variation. Nonlinear Dyn 51:399–407

Matcharashvili T, Chelidze T, Abashidze V et al (2010) Changes in dynamics of seismic processes around Enguri high dam reservoir induced by periodic variation of water level. In: de Rubeis V, Czechowski Z, Teisseyre R (eds) Synchronization and triggering: from fracture to earthquake processes. GeoPlanet: earth and planetary sciences, vol 1. Springer, Berlin/Heidelberg, pp. 273–286

Peng CK, Buldyrev SV, Havlin S et al (1994) Mosaic organization of DNA nucleotides. Phys Rev E 49:1685–1689. https://doi.org/10.1103/PhysRevE.49.1685

Peng CK, Havlin S, Stanley H et al (1995) Quantification of scaling exponents and crossover phenomena in nonstationary heartbeat time series. Chaos 5:82–87

Savage H, Kirkpatrick J, Mori J et al (2017) Scientific exploration of induced seismicity and stress (SEISMS). Sci Drill 23:57–63. https://doi.org/10.5194/sd-23-57-2017

Talwani P (1997) On nature of reservoir-induced seismicity. Pure Appl Geophys 150:473–492

Telesca L, Chelidze T (2018) Visibility graph analysis of seismicity around Enguri high arch dam, Caucasus. Bull Seismol Soc Am 108(5B):3141–3147. https://doi.org/10.1785/0120170370

Telesca L, Matcharasvili T, Chelidze T et al (2012) Relationship between seismicity and water level in the Enguri high dam area (Georgia) using the singular spectrum analysis. Nat Hazards Earth Syst Sci 12(8)

Tibaldi A, Alania V, Bonali F et al (2017) Active inversion tectonics, simple shear folding and back-thrusting at Rioni Basin, Georgia. J Struct Geol 96:33–53. https://doi.org/10.1016/j.jsg.2017.01.005

Varazanashvili O, Tsereteli N, Tsereteli E (2011) Historical earthquakes in Georgia (up to 1900). LTD "MVP", Tbilisi

Varazanashvili O, Tsereteli N, Bonali FL et al (2018) GeoInt: the first macroseismic intensity database for the Republic of Georgia. J Seismol 22(3):625–667

Webber C, Marwan N (eds) (2015) Recurrence quantification analysis: theory and best practices. Springer, Cham/Heidelberg

Webber CL, Zbilut JP (1994) Dynamical assessment of physiological systems and states using recurrence plot strategies. J Appl Physiol 76:965–973

Webber CL, Zbilut JP (2005) Recurrence quantification analysis of nonlinear dynamical systems. In: Riley M, Van Orden G (eds) Tutorials in contemporary nonlinear methods for the behavioral sciences. National Science Foundation, Arlington, pp 26–94

Webber CL, Marwan N, Facchini A et al (2009) Simpler methods do it better: success of recurrence quantification analysis as a general purpose data analysis tool. Phys Lett A 373:3753–3756

Zbilut JP, Webber CL (1992) Embeddings and delays as derived from quantification of recurrence plots. Phys Lett A 171:199–203

Chapter 11
Geohazard Assessment Along the Southern Slope of the Greater Caucasus (Azerbaijan)

Gulam Babayev

Abstract The present work provides an assessment of the seismic hazard across the southern slope of the Greater Caucasus (territory of Azerbaijan). The parameters that have been taken into account include the moment magnitude, seismic energy, peak ground acceleration and site effects. The analysis output is expressed as the spatial distribution of peak ground acceleration (PGA) assessed at the maximum magnitude. The earthquake catalogue from the Republican Center of Seismological Survey (RCSS) at the Azerbaijan National Academy of Sciences (ANAS) was employed in order to perform the research. From the results of the probabilistic analysis, an earthquake with maximum M_W 7.9, is estimated for the central area, corresponding to the worst-case seismic scenario for the region. This is predicted to be the strongest possible earthquake, which can be generated within a collision zone along the Main Caucasus Thrust (MCT) and the West Caspian Fault (WCF). Considering the parameters of two target (scenario) earthquakes, based on the deterministic hazard assessment, the highest PGA values are distributed in the north-western part of the study area: 95–110 Gal (0.95–1.10 m/s^2). This may have a significant impact, especially in terms of the safety of critical facilities in those areas.

Keywords Azerbaijan · Fault · Magnitude · Probabilistic · Catalogue

11.1 Introduction

The research area is situated in a complex zone, in terms of geodynamics and tectonics, included in the Alpine, fold-and-thrust system. The current research has been focused on a region enclosed between 40°N to 42°N and from 46°E to

G. Babayev (✉)
Seismology and geomagnetism Department, Geology and Geophysics Institute, Azerbaijan National Academy of Sciences, Baku, Azerbaijan

© Springer Nature B.V. 2021

F. L. Bonali et al. (eds.), *Building Knowledge for Geohazard Assessment and Management in the Caucasus and other Orogenic Regions*, NATO Science for Peace and Security Series C: Environmental Security,
https://doi.org/10.1007/978-94-024-2046-3_11

50°E. This region hosts a large number of mud volcanoes, linked with the tectonic setting and present-day fault activity (Babayev et al. 2014). Here, we present a new approach for analyzing the possible effects of potential earthquakes in the studied region. In the historical past, several strong and destructive earthquakes struck the study area, causing major damage and economic losses. It is worth pointing out the following earthquakes, with magnitude and year of occurrence: 427, Ms ~ 6.7; 1667, Ms = 6.9; 1669, Ms = 5.7; 1842, Ms ~ 4.5; 1828, Ms = 5.7; 1859, Ms = 5.9; 1872, Ms = 5.7; 1902, Ms = 6.9; 1960, Ms ~ 4.5; 1981, Ms ~ 5.5; 2012, Ms ~ 5.6; 2014, Ms ~ 4.5; 2019, Ms = 5.3 (Telesca et al. 2013; Babayev et al. 2014; Alizadeh et al. 2017; Babayev et al. 2017a; Babayev et al. 2017b; Telesca et al. 2017; Babayev et al. 2019); needless to say, strong historical earthquakes are also reported in other regions in the Greater Caucasus, such as the Republic of Georgia (Varazanashvili et al. 2018). Most earthquakes take place across the eastern sector of the southern slope of the Greater Caucasus: the strongest one, with magnitude M = 6.9, occurred in 1667 near Shamakhi; others, with magnitude M = 4.5 ~ 5.5, occurred near the cities of Sheki and Balaken (Figs. 11.1 and 11.2). The focal depth of those earthquakes varies between 10–15 km.

The goal of this study is to perform an assessment of seismic hazard across the southern slope of the Greater Caucasus (territory of Azerbaijan), both through a probabilistic and a deterministic approach, taking into consideration the earthquakes that hit the area in the 2012–2018 time interval.

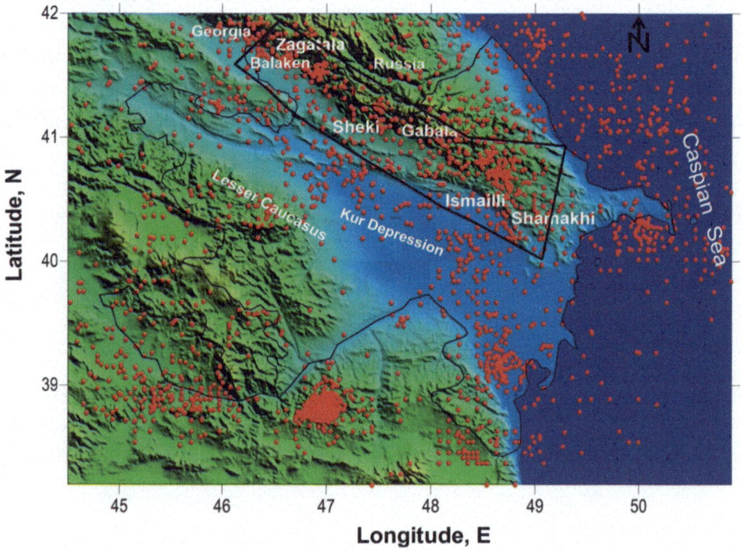

Fig. 11.1 Seismicity of Azerbaijan in the 1963–2018 time interval; Ml range = 3.0–6.6. Note: The location of the study area is highlighted by the black polygon. Reference system: WGS84 / geographic coordinates

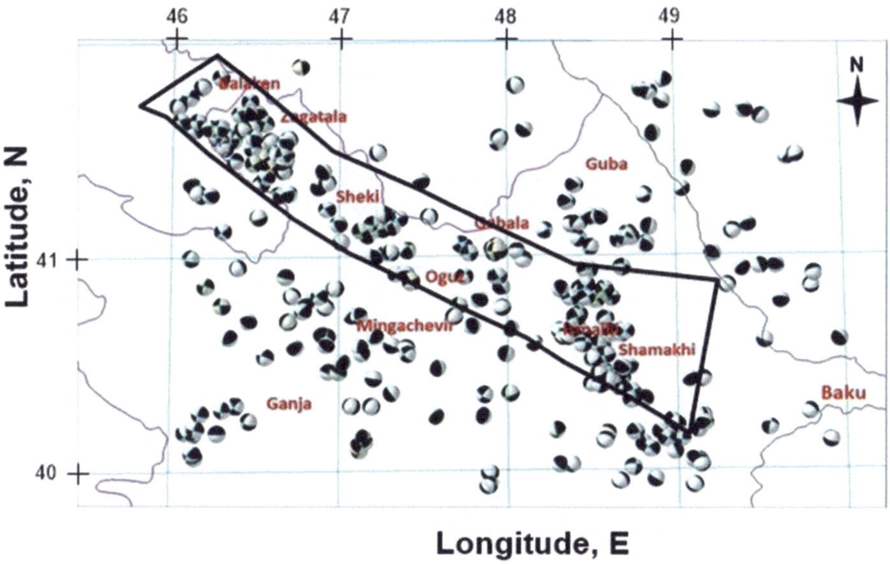

Fig. 11.2 Focal mechanism solutions for the 2003–2018 time interval; earthquakes with Ml > 3. Reference system: (Babayev et al. 2019): WGS84 / geographic coordinates

11.2 The Study Area

The present-day Caucasus range encompasses the southern edge of the Eurasian continent and the northern edge of the central Iranian microcontinent, with the south Azerbaijan segment (Alizadeh et al. 2016). Tectonically, the Greater Caucasus is located in the collision zone between the Arabian and Eurasian plates (Mckenzie 1972; Sengor et al. 1985; Philip et al. 1989; Kadirov 2000; Reilinger et al. 2006; Tibaldi et al. 2017a, b). The seismogenic structures of the region are due to the continuous movement of the Arabian Plate toward the north-northeast (Alizadeh et al. 2016). Azerbaijan belongs to the eastern portion of the Caucasian segment of the Mediterranean fold-and-thrust belt. A characteristic feature of the structure is its tectonic heterogeneity, with a great number of structural units and stratigraphic successions that were subject to a wide gamut of formation and deformation processes (Alizadeh et al. 2016).

11.3 Methodology

The employed methodology consists of a probabilistic and a deterministic component. The main parameters used in both approaches are earthquake instrumental data (magnitude, epicenter distance, focal depth), as well as the physical and dynamic properties of the soil. A correlation between the maximum amplitude of ground shaking and seismic intensity is carried out according to the generally accepted international scale (e.g., Trifunac and Brady 1975). For the study area, with the purpose of performing a probabilistic seismic hazard assessment, the relationship between the magnitude of the earthquake and the length of faults and lineaments has been explored and analyzed, through mathematical formulas. The results obtained reflect the relationship between fault length and the magnitude of strong earthquakes occurring in different regions all over the world. The hereunder performed analysis indicates that these parameters are fully consistent with each other.

The length of the various faults and lineaments, as well as the estimated magnitude based on empirical correlations (Bonilla et al. 1984; Gubin 1974; Riznichenko 1965; Shebalin 1961; Utsi 1961) are shown in Table 11.1, calculated on the basis of the Azerbaijan fault map (Figs. 11.3 and 11.4). In Table 11.1, magnitude (Mw) and energy class (K) data are provided for faults with length L $\geq$ 30 km, assuming that the faults in that length class can produce greater magnitudes. Fault lengths (Fig. 11.3) were calculated through the ArcGIS 10.0 software (ESRI 2011). The rupture lengths along the faults and lineaments were taken as one-third of total fault lengths, according to Baghbani et al. (2016).

Table 11.1 Magnitude (M_W), fault length (L) and energy (K) for various faults estimated by empirical correlations

Fault, №	Length (km), L	Average Mw	K
1	30	6.3	15
2	35	6.5	16
3	88	7.2	17
4	35	6.5	16
5	34	6.4	16
6	30	6.3	15
7	40	6.6	16
8	35	6.5	16
9	40	6.6	16
10	40	6.6	16
11	35	6.5	16
12	200	7.9	18
13	30	6.3	15
14	36	6.5	16
15	46	6.7	16
16	37	6.5	16
17	85	7.2	17
18	70	7.0	17
19	40	6.6	16
20	50	6.8	16
21	65	7.0	17
22	38	6.5	16
23	58	6.9	16
24	30	6.3	15

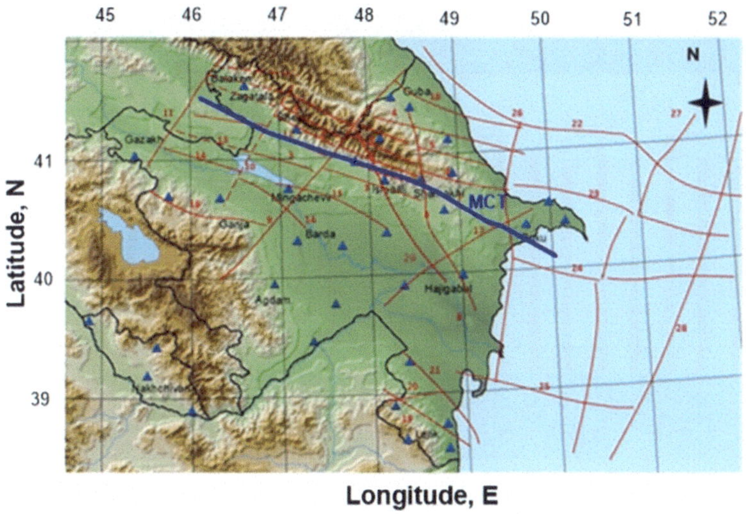

Fig. 11.3 Scheme of active faults in the Azerbaijan territory. Faults: 1-Dashgil-Mudresinsky, 2-Vandam, 3-Geokchai, 4-Siyazan, 5-Zangi-Kozluchay, 6-Germian, 7-Adzhichay-Alat, 8-West-Caspian, 9-Arpa-Samur, 10-Ganjachay-Alazan, 11-Kazakh-Signagi, 12-North Adzhinour, 13-Iori, 14-Kur, 15-Mingachevir-Saatli, 16-Bashlybelsky, 17-Palmiro-Absheron, 18-Akhty-Nyugedi-Kiliazin 19-Talysh, 20-Yardimli, 21-Near Talysh, 22-Central Caspian, 23-Apsheron-Balkhan, 24-Sangachal-Ogurchinsky, 25-Mil-Chikishlyar, 26-Jasper flexure, 26a-Gizilagach, 27–Shakhov-Azizbayov, 28-Karabogaz-Safidrud, 29-Syghyrly-Gabala. Blue line denotes Main Caucasus Thrust (MCT). Triangles point to seismic station locations. Reference system: (Babayev et al. 2019): WGS84 / geographic coordinates

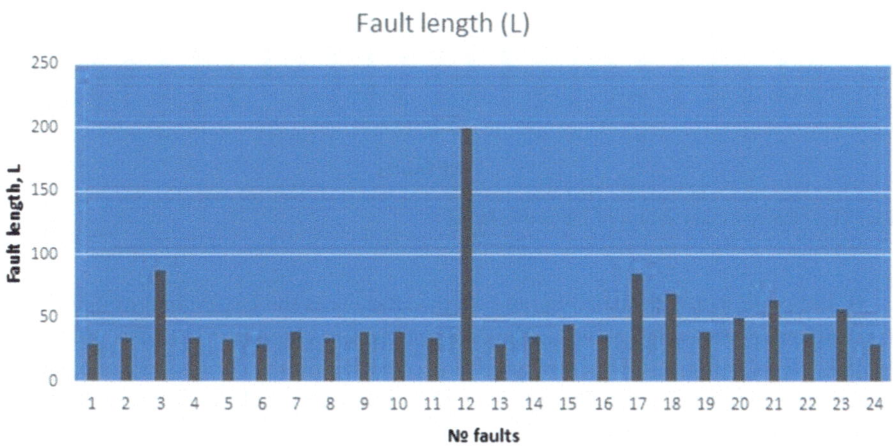

Fig. 11.4 Fault lengths of the active faults, L is expressed in km

Hereunder, the calculated parameters summarized in Table 11.1 are illustrated in Figs. 11.4, 11.5 and 11.6, as the column graphs with the respective values.

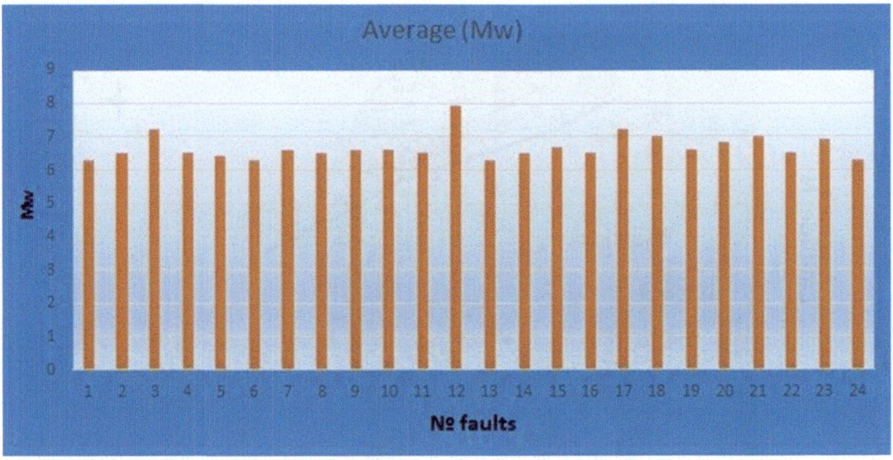

Fig. 11.5 Moment magnitude values of the active faults

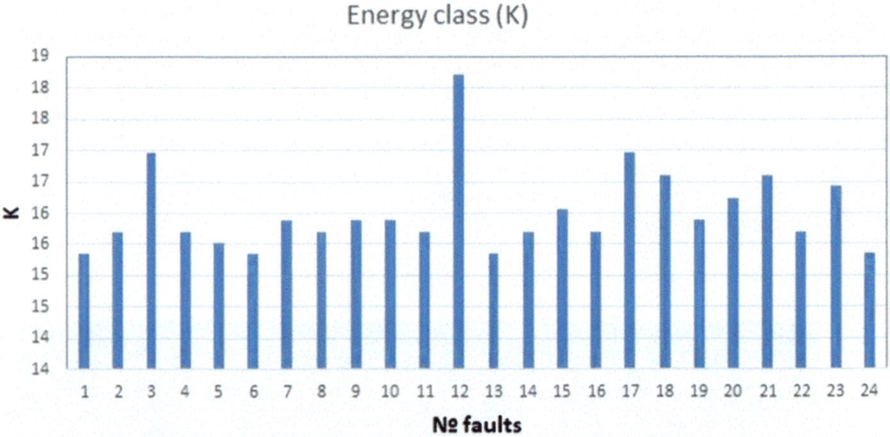

Fig. 11.6 Energy class values of the active faults

Based on this probabilistic analysis, in the study area the largest magnitude (M_W) the fault can generate is 7.9, while the energy class is as high as 18. This is predicted to be a huge earthquake, which may be generated within the collisional zone, along the Main Caucasus Thrust (MCT) (Fig. 11.3, blue line) and the WCF fault (West Caspian Fault) (Telesca et al. 2017).

In the terms of a deterministic assessment, the method includes the consideration of seismogenic elements, which enable determining the distribution of the PGA for each cell of the square grid. This calculation involved 46 square grid cells with an average area of 10×10 km^2, which incorporates the study area. Using the SHAKE program (Ordonez 2003), amplification amplitudes of seismic waves passing through the soil strata were calculated for a typical lithology. The lithology

Table 11.2 Lithological and physical parameters of ground units for the study area

Age	Lithology	Vp, km/s	Vs, km/s	Density, kg/sm^3	Amplification factor>
Holocene (modern deposits)	Argillites, limestone, marlstone, sandstone	1.50	0.35	1.9	1.10
Upper Pleistocene	Thin-layered sandstone	3.30	2.02	2.35	0.53
Lower Pleistocene	Argillites and sandstone	2.10	0.50	2.00	0.95
Neogene	Dolomite, marlstone, sandstone	2.67	0.88	1.98	0.75
Paleogene	Limestone, aleurolite, marlstone	2.23	0.69	1.87	0.83
Cretaceous	Sandstone, aleurolite and clayey shales	3.93	1.63	2.38	0.58
Jurassic	Aleurolite, marlstone	4.18	1.82	2.49	0.55

of the study area consists mainly of Tertiary and Quaternary sandstones, limestones, marlstones, argillites, clayey shales and gabbro rocks. The subsurface models employed in our study are presented in Table 11.2.

In order to calculate the expected maximum amplitude of acceleration of the bedrock, the following formulas were employed for each square grid cell (1), (2) (Aptikayev and Kopnichev 1979):

For far seismic events:

$$Lg\ Ap = 0.80M - 2.3LgR + 0.80\ \text{A} < 160\ \text{sm/s}^2 \qquad (11.1)$$

For near events:

$$LgAp = 0.28M - 0.8LgR + 1.7\ \text{A} \geq 160\ \text{sm/s}^2 \qquad (11.2)$$

Here, Ap is the peak acceleration in Gal, M is the magnitude, and R is the hypocentral distance in km.

Using macroseismic parameters of selected scenario earthquakes, we have calculated peak ground acceleration values at the surface; their distribution is reflected through the plotted numerical cell models.

As scenario earthquakes, two seismic events were chosen from the selective published catalogues (Babayev 2010; Babayev et al. 2010, 2017a; b; Babayev and Telesca 2014): *i)* the Oguz earthquake, which occurred in 2015 ($\varphi = 41.67$; $\lambda = 47.47$) with magnitude Ml 5.9, at a depth (h) of 20 km; *ii)* the Zagatala earthquake, which struck in 2018 ($\varphi = 41.50$; $\lambda = 46.67$) with a magnitude Ml 5.5, at a depth (h) of 10 km. The hypocentral distances are calculated from the epicentral distances assuming the suitable depth to focus (hypocenter) (Baghbani et al. 2016). By using the illustrated methodology, maps of the distribution of peak ground acceleration throughout the study area were produced.

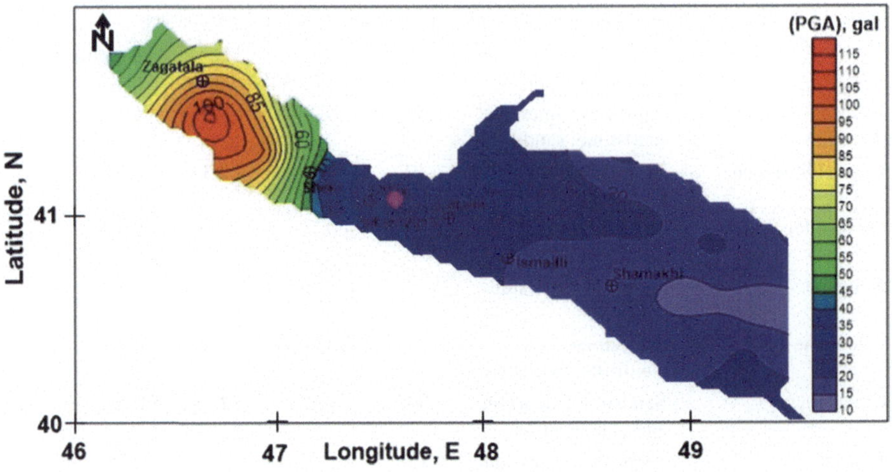

Fig. 11.7 Peak Ground Acceleration (PGA) distribution based on the target (scenario) earthquake of 2015. Note: pink circle point to the epicentre of the scenario earthquake. Reference system: WGS84 / geographic coordinates

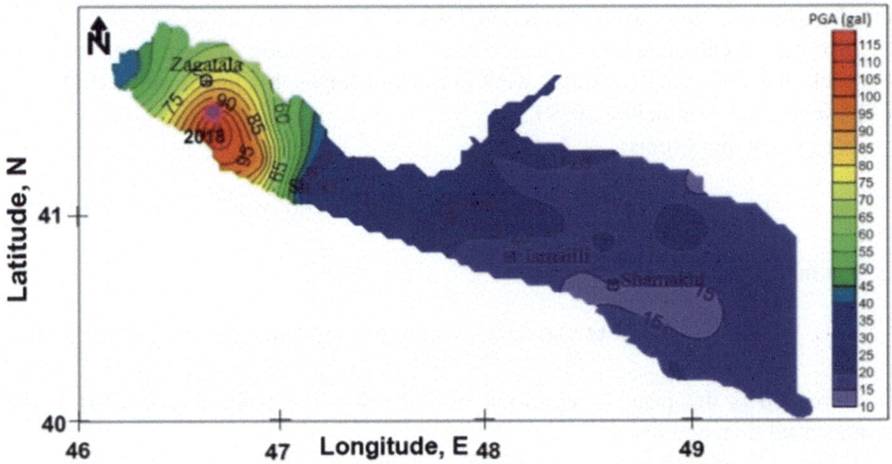

Fig. 11.8 Peak Ground Acceleration (PGA) distribution based on the target (scenario) earthquake of 2018. Note: pink circle point to the epicentre of the scenario earthquake. Reference system: WGS84 / geographic coordinates

Analyzing the two maps (Figs. 11.7 and 11.8), it is clear that the highest PGAs are distributed in the north-western part of the study area with values in the 95–110 Gal (0.95–1.10 m/s^2) range. This may have a significant impacts, especially in terms of the safety of critical facilities in those areas.

11.4 Conclusions

Considering the macroseismic parameters of the 2018, Zagatala earthquake and 2015, Oguz earthquake, the calculated values of the PGA range from 110 to 10 Gal. The sharp decrease to the lowest values can be explained in terms of the effects of soil damping in the study area. As expected, although there are several lineaments and faults in the region, which can generate major earthquakes, site effects play an important role in seismic wave attenuation. This integrated analysis of seismic parameters has enabled hypothesizing the possibility of a major, future earthquake along the MCT and the WCF.

References

Alizadeh AA, Guliyev IS, Kadirov FA, Eppelbaum LV (2016) Geosciences of Azerbaijan. Volume I: geology. Springer, Cham, p 239. https://doi.org/10.1007/978-3-319-27395-2.239

Alizadeh AA, Guliyev IS, Kadirov FA, Eppelbaum LV (2017) Geosciences of Azerbaijan. Volume II: economic minerals and applied geophysics. Springer, Cham, p 340. https://doi.org/10.1007/978-3-319-40493-6

Aptikayev F, Kopnichev Y (1979) Considering focal earthquake mechanism at the prediction of strong motion parameters. Doklady/Transactions of the U.S.S.R. Academy of Sciences 247, pp. 822–825 (in Russian)

Babayev G (2010) About some aspects of probabilistic seismic hazard assessment of Absheron peninsula. Republican Seismic Survey Center of Azerbaijan National Academy of Sciences. Catalogue of Seismoprognosis Research Carried Out in Azerbaijan Territory in 2009, "Teknur", Baku, pp. 59–64 (in Russian)

Babayev G, Telesca L (2014) Strong motion scenario of 25th November 2000 earthquake for Absheron peninsula (Azerbaijan). Nat Hazards 73:1647–1661

Babayev G, Ismail-Zadeh A, Le Mouël J-L (2010) Scenario-based earthquake hazard and risk assessment for Baku (Azerbaijan). Natural Hazards Earth System Science 10:2697–2712. https://doi.org/10.5194/nhess-10-2697-2010

Babayev G, Tibaldi A, Bonali FL, Kadirov F (2014) Evaluation of earthquake-induced strain in promoting mud eruptions: the case of Shamakhi–Gobustan–Absheron areas, Azerbaijan. Nat Hazards 72(2):789–808

Babayev G, Muradi I, Babayev E (2017a) Probabilistic seismic hazard assessment for Absheron peninsula (Azerbaijan). Proceedings of AASSA NAS KR Regional Workshop "Science, State, Structures and Public – Joint Efforts to Reduce the Risks and Consequences of Earthquakes", "Printhouse" Publishing House, Bishkek, Kirgizistan, October, pp 33–38

Babayev GR, Akhmedova EV, Kadirov FA (2017b) Analysis of stress-strain state of Caucasus region (Azerbaijan) on the basis of maximum horizontal stress vectors and world stress map. Application technique. Geophys J 3(39):26–39. (in Russian). https://doi.org/10.24028/gzh.0203-3100.v39i3.2017.104026

Babayev G., Yetirmishli G., Kazimova S., Kadirov F., Telesca L (2020) Stress Field Pattern in the Northeastern Part of Azerbaijan. Pure Appl. Geophys. Vol. 177(1):2739–2751. Springer Nature Switzerland AS. https://doi.org/10.1007/s00024-019-02371-5

Baghbani M, Ebrahim G, Rostami BHR (2016) Probabilistic seismic Hazard analysis for a Dam Siyaho in South Khorasan province (Eastern Iran). Geodynamics Research International Bulletin 4(02):34–48

Bonilla MG, Mark RK, Lienkaemper JJ (1984) Statistical relations among earthquake magnitude, surface rupture length and surface fault displacement, vol 74. Bulletin of Seismology Society of American, pp 2379–2411

ESRI. ArcGIS desktop: release 10. Environmental Systems Research Institute (ESRI), 2011, Redlands

Gubin IE (1974) Seismogenic faults and their meaning for seismic zoning. Geotectonics 6:29–40. in Russian

Kadirov FA (2000) Gravity field and models of deep structure of Azerbaijan. Nafta-Press, Baku, p 112. (monograph in Russian)

McKenzie DP (1972) Active tectonics of the Mediterranean region. Geophys J R Astron Soc 30:239–243

Ordónez GA (2003) SHAKE2000: A computer program for the 1-D analysis of the geotechnical earthquake engineering problem

Philip H, Cisternas A, Gvishiani A, Gorshkov A (1989) The Caucasus: an actual example of the initial stages of continental collision. Tectonophysics 161:1–21

Reilinger RE, McClusky SC, Vernant P, Lawrence S, Ergintav S, Cakmak R, Ozener H, Kadirov F, Guliev I, Stepanian R, Nadariya M, Hahubia G, Mahmoud S, Sakr K, Arrajehi A, Paradissis D, Al-Aydrus A, Prilepin M, Guseva T, Evren E, Dmirotsa A, Filikov SV, Gomez F, Al-Ghazzi R, Karam G (2006) GPS constraints on continental deformation in the Africa-Arabia-Eurasia continental collision zone and implications for the dynamics of plate interactions. J Geophys Res 111:B5. https://doi.org/10.1029/2005JB004051

Riznichenko YV (1965) From the activity of seismic sources to the intensity recurrence at the ground surface. Izv AN SSSR, Fizika Zemli 11:1–12. (in Russian)

Sengor AMC, Gorur N, Saroglu F (1985) Strike slip faulting and related basin formation in zones of tectonic escape: Turkey as a case study. In K. T. Biddle& N. Christie-Blick (Eds.), Strike slip faulting and basin formation. SEPM Special Publication, vol. 37, pp 227–264

Shebalin N (1961) Intensity, magnitude, and source depth of earthquakes. In: Earthquakes in the USSR. USSR Academy of Sciences. p. 126–138

Telesca L, Lovallo M, Babayev G, Kadirov F (2013) Spectral and informational analysis of seismicity: an application to the 1996-2012 seismicity of northern Caucasus-Azerbaijan part of greater Caucasus - Kopet dag region. Physica-A Statistical Mechanics and Its Applications 392:6064–6078. https://doi.org/10.1016/j.physa.2013.07.031

Telesca L, Kadirov F, Yetirmishli G, Safarov R, Babayev G, Ismaylova S (2017) Statistical analysis of the 2003-2016 seismicity of Azerbaijan and surrounding areas. J Seismol 21:1467–1485. https://doi.org/10.1007/s10950-017-9677-x

Tibaldi A, Alania V, Bonali FL, Enukidze O, Tsereteli N, Kvavadze N, Varazanashvili O (2017a) Active inversion tectonics, simple shear folding and back-thrusting at Rioni Basin, Georgia. J Struct Geol 96:35–53

Tibaldi A, Russo E, Bonali FL, Alania V, Chabukiani A, Enukidze O, Tsereteli N (2017b) 3-D anatomy of an active fault-propagation fold: a multidisciplinary case study from Tsaishi, western Caucasus (Georgia). Tectonophysics 717:253–269

Trifunac M, Brady A (1975) On the correlation of seismic intensity scales with the peaks of recorded strong ground motion. Bull Seismol Soc Am 65:139–162

Utsi T (1961) Stastistical study of occurrence of aftershocks. Geophys Mag 30:521–605

Varazanashvili O, Tsereteli N, Bonali FL, Arabidze V, Russo E, Mariotto FP et al (2018) GeoInt: the first macroseismic intensity database for the Republic of Georgia. J Seismol 22(3):625–667

Part IV
Key Studies for Volcanic Hazard Assessment

Chapter 12
Quaternary Volcanic Activity in the Greater Caucasus: A Review of Elbrus, Kazbek and Keli Volcanoes

Giorgi Vashakidze, Anton V. Parfenov, Ketevan Gabarashvili, Manana Togonidze, Tamar Beridze, and Vladimir A. Lebedev

Abstract Our work provides as review of Quaternary volcanic activity in the Greater Caucasus. In particular, we have focused our attention on the Elbrus (northern slope of the main range of the Greater Caucasus), Kazbek (central sector of the Greater Caucasus) and Keli (southern slope of the main range of the Greater Caucasus) volcanoes. After conducting detailed isotope-geochronological, geological and petrological-mineralogical studies regarding the above volcanic centres, we have been able to document multiple phases of volcanic activity: *i)* five phases have been defined for the Elbrus volcanic centre: I – 950-900 ka (andesite-basalts, trachyandesites and dacites), II – 800-700 ka (ignimbrites, rhyodacites, dacites and andesites), III – 225-170 ka (dacites), IV – 110-70 ka (dacites) and V phase – less than 35 ka (dacites); *ii)* four phases have been defined for the Kazbek centre: I – 460-380 ka (basaltic andesites and andesites), II – 310-200 ka (latest phase – andesites and dacites, earliest phase – basaltic trachyandesites, basaltic andesites, and dacites), III – 130-90 ka (latest phase – dacites, earliest phase – andesites and trachyandesites), IV – less than 50 ka (andesites and dacites); *iii)* for the Keli centre, three phases have been defined: I – 245-170 ka (dacites, rhyolites and andesite-dacites), II – 135-70 ka (earliest phase – dacites and rhyolites, latest phase – dacites, rhyolites, andesites and trachyandesites), III – less than 30 ka (dacites and andesites). Owing to the Holocene age of the latest phases of volcanic activity, these volcanic centres ought to be regarded as potentially active (dormant).

G. Vashakidze (✉) · K. Gabarashvili · M. Togonidze · T. Beridze
A. Janelidze Institute of Geology, I. Javakhishvili Tbilisi State University, Tbilisi, Georgia

A. V. Parfenov · V. A. Lebedev
Institute of Geology of Ore Deposits, Petrography, Mineralogy, and Geochemistry, Russian Academy of Sciences, Moscow, Russia

© Springer Nature B.V. 2021

F. L. Bonali et al. (eds.), *Building Knowledge for Geohazard Assessment and Management in the Caucasus and other Orogenic Regions*, NATO Science for Peace and Security Series C: Environmental Security,
https://doi.org/10.1007/978-94-024-2046-3_12

12.1 Introduction and Background

The Greater Caucasus region has been affected by several geohazards during historical times, including earthquakes (Tsereteli et al. 2016; Varazanashvili et al. 2018) and landslides (Tibaldi et al. 2019). Moreover, in Quaternary times the Greater Caucasus region has been subject to major volcanic activity. The Kazbek (Mkinvartsveri), Keli, Elbrus (Fig. 12.1), Kabardjina and Djava (Dzau) neovolcanic centres as well as of some other volcanos, erupted several times; Quaternary volcanic activity is also reported in the Lesser Caucasus (Pasquaré et al. 2011). Previous research (Burchuladze et al. 1976; Chernyshev et al. 2002, 2008; Vashakidze et al. 2019) has enabled to assess Holocene volcanic activity at the Kazbek and Elbrus volcanic centres, and thus the existence of volcanic hazard in the Greater Caucasus. This prompted the need for multidisciplinary geological, geophysical, petrological-geochemical and isotope studies regarding the Quaternary magmatic activity in this region. As a result of field-based investigations and subsequent isotope-geochronological studies of igneous rocks - K-Ar dating of young volcanic rocks was performed through high-sensitivity, low-background techniques elaborated at the Institute of ore geology, petrography, mineralogy and geochemistry IGEM RAS -, the contours of areas of younger volcanic activity have been determined; this has enabled revealing a significant number of new and unknown volcanic vents, which were active in Late Pleistocene-Holocene times. In the present work, we have attempted to assemble all available data on the Kazbek, Keli and Elbrus volcanic centres as well as to make correlations between their pulses of activity, with as much detail as possible.

Fig. 12.1 Location of the three investigated neovolcanic centres in the Greater Caucasus: Elbrus, Kazbek and Keli. Shaded view is based on Aster GDEM v. 2 (Meyer 2011; Tachikawa et al. 2011). Reference system: WGS84 / geographic coordinates

12.2 The Kazbek Neovolcanic Centre

The Kazbek neovolcanic centre (Fig. 12.1) covers one of the largest areas affected by Quaternary volcanic activity in the Greater Caucasus. It is located in the central part of the range, at the Georgia-Russia border, in the Khokhi Range area. This volcanic centre extends across an area of about 250 km^2 in the headwaters of the river Terek (Tergi) and its left tributaries – Devdoraki, Chkheri, Kesia, Mnaisi and Suatisi (Fig. 12.2). The centre is associated with the intersection between the Main Caucasus reverse fault and the Kazbek–Tskhinvali deep-seated fault, which is a part of the left lateral, Agrakhan-Tbilisi Levantian strike-slip fault. The Kazbek centre incorporates the polygenetic stratovolcano Kazbek as well as more than ten satellite volcanoes, which surround it in the form of an arcuate zone from the East, South and West.

We have assessed four phases of volcanic activity, based on detailed fieldwork and petrological-geochemical and isotope-geochronological studies. These are: phase I – 460-380 ka, phase II – 310-200 ka, phase III – 130-90 ka, phase IV – less than 50 ka. Moreover, an early (II^1 – 310-260 ka) and a late (II^2–240-200 ka) episode have been distinguished within phase II. Unfortunately, errors in the obtained isotope dating results for phase III have not enabled us to assess the transition in time between the two episodes. In this case, lava flows have been attributed to early or late episodes of phase III, on the basis of their stratigraphic location and correlations between the chemical compositions of the rocks.

12.2.1 Phase I (Mid-Pleistocene, 460-380 ka)

The first phase of the Kazbek volcanic centre activity was concentrated in its northern part and corresponds to unit 4 in Fig. 12.2. During this phase, three lava flows (Devdoraki, Gveleti and Bloti) with different areal extent were emplaced (Fig. 12.2).

The nearly 7-km-long Devdoraki lava flow covers the Bart-Kort ridge. It is composed of two branches: the northern branch overlains Paleozoic granitoids of the Dariali massif whereas the southern branch rests over Jurassic schists. According to our estimates, the Devdoraki flow comprises five to six basaltic andesite units and, above them, two (in the southern branch) to seven (in the northern branch) andesite units (Lebedev et al. 2018). Remnants of the about 6-km-long Gveleti lava flow crop out on the inaccessible Bagni-Arch-Kort ridge (the rivers Devdoraki and Khreti watershed) and are made up of basalt andesites.

In the past, the Bloti lava flow filled the valley of the same river for an extent of about 5 km; today, it is almost completely eroded: in fact, we found only two minor andesite remnants on the upper terrace of the left bank of the Bloti canyon (west of the Artsivistsveri mountain) (Lebedev et al. 2018).

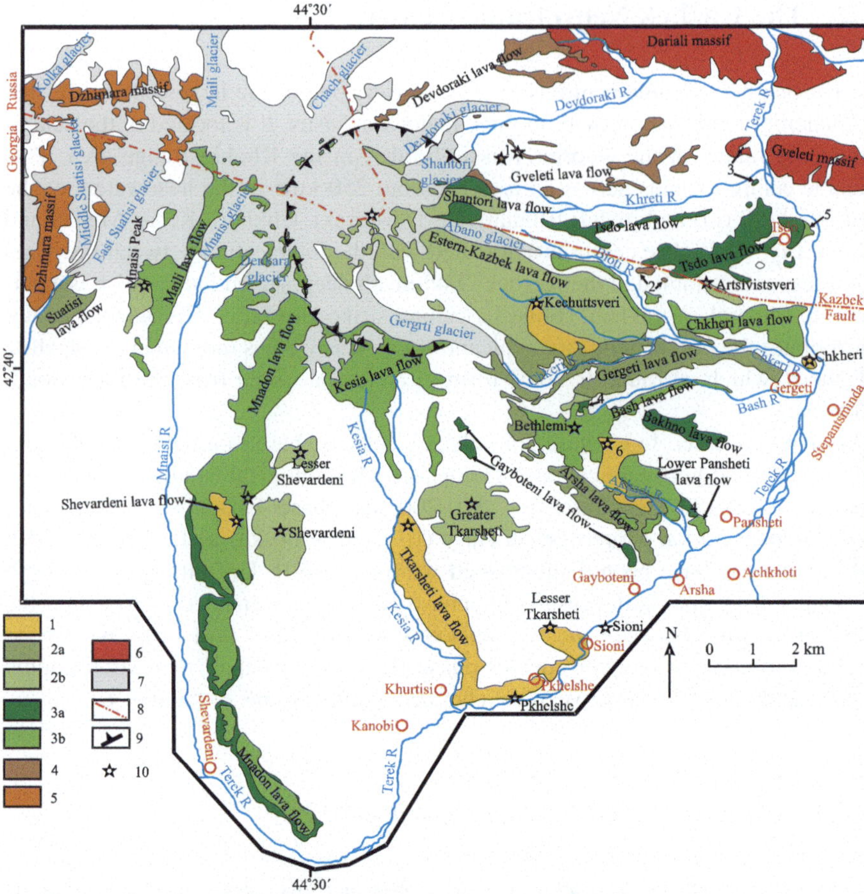

Fig. 12.2 Schematic geological map of the Kazbek neovolcanic centre. (1–4) volcanic rocks of the Kazbek centre of different ages: (1) phase IV, (2a) early episode of phase III, (2b) late episode of phase III, (3a) early episode of phase II, (3b) late episode of phase II, (4) phase I, (5) Pliocene granodiorite and quartz diorite of the Dzhimara massif, (6) granites of the Darial and Gveleti massifs (Middle–Late Carboniferous), (7) glaciers and firn fields (as of 2010), (8) modern active faults, (9) caldera scarp of Paleo-Kazbek Volcano, (10) volcanic apparatuses (lava cones, extrusive domes, and explosive centres). Numerals in the map: (1) dikes and necks of the Devdoraki Gorge (phase I), (2) remnants of the Bloti lava flow (phase I), (3) Gveleti and second Gveleti lava remnants of the Tsdo lava flow (early stage of phase II), (4) remnants of the Akhadi lava flow (early stage of phase II), (5) remnant of the Chkheri lava flow (lower part of lava remnant under the Tsdo settlement, late stage of phase II), (6) Upper Arsha flow of the Bethlemi Volcano (phase IV), (7) neck on the western slope of Shevardeni Volcano (phase IV)

The study of the possible paths of all the above mentioned lava flows indicates that they were formed as a result of the activity of a single volcanic vent, the Proto-Kazbek (apparently a fissure vent), which was located in the source area of the present-day Devdoraki glacier, which in turn, was completely obliterated by erosion and subsequent magmatic activity.

Two minor volcanic edifices have been identified as well. Their remnants, represented by dacitic necks, are located within the deep troughs in the southern part of the Devdoraki gorge. They separate into thin andesite dikes, which crosscut Jurassic shales and siltstones (Lebedev et al. 2016, 2018).

12.2.2 Phase II (Middle Pleistocene, 310-200 ka)

Throughout the second magmatic phase of activity, that corresponds to unit 3a and 3b in Fig. 12.2, the Paleo-Kazbek stratovolcano, which occupied the area of the present-day Kazbek volcano, was the main eruptive site (Dzotsenidze et al. 1985; Lebedev and Vashakidze 2014). The scale of its activity and the volume of the erupted material significantly exceed both those of its predecessor – the Proto-Kazbek volcano, and the present-day edifice. Overall, phase II of activity at the Kazbek centre may be regarded as the most productive, during the whole period of its development.

12.2.2.1 Early Episode of Phase II (II^1–310-260 ka)

During the early episode of phase II, that corresponds to unit 3a in Fig. 12.2, five lava flows with various thickness (Tsdo, Bakhno, Akhadi, Gaiboteni and lower horizons of Mnadon) were formed. They moved towards the east and south.

The lavas of the Tsdo flow spread east of the volcano, along the Khreti River paleo valleys and its tributaries for more than 10 km. At present, they form shields on the watershed ridges between the Bloti and Khreti Rivers. The initial portion of the flow is preserved only as a series of minor isolated remnants of andesites and dacites. In its middle part, the Tsdo flow separates into a southern and a northern (Khreti) branches. The latter branch is less than 3 km long and forms a shield on the watershed between the Khreti River and its right tributary, the Tibaistskali. The major branch of the flow above the Tsdo village was the least affected by erosion. Two horizons of basalt (trachy-) andesites form its lower part. They are overlain by at least four horizons of (trachy-) andesite lavas. Downstream the Terek River, two minor remnants of the Tsdo lava flow can be found. The first one – Gveleti – is located on the rocky spur above the eponymous village and made of basalt trachyandesites, which are similar in composition to the rocks of the lower horizons of the section above the village of Tsdo. Another remnant (Gveleti 2), documented by Lebedev et al. (2018), is located on the left bank of the Terek River, 0.5 km below the village of Gveleti, on the spur built by the Gveleti massif of Paleozoic granitoids. Its andesite lavas are close in composition to the volcanic rocks of the middle part of the section above the Tsdo village. Thus, the termination part of the Tsdo flow has been now almost completely obliterated by erosion.

On the crest of a small ridge south of the Bash creek, opposite Stepantsminda, the 2 km long remnant of the Bakhno andesite lava flow is preserved.

A minor, 1-km-long remnant of basaltic trachyandesites is located to the south, above the Pansheti village. It represents the termination of a small lava flow, which was formed at the initial stage of the Paleo-Kazbek activity.

On the left slope of the Terek River valley, near the Gaiboteni village, a columnar-jointed remnant of the eponymous andesite flow is preserved. Higher upslope, along the watershed between the river Kesia and the creek Chkhati, we found two extra remnants of the upper part of the Gaiboteni flow (Lebedev et al. 2018), which are similar in composition and of the same age.

The most extensive (15 km) of all flows emitted by the centre – the Mnadon flow - is the result of the final eruption of the early episode of phase II activity at the Kazbek centre. During this phase, the lowermost units of the Mnadon flow, of andesite composition, were formed.

12.2.2.2 Late Episode of Phase II (II^2–240-200 ka)

During the late episode of the second phase of activity, corresponding to unit 3b in Fig. 12.2, activity at the Paleo-Kazbek volcano resumed. The Mnadon (upper horizons), Kesia, Maili and Chkheri lava flows, with different thicknesses, were formed (Fig. 12.2).

The Mnadon flow continued its formation: in fact, 2–3 units of andesite composition were generated. In the area of the later extrusive dome Shevardeni, the flow succession is overlain by the sheet of the dacite lava flows with the same age and further towards the sources of the river Mnaisi by at least seven horizons of volcanic rocks of moderately acid composition.

The Kesia lava flow originated on the slopes of Ortsveri peak and produced two small lobes of andesite composition extending southward, which are 2.5–3 km long.

In the springs of the Mnaisi River, on the right bank of its gorge, the 6 km long Maili lava flow of dacite composition is observed. Due to the age, stratigraphic position and composition of igneous rocks dacites of the Maili flow could be correlated with volcanic rocks of the upper part of the Mnadon flow.

In the eastern part, the up-to-10-km-long Chkheri andesite lava flow was formed. It begins near the termination of the Abano glacier and is further distributed along the right bank of the Bloti River gorge, at the base of its slope until the mouth of the Bloti River. Below, the flow is completely dissected by the gorge of the Bloti River and further forms the left bench of the Chkheri River gorge. It should be emphasized that the chemical composition of the Chkheri lavas is similar to the andesites of the same age from the middle part of the Mnadon flow section and upper reaches of the Kesia River.

During this late episode of the II phase, synchronously with the Paleo-Kazbek eruptions, the active polygenetic volcano Bethlemi was formed, producing the Lower Pansheti and Bash lava flows.

The Lower Pansheti flow, up to 3 km long, is composed of lavas of andesitic composition and occupies the slope of the Terek River in the environs of the Pansheti village. Pyroclastic deposits can be observed at the base of the Lower Pansheti flow.

The short (up to 2.5 km long) and narrow flow of the Bashi stream rests at the bottom of the modern trough of this small watercourse and comprises two andesite lavas units. Its termination is observed in the proximity of the Tsminda-Sameba (Holy Trinity) church.

Voluminous eruptions of Paleo-Kazbek during the entire phase II of magmatic activity led to the removal of magma from the shallow reservoirs beneath the volcano. As a result, the subsequent subsidence and collapse of the Paleo-Kazbek cone (apparently during 200–150 ka) took place, followed by the formation of a caldera, about 5 km in diameter. This process was most likely not accompanied by major explosive activity, and this is confirmed by the almost complete absence of Late Pleistocene pyroclastic units in the study area.

12.2.3 Phase III (Late Pleistocene, 130-90 ka)

Detailed field studies of key lava successions in the gorges of the Chkheri and Bloti Rivers, as well as the results of chemical analyses of the rocks and interpretation of satellite images, allowed us to reliably subdivide the magmatic formations of phase III (130-90 ka) into two units, corresponding to units 2a and 2b in Fig. 12.2, corresponding to early and late activity episodes during this time period. It is worth pointing out that all volcanic centres of the early episode have a trachy-andesitic composition, whereas the volcanic centres of the late episode are of dacitic composition. However, K-Ar dating results for all volcanics of the Kazbek centre erupted during phase III are identical, within the realm of analytical uncertainty. Accordingly, there is no possibility to detect the age boundary based on isotope data, which divides volcanism of this phase into early (III^1) and late (III^2) episodes.

12.2.3.1 Early Episode of the Third Phase of Activity (III^1–130-90 ka)

The Gergeti, Arsha and Eastern Kazbek lava flows (Fig. 12.2) characterized by different thicknesses and lengths and by andesitic to trachyandesitic composition, were emplaced during the early episode of phase III of the Kazbek centre. This phase corresponds to unit 2a in Fig. 12.2. They were erupted by the Kazbek volcano, which started its activity within the Paleo-Kazbek caldera, and by the Mnaisi Peak volcano (Fig. 12.3), on the right bank of the Mnaisi River, near the eponymous glacier.

The Gergeti is the most extensive lava flow (up to 8 km) formed during the early episode of the third phase of volcanic activity. It begins at the termination of the eponymous glacier, forms the Kvenetmta ridge and terminates on the left bank of the Terek River beneath the Gergeti village.

Fig. 12.3 The Mnaisi Peak volcano (Photo by V.A. Lebedev); for its location, see Fig. 12.2

The 6-km-long Arsha flow is split from the Gergeti flow, in the sources of the river Chkheri south of the Gergeti (Ortsveri) glacier. It bypassed the Bethlemi volcano cone to the west, filled the Chkhati River paleo-valley and reached the Terek River in the area of modern Gaiboteni and Arsha villages.

The Eastern Kazbek flow was mapped for the first time by the authors of this publication (Lebedev et al. 2018). It begins at the eastern base of the modern Kazbek cone and covers most of the area between the Chkheri and Bloti Rivers. Trachyandesites in the lower part of its section crop out only above the Abano glacier, on the right side of its valley as well as NW of the Kechuttsveri volcano, at the base of a small lava ridge. Apparently, the length of the trachyandesite flows did not exceed 3 km.

During this episode of activity, the small Mnaisi Peak volcano was active in the upper reaches of the Suatisi and Mnaisi rivers, located on their watershed and firstly described by Lebedev et al. (2018). As a result of its eruptions, the Suatisi flow, 3 km long, was formed and rests on the Suatisi riverbed at its source.

The resumption of volcanic activity in the region about 130 ka with the eruption of more basic lavas in comparison with the end of phase II, was due most likely to the inflow of a new portion of deep basic magma melts (replenishment) into the magma chamber. This event led to an abrupt change of the geochemical features of the magma, and to subsequent, large-scale outpourings of superheated, hybrid andesite melts.

12.2.3.2 Late Episode of Phase III (III^2–130-90 ka)

Sustained eruptions at the Kazbek volcano and the build-up of its edifice continued during the late episode of phase II, corresponding to unit 2b in Fig. 12.2; however, the composition of the erupted products became more acid. The modern Kazbek cone is made of alternating layers of dacite lavas, lava-breccias and agglomerates. It produced two lava flows, the Eastern Kazbek (upper part) and the Shantori (Fig. 12.2). The Mnaisi volcanic peak was back in activity as well. Small volcanic edifices, composed of dacite lavas, were formed: the Artsivistsveri, Greater Tkarsheti, Shevardeni, Lesser Shevardeni, Pkhelshe and Sioni (Fig. 12.2).

Continuing activity at the volcano led to the formation of dacite lava sheets in the upper part of the Eastern Kazbek flow. As a result of glacial erosion, the flow body split up into two separate ridges – a thicker, northern flow and a minor, southern flow with two small rocky lava remnants in between. The frontal part of the southern branch was subsequently, completely eroded by the Chkheri River. Only a small remnant was preserved, on the right bank of this river, at the northern foot of the Bethlemi volcano (Sabertse Pass).

The 1.5-km-long, Shantori lava flow is located in a remote mountainous area at an altitude of 3600–4000 m at the northeast foot of the volcano on the watershed of the Abano and Shantori glaciers.

The small Peak Mnaisi volcano resumed its activity. The dacitic cone of the volcano and a small lava flow were formed. The latter moved down south, along the roof of the Maili flow (related to phase II^2) at a maximum distance of 1 km. As a result of subsequent glacial processes, the cone was intensively eroded and has survived until the present day (Fig. 12.3) in the form of an up to 40-m-high, pillar-like remnant (neck).

The Artsivistsveri volcano, identified for the first time by Lebedev et al. 2018, is located on the eastern slope of the eponymous mountain peak (2540 m of elevation). This volcano was formed as a result of dacitic magma rising along the fault and, at present, it is a narrow (up to 150 m) and about 1-km- long lava ridge.

The polygenetic, Greater Tkarsheti volcano (Fig. 12.4), is located in the glacier cirque at the springs of the Kesia River. Its slopes are covered by rocky debris. According to the stratigraphic position (Tkarsheti lavas overlap volcanics of the

Fig. 12.4 Panoramic view of the Greater Tkarsheti volcano (Photo by V.A. Lebedev); for its location, see Fig. 12.2

end of phase II, and the cone is injected by a flank eruption of phase IV) and the composition of its rocks, the dome of the Greater Tkarsheti can be reliably assigned to the late formations belonging to the third activity phase of the Kazbek centre.

The Shevardeni and the Lesser Shevardeni dacitic volcanoes, recently identified by Lebedev et al. (2018), are located close to each other in the valley of the Mnaisi River, on its left bank above the cliff made of the Mnadoni lava flows (related to phase II), at elevations between 3100 m and 3400. The Shevardeni is a polygenetic volcano with an oval shape in plan view. The Lesser Shevardeni, monogenetic extrusive dome, has an irregular shape, is located 0.5 km to the north. The slopes of both volcanoes are covered by rock debris.

The minor Pkhelshe and Sioni explosive volcanoes are located directly in the river Terek valley near the eponymous villages. They are composed of boulders of dacite lavas and loose pyroclastic material, which share almost similar compositions and petrographic features. Presently, the Sioni cone is almost completely carved by fluvial erosion and its remains can be observed on a cliff on the right bank of the Terek River.

12.2.4 Phase IV (Late Pleistocene-Holocene, <50 ka)

The magmatic activity of phase IV had a pronounced areal character, lacking a dominant volcanic apparatus; it corresponds to unit 1 in Fig. 12.2. During this period, the Kechuttsveri and Lesser Tkarsheti volcanoes erupted, fissure eruptions of andesite-dacite lavas took place on the slope of the Bethlemi cone, the extrusive domes of Shevardeni and Greater Tkarsheti formed (Fig. 12.2). The formation of the Chkheri explosive volcano dates back to this phase as well.

The small Kechuttsveri volcanic cone is located in the area between the Chkheri and Bloti rivers. Its eruptions at the Pleistocene-Holocene transition led to the formation of a 1.5-km-long andesite flow.

Eruption of andesite lavas from the lateral vent on the SE slope of the volcano (the Upper Pansheti flow) during this phase is related to the Bethlemi polygenetic volcano. The lava flows form two small tongues: the northern one is 1 km long, whereas the southern one is up to 2 km long.

The later, dacitic lava flow of the Shevardeni volcano, was erupted in the western part of this dome from underneath its base and rests on the lavas of the upper horizons of Mnadon flow (related to phase II).

We found a small neck of hyalodacites in a small ravine 450 m north of the beginning of the Shevardeni flow (Lebedev et al. 2018), which cuts across the lavas of the upper horizon of the Mnadon flow and most likely was formed during this phase. The rocks composing the neck are heavily oxidized and altered. This made it impossible to perform their isotope dating.

At the base of the Greater Tkarsheti volcano, on its SW slope, a lateral apparatus was formed, which produced the major (up to 5.5 km long) andesitic-dacitic Tkarsheti flow. The latter filled the Kesia River paleo-valley, reached the Terek

River valley and formed there an up to 2.5-km-long lobe (environs of the Khurtisi-Pkhelshe-Goristsikhe villages). At present, the main body of the flow is a small lava ridge in-between the Kesia and Khevia river gorges.

A small cone of the Lesser Tkarsheti monogenetic volcano is located in the valley of the Terek River at an altitude of about 400 m above the Goristsikhe village. A single eruption of this apparatus led to the formation of an andesite lava flow that is up to 1 km long; its levees and a fresh rubble surface can be easily recognized. The lava flow blocked the valley of the Terek River in its narrow part, forming a small dam. Subsequently, a lake was formed upstream. To date, the lava flow, towards its termination, has been cut by the Terek River, to a depth of 10 m; the reservoir water level has lowered and lacustrine clay sediments crop out in the floodplain section on the right bench of the river bed. Abundant carbonized wood remnants were found and dated: their age, according to ^{14}C datings, equals approximately to 6000 years BP (Burchuladze et al. 1976; Chernyshev et al. 2002).

The small Chkheri explosive centre is located at the mouth of the eponymous river on the left bench of the Terek River valley and adjoins the frontal part of the Chkheri lava flow (related to phase II^2). Its pyroclastic rocks are andesitic in composition. At the foot of the edifice, alluvial sediments are observed. According to K-Ar datings, the formation of the Chkheri volcano took place less than 20 ka.

Magmatic activity of the Kazbek centre during this phase was not significant and was related to eruptions (most likely different in time) of several minor, mainly monogenetic vents at its peripheries. This continued the trend in the development of volcanism, outlined for the end of phase III and expressed by the gradual "extinction" of the main apparatus and the transition to an areal type of volcanic activity.

12.3 The Keli Neovolcanic Centre

The Keli Highland (Fig. 12.1) is located on the southern slope of the Main Caucasus Range west of the Cross (Jvari) Pass and covers the area of the White Aragvi, Ksani, Greater and Lesser Liakhvi rivers. It is one of the largest areas of Late Quaternary magmatism in the Greater Caucasus. The distinctive feature of such neovolcanic centre is the major development of areal, moderately acidic volcanism: across a relatively small area (30 × 25 km) and during a very short time, more than thirty mainly monogenetic volcanoes were active. Geomorphologically, the Keli Highland is subdivided into three zones: southern, northern and western. Besides, K-Ar dates strictly fall into three age groups –phase I– 245-170 ka, phase II – 135-70 ka and phase III – less than 30 ka. In addition, early (II^1–135-100 ka) and late (II^2–100-70 ka) activity episodes were identified within phase II.

12.3.1 Phase I (Middle Pleistocene, 245-170 ka)

During the first phase of activity in the western zone of the Keli plateau, that corresponds to unit 4 in Fig. 12.5, the formation of the Fidarkhokh, Ermani, Knogo II extrusive domes took place, as well as the edification of an unnamed volcano (2997 m), the remains of which are preserved on the northern bank of the lake Kelistba and are overlain by Late Pleistocene lavas of the Keli plateau (Fig. 12.5). We suppose that two very small volcanoes of the western zone were formed in the sources of the Ermani River and north of the Gozo village ruins during this phase as well. The rocks composing the above listed volcanoes have similar chemical composition, being felsic porphyries or, in rare cases, calc-alkaline aphyric dacites. The Fidarkhokh, Ermani, and Knogo II volcanoes have been strongly affected by glacial erosion.

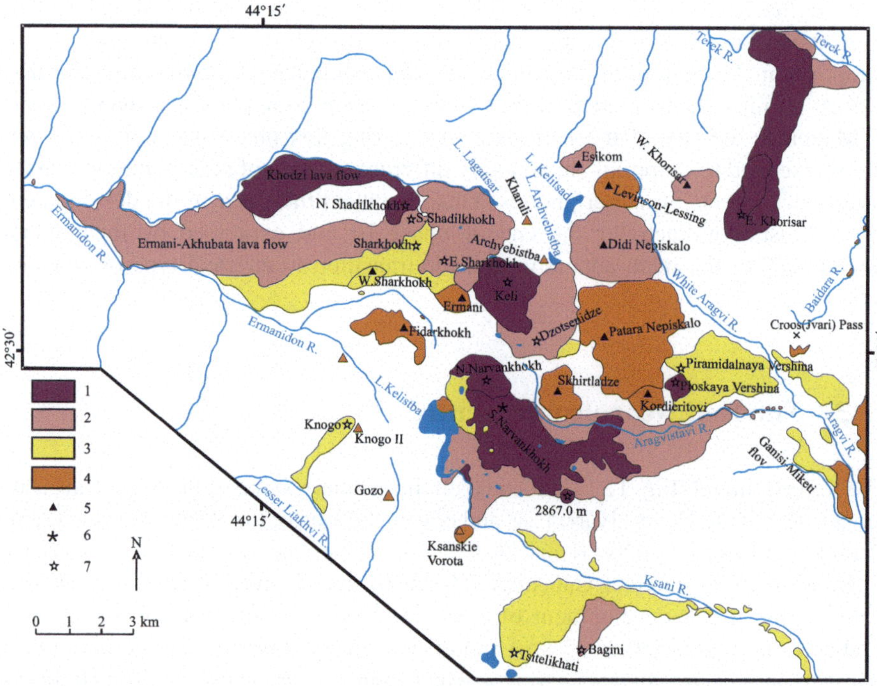

Fig. 12.5 Map showing the distribution of products of Quaternary volcanism at the Keli Highland, Greater Caucasus (prepared by V.A. Lebedev). (1) Volcanic rocks of activity phase III (Late Neopleistocene–Holocene); (2) volcanic rocks of the end of phase II (Late Neopleistocene); (3) volcanic rocks of the beginning of phase II (Late Neopleistocene); (4) volcanic rocks of phase I (Middle Neopleistocene); (5) stratovolcanoes and extrusion domes; (6) cinder cones; (7) lava volcanoes

Fig. 12.6 Panoramic view of the Levinson-Lessing volcano (Photo by V.A. Lebedev); for its location, see Fig. 12.5

Three huge (Patara-Nepiskalo, Skhirtladze, Levinson-Lessing, Fig. 12.6) and two minor extrusive domes were formed in the northern zone (Fig. 12.5), described for the first time by Lebedev et al. (2011b) at the Kharuli pass and on the western bank of Lake Archvebistba. The Patara-Nepiskalo volcano is one of the biggest edifices in the Keli highland, and has an extrusive dome shape, which did not produce lava flows. Some researchers (Skhirtladze 1958) consider the bowl shape of the Patara – Nepiskalo volcano massif as an indicator of a small collapse caldera formation at the final stage of its activity. However, we have found no field evidences to support this assumption. In the Aragvistavi River valley, stratigraphic correlations between the lavas emitted by the Skhirtladze volcano and the younger andesite flows of the Keli plateau were performed. The chemical compositions of the rocks which compose all extrusive volcanoes of the northern zone correspond to dacite. The only exception is Patara-Nepiskalo, made of rhyolites.

Within the southern zone, the Kordierituli and Ksanis Karibche (Ksani Gates) extrusive volcanoes erupted, and the same did the Levinson-Lessing volcano at the Cross Pass (Fig. 12.5). The Kordierituli extrusive dome is made of andesite-dacites with garnet and cordierite (osumilite); these features are unique of the Keli plateau.

12.3.2 *Phase II (Late Pleistocene, 135-70 ka)*

12.3.2.1 Early Episode of the Phase II (II^1–135-100 ka)

The early episode of the second phase of activity at the Keli Highlands is mainly associated with eruptions of volcanoes producing lava flows and it corresponds to unit 3 in Fig. 12.5.

In the western part of the Keli centre, the most significant eruptions are related to the activity of the Sharkhokh volcano – the southernmost volcano on the Kaidon ridge (Fig. 12.5). The volcanic edifice is represented by three accreted peaks (elevation 3254 m, Sharkhokh and Sirkhokh) which form an arcuate ridge, open to the NW. It can be assumed that the volcano had a crater, the western part of which was broken as a result of large-scale outpourings of lavas. It is worth noting that, in the past, the peak of Syrkhokh, adjacent to the main peak of the Sharkhokh volcano from the south, was considered a separate volcanic apparatus. According to our observations, it is composed of dacites, identical in appearance to the rocks of other parts of the volcano and, undoubtedly, it belongs to the Sharkhokh volcanic edifice (Lebedev et al. 2011). The Sharkhokh volcano is the centre of the eruption that produced the extended (up to 6 km) Lower Ermani-Akhubat lava flow, which begins on its western slope and extends to the area of the Khoji and Ermani villages. The Sharkhokh dacitic lavas also spread southward: on the right side of the Ermani river valley between the Ermani and West Sharkhokh volcanoes, there is an up to 50 m thick lava scarp composed of aphyric rocks, close in composition to the effusive rocks constructing the Sharkhokh volcanic structure.

On the right bank of the Ermani River valley, 1 km west of the Sharkhokh volcano, the West Sharkhokh extrusive dome (2975 m) is exposed. According to our observations, lavas of the Lower Ermani-Akhubati flowed around the base of volcano Sharkhokh and descended into the Ermany River valley.

The Knogo volcano was formed in the Late Neopleistocene and is located west of the destroyed extrusive dome Knogo II, in the source of the river Malaya (Small) Liakhvi (Fig. 12.5). SW of the volcano Knogo 3 km long lava flow passes into the river Malaia Liakhvi valley.

On the eastern bank of Lake Kelistba, as high as 140 m, a lava cliff crops out, made of dacitic lavas upwards, overlain by andesite flows of the South Narvankhoh volcano. The dacites of the same kind are exposed from under the younger lavas in single outcrops south of the South Narvankhoh volcano. This lava layer forms the lower lava tier of the Keli plateau. We suppose that in the same place where North Narvankhoh volcano is located (which was formed later during phase III), another volcanic edifice existed. From this edifice, lava flows (up to 3 km long) were produced and mainly moved in western and northern directions, forming the lower tier of the Keli plateau (Lebedev et al. 2011b). To date, the predecessor of the North Narvankhokh volcano is buried under the lavas of the latter.

In the small gorge between the Patara (Small) Nepiskalo volcanoes and Dzotsenidze volcano, small exposures of columnar jointed lavas are preserved. Apparently, most of the lava flow located here was destroyed and transformed into an accumulation of columnar-shaped boulders. The beginning of the flow is buried underneath younger effusive rocks of the volcano Dzotsenidze. In our opinion, this flow is linked to earlier eruptions of the Dzotsenidze volcano, which was mainly active at the end of phase II of the Keli centre magmatic activity.

The Pyramidal Peak volcano was the only active volcano in the northern zone (Fig. 12.5). It adjoins from the southeast the Patara (Small) Nepiskalo circular ridge and is the eruption centre of the lavas of Ganisi-Miketi flow. The latter stretches along the right bench of the Tetri (White) Aragvi River modern valley. The products of the Pyramidal Peak are represented by aphyric dacites where xenoliths of mafic aphyric lavas of andesite composition can be observed. The chemical composition of the xenoliths is totally different from that of their host rocks. The petrography of aphyric dacites of the Pyramidal Peak volcano is similar to lavas of the Patara Nepiskalo volcano, which was active nearby during phase I.

In the southern area, only the Tsitelikhati volcano (Fig. 12.5) was active and produced an up to 12 km long lava flow. The flow remains are distributed mainly on the right bank of the Ksani River; in the area of Bagini village they are overlain by younger andesite lavas of the Bagini volcano, made up of rhyolitic lavas.

12.3.2.2 Late Episode of the Phase II (II^2–100-70 ka)

The late episode of phase II at Keli highland was the most intensive during the Keli centre development period and it corresponds to unit 2 in Fig. 12.5. Volcanoes producing lava flows as well as extrusive domes were active.

In the western zone of the Keli centre, the South Shadilkhokh, East Sharkhokh and Dzotsenidze volcanoes were active (Fig. 12.5). In addition, at that time, apparently, an independent apparatus/vent at the place of the later formed Keli volcano was active. All above mentioned volcanoes are distributed in the NE periphery of the western zone along the both slopes of the along-strike Kharuli range.

The South Shadilkhokh Volcano is located in the central part of the Kai-don ridge and is the eruption centre of the Upper Emani-Akhubati flow dacite lavas (Fig. 12.5). The flow is 12 km long and extends to the west. In its upper part, the Upper Ermani-Akhubati flow rests upon the lavas of the Sharkhokh volcano.

The East Sharkhokh volcano is located between the Kai-on ridge and Keli volcano. From the North the plateau built up by up to 2, 5 km long lava flows of dacite composition. To the South, they lie over the Sharkhokh volcano lavas and to the north near the Kai-don ridge, they rest on the Shadilkhokh volcano lavas. Apparently, the East Sharkhokh was active a bit later than the South Shadilkhokh volcano, which was also active during the late episode of phase II at the Keli centre. It should be noticed that the surface of the volcano lava plateau, despite its height (3100–3250 m) has not been significantly affected by glacial erosion, while the lavas lying in the immediate vicinity of the South Shadilkhokh (east of this volcano) were affected by glacial erosion.

West of the Kai-don ridge, in the depression between the Ermani, East Sharkhokh and Keli volcanoes, a small lava plateau with an area of 500 × 700 m is located. It is separated from the East Sharkhokh lava plateau by a 50 m deep gorge. In the southern part, the plateau lavas rest on the slope of the Ermani dome and in the eastern part they disappear beneath the younger edifice of Keli volcano. The surface of the plateau, in contrast with the adjacent lavas of the Sharkhokh volcano, was heavily eroded by glaciers. We suggest that the eruption centre of these lavas was the apparatus/vent located in the place of Keli volcano, which was later buried under the lavas of the latter (Lebedev et al. 2011b).

The Dzotsenidze Volcano, which became active during the early episode of phase II, was formed as a minor (relative height 100 m) rise at the late episode of its activity. It was the eruption centre of dacite lava flows which generated a plateau between the Kharuli range and the Patara Nepiskalo volcano. To the SE, the lava flows of the Dzotsenidze volcano cover the columnar-jointed lava remnant which apparently is related to the first impulse of the volcano activity. To the East, the flows adjoin the Patara Nepiskalo edifice and to the West they subside under the Keli volcano lavas. The Dzotsenidze volcano lava plateau surface has no signs of glacial erosion.

In the northern zone of the Keli centre's Esikom and West Khorisar extrusive domes, the East Khorisar volcano and the large Didi Nepiskalo stratovolcano were active (Fig. 12.5).

The Didi Nepiskalo Volcano (Fig. 12.7) is made up of massive and brecciated rhyolite lavas. The latter shows the alternation of effusive and explosive eruption periods during its activity.

The East Khorisar polygenetic volcano produced a 7 km long dacite lava flow. The flow entered the Terek River canyon, dammed it and formed a small lobe downstream (in proximity of Nogkau village). The East Khorisar lava flow, along almost its entire length (with the exception of its termination part) was overlain by the products of later eruptions of this volcano.

Fig. 12.7 Field view of the Didi Nepiskalo volcano (Photo by V.A. Lebedev); for its location, see Fig. 12.5

In the southern part of the Keli centre, effusive eruptions of the Bagini and South Narvankhokh volcanoes took place (Fig. 12.5). Bagini was the eruption centre for a 2 km long small andesite lava flow with a fresh surface relief and fairly well-expressed levees. In the area of the Bagini village ruins, the volcano lavas rest on the Tsitelikhati volcano rhyolite lava flow.

During this period of time, the South Narvankhokh volcano erupted in effusive style and produced extended trachyandesite-andesite flows distributed towards the east and south. Lava flows filled and levelled the glacial cirque at the source of the Aragvistavi River and formed the middle tier of the Keli plateau. As a result of this damming, Lake Kelistba was generated, the Aragvistavi River changed its source and started to cut a new riverbed along the northern contact of the Keli plateau lavas with the host rocks in the area of Ksani Gates pass. To the east, the South Narvankhokh lava flows extended as far as the confluence between the Aragvistavi and Tetri (White) Aragvi Rivers. Today, lava remnants in this area are observed on lower terraces along both banks of the Aragvistavi River. A small portion of lavas produced by this volcano passed over the Arkh ridge after the Keli plateau formation and generated a thin and long flow in the river Goristskali paleo bed.

12.3.3 Phase III (Late Pleistocene-Holocene, <30 ka)

During phase III at the Keli highland, which corresponds to unit 1 in Fig. 12.5, both effusive and extrusive volcanoes were active. In the western zone of the Keli centre, magmatic activity is linked to effusive eruptions from three volcanoes – the North Shadilkhokh, North Narvankhokh and Keli (Fig. 12.5). The North Shadilkhokh volcano is located at the northernmost termination of the Kai-don ridge. From the peak of this volcano to the NW, and in lesser extent to the SW, two dacitic lava flows moved downslope. The 50-km-long Khodzi flow filled the Deksokhirdon river paleocanyon in the proximity of the Khodzi and Kabuzta villages, on the right flank of the Ermani-Akhubati flow. The Khodzi flow is completely fresh and not affected by glacial erosion.

The North Narvankhokh volcano is located NE of Lake Kelistba and produced a small dacitic lava flow, which extends in northern, western and southwestern directions.

The Keli volcano is located in its central part on the crest of the Kharuli range. South of the volcano a rather short (1–1.5 km) but thick dacite lava flow is found over the Dzotsenidze volcano dacites. To the west, as mentioned above, the lava plateau remnant emerges from under the Keli volcano, made up of andesites, which we believe were formed as a result of the eruptions from the predecessor vent of this volcano. In the northern zone of the Keli centre, magmatic activity is related to eruptive activity of the Easy Khorisar volcano, and the Flat Peak extrusive dome (Fig. 12.5).

Eruptions at the East Khorisar volcano took place twice at the end of late Neopleistocene or Holocene. The first eruption led to the formation of an extended lava flow (up to 7 km) of dacitic composition. The flow moved down towards the Terek riverbed along the surface of older flow and turned in its lower part up the valley due to formed before lava dam. The flow has fairly well-expressed levees and the surface is not affected by glacial action. The successive and last eruption of the East Khorisar Volcano occurred without any significant interruption. As a result, the northern wall of the crater was breached and the eruption of a minor (up to 2 km long) dacite lava flow took place. The latter covered the products of earlier eruptions.

The Flat Peak extrusive dome is located SE of the Patara Nepiskalo volcano, on the left bank of the Aragvistavi river. The rocks constructing the volcano are weakly porous aphyric dacites. On the southern slope of the volcanic edifice, we have found fragments of glassy lavas, apparently volcanic bombs. Their composition is markedly different from the effusive rocks that make up the extrusive dome and that correspond to felsic andesites. It may be supposed that the andesites are remelted material of xenogenous origin, erupted at the early stage of the Flat Peak volcano activity.

In the southern zone of the Keli centre, magmatic activity is related to the South Narvankhokh volcano and an unnamed cinder cone on the Arkh range (2867 m) (Fig. 12.5).

The South Narvankhokh Volcano resumed its activity and produced the 4–5 km long andesite lava flow which formed the upper tier of the Keli plateau. At the final stage, the volcano generated a cinder cone which is, at present, a modern edifice of the volcano. Lava flows of South Narvankhokh are fresh and not affected by glaciers rubbly surfaces.

12.4 Elbrus Neovolcanic Centre

The Elbrus volcanic centre (Fig. 12.1) is located on the northern slope of the Main Range of the Greater Caucasus on the watershed of Malka, Baksan and Kuban rivers. It includes the Elbrus volcano as well as several small volcanic apparatus along its periphery (Paleo-Elbrus, Chuchkhur, Chomart-Kol, Siltran, Tash-Tebe, Tashlisirt, Tizil).

On the basis of fieldwork, petrological-geochemical and isotope-geochronological studies of selected rock samples, 5 phases of volcanic activity could be distinguished at the Elbrus volcanic centre: phase I – 950-900 ka, phase II – 800-700 ka, phase III – 225-170 ka, phase IV – 110-70 ka, phase V – less than 35 ka.

12.4.1 Phase I (End of the Eopleistocene, 950–900 ka)

During this time window, a number of volcanoes located east of modern Elbrus became active: Tizil, Tashlisirt and Siltran (Fig. 12.8). It should be noted that all of them are arranged along an extended linear submeridional zone, which apparently

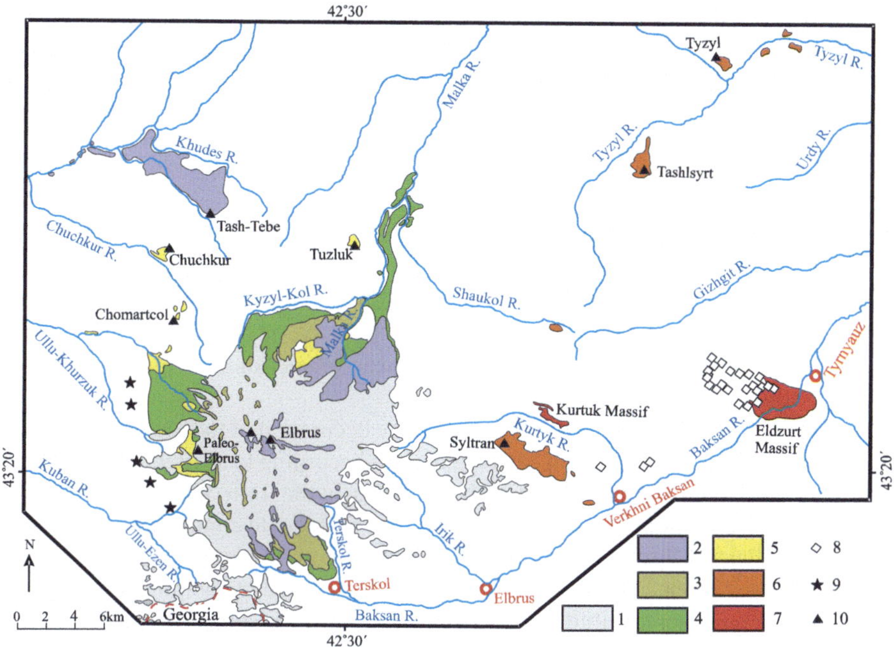

Fig. 12.8 Schematic geological map of the Elbrus volcanic centre. (1) glaciers and firn fields; (2) Late Neopleistocene-Holocene, <35 ka (Phase V); (3) Late Neopleistocene, 110-70 ka (Phase IV); (4) Middle Neopleistocene, 225-170 ka (Phase III); (5) Early Neopleistocene - 800-700 ka (Phase II); (6) end of the Eopleistocene, 950-900 ka (Phase I); (7) Late Pliocene granitoids; (8) necks, dikes, and remnants of Late Pliocene volcanoes (rhyolites, dacites); (9) necks, dikes, and remnants of Quaternary volcanoes (dacites); (10) Quaternary volcanoes

traces a deep-seated fault, active in Quaternary times (Lebedev et al. 2010). This phase corresponds to unit 6 in Fig. 12.8.

In the Tizil river valley, at the mouth of its left tributary Kinzhal river, a remnant of the Tizil lava flow can be observed. It is made of andesite-basalts (lower part of the section) and trachyandesites (upper part of the section). Below, downstream of the Tizil, minor remnants of this flow could be observed.

The Tashlistir Volcano is located on the northern slope of the eponymous peak, in the springs of the Tizil River, between its right tributaries Tallikol and Tashlisirt. According to our observations, the initial stage of its activity was related to the eruption of a thin flow of dacite lavas which spread to the north at about 5 km from the centre. Subsequently, the lava volcano was replaced by a secondary cinder cone, which later on was almost completely erased and is hardly distinguished in the present-day topography.

The Siltran Volcano is located on the Siltran range between the Kirtik and Siltransu Rivers (the Baksan river basin). At present, this sublatitudinal range comprises four peaks, intergrown at their base. It is made of dacite volcanic rocks. We suppose that the eruption centre of the dacites is located in the western, higher part of the range, where the flows spilled down to the east, following the paleorelief slope towards the Baksan river valley. The termination part of the volcano Siltran flow is preserved on the watershed of the Siltransu and Baksan river gorges, as an isolated, columnar jointed remnant.

12.4.2 Phase II (Early Neopleistocene – 800-700 ka)

In this time interval, a few volcanoes west of Elbrus became active: Paleo-Elbrus, Chomartkol, Chuchkhur and Tash-Tebe (Fig. 12.8). We need to point out that all are located within the up to 20–25-km-long linear submeridional zone. This phase corresponds to unit 5 in Fig. 12.8.

Remnants of the Paleo-Elbrus volcanic edifice occur west of the modern Elbrus in the heads of the Kukurtlu and Ullu-Kam glaciers. According to available K-Ar dates, activity at the Paleo-Elbrus took place about 800 ka, in the form of a fissure eruption, which produced ignimbrites deposits and lavas of rhyodacite composition. At about 720 ka, the Paleo-Elbrus erupted once again. This eruption was apparently of the central type and produced lava sheets of dacite and rhyodacite composition. Around the same time, a minor satellite volcano was active. Its neck-shaped remnants, which can be seen on the watershed between the sources of the Kukurtlu and Ullu-Kam rivers, are composed of rhyodacites and dacites (Lebedev et al. 2011a).

At the springs of the Chomart-Kol River, 10 km north of Paleo-Elbrus, several remnants of rhyodacite tuff can be noticed, indicating the presence of an earlier explosive volcanic vent located in the area. Further north, a minor ignimbrite sheet exists in the springs of the Chuchkhur river which was produced by the independent Chuchkhur volcano.

Located at the springs of the Khudes River, the Tash-Tebe volcano is the northernmost Early Neopleistocene volcano of phase II. Lavas outpoured from this volcano are andesitic in composition.

In the Khudes River mouth, on the watershed, the remnant of the Khudess lava flow can be observed. These lavas have a petrography and a chemical composition that are similar to Tash-Tebe volcano. In addition, K-Ar dates performed by the authors coincide with the data on Tash-Tebe within the error limits (Lebedev et al. 2011a). As a result of the obtained data analysis, it can be assumed that volcanoes from the Khudes River's mouth are preserved, termination portions of the lava flow formed as a result of the Tash-Tebe volcano eruptions. The Khudess flow, up to 30 km long, was the longest among the similar Late Cenozoic formations in the Greater Caucasus.

12.4.3 *Phase III (Middle Neopleistocene, 225-170 ka)*

During this phase, which corresponds to unit 4 in Fig. 12.8, Elbrus volcano originated, and produced dacitic lavas. The formation of the lower tier of lavas in the Azau River valley (southern slope), as well as most of the lava flows of the Biitik-Tebe River canyon (western slope) and the springs of the Birdjalisu River (northern slope) (Lebedev et al. 2005) is related to this phase of activity.

12.4.4 *Phase IV (Late Neopleistocene, 110-70 ka)*

During this phase, Elbrus continued its volcanic activity, producing dacitic lavas (Fig. 12.8 – unit 3). On the southern slope of the volcano, flows were generated, which constructed the second, third and fourth lava tiers of the Azau River valley. On the northern slope, during this phase of magmatic activity, the "Sultan Waterfall" flow was emitted, which dammed the Kizilkol River and froze in the area of the Zhelasu mineral springs (Lebedev et al. 2005).

12.4.5 *Phase V (Late Neopleistocene-Holocene, <35 ka)*

During this phase of activity, Elbrus volcano resumed its activity again (Fig. 12.8 – unit 2). Lavas belonging to this phase can be observed on the southern and northern slopes of Elbrus and are represented by dacites. In the Azau River valley, they form lava flows near the Irik glacier in the area of Ice Base, watersheds of the Garbashi, Small Azau and Small Azau-Bol rivers. In the Malkha River basin, they cover the greater part of the Elbrus northern slope (Lebedev et al. 2005).

It is worth noting that, by using the microseismic sounding technique, the presence of two deep volumes with similar properties and morphology under the Caucasus mountain range was assessed. These bodies are subvertical and marked by low (L-I) and high (H-II) seismic velocities. The L-I body is located mainly beneath the northern limb and axial part of the Greater Caucasus meganticlinorium; the H-II body is mostly located beneath the southern limb (Gorbatikov et al. 2015).

12.5 Final Remarks

With the purpose of summarizing the results of our study focused on the Quaternary magmatic activity in the Greater Caucasus, we have combined the activity phases of all three centres into Table 12.1. As can be seen from the table, during the late

Table 12.1 Correlation of Quaternary volcanic activity phases and stages in the Greater Caucasus

Activity phases	Activity phases	Activity episodes	Activity phases	Activity episodes
Elbrus Centre	Kazbek Centre		Keli Centre	
I – 950-900 ka				
II – 800-700 ka				
	I – 460–380 ka			
	II – 310–200 ka	II^1–310–260 ka		
III – 225–170 ka		II^2–240-200 ka	I – 245-170 ka	
IV – 110-70 ka	III – 130-90 ka	III^1	II – 135-70 ka	II^1–135-100 ka
		III^2		II^2–100-70 ka
V – $\leq$ 35 ka	IV – $\leq$ 50 ka		III – $\leq$ 30 ka	

episode of the second phase of activity of the Kazbek centre, the Elbrus and Keli centres also became active.

The nature of the recent magmatism that took place at the Elbrus, Kazbek and Keli volcanic centres, when the voluminous volcanic activity repeatedly resumed after long interruptions, as well as the Holocene age of the latest known eruption (Burchuladze et al. 1976; Chernyshev et al. 2002, 2008; Vashakidze et al. 2019), may be evidence of the possible incompleteness of the last phase of magmatism. This is indicated by the presence of fumaroles at the top of the Elbrus volcano and thermal springs along the periphery of the Elbrus and Kazbek volcanoes as well. The above observations enable us to conclude that there may be the possibility of renewed activity at all three volcanic centres.

References

Burchuladze AA, Dzhanelidze CG, Togonidze GI (1976) Application of radiocarbon method for solution of some problems of Pleistocene and Holocene paleogeograophy of Georgia. In: Aktual'nye voprosy sovremennoi geokhronologii (Current problems of geochronology). Nauka, Moscow, pp 238–243

Chernyshev IV, Lebedev VA, Bubnov SN et al (2002) Isotopic geochronology of quaternary volcanic eruptions in the greater Caucasus. Geochem Int 40(11):1042–1055

Chernyshev IV, Lebedev VA, Bubnov SN et al (2008) The youngest volcanism of Caucasus and regularities of its manifestation in time and location: evidence from geochronological study (in Russian). In: Changes in the environment and climate: natural and related to them technogenic hazards, vol 2. IFZ RAN, Moscow, pp 206–234

Dzotsenidze NM, Khaburzaniya IA, Bagratishvili TD (1985) Otchet "Geologicheskoe i paleomagnitnoe izuchenie lavovykh nakoplenii Kazbegskogo raiona B. Kavkaza s tsel'yu ikh korrelyatsii, utochneniya vozrasta i opredeleniya vozmozhnoi rudonosnosti" za 1983–1985 gg (Report on the geological and Paleomagnetic study of lava piles of the Kazbek area in the greater Caucasus aimed at their correlation, refinement of the age and determination of possible ore potential). KIMS, Tbilisi

Gorbatikov AV, Rogozhin EA, Stepanova MY et al (2015) The pattern of deep structure and recent tectonics of the greater Caucasus in the Ossetian sector from the complex geophysical data. Izv Phys Solid Earth 51(1):26–37. https://doi.org/10.1134/S1069351315010073

Lebedev VA, Vashakidze GT (2014) The catalogue of quaternary volcanoes of the greater Caucasus based on geochronological, Volcanological and Isotope_Geochemical data. J Volcanol Seismol 8(2):93–107. https://doi.org/10.1134/S0742046314020043

Lebedev VA, Chernyshev IV, Bubnov SN et al (2005) Chronology of magmatic activity of the Elbrus volcano (Greater Caucasus): evidence from K–Ar isotope dating of lavas. Dokl Earth Sci 405A(9):1321–1326

Lebedev VA, Sakhno VG, Yakushev AI (2010) Total duration and spatial migration of quaternary volcanism in the El'brus Region, Greater Caucasus. Dokl Earth Sci 430(1):80–85. https://doi.org/10.1134/S1028334X10010186

Lebedev VA, Bubnov SN, Yakushev AI (2011a) Magmatic activity within the Northern Caucasus in the early Neopleistocene: active volcanoes of the Elbrus center, chronology, and character of eruptions. Dokl Earth Sci 436(1):32–38. https://doi.org/10.1134/S1028334X11010065

Lebedev VA, Vashakidze GT, Arutyunyan EV et al (2011b) Geochronology and evolution of quaternary volcanism at the Keli Highland, greater Caucasus. Geochem Int 49(11):1120–1144. https://doi.org/10.1134/S0016702911090035

Lebedev VA, Chugaev AV, Vashakidze GT et al (2016) Formation stages and ore matter sources of the Devdoraki copper deposit, Kazbek volcanic center, the greater Caucasus. Geol Ore Deposit 58(6):465–484. https://doi.org/10.1134/S1075701516060040

Lebedev VA, Parfenov AV, Vashakidze GT et al (2018) Chronology of magmatic activity and petrologic–mineralogical characteristics of lavas of Kazbek quaternary volcano, greater Caucasus. Petrology 26(1):1–28. https://doi.org/10.1134/S086959111801006X

Meyer D (2011) ASTER-GDEM-validation-team ASTER global digital elevation model version 2-summary of validation results. Jpn Syst 7

Pasquarè FA, Tormey D, Vezzoli L et al (2011) Mitigating the consequences of extreme events on strategic facilities: evaluation of volcanic and seismic risk affecting the Caspian oil and gas pipelines in the Republic of Georgia. J Environ Manag 92(7):1774–1782

Skhirtladze NN (1958) Postpaleogenovyi effuzivnyi vulkanizm Gruzii (Post-Paleogene Volcanism of Georgia). Izd Akad Nauk GruzSSR, Tbilisi

Tachikawa T, Hato M, Kaku M et al (2011) Characteristics of ASTER GDEM version 2. In: Proceedings of the 2011 IEEE international geoscience and remote sensing symposium, Vancouver, 2011. IGARSS-IEEE, pp 3657–3660

Tibaldi A, Oppizzi P, Gierke JS et al (2019) Landslides near Enguri dam (Caucasus, Georgia) and possible seismotectonic effects. Nat Hazard Earth Sys 19:71

Tsereteli N, Tibaldi A, Alania V et al (2016) Active tectonics of Central-Western Caucasus, Georgia. Tectonophysics 691:328–344

Varazanashvili O, Tsereteli N, Bonali FL et al (2018) GeoInt: the first macroseismic intensity database for the Republic of Georgia. J Seismol 22(3):625–667

Vashakidze G, Goguitchaichvili A, Garcia-Redondo N et al (2019) Magnetic dating of the Holocene monogenetic Tkarsheti volcano in the Kazbeki region (Great Caucasus). Earth Planets Space 71(1):1–7. https://doi.org/10.1186/s40623-019-1109-4

Chapter 13
Tectonic Control Over the Abuli Samsari Volcanic Ridge, Lesser Caucasus, Georgia

Federico Pasquaré Mariotto, Fabio L. Bonali, and Corrado Venturini

Abstract The purpose of the present work is to integrate previous research focused on the Abuli Samsari Volcanic Ridge, situated in the Javakheti Highland, Georgia. Through a synergic approach, consisting in the collection and analysis of field and satellite data, combined with the results of previously published research, we have been able to define the overall structure of the volcanic ridge, which, on its northern sector, is cut across by two parallel pipelines, carrying oil and gas from the Caspian Sea to the western countries. Despite the likelihood of seismic or volcanic events in the area, geohazard assessment had never been adequately performed for this section of the pipelines' route across Southern Georgia. The most relevant outcomes of our effort, aimed at filling this critical gap, consist in: the identification and mapping of eruptive centers and tectonic lineaments; the reconstruction of magma pathways; the definition of the expected moment magnitude for possible earthquakes; the assessment of orientation of the maximum horizontal stress from the Late Miocene to the present day. We have used these results to evaluate the current seismic and volcanic hazards affecting the Abuli Samsari Volcanic Ridge, which may have major impacts on the security of the pipelines. The calculated, about N-S directed maximum horizontal stress may play a key role in volcanic reactivation, which might occur in the form of fissure eruptions and the formation of new vents and monogenetic as well as composite volcanoes. As the track of the pipelines lies just

F. Pasquaré Mariotto (✉)
Department of Human and Innovation Sciences, Insubria University, Como, Italy
e-mail: pas.mariotto@uninsubria.it

F. L. Bonali
Department of Earth and Environmental Sciences, University of Milan Bicocca, Milan, Italy

CRUST- Interuniversity Center for 3D Seismotectonics with Territorial Applications, Chieti Scalo, Italy

C. Venturini
Department of Biological, Geological and Environmental Sciences, Bologna University, Bologna, Italy

© Springer Nature B.V. 2021

F. L. Bonali et al. (eds.), *Building Knowledge for Geohazard Assessment and Management in the Caucasus and other Orogenic Regions*, NATO Science for Peace and Security Series C: Environmental Security,
https://doi.org/10.1007/978-94-024-2046-3_13

north of the younger volcanic edifice in the ridge we suggest that these lifelines could be severely affected by possible future volcanic eruptions, which might bring about a major interruption in oil delivery from the Caspian Sea towards the west.

Keywords Pipeline · Abuli · Samsari · Magma pathway · Earthquake · Dyke

13.1 Introduction

Georgia,situated in the central part of the Caucasian region, between the Greater and Lesser Caucasus, is the natural oil and gas transportation and pipeline corridor from the Caspian region to the West. The Baku-Supsa (BS) and the Baku-Tbilisi-Ceyhan (BTC) oil pipelines, as well as the Baku-Tbilisi-Erzurum gas pipeline (BTE), all run across the Caspian region through Georgia (Fig. 13.1). The BTC and BTE pipelines run parallel, separated by approximately 20 m, for the 248-km section across the Georgian territory. For 70 km of this stretch, the pipelines lie within the northern part of the Javakheti volcanic highland in Southern Georgia. In this geologically-

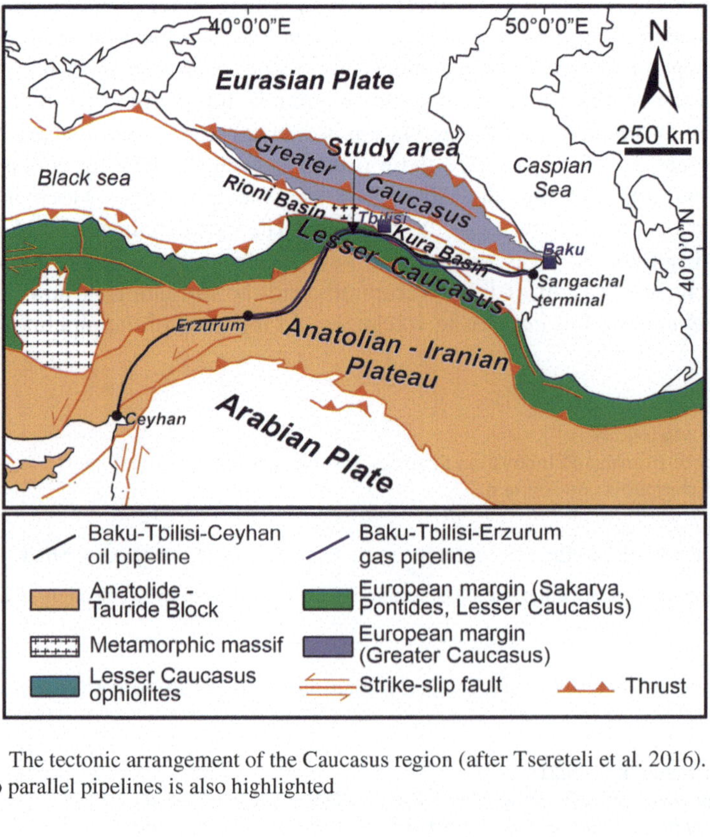

Fig. 13.1 The tectonic arrangement of the Caucasus region (after Tsereteli et al. 2016). The route of the two parallel pipelines is also highlighted

active area, there is major possibility that volcanic and seismic events may put these crucial energy links in danger. Despite the likelihood of these geohazard-related events, the nature and magnitude of the volcanic and seismic threats to these lifelines had not been envisioned, as far as the Georgian section is concerned. Also, possible mitigation measures to be planned beforehand had not been developed prior to pipeline installation. In the past, Pasquaré et al. (2011) attempted to fill this gap, through an international scientific cooperation aimed at assessing the possible geo-related impacts to the pipelines across their stretch across the Javakheti plateau. Their work enabled to highlight and quantify the potential volcanic, seismic, and related geohazards that affect the area crossed by the pipelines, which were not specifically considered in the original design of these facilities.

The present work is aimed at integrating and completing the effort by Pasquaré et al. (2011), on the basis of new field studies, seismic data, and the interpretation of satellite images. Through the above approach, we have been able to locate all volcanic centers in the area crossed by the pipelines, also the minor ones that had not been identified so far and to define the most relevant tectonic structures in the area. Most crucially, we have carried out a meticulous magma pathway reconstruction and field stress assessment, which, in turn, has allowed us to evaluate the seismic hazard in the area as well as to identify the preferential directions along which future volcanic reactivation may take place.

13.2 Geological Setting

The Caucasus is an orogenic system that formed in the Cenozoic following the Arabian-Eurasian plate collision. It is subdivided into a Greater and Lesser Caucasus: the former has developed though the formation of mainly WNW-ESE trending, folds and faults, which resulted in a rather rectilinear mountain front in plan view, especially along its southern side (Fig. 13.1). The Transcaucasian intermontane valley, separating the Greater Caucasus to the north from the Lesser Caucasus to the south, is composed of: (i) the Rioni foreland fold-and-thrust belt and the Rioni foreland basin to the west (Adamia and Gujabidze 2004; Forte et al. 2004); (ii) the Dzirula massif in the centre and, (iii) the Kura Basin to the east.

The Lesser Caucasus (Fig. 13.1) represents a major portion of the Tethyan belt; it extends from the Black Sea to the Caspian Sea, across Georgia, Armenia, and Azerbaijan, and was built due to of convergence and collision of Eurasia, Gondwana-derived terranes, and Arabia (Moritz et al. 2016; Avagyan et al. 2005). In terms of tectonic structures on a larger scale, the dominant structure in the Lesser Caucasus is the Pambak-Sevan Fault Zone, a 300-km long, 10-km wide, NW-SE trending, right lateral strike-slip fault zone located along the southern front of the range (Philip et al. 2001). One of the most important and active compressional structures in the Lesser Caucasus is the 20-km long Spitak reverse fault, that triggered the eponymous seismic event in December 1988. The rupture that accompanied the earthquake, according to Trifonov et al. (1990), was 37 km long, with a staggering vertical offset of about 1.8 m.

In the central part of the Lesser Caucasus is located the 2500-km^2 wide, 1500–2000-m high Javakheti volcanic highland (or plateau), mainly in South Georgia as well as on the adjacent territories of Armenia and Turkey. The plateau is one of the largest areas affected by major Neogene-Quaternary, mainly mafic magmatism in the Lesser Caucasus (Lebedev et al. 2019) and hosts a great number of major and minor volcanic centers (Lebedev et al. 2003; Chernyshev et al. 2006). The youngest volcanic system in the Javakheti Highland is the centrally-located, 40 km long, Abuli-Samsari Volcanic Ridge (ASVR), extending in a roughly N–S direction from the southern slopes of the Trialeti Ridge to the Paravani River valley. According to Lebedev et al. (2003), the most important edifices in the ASVR are Tavkvetili, Beberiklde (Egoisar), Shavnabada, Samsari, Kerogly, Godorebi, Didi-Abuli, Patara-Abuli, and Eshtia. The most prominent ones are Didi Abuli (3300 m asl) and Samsari (3284 m asl).

The basement of the AVSR is made of Cretaceous–Paleogene volcanosedimentary rocks locally exposed by erosion (Fig. 13.2). Before the work of Lebedev et al. (2003), almost all volcanics of the AVSR were attributed to the upper Miocene and Pliocene. Alongside these formations, the ASVR is marked by younger volcanic deposits testifying to Quaternary volcanism (Fig. 13.2). Lebedev et al. (2003), on the basis of K-Ar datings, documented four main activity intervals in the volcanic ridge: (i) 800–700 ka; (ii) 400 ka; (iii) 320–170 ka, and (iv) Late Pleistocene-Holocene (less than 30 ka).

13.3 Methods

Field methods used during the research leading to the present work comprised identifying, tracking and studying faults, by measuring fault plane attitude and slickensides, whenever this was possible; dykes and scoria cones were also studied in the field, by assessing dyke attitude and determining the morphometric parameters of monogenetic cones. Fieldwork was integrated by a number of further methods, as detailed in Sect. 13.3.1.

In order to assess the volcanic hazard affecting a defined area, such as the considered ASVR, it is paramount to define the possible direction of magma pathways, considering the present stress field, active and inactive tectonic structures through which magma can reach the surface; the assessment of magma pathway directions depends on various factors, as detailed hereunder. First, it is well known that intrusive dyke swarms are usually expressed at the surface by aligned pyroclastic centers and swarms of eruptive and dry fissures and faults (Johnson and Harrison 1990; Strecker and Bosworth 1991; Annen et al. 2001; Walter and Schmincke 2002; Tibaldi et al. 2013; Tadini et al. 2014). Dykes, along with inclined sheets are among the most important sub-volcanic features feeding magma to the surface (e.g. Pasquaré and Tibaldi 2007; Tibaldi et al. 2008a, 2011; Tibaldi and

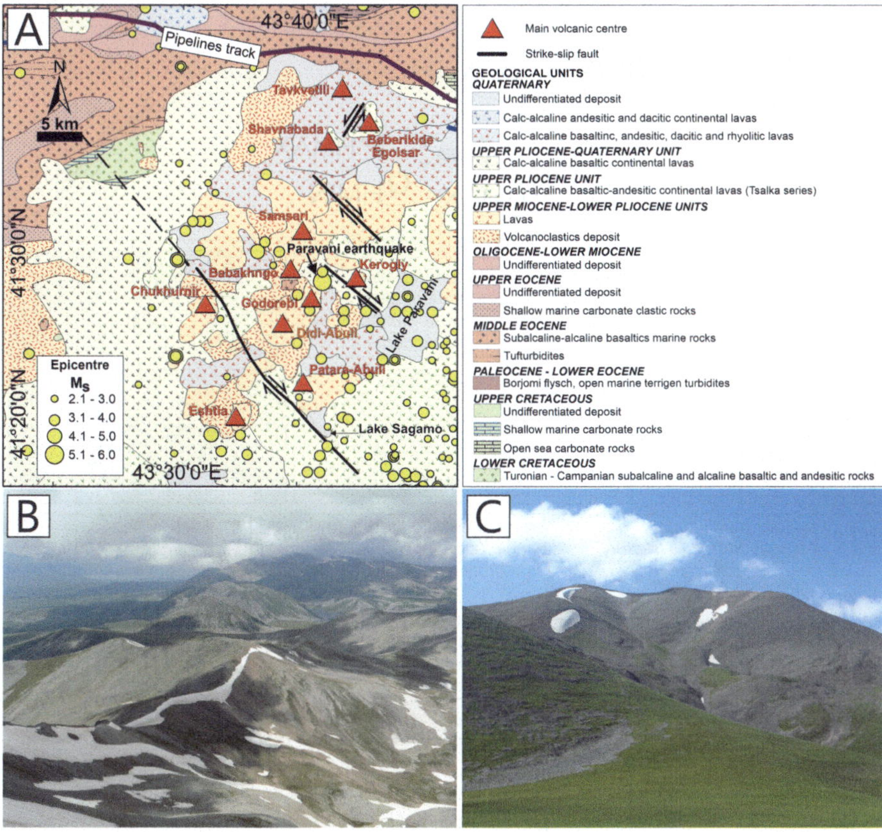

Fig. 13.2 (**a**) Geological map of the Abuli-Samsari Volcanic Ridge. Geological units are from Adamia and Gujabidze (2004). Post-1900 AD Earthquakes are from Zare et al. (2014) and Tsereteli et al. (2016). Main Volcanic Centres are from the Global Volcanism Program (2013). Strike-slip faults are from Rebai et al. (1993) and Pasquaré et al. (2011). (**b**) The AVSR, as seen looking North from the top of Didi Abuli Volcano. (**c**) The southern slope of Didi Abuli Volcano

Pasquaré 2008; Bistacchi et al. 2012); they are also capable of destabilizing the flanks of volcanoes (e.g. Tibaldi et al. 2008b). Dykes intercepting the surface result in the formation of aligned volcanic centres; a great number of parallel dykes result in the emplacement of tens of pyroclastic cones. By statistically studying the maximum concentration of volcanic vents it is possible to assess the trend of the underlying dyke swarm (e.g. Nakamura 1977; Nakamura and Uyeda 1980; Fig. 13.3a, b). Several statistical methods have been proposed over the years, aimed at defining vent alignment patterns (Wadge and Cross 1988; Connor 1990; Lutz and Gutmann 1995; Germa et al. 2013); one of the most critical factors for ascribing a number of aligned vents to a single feeder-dyke, is the identification of coeval vents.

In the present work, magma pathway reconstruction and field stress assessment have been meticulously carried out through several steps. Considering the extent

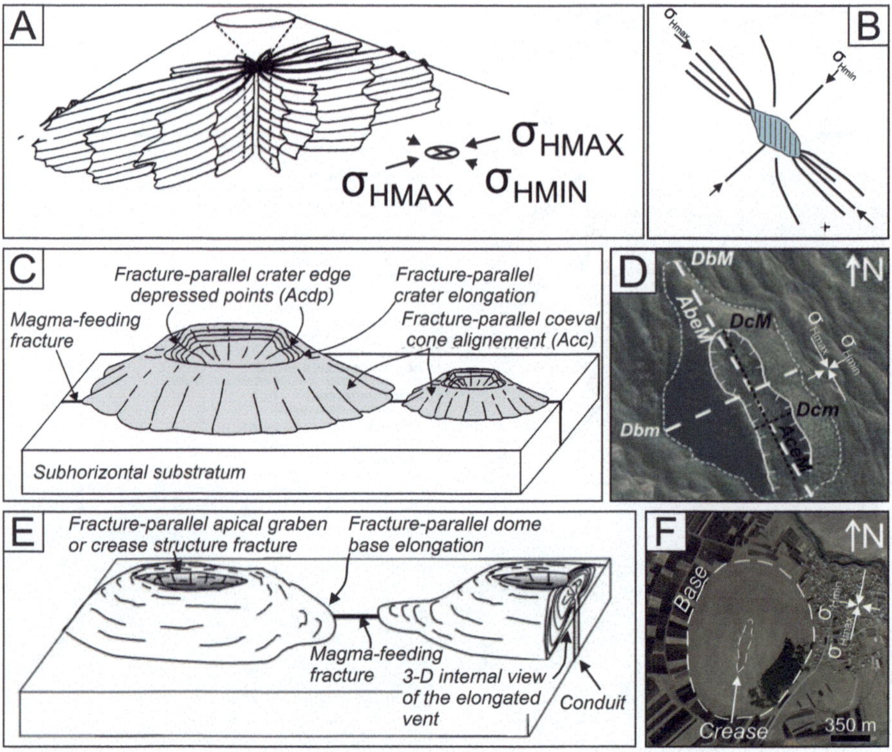

Fig. 13.3 (**a**) Sketch 3-D view of radial dykes near the central conduit that, with distance, are reoriented parallel to the horizontal greatest principal stress (σ_{Hmax}); vents can be noticed on either flank of the edifice (redrawn after Nakamura 1977). (**b**) Theoretical dyke geometry, under a differential horizontal stress, that becomes mostly parallel to the σ_{Hmax} (after Nakamura and Uyeda 1980). (**c**) Morphometric parameters of pyroclastic cones and their relationship with the underlying feeding fracture (modified after Tibaldi 1995). (**d**) Textbook example of a pyroclastic cone with indication of the trend of morphometric parameters (from Réunion Island). AceM, AbeM: fracture-parallel crater and base elongation; Dbm, DbM: minimum and maximum diameter of cone base; Dcm, DcM: minimum and maximum crater diameter (modified after Bonali et al. 2011). (**e**) Sketch of volcanic domes that show features such as apical grabens (crease structures), elongation of the base and alignment of more edifices. Modified after Pasquaré and Tibaldi (2003). (**f**) Outstanding example of a NNE-SSW elongated dome belonging to the study area (41°17′45.06″N – 43°36′25.92″E), with a fracture-parallel apical graben/crease structure

of the area, the harsh logistic conditions that prevented from carrying out a very detailed fieldwork as well as the presence of recent deposits that can hide structures, for most of the centres we inferred magma pathway geometry by applying indirect methods related to volcano morphology, volcanic vent distribution, scoria cones and dome morphology, as explained hereunder in detail. Several works have been dedicated to this interesting topic in the recent past, confirming the relevance of this approach in settings such as entire volcanic arcs (Bonali et al. 2013, 2015a; Tibaldi and Bonali 2017; Tibaldi et al. 2017), single volcano or single volcanic complex

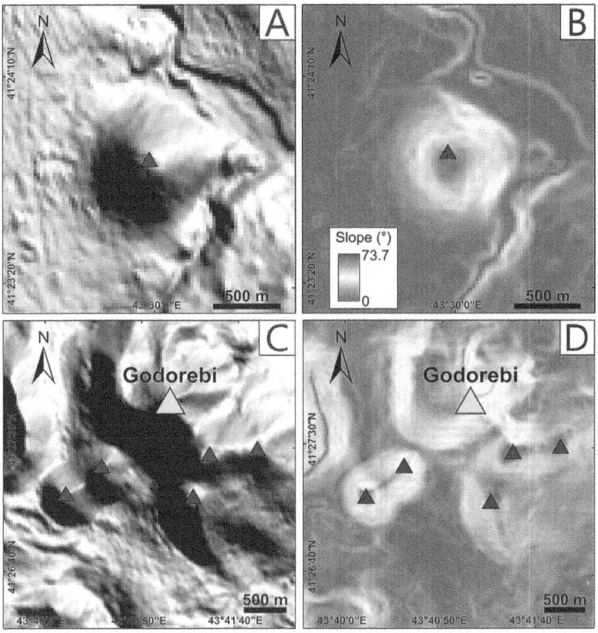

Fig. 13.4 Example of unmapped volcanic centers here identified on ALOS GDEM shaded (**a**) and slope (**b**) view (https://www.eorc.jaxa.jp/ALOS/en/aw3d30/index.htm). (**c–d**) Unmapped volcanic centers located nearby Godorebi volcano on ALOS GDEM shaded (**c**) and slope (**d**) view; it is worth noting that two centers to the west show a clear alignment. The yellow triangle represents the volcano that is already cited in the literature; blue triangles represent newly-identified volcanic centers

(Bonali et al. 2011, 2012, 2015b, Bonali 2013) and rift zones (Tibaldi 1995; Tadini et al. 2014; Tibaldi et al. 2013).

13.3.1 Volcanic Centres and Tectonic Lineaments

The first stepof our research has been gathering all available information in the literature about the study area, including scientific papers (e.g. Lebedev et al. 2003; Pasquaré et al. 2011), geological maps e.g. (Adamia and Gujabidze 2004) and online resources (e.g. the Global Volcanism Program – http://volcano.si.edu/, Global Volcanism Program 2013). Along with this, we also used topographic maps of the area at the 1:25,000 scale, the freely available satellite images provided by Google EarthTM and ArcGIS Online basemaps service, free ALOS Global Digital Surface Model (GDEM) with a pixel size of 30 m provided by provided by JAXA (https://www.eorc.jaxa.jp/ALOS/en/aw3d30/index.htm; Tadono et al. 2016); this was aimed at identifying and locating all unmapped volcanic centres (e.g. Fig. 13.4c–d) and possible tectonic lineaments in the study area. On satellite images and/or shaded and slope views of GDEM (e.g. Fig. 13.4a, b), the volcanic centres have been identified by individuating the presence of craters, calderas, volcano deposits, and recognizing the typical morphological features of the edifices. Tectonic lineaments have been located on satellite images, to be compared with inferred magna pathways, through well-established methods (e.g. Marino and Tibaldi 1988; Tibaldi and Ferrari 1991).

Lineaments have been also interpreted on different shaded views and slope models of ALOS GDEM (e.g. Koike et al. 1998; Jordan 2003; Silva et al. 2007). The collected data are discussed in the section dedicated to results, also considering focal mechanism solutions from Tsereteli et al. (2016) and the maximum horizontal stress (σ_{HMAX}) data from the World Stress Map 2016 (http://www.world-stress-map.org/; Heidbach et al. 2016, 2018).

13.3.2 Areal Density of Volcanic Centres

Areal density analysis is a helpful technique in the framework of the spatial analysis methodology, in that it results in a continuous surface marked by different density values across a landscape (Silverman 1986). This effective technique takes into account the different elongation values of the areas characterized by higher density of volcanic activity and correlates them with the azimuth of the feeder dykes (Bonali et al. 2011; Tadini et al. 2014). In agreement with Connor (1990) and Bonali et al. (2011), we used all the identified vents as points from which to project a 675-m search radius, based on the empirical spacing between coeval vents (1.350 m) documented by Tibaldi (1995) for the Canary Islands (other values of spacing were provided by the same author for further, selected volcanic areas). Using half of this distance as search radius allows for the identification of different overlapping areas and their elongation; the latter can then be compared with the strike of eruptive fissures, as well as faults. The importance of the choice of the appropriate search radius value is clear when it comes to comparing elongations with feeder dyke orientations, as it is likely that many elongations could represent eruptive fissures linking coeval vents to each other. For the purpose of our analysis, we used the kernel method (the "calculate density" option in the ArcGIS® software), by dividing isodensity areas into four classes by equal intervals with respect to the maximum value (Bonali et al. 2011): this choice was made with the purpose of better visually highlighting elongations and alignments of overlapping areas in the resulting map.

13.3.3 Inferring Magma Pathway Azimuth

A detailed reconstruction of shallow magma paths can be accomplished by studying a number of morphometric parameters of main craters and parasitic vents (Fig. 13.3c; Tibaldi 1995; Corazzato and Tibaldi 2006; Bonali et al. 2011), possibly integrated by seismological data (e.g. Tibaldi and Bonali 2017; Russo et al. 2017). When a dyke intercepts the slope of a volcano, this leads to the formation of scoria ramparts that are parallel to the dyke, or pyroclastic cones with an elongated base and crater (e.g. Figure 13.3c–d) whose azimuth is controlled by the strike of the magma path (Tibaldi 1995). The azimuths of cone base and crater (AbeM and DcM in Fig. 13.3d) can be used, provided that the pyroclastic cone developed on a substrate with an inclination lower than 7°; otherwise, the elongation of the cone was affected by slope gravity control. Particularly attention has also been paid to the presence of lava flows younger than the pyroclastic cones, which can mask the real shape of the cone base. Crater rims can also have symmetric higher and lower points: the line connecting two opposite lower points (Acdp in Fig. 13.3c) is regarded to be parallel to the underlying magma fissure (Tibaldi 1995) as also shown in Fig. 13.3d. Fissure eruptions can also give rise to closely-spaced vents (e.g. Acc in Fig. 13.3a) resulting in composite cones marked by different degrees of coalescence (Corazzato and Tibaldi 2006). All the described features are the surface expression of shallow dykes; they tend to be perpendicular to the horizontal least principal stress (σ_{Hmin}) and parallel to the horizontal greatest principal stress (σ_{Hmax}). Usually the σ_{Hmin} coincides with the least principal stress (σ_3), whereas the σ_{Hmax} can be the greatest principal stress (σ_1) or the intermediate principal stress (σ_2). Fissures and faults can also be directly identified across the substrate surrounding volcanic edifices, hence enabling to gain insights into the style and geometry of the fracture field. Volcanic domes can also provide key information, in terms of their morpho-structural features (Fig. 13.3e), such as apical depressions (Fig. 13.3f), elongation of the base and alignment of coeval edifices, which may be the expression of the strike of the underlying fractures through which magma is transported to the surface (Pasquaré and Tibaldi 2003).

Geophysical data recorded during magma injection events can also be helpful: among these, the precise location of volcano-tectonic earthquakes and related focal mechanisms that, in turn, can be used to come up with the dominant kinematics/geometry of magma paths. Published interferometric data have been also studied especially whenever they were related to dyke injection processes (e.g. Tizzani et al. 2015). During the present research, all available published geological maps and papers containing field data have also been used to infer the most likely, hidden eruptive fissures to all identified volcanic centers.

In summary, we performed the reconstruction of the azimuth of magma paths has here been based on the following: (*i*) alignment of closely-spaced vents; (*ii*) alignment of composite craters; (*iii*) preferred distribution of swarms of parasitic and flank cones; (*iv*) elongation of volcano centers or the whole volcanic chain; (*v*) trend of the base of elongated pyroclastic cones or volcanic domes; (*vi*) trend of crater rim depressed points; (*vii*) trend of dome apical grabens.

13.4 Results

13.4.1 Volcanic Centres and Lineaments Identification

We have mapped a total of 135 volcanic centres, 11 of which are classified as main volcanic centers in the literature and belong to the Abuli Samsari Volcanic Ridge (e.g. Lebedev et al. 2003; Pasquaré et al. 2011; Global Volcanism Program 2013). 57

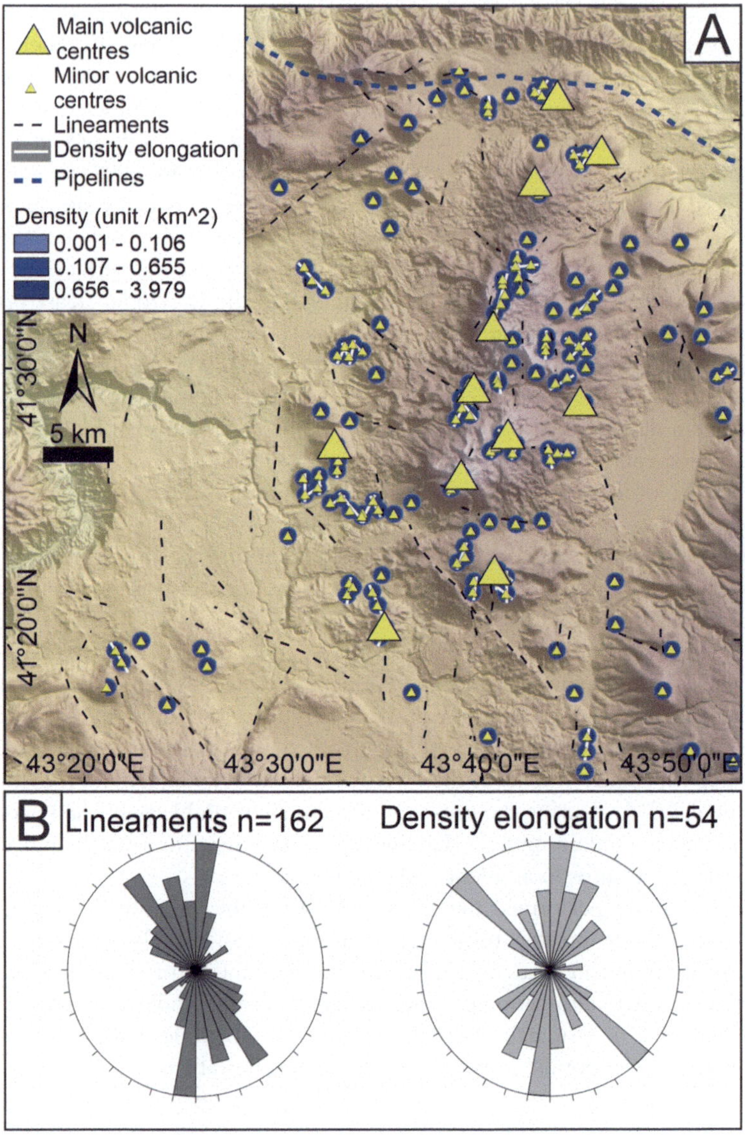

Fig. 13.5 (**a**) All the mapped volcanic centers, together with mapped lineaments and calculated density areas. (**b**) Rose diagrams for lineaments and elongation of density area

centres are Quaternary, 15 are of Pliocene-Quaternary age, 36 are related to Pliocene units and 27 to Miocene-Pliocene ones. At a general level, they are scattered across an area with an about N-S overall elongation. We have also mapped 162 lineaments, with represented trends ranging from N-S to NW-SE and NNW-SSE; few structures strike NE-SW as well (Fig. 13.5).

13.4.2 Areal Density of Volcanic Centers

As far as isodensity areas are concerned (Fig. 13.5), they were used for information related to alignments, by measuring all 54 elongation directions provided in the rose diagram of Fig. 13.5b. The evaluation of all the possible elongations of overlapping areas was performed by tying all the recognized vents within the three highest classes of density and by measuring the azimuths of the lines connecting two or more vents (Fig. 13.5a). The rose diagrams show a dominance of N-S and NW-SE trending elongations, along with subordinate NNE-SSW, and NE-SW trends. The prevailing, N-S elongation direction is in agreement with the general trend of the ASVR, whereas the NW-SE direction is consistent with the general trend of the major fault zones reported by previous authors (Rebaï et al. 1993; Pasquaré et al. 2011). The N-S to NNE direction agrees with the principal stress trajectories suggested by Tsereteli et al. (2016), based on focal mechanism data.

13.4.3 Inferring the Direction of Magma Pathways

By applying the above mentioned methods, we were able to define 66 different directions for magma pathways, as shown in Fig. 13.6. As highlighted in the rose diagram (Fig. 13.6), the greater number of magma pathways strike mostly N-S (24), with a minor population (12) of about NW-SE and NNE-SSW (10) directions. The N-S and NE-SE trends seem to be distributed across the whole area; the NW-SE direction appears in proximity of the NW-SE trending, major fault zones. Furthermore, we also considered the age volcano deposits related to the inferred magma paths, as displayed in Fig. 13.7. Regarding Quaternary times, the most represented directions are N-S, WNW-ESE and NE; for Pliocene-Quaternary times, the most represented directions are N-S and NE; for Pliocene-Quaternary times, the most represented directions are N-S, NW-SE and NE; for Miocene-Pliocene times, the most frequent directions are N-S, followed by NW-SE and NE-SE. The N-S direction is always present over the whole considered time interval.

13.4.4 Field Data

Here, we present new field observations, which integrate data provided by Rebaï et al. (1993), Koçyigit et al. (2001) and Pasquaré et al. (2011). Koçyigit et al. (2001) documented a strike-slip-dominated compressional extensional regime in the area, which dates back to Pliocene times. According to these authors, the dominant tectonic and volcanic features across the ASVR are NW-SE to NE-SW trending conjugate sets of strike-slip faults, N-S to NNW-trending normal faults, fissures, and a number of Plio-Quaternary volcanoes.

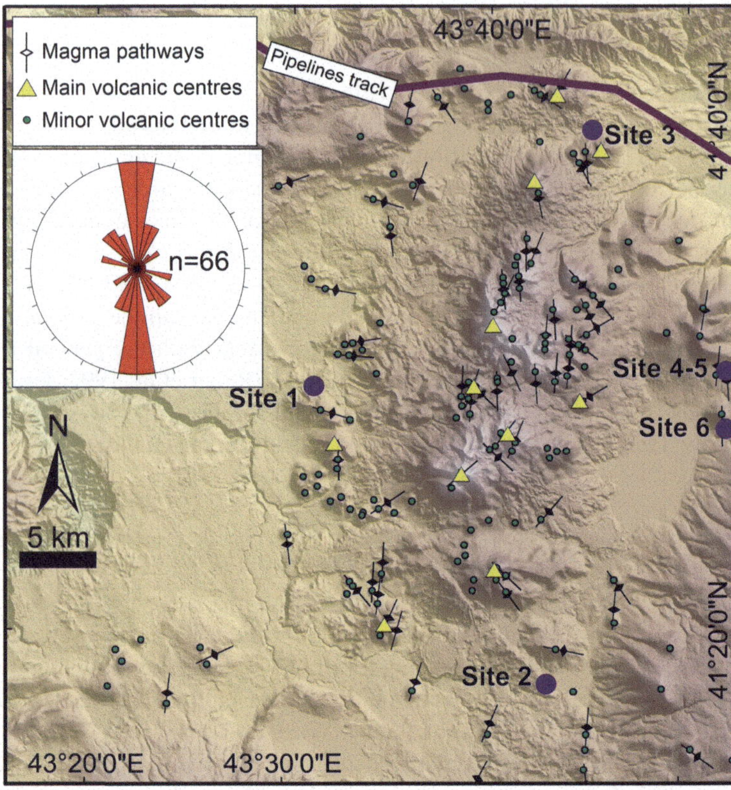

Fig. 13.6 Direction and location of the 66 inferred magma pathways as well as the related rose diagram

Pasquaré et al. (2011), besides analysing recent volcanic edifices which are the closest to the pipeline track, devoted particular attention to a major, NW-SE striking fault which extends for about 25 km, as shown in Fig. 13.2. This fault had been first recognized on SPOT images by Rebaï et al. (1993) who observed, close to the south-eastern tip of the fault, the displacement of two river channels, revealing a major right-lateral, strike-slip component. However, as Rebaï et al. (1993) had not carried out detailed structural surveys of the fault, Pasquaré et al. (2011) performed detailed structural field surveys along the fault. They made measurements at several locations (continuous observation of the fault was made impossible by the thick vegetative cover); site 1 (see Fig. 13.6) displays an outstanding example of a very steep fault plane (Fig. 13.8a), with slicknesides revealing two components of movement: an older normal component (with striation pitches close to 90°), upon which younger striations are superimposed, indicating almost pure strike-slip movements on the fault plane. The resulting horizontal σ_{Hmax} strikes N12°.

Another example of the appearance of the fault in the field can be observed at site 2 (Fig. 13.8b). According to the freshness of the slickensides, also this

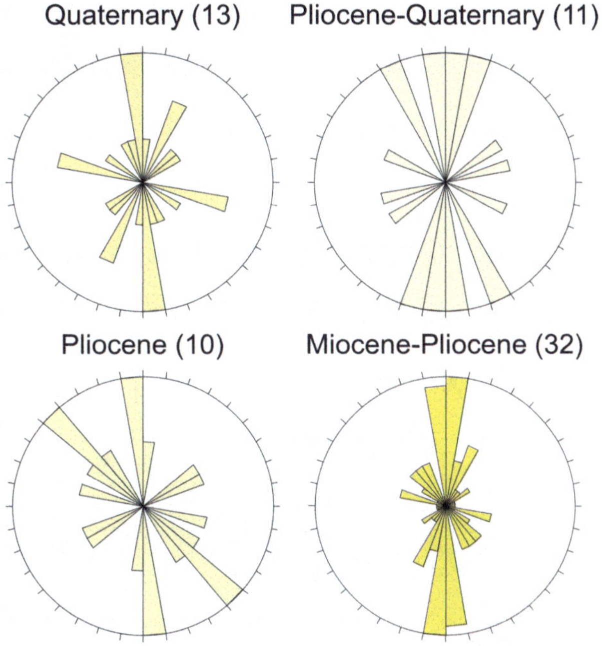

Fig. 13.7 Rose diagrams of the inferred magma pathways, subdivided according to the age of the related volcanic deposits

outcrop reveals the fault's recent activity. Stress tensor computations indicate an approximately N-S directed σ1, as reported in Fig. 13.6. The resulting σ_{Hmax} strikes N8°. Furthermore, in the Tavkvetili volcano area, closer to the pipeline track, Pasquaré et al. (2011) performed measurements along a NE-SW striking strike-slip fault, which had already been identified by Rebaï et al. (1993) based on the observation of SPOT images. The analysis of steep fault planes (site 3 in Fig. 13.6) with clear slickensides, clearly suggest a left-lateral, strike-slip kinematics for this fault. The related horizontal compressive stress (σ1) is in very good agreement with the results from the NW-SE directed, right-lateral fault. The resulting horizontal σ_{Hmax} strikes N10.5°.

We have also identified a number of dykes within a few scoria cones that have been dug for mining purposes; where quarries are abandoned, it has been possible to access them and carry out structural fieldwork (their location is provided in Fig. 13.6). At site 4, we identified a textbook example of a 1.2-m thick, N-S striking vertical dyke (Fig. 13.8c). At site 5, a few tens of meters away to the NW within the same scoria cone, we found another dyke, striking NE and with a thickness of about 1 m (Fig. 13.8d). This thickness is typical of feeder dykes related to monogenetic cones (e.g. Tibaldi et al. 2013). At site 6, located within a recently-dug scoria cone as well, we found a 1.5-m-thick dyke and a pervasive set of sub-vertical structures, all of which are striking NE-SW (Fig. 13.8e).

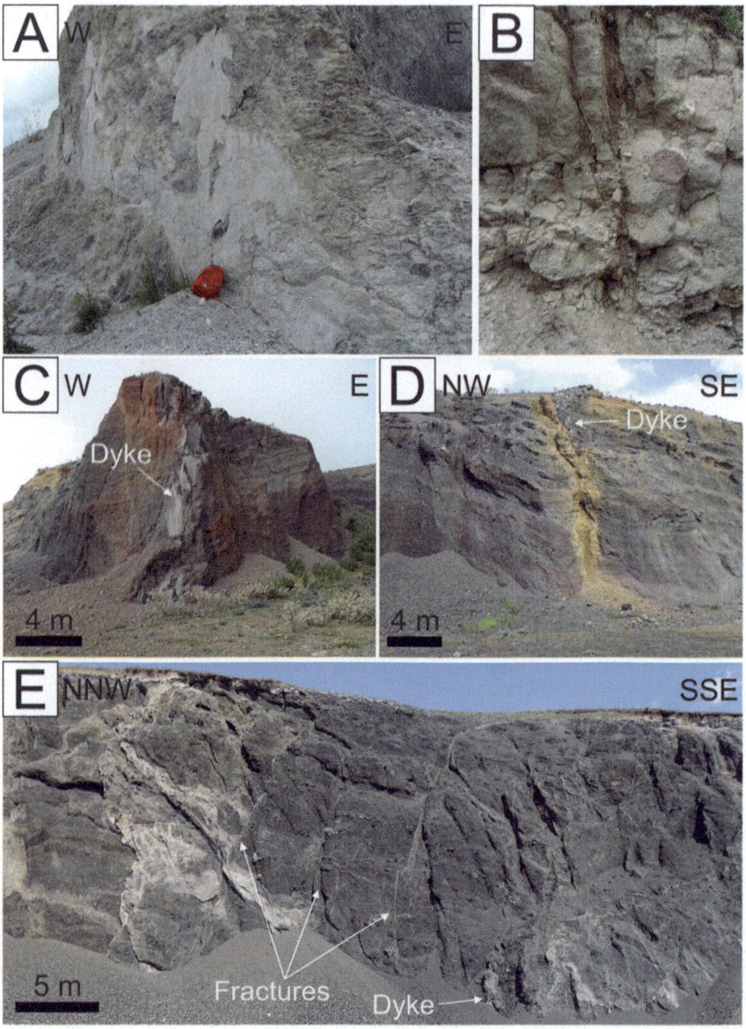

Fig. 13.8 (**a**) Fault zone studied at site 1; (**b**) Fault zone studied at site 2; (**c**) N-S trending dyke surveyed at site 4; (**d**) NE-SW dyke surveyed at site 5; (**e**) NE-striking dyke and NE-striking facture zone studied at site 6

13.5 Discussion

13.5.1 Geological-Structural Features

The study area, prior to this research effort, had been the focus of very few studies aimed at studying the structural factors guiding the ascent of magma to the surface, and at defining the dominant stress field in the area, which may lead to future earthquakes. With the aim of contributing to the above goals, we have taken into consideration: (*i*) field observations and data from Koçyigit et al. (2001) as well as Pasquaré et al. (2011); (*ii*) seismological data from Tsereteli et al. (2016); (*iii*) new field data collected for the purpose of the present work; (*iv*) new data related to the orientation of magma pathways. Koçyigit et al. (2001) proposed that the neotectonic regime of the area is characterized by a strike-slip-dominated regime. Pasquaré et al. (2011) performed structural field surveys aimed at recognizing the evolution of the kinematics of the main strike-slip fault in the study area, and deriving stress tensor orientations. Their results confirmed the presence of dextral strike-slip faulting in the area at sites 1 and 2 and of left-lateral strike-slip faulting at site 3 (Fig. 13.2a and Fig. 13.6). Stress inversion on data from the three sites indicates that σ_1 (assumed as σ_{HMAX}) respectively strikes N12°, N8° and N10.5°, therefore suggesting an about N-S direction of compression. Also the trend of dykes measured at site 4 support this hypothesis. Regarding seismological data, σ_1 trajectories (Fig. 13.9) proposed by Tsereteli et al. (2016), based on focal mechanism data, document a direction of N 7°; data from the World Stress Map provide more directions of the σ_{HMAX} but the one related to the M_S 5.6,1986 Paravani earthquake suggest a direction of N 12°, consistent with all our field data. By combining all the above, we claim that the about N-S to N10°E trend can be suggested as the predominant compressional direction in the area, as shown in Fig. 13.9. In addition to field data, we also performed, for the first time, an *ad-hoc* analysis aimed at reconstructing the direction of magma pathways in the area. To do so, we first identified a total of 135 volcanic centres, 11 of which are classified as main volcanic centers from the literature and belong to the Abuli Samsari Volcanic Ridge (e.g. Lebedev et al. 2003; Pasquaré et al. 2011; Global Volcanism Program 2013). These centres are placed across an area with an overall, about N-S trending elongation. We identified 66 different directions and location for magma pathways as shown in Fig. 13.6. As highlighted in the rose diagram, the greater number of magma pathways strike mostly N-S (24), with a minor population the strike about NW-SE (12) and NNE-SSW (10). The N-S and NE-SE trends seem to be distributed across the whole area. Furthermore, according to the age of volcano deposits related with the inferred magma pathways (Fig. 13.7), the most represented directions for Quaternary times are N-S, WNW-ESE and NE; for Pliocene-Quaternary times the most represented direction are N-S and NE; for Pliocene-Quaternary time the most represented directions are N-S, NW-SE and NE; for the Late Miocene-Pliocene the most represented directions are N-S, followed by NW-SE and NE-SE. The N-S direction is consistently present over time, and this suggests that it can be related to the σ_{HMAX}; such direction is still predominant to the present day, as suggested by data from Tsereteli et al. (2016) and Pasquaré et al. (2011). The NW-SE direction of magma pathways can be related to major NW-SE strike-slip faults and other unmapped weakness zone in the area. The NE-SE direction is consistent with dykes and fractures discovered at Site 5 and 6, as well as with the presence of faults, as suggested by Pasquaré et al. (2011). Also the lineaments in the whole area enable to assess that the more represented directions are N-S, NW-SE and NNW-SSE; few also strike NE-SW (Fig. 13.5). The rose diagram related to elongation of density areas (Fig. 13.5b) shows a marked

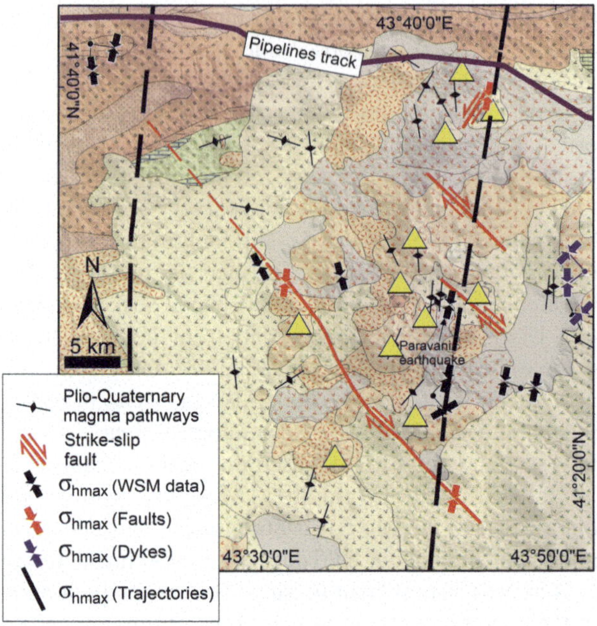

Fig. 13.9 The directions of the σ_{HMAX} across the whole study area, based on WSM data, faults, dykes. Horizontal stress trajectories are also provided, based on Tsereteli et al. (2016)

prevalence of N-S and NW-SE trending elongations, along with subordinate NNE-SSW, and NE-SW trends. In Fig. 13.9, we have summarized all available data, which enable documenting the direction of the σ_{HMAX}, indicated as trajectories in the study area.

13.5.2 Seismic and Volcanic Hazard Assessment

Regarding seismic hazard, the Georgian catalogue (historical times) reports 244 single events in the area for a selected depth between 1 and 15 km; 118 earthquakes have a magnitude of 2.1–3.0 M_S, 113 have a magnitude of 3.1–4.0 M_S, 12 have a magnitude of 4.1–5.0 M_S and only one has a magnitude between 5.1 and 6 M_S. They are plotted in Fig. 13.2a and their distribution suggest possible trend of epicentres following NW-SE, NE-SW and N-S directions.

The strongest historical event in the Abuli Samsari area hit on May 13th, 1986 at 08:44 UTC, with a given magnitude of 5.6 M_S (5.8 M_w) and an hypocentral depth of 12 km (Tsereteli et al. 2016); it was produced by a strike-slip tectonic regime (WSM data, Heidbach et al. 2008); the intensity data points related to this event range from 3 to 7–8 (MSK) and are all located in Georgia, Armenia and Azerbaijan (Varazanashvili et al. 2018). The triggering fault for this event has not been defined yet. However, we believe that a structure belonging to the major fault system we documented (Figs. 13.2a and 13.8b) might be regarded as the cause for the seismic event of 1986, based on the following considerations: (*i*) the Paravani lake area, where the earthquake and its aftershock (Fig. 13.2) were recorded, is less than 5 km from the main fault segment we identified in the field; (*ii*) no other fault in the area is comparable to this system both in terms of persistence in length and freshness of the fault planes (Fig. 13.8a). Another two, lesser fault segments are reported (Figs. 13.2 and 13.9), close to the Paravani Lake, which were identified by Rebaï et al. (1993) on SPOT images and interpreted as right-lateral, strike-slip faults. These might also be regarded as possible triggering structures for the 1986 Paravani event; if this was the case, their trend and kinematics are still compatible with an about N-S directed σ_1.

By using the Regressions of Rupture Length and Magnitude for strike-slip faults, as indicated by Blaser et al. (2010) for continental settings, we are able to calculate the expected moment magnitude for possible earthquakes triggered by the main four faults (Figs. 13.2a and 13.9): the northeastern one, striking about NE-SW, can produce an earthquake of M_w 5.1; the first fault from the North, striking NW-SE, can produce an earthquake of M_w 5.7, the second (from the North) NW-SE fault can produce an earthquake of M_w 5.5, and finally, the southernmost, longer one is capable of triggering an earthquake of M_w 6.8. All the above suggests that the area is affected by significant seismic hazard and this, as already underscored by Pasquaré et al. (2011), may be a threat for the pipeline route that runs across the northernmost portion of the volcanic range. In this case, as the faults do not crosscut the pipeline track, the risk posed to the lifeline would by represented by horizontal ground acceleration and oscillatory motions produced by a possible earthquake, and not by direct rupture.

As regards volcanic hazard, at the northernmost edge of the range, extremely close to the pipeline corridor, particularly notable are Tavkvetili and Shavnabada volcanoes (Fig. 13.2a). Tavkvetili volcano is a scoria cone, up to 2582 m asl in elevation, with a well-preserved summit crater, 200 m in diameter. Several lava flows were outpoured from the vent and flowed northward and southward as far as 4 km away from it. Shavnabada volcano is located 6 km south of Tavkvetili and is composed of two vents. The northern one produced a scoria cone, up to 2929 m asl in elevation. Activity at the southern vent resulted in the formation of a small shield cone with a well-preserved summit crater and lava flow field. Tavkvetili and Shavnabada are among the youngest volcanoes in the ASVR (Lebedev et al. 2003); this is attested also by our geomorphologic observations, pointing to the absence of periglacial activity on slopes; the observation of well-preserved summit craters (Pasquaré et al. 2011) suggest that volcanic activity here probably postdates the last glacial retreat (<10,000 yr BP).

Based on the results of our research, the about N-S directed σ_1 may be crucial also in terms of volcanic reactivation, which might take place through the opening of new fissures and vents, as well as through the formation of monogenetic and composite volcanoes. As the track of the pipelines lies just north of Tavkvetili Volcano, it could be affected by a reactivation of volcanic activity, and this, in turn, might bring about a major interruption in oil delivery from the Caspian Sea towards the west.

13.6 Conclusions

On the basis of new field studies, seismic data, and the interpretation of satellite images, we have been able to conduct a thorough investigation focused on the Abuli Samsari Volcanic Ridge, whose northern portion is crossed by two major pipeline tracks. The most relevant outcomes of our work are summarized hereunder. In detail, we have: (*i*) mapped a total of 135 volcanic centres, scattered across an area with an about N-S overall elongation; (*ii*) mapped 162 lineaments, with trends ranging from N-S to NW-SE and NNW-SSE; (*iii*) measured 54 elongation directions, with a dominance of N-S and NW-SE trending elongations, along with subordinate NNE-SSW, and NE-SW ones; (*iv*) defined 66 different directions for magma pathways, the greater number of which strike mostly N-S with subordinate, about NW-SE and NNE-SSW trends; (*v*) assessed the direction of the σ_{HMAX} from the Late Miocene to the present day; in this regard, we suggest the about N-S direction as the presently dominant one; (*vi*) estimated the expected moment magnitude for possible earthquakes triggered by the most relevant faults identified in the area. All our results point to the possibility that earthquakes may hit the area, induced by the dominant, about N-S directed $\sigma_{HMAX.}$ Moreover, the likelihood of a resumption of volcanic activity, guided by the present-day σ_{HMAX} cannot be ruled out and should be taken in consideration.

Further studies are needed, which should enable to identify, with as much detail as possible, the sections of the Abuli Samsari Volcanic Ridge that are the most prone to host a reactivation of volcanism, as well as to come up with suitable remediation measures aimed at ensuring security of the strategic pipelines that run across the area.

References

Adamia SH, Gujabidze G (2004) Geological Map of Georgia 1:500,000 Scale. Georgian Department of Geology (ISTC Project)

Annen C, Lénat JF, Provost A (2001) The long-term growth of volcanic edifices: numerical modelling of the role of dyke intrusion and lava flow emplacement. J Volcanol Geotherm Res 105:263–289

Avagyan A, Sosson M, Philip H, Karakhanian A, Rolland Y, Melkonyan R, Rebaï S, Davtyan V (2005) Neogene to Quaternary stress field evolution in Lesser Caucasus and adjacent regions using fault kinematics analysis and volcanic cluster data. Geodin Acta 18:401–416

Bistacchi A, Tibaldi A, Pasquaré Mariotto F, Rust D (2012) The association of cone-sheets and radial dykes: data from the Isle of Skye (UK), numerical modelling, and implications for shallow magma chambers. Earth Planet Sci Lett 339–340:46–56

Blaser L, Krüger F, Ohrnberger M, Scherbaum F (2010) Scaling relations of earthquake source parameter estimates with special focus on subduction environment. Bull Seism Soc Am 100(6):2914–2926

Bonali FL, Corazzato C, Tibaldi A (2012) Elastic stress interaction between faulting and volcanism in the Olacapato–San Antonio de Los Cobres area (Puna plateau, Argentina). Glob Planet Change 90:104–120

Bonali FL, Corazzato C, Tibaldi A (2011) Identifying rift zones on volcanoes: an example from La Réunion island, Indian Ocean. Bull Volcanol 73(3):347–366
Bonali FL (2013) Earthquake-induced static stress change on magma pathway in promoting the 2012 Copahue eruption. Tectonophysics 608:127–137
Bonali FL, Tibaldi A, Corazzato C, Tormey DR, Lara LE (2013) Quantifying the effect of large earthquakes in promoting eruptions due to stress changes on magma pathway: the Chile case. Tectonophysics 583:54–67
Bonali FL, Tibaldi A, Corazzato C (2015a) Sensitivity analysis of earthquake-induced static stress changes on volcanoes: the 2010 Mw 8.8 Chile earthquake. Geophys J Int 201(3):1868–1890
Bonali FL, Corazzato C, Bellotti F, Groppelli G (2015b) Active tectonics and its interactions with Copahue volcano. In: Copahue Volcano. Springer, Berlin/Heidelberg, pp 23–45
Chernyshev IV, Lebedev VA, Arakelyants MM (2006) K-Ar dating of quaternary volcanics: methodology and interpretation of results. Petrology 14:62–80
Connor CB (1990) Cinder cone clustering in the TransMexican Volcanic Belt: implications for structural and petrologic models. J Geophys Res Solid Earth 95(B12):19395–19405
Corazzato C, Tibaldi A (2006) Basement fracture control on type, distribution, and morphology of parasitic volcanic cones: an example from Mt. Etna, Italy. In: Tibaldi A, Lagmay M (eds) Interaction between volcanoes and their basement. J Volcanol Geotherm Res (Special issue) 158:177–194
Forte AM, Eric C, Whipple KX (2014) Transition from a singly vergent to doubly vergent wedge in a young orogen: the greater Caucasus. Tectonics 33:2077–2101
Germa S, Connor LJ, Cañon-Tapia E, Le Corvec N (2013) Tectonic and magmatic controls on the location of post-subduction monogenetic volcanoes in Baja California, Mexico, revealed through spatial analysis of eruptive vents. Bull Volcanol 75(12):1–14
Global Volcanism Program (2013) In: Venzke E (ed). Volcanoes of the World, vol 4.8.5. Smithsonian Institution. https://doi.org/10.5479/si.GVP.VOTW4-2013. Accessed 02 Jan 2020
Heidbach O, Rajabi M, Cui X, Fuchs K, Müller B, Reinecker J, Reiter K, Tingay M, Wenzel F, Xie F, Ziegler MO, Zoback ML, Zoback MD (2018) The World Stress Map database release 2016a: crustal stress pattern across scales. Tectonophysics 744:484–498
Heidbach O, Rajabi M, Reiter K, Ziegler M and WSM Team (2016) World stress map database release 2016. GFZ Data Services
Heidbach O, Tingay M, Barth A, Reinecker J, Kurfeß D, Müller B (2008) The 2008 release of the world stress map. Available online at: www.world-stress-map.org
Johnson CA, Harrison CGA (1990) Neotectonics in central Mexico. Phys Earth Planet Int 64:187–210
Jordan G (2003) Morphometric analysis and tectonic interpretation of digital terrain data: a case study. Earth Surf Processes Landf J British Geomorphol Res Group 28(8):807–822
Koçyigit A, Yılmaz A, Adamia S, Kuloshvili S (2001) Neotectonics of east Anatolian plateau (Turkey) and Lesser Caucasus: implication for transition thrusting to strike-slip faulting. Geodin Acta 14:177–195
Koike K, Nagano S, Kawaba K (1998) Construction and analysis of interpreted fracture planes through combination of satellite-image derived lineaments and digital elevation model data. Comput Geosci 24(6):573–583
Lebedev VA, Chernishev IV, Dudauri OZ, Arakelyants MM, Bairova ED, Gol'tsman YV, Chugaen AV, Vashakidze GT (2003) The Samsari volcanic center as an example of recent volcanism in the Lesser Caucasus: K-Ar geochronological and Sr-Nd isotopic data. Doklady Earth Sci 393:1323–1328
Lebedev VA, Vashakidze GT, Parfenov AV, Yakushev AI (2019) The Origin of Adakite-Like Magmas in the Modern Continental Collision Zone: Evidence from Pliocene Dacitic Volcanism of the Akhalkalaki Lava Plateau (Javakheti Highland, Lesser Caucasus). Petrology 27:307–327
Lutz T, Gutmann JT (1995) An improved method for determining and characterizing alignments of point-like features and its implications for the Pinacate volcanic field, Sonora, Mexico. J Geophys Res 100(B9):17659–17670

Marino CM, Tibaldi A (1988) Use of Landsat and Seasat data as a tool in kinematic analysis. Int J Remote Sens 9(10–11):1659–1673

Moritz R, Melkonyan R, Selby D, Popkhadze N, Gugushvili V, Tayan R, Ramazanov V (2016) Metallogeny of the Lesser Caucasus: from arc construction to post-collision evolution. Soc Econ Geologists (Spec Publ) 19:157–192

Nakamura K (1977) Volcanoes as possible indicators of tectonic stress orientation—principle and proposal. J Volcanol Geotherm Res 2(1):1–16

Nakamura K, Uyeda S (1980) Stress gradient in arc–back arc regions and plate subduction. J Geophys Res Solid Earth 85(B11):6419–6428

Pasquaré FA, Tibaldi A (2003) Do transcurrent faults guide volcano growth? The case of NW Bicol Volcanic Arc, Luzon, Philippines. Terra Nova 15(3):204–212

Pasquaré F, Tibaldi A (2007) Structure of a sheet-laccolith system revealing the interplay between tectonic and magma stresses at Stardalur Volcano, Iceland. J Volcanol Geotherm Res 161:131–150

Pasquaré FA, Tormey D, Vezzoli L, Okrostsvaridze A, Tutberidze B (2011) Mitigating the consequences of extreme events on strategic facilities: Evaluation of volcanic and seismic risk affecting the Caspian oil and gas pipelines in the Republic of Georgia. J Environ Manag 92(7):1774–1782

Philip H, Avagyan A, Karakhanian A, Ritz J-F, Rebai S (2001) Estimating slip rates and recurrence intervals for strong earthquakes along an intracontinental fault: example of the Pambak-Sevane-Sunik fault (Armenia). Tectonophysics 343:205–232

Rebaï S, Philip H, Dorbath L, Borissoff B, Haessler H, Cisternas A (1993) Active tectonics in the Lesser Caucasus: coexistence of compressive and extensional structures. Tectonics 12:1089–1114

Russo E, Waite GP, Tibaldi A (2017) Evaluation of the evolving stress field of the Yellowstone volcanic plateau, 1988 to 2010, from earthquake first-motion inversions. Tectonophysics 700:80–91

Silva CL, Morales N, Crósta AP, Costa SS, Jiménez-Rueda JR (2007) Analysis of tectonic-controlled fluvial morphology and sedimentary processes of the western Amazon Basin: an approach using satellite images and digital elevation model. An Acad Bras Cienc 79(4):693–711

Silverman BW (1986) Density estimation for statistics and data analysis, vol 26. CRC Press, Boca Raton

Strecker M, Bosworth W (1991) Quaternary stress-field change in the Gregory Rift, Kenya. EOS Trans Am Geophys Union 72:17–22

Tadini A, Bonali FL, Corazzato C, Cortés JA, Tibaldi A, Valentine GA (2014) Spatial distribution and structural analysis of vents in the Lunar Crater Volcanic Field (Nevada, USA). Bull Volcanol 76(11):877

Tadono T, Nagai H, Ishida H, Oda F, Naito S, Minakawa K, Iwamoto H (2016) Initial Validation of the 30 m-mesh Global Digital Surface Model Generated by ALOS PRISM, The International Archives of the Photogrammetry, Remote Sensing and Spatial Information Sciences – ISPRS XLI-B4:157–162

Tibaldi A (1995) Morphology of pyroclastic cones and tectonics. J Geophys Res Solid Earth 100(B12):24521–24535

Tibaldi A, Ferrari L (1991) Multisource remotely sensed data, field checks and seismicity for the definition of active tectonics in Ecuadorian Andes. Remote Sens 12(11):2343–2358

Tibaldi A, Pasquaré F (2008) A new mode of inner volcano growth: the "flower intrusive structure". Earth Planet Sci Lett 271:202–208

Tibaldi A, Bonali FL (2017) Intra-arc and back-arc volcano-tectonics: magma pathways at Holocene Alaska-Aleutian volcanoes. Earth-Sci Rev 167:1–26

Tibaldi A, Vezzoli L, Pasquaré Mariotto F, Rust D (2008a) Strike-slip fault tectonics and the emplacement of sheet-laccolith systems: the Thverfell case study (SW Iceland). J Struct Geol 30:274–290

Tibaldi A, Pasquaré Mariotto F, Papanikolaou D, Nomikou P (2008b) Discovery of a huge sector collapse at the Nisyros volcano, Greece, by on-land and offshore geological-structural data. J Volcanol Geotherm Res 177:485–499

Tibaldi A, Pasquaré Mariotto F, Rust D (2011) New insights into the cone–sheet structure of the Cuillin Complex, Isle of Skye, Scotland. J Geol Soc 168:689–704

Tibaldi A, Bonali FL, Pasquaré FA, Rust D, Cavallo A, D'Urso A (2013) Structure of regional dykes and local cone sheets in the Midhyrna-Lysuskard area, Snaefellsnes Peninsula (NW Iceland). Bull Volcanol 75(11):764

Tibaldi A, Bonali FL, Corazzato C (2017) Structural control on volcanoes and magma paths from local-to orogen-scale: The central Andes case. Tectonophysics 699:16–41

Tizzani P, Battaglia M, Castaldo R, Pepe A, Zeni G, Lanari R (2015) Magma and fluid migration at Yellowstone Caldera in the last three decades inferred from InSAR, leveling, and gravity measurements. J Geophys Res Solid Earth 120(4):2627–2647

Trifonov VG, Karahanian AS, Kozhurin AI (1990) The Spitak earthquake as recent tectonic activity manifestation. Geotektonika 6:46–60. (in Russian)

Tsereteli N, Tibaldi A, Alania V, Gventsadse A, Enukidze O, Varazanashvili O, Müller BI-R (2016) Active tectonics of central-western Caucasus, Georgia. Tectonophysics 691:328–344

Varazanashvili O, Tsereteli N, Bonali FL, Arabidze V, Russo E, Pasquaré Mariotto F, Oppizzi P (2018) GeoInt: the first macroseismic intensity database for the Republic of Georgia. J Seismol 22(3):625–667

Walter TR, Schmincke H-U (2002) Rifting, recurrent landsliding and Miocene structural reorganization on NW-Tenerife (Canary Islands). Int J Earth Sci 91:615–628

Wadge G, Cross A (1988) Quantitative methods for detecting aligned points - an application to the volcanic vents of the Michoacan-Guanajuato volcanic field, Mexico. Geology 16:815–818

Zare M, Amini H, Yazdi P, Sesetyan K, Demircioglu MB, Kalafat D, Erdik M, Giardini D, Khan MA, Tsereteli N (2014) Recent developments of the Middle East catalog. J Seismol 18(4):749–772

Part V
Key Studies for Hydrological, Landslide and Coastal Hazard Assessment

Chapter 14
Landslide and Mudflow Hazard Assessment in Georgia

Merab Gaprindashvili, Emil Tsereteli, George Gaprindashvili, and Otar Kurtsikidze

Abstract Geological hazard posed by landslides, debris flows, rock avalanches and mudflows has always been and still represents a major threat for communities all over the world, causing extensive damage and often times the destruction of infrastructures and facilities. Over the last decades, the protection of the population from geohazards and the safe operation of infrastructures have become a significant priority for most countries in the world. Geological-related, adverse phenomena are more frequent in mountainous countries with high rainfall amounts and complicated geological settings. The above-mentioned geological hazards can also be favoured by climate change, earthquakes, as well as by pervasive human activities. Georgia belongs to one of the most complicated regions among the world's mountainous countries: thousands of settlements, buildings, roads, oil and gas pipelines, high-voltage power lines are prone to geohazards, that can trigger disasters and lead to widespread losses of life and property. Consequently, geohazard assessment is an important step towards the management and mitigation of adverse, natural events. In the present work, we introduce a set of new hazard maps for geological hazard assessment, compiled by employing methods applied at the national level, with a particular focus on landslides and mudflows.

Keywords Georgia · Geohazard · Landslide · Debris flow · Caucasus

M. Gaprindashvili (✉) · O. Kurtsikidze
Department of Geology, LEPL National Environmental Agency, Tbilisi, Georgia

E. Tsereteli · G. Gaprindashvili
Department of Geology, LEPL National Environmental Agency, Tbilisi, Georgia

Ivane Javakhishvili Tbilisi State University, Vakhushti Bagrationi Institute of Geography, Tbilisi, Georgia

© Springer Nature B.V. 2021

F. L. Bonali et al. (eds.), *Building Knowledge for Geohazard Assessment and Management in the Caucasus and other Orogenic Regions*, NATO Science for Peace and Security Series C: Environmental Security,
https://doi.org/10.1007/978-94-024-2046-3_14

14.1 Rationale

Georgia is one of the most mountainous countries on Earth, bordered by the Greater Caucasus to the North and the Lesser Caucasus to the South (Fig. 14.1). It is worth pointing out that the country is also remarkable in terms of the development of different types of geological hazards, including landslides (Tibaldi et al. 2019) and earthquakes (Tibaldi et al. 2017a,b; Varazanashvili et al. 2018), with the exception of historic volcanic eruptions, though volcanoes with Holocene activity are present in the Lesser Caucasus (Pasquarè et al. 2011). Georgia hosts a great variety of morphological and climate conditions - from the humid, subtropical coastlines to highland-alpine-nival areas; whenever natural disasters strike the country, this usually leads to a great number of casualties, also due to very populated human settlements all across Georgia (Tsereteli et al. 2012; Tsereteli et al. 2014; Gaprindashvili et al. 2014; Gaprindashvili and Westen 2016; Gaprindashvili et al. 2018). A few examples of recent landslides, rock avalanches and debris flows are shown in Fig. 14.2; their location is highlighted in Fig. 14.1, which testifies to the large aerial distribution of such natural phenomena, from the Greater to the Lesser Caucasus.

Considering that geological disasters, as well as their adverse effects, have been on the rise in Georgia since the last century, based on the available data, the national government adopted a special decree (No. 967) in the 1960s, requiring the State Geology Department to develop a methodology for carrying out engineering

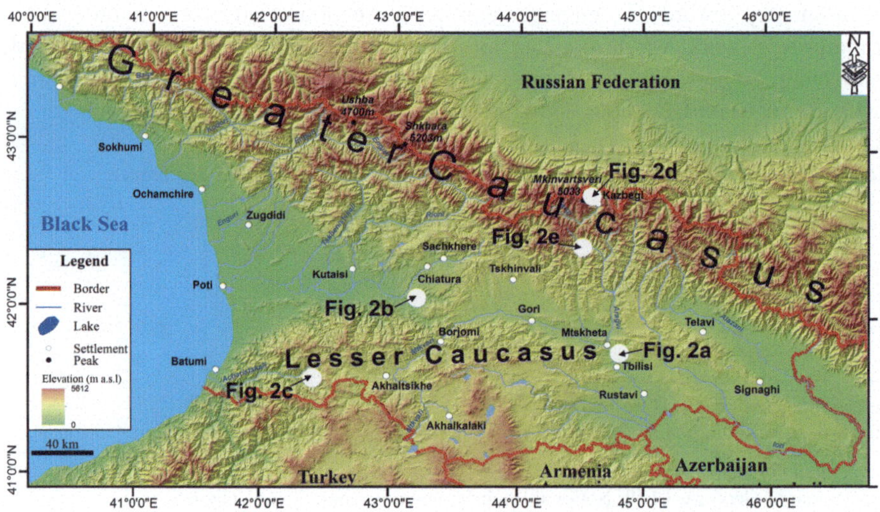

Fig. 14.1 Elevation map of the study area (Georgia), with indication of major mountain peaks and settlements, country borders, rivers and lakes (redrawn after Gaprindashvili and Westen 2016). The location of the events reported in Fig. 14.2 is also provided. Elevation data are from Aster GDEM v. 2 (Meyer 2011; Tachikawa et al. 2011); reference system: WGS84/geographic coordinates

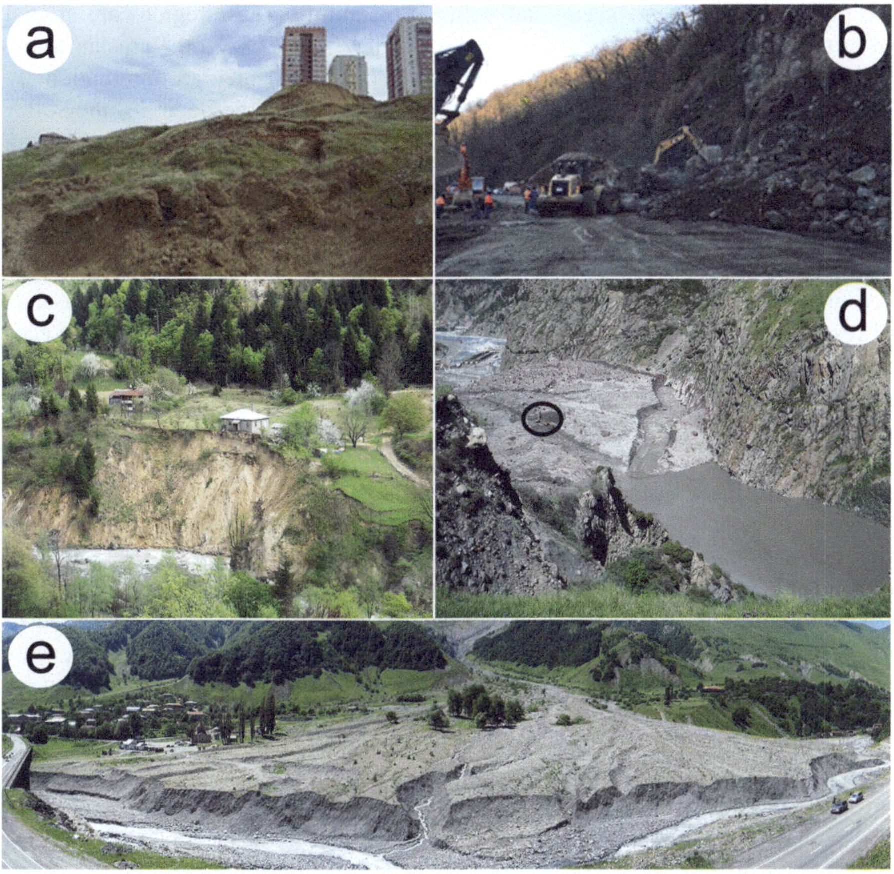

Fig. 14.2 (**a**) A major landslide occurred in Tbilisi in 2016 (Sheshelidze str.), buildings for scale; (**b**) a rock avalanche took place in 2011 at Rikoti pass (scraper for scale); (**c**) a landslide hit the Khulo municipality in 2013 (house for scale); (**d**) the Amali-Devdoraki rock avalanche and debris flow, 2014 (Kazbegi) (scraper for scale); (**e**) the Mleta debris flow in the Dusheti municipality, 2014 (cars for scale)

geodynamic studies. These comprised field geological surveys, the collection of all historical data and stationary observations; the coordination and monitoring of these activities were managed by the Ministry of Environmental Protection and Natural Resources.

By the end of the XX century in Georgia, scientists had performed geological surveys at the 1: 200,000 and 1: 50,000 scales across the whole country, and at the 1: 25,000 at 1: 10,000 scales across 45% of the territory, resulting in the production of several maps aimed at geohazard assessment.

These studies revealed that more than 70% of the Country's total area, including relevant settlements and major facilities, are prone to different geohazards. In this regard, the most critical areas are those in mountainous regions, also because these are the most affected by social and political instability. These regions are crucial from the geopolitical and economic point of views, as: *i)* several Eurasian transport facilities run through Georgian mountain passes; *ii)* they are internationally recognized landscapes with a unique historical and cultural heritage; *iii)* they offer great opportunities for mining and tourism, including trekking, skiing and climbing activities.

In Georgia, from 1967 to 2019 A.D., a total of about 3000 settlements (63% of the total) have been classified as located in areas potentially affected by natural disasters; as a consequence of this, about 60,000 families have been displaced as eco-migrants. Presently, as a result of the exposure to geological hazards, thousands of residents may have to be relocated to safer settlements, thus making them eco-migrants and causing them stress and resentment.

It is worth noting that, under the General Scheme of Anti-Erosion Measures (General Scheme of Anti-Erosion Measures 1988), geological disaster management funds were set at 1.3 billion GEL. It also needs to be underscored that, based on our data, the consequences to the population and the damage produced to infrastructures and facilities, as well as the number of human casualties, are three times greater than in the Southern Caucasus countries. As of today, the total number of landslide-gravity events exceeds 50,000, whereas those that directly affected the population and facilities/infrastructures are almost 3000. In the past, due to the more limited number of studies focused on landslide assessment, fewer events were recognized and mapped. The clearest confirmation of this was provided by the National Environmental Agency of the Ministry of Environment Protection and Agriculture of Georgia: engineering surveys conducted in Tbilisi area in 2016–2019 resulted in the identification of more than 500 landslides (Gaprindashvili et al. 2019), whereas only 60 landslides had been included in the Bulletin of Hazardous Geological Processes of the State Department of Geology in 2000. This is also confirmed by the analysis of the results of the geo-monitoring and disaster survey of the National Environment Agency in 2009–2018 (Table 14.1), which are aligned with those of the dynamic activity of landslide-gravitational and mudflow events (Fig. 14.3).

The spatial distribution of landslide and mudflow events over the territory encompasses both the seaside and the highlands. Out of the 70 municipalities regarded as affected by landslide hazards, 29 are included in the average-hazard zone (41%) and 35 (50%) in the high-hazard zone. According to recent data (2011–2018) from 8229 families (Source: Department of Geology of Georgia 2019), as a result of this monitoring activity, 1545 families were given the recommendation to relocate to safer houses/settlements (Department of Geology 2019). In view of the above, we hereunder introduce a set of new landslide and debris/mudflow hazard maps, compiled through the application of a national-scale approach.

Table 14.1 Number of landslides and mudflows with associated casualties and economic losses in Georgia (2009–2018)

Year	Landslide			Debris/Mudflow			Total casualties	Total economic loss ($ mln)	Affected objects	
	Re-activated and new	Human casualties	Approx. economic loss ($ mln)	Re-activated and new	Human casualties	Approx. economic loss ($ mln)			Settlements	Houses
2009	323	1	27.6	193	3	7.2	4	34.8	285	2696
2010	250	3	8.7	81	2	2.2	5	10.9	295	822
2011	94	3	5.6	37	8	3.9	11	9.5	309	463
2012	325	1	11.8	88	5	21.7	6	33.5	350	845
2013	336	0	19.5	93	0	20.3	0	39.8	472	1269
2014	727	0	26.7	141	10	65.2	10	91.9	845	962
2015	936	4	29.8	167	19	108.7	23	138.5	931	1014
2016	780	0	17.6	208	0	8.9	0	26.5	934	1084
2017	845	0	21.3	165	0	10.6	0	31.9	1042	1353
2018	702	1	19.6	122	0	16.8	1	36.4	1057	1245
Total	**5318**	**13**	**188.2**	**1295**	**47**	**265.5**	**60**	**453.7**		**11,753**

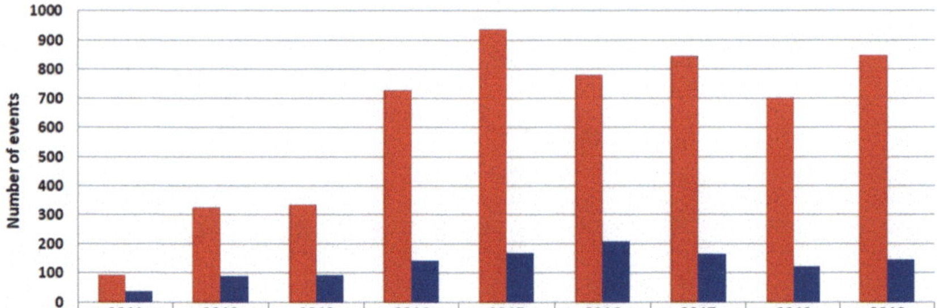

	2011	2012	2013	2014	2015	2016	2017	2018	2019
Landslide	94	325	336	727	936	780	845	702	849
Debris/Mudflow	37	88	93	141	167	208	165	122	146

Fig. 14.3 Number of landslides and debris/mudflows that occurred in Georgia (2009–2019). (Source: Department of Geology of Georgia)

14.2 Morphological Aspects of the Territory

The magnitude of landslide-gravitational and mudflow processes and events and the associated hazards depend on factors such as rock and terrain energy potential, together with external environmental factors such as meteorological events, earthquakes and anthropogenic contributions.

Studies focused on the development and triggering causes of landslides and mudflows carried out in Georgia in the last tens of years, have been considered of relevant importance by UNEP- UNESCO, and were summarized in the monograph "Landslides and Debrisflow" (Tsereteli 1984). Based on this experience, the landslide and mudflow hazard has been assessed, providing new hazard maps and recognizing the geological units and characteristics that are more prone to potentially develop landslides and mudflows. Among the major constituent elements of the geological environment are rocks and soils, which represent not only the foundation on which human-engineering structures and facilities are built, but are also crucial in terms of the activation of geological processes, primarily landslides and mudflows. The role of such components is particularly relevant, if the Alpine-Himalayan-type geological environment of the Georgian central segment of the Caucasus is considered. In Georgia, regional studies analysing the exogenous conditions that led to the development of gravitational events, have identified the categories of terrains and slopes that are more prone to this type of geohazard (Gaprindashvili 2016; Gaprindashvili et al. 2016a, b, 2019).

The first aspect we need to consider is the relation between the terrain slope and exogenous processes, here referred to as the "energy potential of the terrain"; this is provided considering the following classes of terrain slope: up to 3°; 3–8°; 8–15°; 15–25°; 25–35°; 35–45°; 45–65°; 65–90° (Fig. 14.4).

For each class of "terrain energy potential", that corresponds to specific areas in the Georgian territory (see Fig. 14.4), all the types of dominant exogenous processes have been identified and reported in Fig. 14.5. In addition, four different

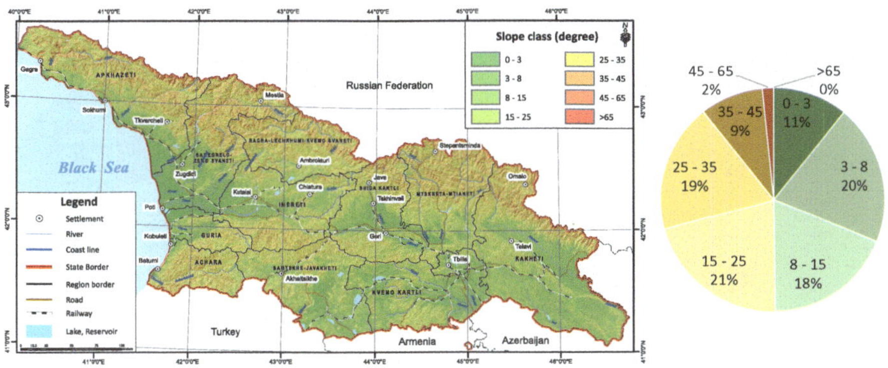

Fig. 14.4 Terrain slope map of Georgia calculated on Aster GDEM V.2 (pixel size is approximately 30 meters; Meyer 2011; Tachikawa et al. 2011); reference system: WGS84/geographic coordinates. Terrain slope classes are reported in percentage of frequency

Terrain slope (°)	Geological processes and related hazards
>65	Rocky terrains, with deep erosion and gravitational processes in the form of rock avalanches and accumulation of thick colluvium in debris/mudflow source areas.
45-65	Major gravitational processes, such as rock avalanches, rockfalls, rockslide-type landslides, debris/mudlflows, river bank erosional phenomena.
35-45	Rockfalls and rock avalanche-type landslides, river erosion and debris/mudflow occurrence.
25-35	Tectonically active, highly-weathered areas; denudation-gravitational events, with sediment accumulation and debris/mudflows as well as landslides; development of erosional processes in gullies.
15-25	Erosional processes in gullies, accompanied by debris/mudflows and landslides.
8-15	Active processes of rainfall-induced denudation and accumulation of deluvial sediments. High "moisture effects" and development of plastic-flow landslides; active processes of suffosion and erosion in gullies with subsequent mudflows; climatogenic landslides.
3-8	Suffosion processes over plain areas; weathering in mountain pediment areas and rainfall-induced denudation, with formation of deluvial sediments; gully erosion and development of shallow climatogenic landslides.
0-3	Mostly resistant to geological processes; swamping due to rising groundwater, suffosion.

Fig. 14.5 Possible geological processes and related hazards for the different classes of terrain slope reported in Fig. 14.4

areas, defined as "morphological zones" have also been defined, based on different altitudes above the sea level, and have been mapped in Fig. 14.6 and classified by percentage value. Such classes are: *inter-mountain areas* (less than 600 m a.s.l.); *lowland areas* (600–1000 m a.s.l.); *mid-altitude areas* (1001–2000 m a.s.l.); *highlands* (above 2000 m a.s.l.). The "morphological zones" have been considered for geohazard assessment together with those representing the terrain energy potential".

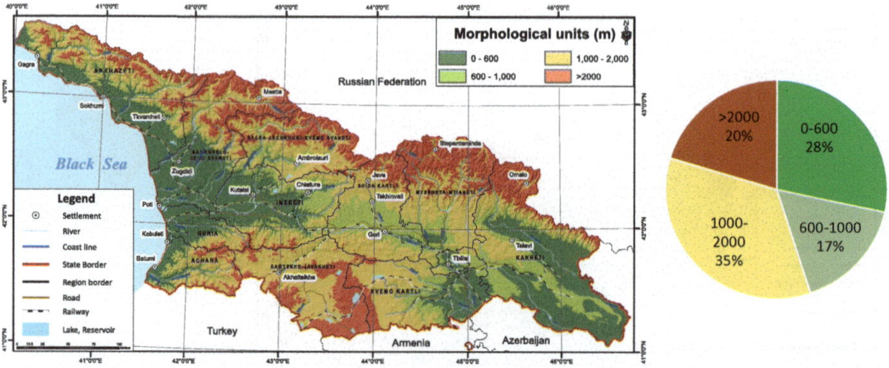

Fig. 14.6 Morphological zones in Georgia; their classification is based on Aster GDEM V.2 (pixel size is approximately 30 meters; Meyer 2011; Tachikawa et al. 2011); reference system: WGS84/geographic coordinates. Morphological zones distribution in Georgia, classified by percentage values

14.3 The Influence of Climate and Weather

Regarding the contribution of climate, the long-term weather regime for a given geographical environment should be considered as a main background for some geological processes (landslides, debris flows), whereas meteorological factors (temperature, precipitation, humidity) are the main determinants for landslide and mudflow events. Consequently, landslides and mudflows may be linked to atmospheric precipitation deviations from the average, in terms of the hereunder provided conditions:

1. **Low background activation** - when annual rates of atmospheric precipitation and “moisture effect” are below the threshold of the annual multiannual standard, and landslides are “stable”;
2. **Background (medium) activation -** when the amount of atmospheric precipitation during the year and the “moisture effect” of the deforming horizon fall within the limits of the multi-annual norm; in this case, landslide processes occur at the background level;
3. **Stressful -** when atmospheric precipitation during the year exceeds the average perennial up to 200 mm. In this case, slopes with rocks characterized by optimal receptor properties tend to activate landslide processes;
4. **Extreme -** when landslide activation begins in events of precipitation above 400 mm above the average perennial norm. In this case, almost all landslide bodies are reactivated, and new landslides processes are triggered;
5. **Paroxysmal -** Regardless of what kind of circulatory regime is present in the atmosphere, when rainfall is above (up to 400–600 mm) the perennial average, the “humidity effect” will be increased to the limit, and all new landslides in stable conditions will be activated.

Thus, in terms of regional development of landslides in Georgia and their dynamic regime, there are three main stages:

1. Maximum reactivation of landslide processes, caused by paroxysmal extremes in atmospheric precipitation (400–600 mm or more);
2. Periods of average activation of landslide processes, mainly involving the intervals between periods of intensified activation of landslide events;
3. Baseline period of landslide processes. Landslides in this kind of dynamics precede the two previous stages.

If the internal annual deviations of atmospheric precipitations and the "moisture effect" play a key role in the dynamic mode of landslide processes, the main determinants of debris/mudflow events are the amount and intensity of daytime rainfall and precipitation (that in Georgia is the warmest period of the year - months April to October).

For the mountainous territory of Georgia, we have established that the background activation of mudflow processes begins with daily precipitation ranging 30–40 mm. Extreme mudflows with high energy and volumes start in the range of 50–80 mm. In the event of precipitation of 80–100 mm or more, there will be catastrophic debris/mudflows (Gaprindashvili et al. 2016a, b; Tsereteli et al. 2019). The classic site for these phenomena is the central part of the Greater Caucasus, with the tributaries of Aragvi, Tergi, Rioni, Tskhenistskali, Enguri, Kodori and almost all the rivers on the southern slope of the Kakheti Caucasus.

As regards the complications due to human activity in Georgia, in terms of reactivation of exogenous processes and events, two categories of adverse human impacts on the geological environment can be clearly identified:

1. Natural-anthropogenic, in which the development of geo-environmental changes and exogenous processes are related to human-engineering activities, but the natural development dominates over the human effects.
2. Anthropo-technogenic adverse effects on the geological environment, when exogenous processes are strongly enhanced by human activities.

14.4 Classification of Landslide Types in the Territory of Georgia

Due to the complexity of landslide processes and events, their geological nature, multifactorial features, morphogenetic features, depth deformation, volumes, slope and many other features are so diverse that the number of classification characteristics exceeds the hundreds (Cruden and Varnes 1996; Hungr et al. 2014; Varnes 1978). All, to varying degrees, reflect the essence of landslide processes and the peculiarities of events. It is recommended that existing landslide bodies and their possible emergence areas should be assessed in terms of the characteristics of the deformable slopes, their structure, movement mechanisms, and the nature of recurring processes characterizing these types of geohazards. According to the above, two landslide classes can be established:

Class A - Landslides in lithified, although tectonically strongly disrupted rocky and semi-rocky substrate rocks. As a rule, rocks of this class are characterized by maximum deformation capacity, and a more or less uniform structure and movement mechanisms;

Class B - Landslides bound lithic rocks and quaternary slope sediments and active zones of weathering. Landslides developed in these sediments are characterized by a great heterogeneity in origin, movement mechanisms, kinematics, morphology and other features.

As far as the landslide hazard factors in the territory of Georgia are concerned, the following groups of landslides can be individuated:

- **Coastal landslides -** the origin of which is associated with erosion of riverbanks, abrasive transformation of seabeds and associated reservoirs, rising water levels, and hydrostatic and hydrodynamic regime fluctuations;
- **Climate-related landslides -** mainly occurring in slope sediments made of weak clay rocks. Associated with quantitative indices of atmospheric precipitation on an annual basis;
- **Tecto-seismogenic landslide-gravitational events -** characterized by possible earthquake triggering and localized in zones affected by tectonic dislocations;
- **Karst landslides -** quite specific in the Caucasus region, and scientific data on them is scarce;
- **Landslides influenced by anthropogenic processes -** they occupy an important place in Georgia, with little attention paid to the study of problems related to them in the recent past;
- **Cryogenic landslides -** characteristic of highland alpine-nival zone terrains in the Caucasus region.

14.5 New Hazard Maps Calculation

After relevant studies were carried out nationwide and the major factors behind the occurrence of landslide-gravitational and mudflow processes and events were identified, the next stage was to identify threats, vulnerabilities, and risks to both territorial and individual municipalities. In this regard, landslide and mudflow events were zoned according to the level of vulnerability of the potentially affected areas, recurrence intervals, and characteristics of the infrastructures. Accordingly, we have come up with different categories of hazard, which could be re-arranged in a different amount of hazard classes, depending on need. In this work, we have provided landslide and debris/mudflow hazard zoning maps grouped into 7 classes for landslide events (Fig. 14.7) and 9 classes for debris/mudflow events (Fig. 14.8).

Fig. 14.7 Landslide hazard zoning map of Georgia; reference system: WGS84 / geographic coordinates

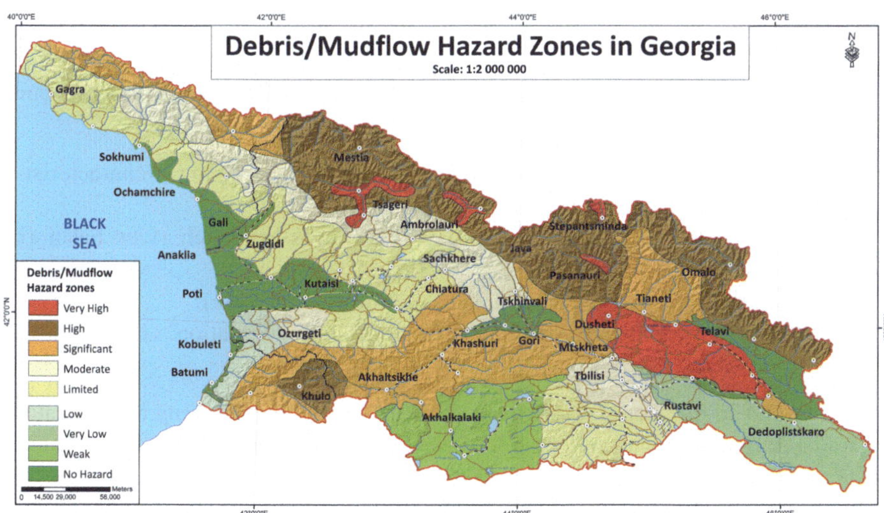

Fig. 14.8 Debris/mudflow hazard zoning map of Georgia; reference system: WGS84 / geographic coordinates

A simplified classification can consider four classes: high, moderate, low and not dangerous; this will be applied in the future using the same methodology.

Specifically, for landslide-gravitational hazards, we use the notion of "landslide potential", which is expressed as process intensity (i.e. hazard severity) and activity (frequency, return period), generated in a uniform geological environment and climate, and in a "moisture effect". The estimated coefficient of these events is estimated using four indicators:

1. The extent of the area occupied by all-time landslide bodies ($F_{cond.}$/km^2) the total area of the geological environment (F_{lands}. km^2);
2. The total area sum of landslide bodies (N area km^2) relative to the total area of geological environment;
3. The homogeneous geological environment space (F_{area}.km^2), associated with newly formed landslides (Kn $_{lands}$.) and earlier landslides (K_N old lands), which allows us to determine the potential for landslide development dynamics;
4. The ratio of areas prone to landslide processes potentially in the state of homeostatism (fkm^2) to the total study area (F Geol.km^2).

As regards mudflows, the mudflow hazard ratio (Ks) has been used. These ratios are:

1. Comparison of the total number of mudflats in the given river basins with rivers in the same basin where no mudflows have been recorded (Ks = $n_1 n_2$);
2. Comparison of the mudflow areas of the relevant geological environment with the total area of the river basin (Ks = f km^2 / F km^2);
3. Comparison of the areas of active nutrient depletion with solid mineral product to the total catchment area of the given river basin (Ks = fmF) (Ks = f mudfl. Source km^2 / F geol.km^2).

In addition to the listed coefficients, the following shall be taken into account when determining the risk of mudflows:

4. Frequency of the transformation of the mudflows originating in the characteristic mudflats into the basin in time;
5. Single maximum volumes of mudflow flows in a mud-rich river basin (in thousand m^3);
6. Objects in the immediate danger area of mudflow events (Kc);
7. Mudflow events that pose a direct threat to the population and important engineering sites.

We use the coefficient of 0 to 1 to estimate the hazard by summing the indicators of landslide-gravitational and mudflow events, where 0 means no hazard and 1 means highest hazard.

14.6 Results

The final weights of the resulting maps ranged from 0 to 1. Landslide and debris/mudflow hazard zoning maps here presented were grouped into 7 classes for landslide events (Fig. 14.7) and 9 classes for debris/mudflow events (Fig. 14.8), as listed hereunder.

Regarding the landslide hazard zoning map (Fig. 14.7), we have broken down the most relevant areas in terms of the level of hazard:

- **Very high hazard zone**, including the mountainous Adjara, Imereti and Racha-Lechkhumi and Lower Svaneti regions, as well as the Black Sea coast of Apkhazeti;
- **High hazard zone**, including the Guria foothills, Mtskheta-Mtianeti region, Okriba foothill zone and the western part of Adjara;
- **Significant hazard zone**, including the northern slopes of the Trialeti ridge with the Akhaltsikhe depression, the Samegrelo region (the hills of the Kolkheti North side) and the Tbilisi-Asureti area;
- **Medium hazard zone**, encompassing the Trialeti ridge, eroded terrains of the medium and high mountain valleys of the Greater Caucasus, Upper Svaneti, the part of the Caucasus Ridge in Kakheti region, the peripheries of the Shida Kartli depression, and foothills of the Tsivi- Gombori Ridge;
- **Low hazard zone,** including Mesozoic units of the southern slope of Caucasus constructed with volcanogenic and carbonate formation, river Iori plateau, the left side of river Iori characterized with hilly terrain, the adjacent territories of Khrami and Loki massifs;
- **Very low hazard zone**, encompassing the Javakheti volcanogenic highland, the axial zone of the Caucasus, the massifs of Khrami, Loki and Kelasuri, the terraces of the river Mtkvari and Khrami rivers, in the Marneuli -Gardabani plain area;
- **No hazard zone**, including the Kolkheti lowland, Shida Kartli, Gardabani, Alazani, and Iori areas presented by low plain-terraces.

Regarding the debris/mudflow hazard zoning map (Fig. 14.8), we have shown the most relevant areas subdivided by level of hazard:

- **Very high hazard zone**, that encompasses Kakheti (Alazani-Iori basins), Racha-Lechkhum-Kvemo Svaneti, Samegrelo-Zemo Svaneti and Mtskheta-Mtianeti regions;
- **High hazard zone**, which includes the Alpine zones of the medium and high mountain areas on the northern and eastern slopes of the Central and Eastern Greater Caucasus, as well as the high mountainous part of the lesser Caucasus, in the Adjara region;
- **Significant hazard zone**, that includes the Trialeti and Meskheti ridges, medium and upper parts of the Kodori and Bzifi River basins of the Western Caucasus, and the western part of Adjara;
- **Medium hazard zone**, encompassing the Rioni, Tskhenistskali, Enguri and Kodori river basins, headwaters of the Kvirila River basin, the low mountain and middle mountain zone of the Caucasus Range in Apkhazeti, Algeti river basin, Trialeti ridge in the vicinity of Tbilisi, and the low-mountain area of Adjara;
- **Limited hazard zone** that includes the lowland zones of raised horst of the Dzirula, Khrami, and Loqi, Foothills of Guria and Imereti regions, Psou, Sandripshi, and Zhoekvara rivers in Aphkhazeti;

- **Low hazard zone** that includes areas made of carbonate rocks in the small and middle-size mountains of Racha, Askhi, Khvamli, Arabika, and the Bzipi and Kodori rivers, foothills of Guria and Adjara;
- **Very low hazard zone** that encompasses the Iori Plateau and part of the downstream of river Mtkvari;
- **Weak hazard zone** that includes the Akhalkalaki plateau and volcanogenic highland of Javakheti;
- **No hazard zone** that encompasses the Kolkheti lowland and Black Sea coast, plains of Eastern Georgia, Kartli and Alazani.

14.7 Final Remarks and Future Developments

The hazard maps of landslide-gravitational and mudflow events have been compiled based on multi-criteria assessment methods, and have been grouped into 7 classes for landslide events and 9 classes for debris/mudflow events (Figs. 14.7 and 14.8). In a future perspective, we intend to provide the related classification subdivided in four homogenous hazard levels: 1. High; 2. Moderate; 3. Low; 4. Not in danger.

The four homogenous hazard levels will be classified as follow:

1. Highly hazardous areas are those in which the components of the geological environment are likely to generate any kind of geological process and are characterized by high vulnerability coefficients (greater than 0.5) and extreme activation intensities averaging 8–11 years;
2. Moderate-hazard areas are those where geological environment is susceptible to landslide and mudflow events within the range of 0.1–0.5 and the probability of activation of extreme processes is up to 10%;
3. Areas of low hazard are those where the probability of landslide and mudflow events does not exceed 1%;
4. Non-hazardous areas are those in which additional unforeseen processes, without triggering agents, shall not occur.

Acknowledgements The authors would like to thank the National Environmental Agency of Ministry of Environment Protection and Agriculture of Georgia for providing the various datasets used in this study. Also sincere gratitude to the editors of this issue for their support.

References

Cruden DM, Varnes DJ (1996) Landslide types and processes. In: Turner AK, Schuster RL (eds) Landslides: investigation and mitigation, Transportation Research Board, Special Report 247. US National Research Council, Washington DC, pp 36–75

Department of Geology, National Environmental Agency of Georgia (NEA) (2019) Information Geological Bulletins (2000–2019)

Gaprindashvili G (2016) Geodynamic processes and expected geo-ecological complications in Southern Highlands of Georgia. Dissertation, Ivane Javakhishvili Tbilisi State University, Tbilisi, Georgia

Gaprindashvili G, Westen CV (2016) Generation of a national landslide hazard and risk map for the country of Georgia. Nat Hazards 80(1):69–101. https://doi.org/10.1007/s11069-015-1958-5

Gaprindashvili G, Guo J, Daorueang P et al (2014) A new statistic approach towards landslide Hazard risk assessment. Int J Geosci 5:38–49. https://doi.org/10.4236/ijg.2014.51006

Gaprindashvili G, Gaprindashvili M, Tsereteli E (2016a) Natural disaster in Tbilisi City (Riv. Vere Basin) in the year 2015. Int J Geosci 7:1074–1087. https://www.scirp.org/journal/paperinformation.aspx?paperid=70622

Gaprindashvili M, Tsereteli E, Gaprindashvili G (2016b) Geological Report of the River Gldanis Khevi Basin. In: Strengthening Urban Risk Management of Tbilisi, UNDP Project. National Environmental Agency, Tbilisi, p 62

Gaprindashvili G, Tsereteli E, Gaprindashvili M (2018) Landslide hazard assessment methodology in Georgia. In: proceedings of the 16th Danube – European conference on. Geotech Eng 2(2–3):217–222. https://doi.org/10.1002/cepa.674

Gaprindashvili M, Tsereteli E, Gaprindashvili G (2019) Engineering-Geodynamic conditions and Geological Hazard assessment of Tbilisi city. Department of Geology, National Environmental Agency, Tbilisi,

Hungr O, Leroueil S, Picarelli L (2014) The Varnes classification of landslide types, an update. Landslides 11(2):167–194

Meyer D (2011) ASTER-GDEM-validation-team ASTER global digital elevation model version 2-summary of validation results. Jpn Syst 7

Pasquarè FA, Tormey D, Vezzoli L et al (2011) Mitigating the consequences of extreme events on strategic facilities: evaluation of volcanic and seismic risk affecting the Caspian oil and gas pipelines in the Republic of Georgia. J Environ Manag 92(7):1774–1782

State Department of Geology of Georgia (1988) General Scheme of Anti-Erosion Measures. Tbilisi, Georgia (in Russian)

Tachikawa T, Hato M, Kaku M et al (2011) Characteristics of ASTER GDEM version 2. In: proceedings of the 2011 IEEE international geoscience and remote sensing symposium. IEEE:3657–3660

Tibaldi A, Alania V, Bonali FL et al (2017a) Active inversion tectonics, simple shear folding and back-thrusting at Rioni Basin, Georgia. J Struct Geol 96:35–53

Tibaldi A, Russo E, Bonali FL et al (2017b) 3-D anatomy of an active fault-propagation fold: a multidisciplinary case study from Tsaishi, western Caucasus (Georgia). Tectonophysics 717:253–269

Tibaldi A, Oppizzi P, Gierke J et al (2019) Landslides near Enguri dam (Caucasus, Georgia) and possible seismotectonic effects. Nat Hazard Earth Sys 19:71

Tsereteli E (1984) Experience for landslide and Debrisflow research in Georgia. In: Kazlovski E (ed) Landslides and Debrisflows, vol 2. UNESCO, UNEP, pp 6–17

Tsereteli E, Gaprindashvili M, Gaprindashvili G et al (2012) Problems of natural and anthropogenic disasters in Georgia. J Georgian Geophys Soc Phys Solid Earth 15A:14–22

Tsereteli E, Gobejishvili R, Gaprindashvili G et al (2014) Challenges of geologic catastrophes in Georgia and ways of their reduction. In: Engineering geology for society and territory, vol 2. Springer, Cham, pp 1767–1773. https://doi.org/10.1007/978-3-319-09057-3_313

Tsereteli E, Gaprindashvili G, Gaprindashvili M et al (2019) Hazard risk of debris/mud flow events in Georgia and methodological approaches for management. In: Shakoor A, Cato K (eds) Proceedings of the IAEG/AEG annual meeting, San Francisco, California, 2018,vol 5. Springer, Cham. https://doi.org/10.1007/978-3-319-93136-4_19

Varazanashvili O, Tsereteli N, Bonali FL et al (2018) GeoInt: the first macroseismic intensity database for the republic of Georgia. J Seismol 22(3):625–667

Varnes DJ (1978) Slope movement types and processes. In: Schuster RL, Krizek RJ (eds) Landslides: analysis and control. Transportation Research Board, special report 176. National Academy of Sciences, Washington DC, pp 11–33

Chapter 15
Significance of the Spatial Resolution of DEM in Regional Slope Stability Analysis Enguri Dam, Republic of Georgia

Ian Nichols, John S. Gierke, Thomas Oommen, and Maria Diletta Acciaro

Abstract Landslides are the most common natural hazard in mountainous terrains and have a high potential to disrupt human activities and damage infrastructure when they occur in populated areas. The study area encompasses a portion of the Enguri Hydroelectric Dam (and reservoir), along with the steep, clay-rich slopes directly to the east of the reservoir. A large landslide in this area would have far-reaching negative consequences, as the Enguri Hydroelectric facility provides 50% of domestic energy production and serves as an anchor for socio-economic stability in the region. In order to reduce the risks associated with these slope failures, it is essential to improve our ability to forecast future landslides and determine which slope areas should be targeted for mitigation practices. This study will view the significance of input Digital Elevation Model (DEM) spatial resolution (30 m vs. 12 m) within a physics-based numerical model for regional slope stability analysis. Unlike the previous researches that have viewed the significance of input DEM spatial resolution within statistical slope stability models, we found that there is considerable improvement in the quality of output landslide susceptibility maps when utilizing DEMs with a higher spatial resolution. At 12 m spatial resolution, there is more distinct delineation of the reservoir boundary adjacent to the base of the hazardous slope. More importantly, the trials utilizing the 12 m DEM accounted for 0.22 km^2 of the new slope area that is considered "unstable" (FS < 1.25). These important details were completely missed at a spatial resolution of 30 m.

I. Nichols · J. S. Gierke · T. Oommen (✉)
Department of Geologic & Mining Engineering & Sciences, Michigan Technological University, Houghton, MI, USA
e-mail: toommen@mtu.edu

M. D. Acciaro
Department of Geologic & Mining Engineering & Sciences, Michigan Technological University, Houghton, MI, USA

Department of Earth and Environmental Sciences, University of Milan Bicocca, Milan, Italy

© Springer Nature B.V. 2021

F. L. Bonali et al. (eds.), *Building Knowledge for Geohazard Assessment and Management in the Caucasus and other Orogenic Regions*, NATO Science for Peace and Security Series C: Environmental Security,
https://doi.org/10.1007/978-94-024-2046-3_15

The findings from this study highlight the advantages of utilizing high spatial resolution DEMs when using a physics-based numerical model for a regional slope stability analysis. Landslide susceptibility maps with higher spatial resolution reveal important slope details that could drastically improve the efficiency of landslide mitigation practices, and aid in the forecasting of potentially catastrophic slope failures.

15.1 Introduction and Background

Regional slope stability is analyzed using susceptibility models that predict where the landslides are likely to occur (Brabb 1984; Guzzetti et al. 2005). Different approaches are used for regional slope stability analysis depending on the availability of the data, including geomorphological mapping, landslide inventory analysis, heuristic approach, physics-based numerical modeling and statistical-based empirical classification (Aleotti and Chowdhury 1999; Corominas et al. 2014; Guzzetti et al. 1999; Naidu et al. 2018; Weidner et al. 2018). One dataset that is used for the regional slope stability analysis irrespective of the approach is the Digital Elevation Model (DEM) (Guzzetti and Tonelli 2004; Oommen et al. 2018). The DEM provides a valuable dataset for computing the terrain slope within the study area (Bouali et al. 2016; Chau et al. 2004; Guzzetti et al. 1994; Van Westen et al. 2008; Vishnu et al. 2019). With the advent of remote sensing, often different spatial resolution terrain models are available for the study sites, which leads to the question of which DEM to use for regional slope stability analysis?

Researchers have tried to address this question by testing different spatial resolution DEMs with statistical models (Guzzetti et al. 1999; Chang et al. 2019; Claessens et al. 2005; Dietrich et al. 2001; Manzo et al. 2013; Paudel et al. 2016). Guzzetti et al. (1999) recommended the use of multiple spatial-resolutions DEMs and to determine the optimal resolution for the final susceptibility mapping. Dietrich et al. (2001) verified the applicability of DEMs of different spatial resolutions (i.e., 30 m, 10 m, 6 m, and 2 m) for shallow landslides. It was concluded that for moderate landslide susceptibility classes, the results were similar for coarse and fine-scale DEMs. However, for high and low susceptibility classes, the finer spatial resolution DEM provides improved identification of the local ridges and valleys (Dietrich et al. 2001). Many of these studies concluded that for statistical modeling of landslide susceptibility, finer spatial resolution does not necessarily lead to higher accuracy in terms of landslide susceptibility (Claessens et al. 2005; Manzo et al. 2013; Paudel et al. 2016). The reasoning was that the DEM with the spatial resolution closer to the mean size of landslide could provide the best results as it represents a meaningful mapping unit to characterize the different variables that control the landslide susceptibility. However, to the best of our knowledge, no studies have evaluated the impact of the spatial resolution of DEMs on physics-based models for regional landslide susceptibility analysis. Therefore, considering the fundamental difference in modeling principles between statistical and physics-based models, it is important to evaluate how the spatial resolution of DEM will impact physics-based models.

The purpose of this study is to use a physics-based model to assess the slope stability of the region in the vicinity of the Enguri Hydroelectric Dam, and evaluate how the spatial resolution of the DEM influences the output quality of the regional slope stability analysis. The slope stability analysis is performed by the infinite slope analysis method using the Geographic Information System Tool for Infinite Slope Stability Analysis (GIS-TISSA) (Escobar-Wolf et al. 2020). To accomplish this goal, two DEMs of different spatial resolution (30 m and 12 m) were used while soil and vegetation parameters were kept constant. This allowed for the independent testing of how increasing DEM resolution enhances output map quality. While soil and vegetation parameters were kept constant, the resolution of the subsequent shapefiles was altered to match that of the input DEM. This is a crucial step, as the minimum input resolution controls the output map spatial resolution.

15.2 Study Area

The study area for this research lies adjacent to the Enguri Hydroelectric Dam reservoir, located in the western part of the Republic of Georgia (Fig. 15.1a), where a catastrophic slope failure could infill large portions of the reservoir or even topple the dam (Telesca et al. 2012). The Enguri Hydroelectric Dam was completed in 1987 and is the second-highest arch dam in the world with a height of 271.5 m. The approximately 15 square kilometer reservoir holds 1.1 billion cubic meters of water. The region in the vicinity of the Enguri Dam was selected for this study for multiple reasons. Primarily, the Enguri hydroelectric facility provides approximately 50% of all electricity domestically produced by Georgia (Telesca et al. 2015). A landslide causing a sudden dam failure would severely cripple the country's electrical supply, increase the country's reliance on foreign electricity, and possibly lead to civil unrest within the country. The Enguri reservoir also forms an international border between Georgia and the separatist region of Abkhazia, where the hydroelectric power station is located (Matcharashvili et al. 2010). There is already a tense political relationship between the two regions, so any threat to energy distribution would surely aggravate the situation. While these are serious problems that could potentially impact many people, the local communities cannot be forgotten. Any problems with the dam would seriously threaten the nearby settlements of Jvari and Potshko Etersi (Abashidze et al. 2010; Chelidze et al. 2013; Tibaldi et al. 2019). Approximately 8000 people would be in the direct path of a catastrophic flood caused by a dam failure. The Greater Caucasus mountains that encompass the Enguri Dam and its surroundings have very steep slopes with active faults. Seismic activity and heavy rainfall are also common, both of which significantly increase the likelihood of a landslide occurring on steep slopes. Slope failures have already been documented on the slopes east of the reservoir. The Khoko Landslide area is an active failure on the reservoir's eastern margin and is only a few kilometers away from the dam, and moves downslope (toward the reservoir) at a rate of 5–10 cm/year (Tibaldi et al. 2019). This is one of several previous landslides that have been documented in the area. As a result, the eastern side of the reservoir was chosen as the study area, and a regional landslide susceptibility map was produced.

The area selected for the extent of the GIS-TISSA simulations is a rectangle measuring approximately 7 × 3 km, elongated in the N-S direction (Fig. 15.1b). Field data used for parameterizing model inputs were only collected on the Georgian (eastern) side of the reservoir due to the political instability on the Abkhazia (western) side making field surveys there too dangerous. The data from this fieldwork revealed that the primary lithological units within the study area are basalt, limestone, shale, and clay. Due to the known slope failures already occurring in the Khoko landslide area, this location was selected for more intensive soil testing. Borehole data from 8 sites within the Khoko landslide area show that the area is comprised of a red clay unit that is over 45 m thick. Basaltic bedrock is assumed to be underneath the clay, but borehole data do not reach deep enough to confirm. These clay-rich soils can be described as very heavily weathered with many gypsum inclusions. A small mine that extracts this gypsum is located at the top of the slope, to the southeast of the Khoko landslide area. It is possible that mining activities may be influencing slope stability in the area surrounding the mine. Road cuts within the study area crosscut most of the lithological units described above and host numerous springs draining water from the steep slopes above. These road cuts also experience numerous rockfalls every day, displaying the active mass wasting experienced throughout the area.

Recent and considerable tectonic activity is very prevalent within the study area, as indicated by exposed fault scarps and folding of rock layers. Large scale uplift of topography has increased erosion in some of the highlands as well. As a result, new fluvial deposits have been created in the lowlands (Tibaldi et al. 2017, 2018, 2019). Seismic records from the past years indicate that this is still a very active tectonic setting (Varazanashvili et al. 2018). Strong seismic events greatly increase the potential for triggering landslides, especially in areas that are already prone to such events.

15.3 Data and Method

15.3.1 Model Inputs

For the physics-based regional slope stability analysis, it was first necessary to gather and format all required model inputs. The GIS – TISSA tool utilizes a total of 5 input files to create the output map of landslide susceptibility. (1) A DEM provides an elevation for every pixel within the study area and allows for the calculation of slope magnitude. (2) A soil shapefile defines the boundaries of each general soil type found in the study area. (3) The influence of vegetation on slope stability is

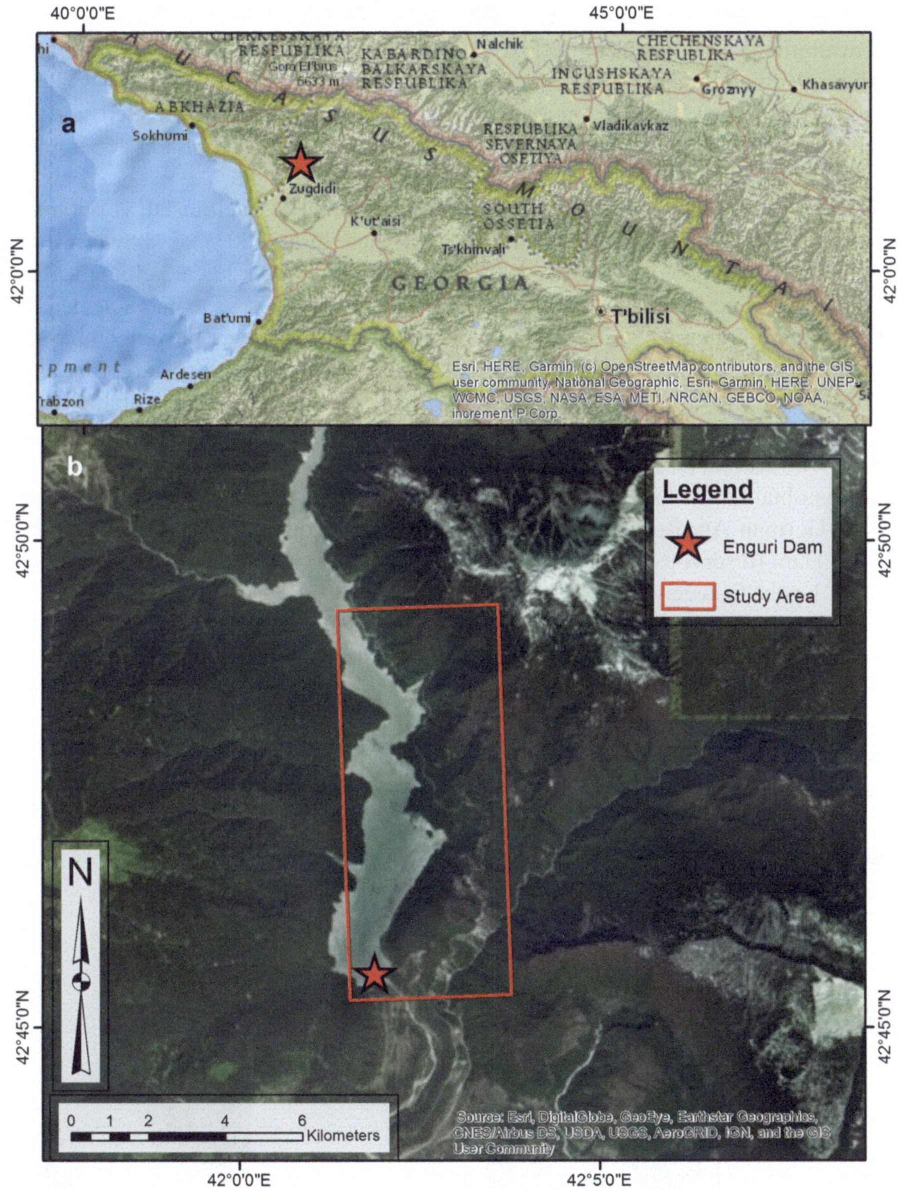

Fig. 15.1 (**a**) Location of Enguri Dam in Georgia is shown with a red place marker. (**b**) Satellite image of the Enguri Dam reservoir. The study area used for slope stability calculations is shown inside the red rectangle. Reference system: WGS84/geographic coordinates

addressed by a vegetation boundary layer. These three layers were all projected to the WGS 1984 UTM Zone 38 N coordinate system and clipped to the study area to ensure proper layer alignment. It is important to note that the input shapefiles only delineate the spatial extent of each soil/vegetation unit. (4–5) The numerical value of the soil/vegetation properties are provided by comma-separated value (.csv) files. Within the .csv files, the distribution type for each parameter can be selected, and the statistical data for each unit can be entered. Reference system: WGS84 / geographic coordinates.

The first DEM used in the study was a 30 m ASTER satellite DEM obtained from NASA EarthData (earthdata.nasa.gov) (Hirano et al. 2003), as shown in Fig. 15.2a. This is an open data source, so the only requirement for downloading the data was creating a personal EarthData account. The desired DEM tile was selected by simply defining an area within the built-in EarthData web viewer that encompassed the entire study area. After downloading, it was necessary to perform a reprojection to the projected coordinate system defined above. The final step was to clip the DEM to the exact extent of the study area. After this, the file was ready for input to the model. The second DEM tested, shown in Fig. 15.2b, has a spatial resolution of 12 m and was obtained from TanDEM-X Science Service System and required approval by the German Aerospace Center (GAC). The data order for a tile containing the Enguri Dam Reservoir was considered sensitive by the GAC, and as a result, it took almost 1 year to receive the DEM (Krieger et al. 2007). Once the correct tile was attained, reprojection and clipping to the study area were performed in the same manner as the ASTER DEM to ensure compatibility with all other inputs.

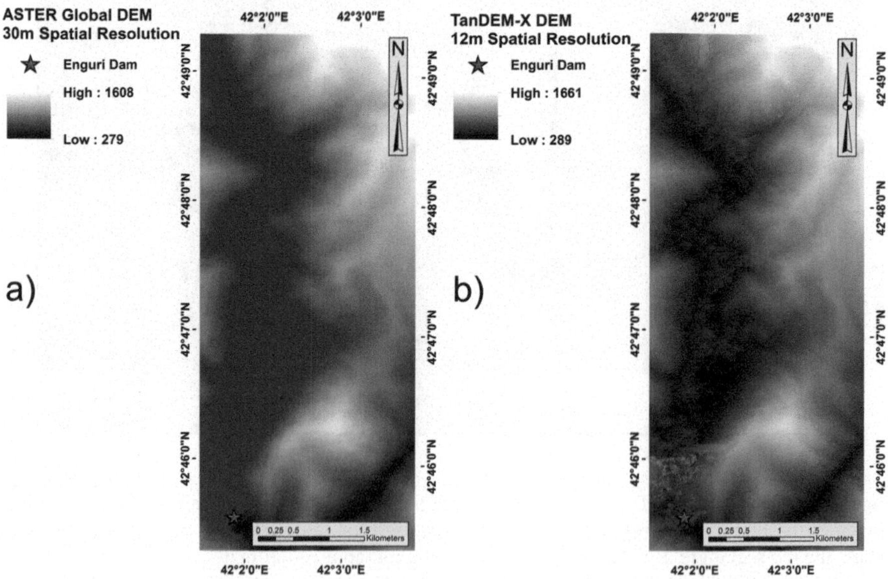

Fig. 15.2 (**a**) ASTER DEM, 30 m spatial resolution. (**b**) TanDEM-X DEM, 12 m spatial resolution. The highest and lowest elevations are shown for each DEM. Reference system: WGS84/geographic coordinates

Soil unit boundaries were defined by using a shapefile that matches the DEM extent (Fig. 15.3a). The shapefiles used in both the 12 m and 30 m trials were derived from the digitization of a geologic map of the Caucasus Mountain Region and contain three distinct soil units. For proper functionality within the TISSA tool, each soil unit was assigned an integer: Limestone (1), Basalt (2), and Clay (3). Vegetation unit boundaries were also defined by using a shapefile with an extent identical to that of the input DEM (Fig. 15.3b). However, in this case, the vegetation data was derived from LANDSAT8 raster data. Vegetation cover was determined by calculating the Normalized Vegetation Index (NDVI), analogous to the total density of vegetation. The vegetation root strength (root cohesion) and total mass could then be determined through the relationship presented by Huang et al. (2006)

$$C_{R(x,y)} = C_{min} + C_{interval} \frac{NDVI_{(x,y)} + 1}{2}$$

where $C_{R(x,y)}$ is the root cohesion, C_{min} and $C_{interval}$ are the minimum cohesion and cohesion interval respectively assuming a uniform distribution. The raster had to be then converted to a shapefile for proper functionality within the TISSA tool. Vegetation unit integers were also assigned. All soil/vegetation boundaries and units remained identical for each trial, with only the spatial resolution changing. This was achieved by creating duplicate shapefiles at the time of raster file conversion, and changing the cell size of the shapefile. This ensured all soil/vegetation boundaries remained constant, and that the cell sizes of all input files were compatible with the

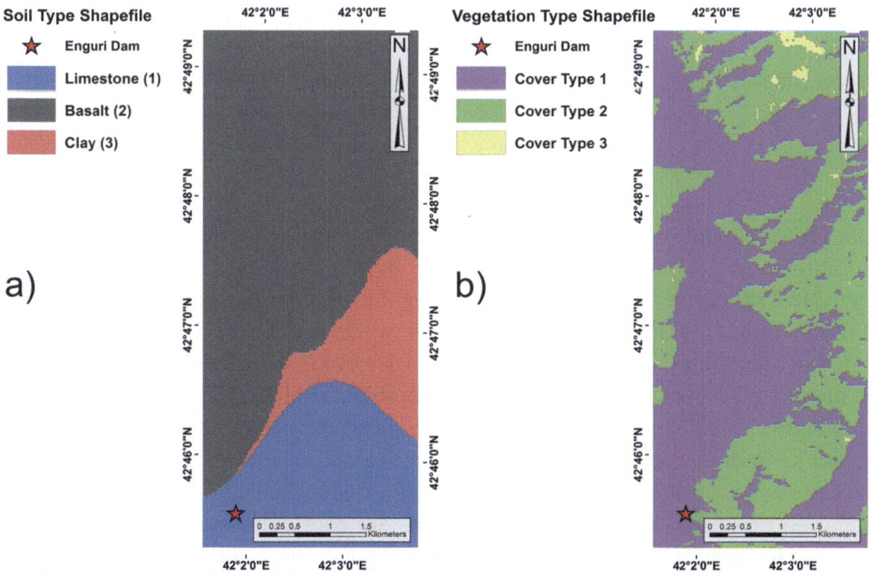

Fig. 15.3 (**a**) Soil type shapefile. (**b**) Vegetation type shapefile. Reference system: WGS84/geographic coordinates

Table 15.1 Soil properties used to parameterize soil type shapefile; internal friction angle (phi), cohesive strength (c, Pa), saturated unit weight (suw, N/m^3), moist unit weight (muw, N/m^3), soil thickness (D, m), pore pressure coefficient (Hw)

Unit	Param	Dist	stat1	stat2	stat3
1	phi	Normal	25.57	1.97	0
2	phi	Normal	40.32	1.43	0
3	phi	Normal	33.73	4.36	0
1	c	Normal	2,450,000	680,000	0
2	c	Normal	6,840,000	1,870,000	0
3	c	Normal	27,830	8020	0
1	suw	Normal	24,000	0	0
2	suw	Normal	30,000	0	0
3	suw	Normal	20,000	0	0
1	muw	Normal	22,000	0	0
2	muw	Normal	28,000	0	0
3	muw	Normal	18,000	0	0
1	D	Uniform	4	7	0
2	D	Uniform	3	5	0
3	D	Uniform	3	5	0
1	Hw	Uniform	0.1	0.5 (1.0)[a]	0
2	Hw	Uniform	0.1	0.5 (1.0)[a]	0
3	Hw	Uniform	0.1	0.5 (1.0)[a]	0

[a] A maximum pore pressure coefficient of 0.5 was used to model unsaturated soil conditions, while a coefficient of 1 was used to model saturated conditions

Table 15.2 Vegetation properties used to parameterize vegetation type shapefile; root cohesive strength (Cr, Pa), tree surcharge (qt, Pa)

Unit	Param	Dist	stat1	stat2	stat3
1	Cr	Normal	4788	2000	0
2	Cr	Normal	4788	2000	0
3	Cr	Normal	4788	2000	0
1	qt	Normal	1197	200	0
2	qt	Normal	1197	200	0
3	qt	Normal	1197	200	0

input DEMs' spatial resolution. It also allowed for the discrete testing of how the resolution quality of input DEM files impacts output landslide susceptibility maps.

Physical properties used for parameterizing the input soil shapefile were acquired from data collected during a field campaign to the study site (Table 15.1). These properties include the internal friction angle (degrees), cohesive strength (pascals), soil thickness (meters), and pore pressure coefficient (unitless). Saturated and moist unit weights (N/m^3) of all units had to be estimated from common reference values for similar rock/soil types. As mentioned previously, vegetation properties were interpreted from remotely sensed LANDSAT8 raster data (Table 15.2). The two vegetation properties used in the calculations are root cohesive strength (pascals) and tree surcharge (pascals). The tables below have columns that correspond to: (a) the soil/vegetation unit, (b) parameter, (c) distribution, (d) statistic 1, (e) statistic 2, (f) statistic 3.

15.3.2 GIS – TISSA

To create the map of the regional landslide susceptibility, the custom GIS – TISSA toolset was used (Escobar-Wolf et al. 2020). The TISSA toolset utilizes the "infinite slope" method for its slope stability analysis. This physics-based method models the landslide as translational, which means the failure surface is relatively shallow and parallel to the slope of the ground surface. The infinite slope method outputs a numerical value known as the "factor of safety" (FS). This FS value is simply a ratio between the forces acting to stabilize the slope (the numerator) and the forces acting to destabilize the slope (the denominator). As a result, FS = 1 means all forces are equal, and the slope is in a state of equilibrium. FS > 1 implies the stabilizing forces are greater than the destabilizing forces, and the slope is stable. On the other hand, FS < 1 implies the destabilizing forces are the greatest, and the slope is not stable (slope failure will occur).

The FS equation (Escobar-Wolf et al. 2020) is outlined below with all important variables, along with a diagram that is useful in understanding how each of the variables contribute to a slope's strength (Fig. 15.4):

$$FS = \frac{Cr + Cs + [q_t + \gamma_m D + (\gamma_{sat} - \gamma_w - \gamma_m)\, H_w D] \cos^2\beta tan\varphi}{[q_t + \gamma_{mD} + (\gamma_{sat} - \gamma_w - \gamma_m)\, H_w D]\, sin\beta cos\beta}$$

Cr = contribution to soil cohesive strength from vegetation roots (when present)
Cs = soil cohesive strength
q_t = weight of vegetation added to the slope
Υ_m = unsaturated soil unit weight (moist, above the phreatic surface)
Υ_{sat} = saturated soil unit weight (under the phreatic surface)

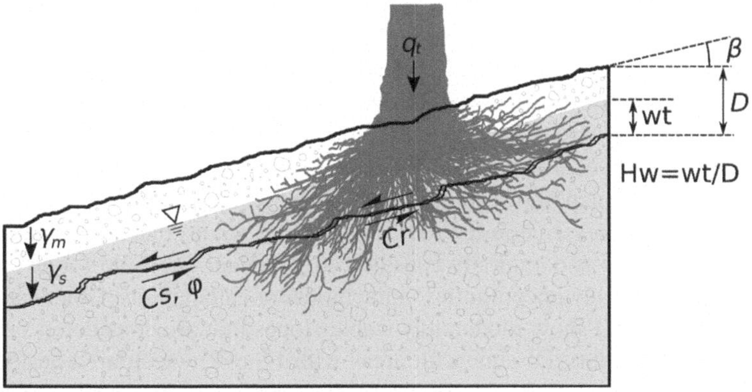

Fig. 15.4 Inputs to the infinite slope model (Escobar-Wolf et al. 2020)

Υ_w = water unit weight (constant equal to 9810 N/m^3 or 62.4 lb./ft^3)
D = depth to failure surface
H_w = depth of phreatic surface
β = terrain slope
φ = internal friction angle of soil

The inputs required for each slope stability calculation were a single DEM raster file in .geotiff format, two ESRI shapefiles for delineating the spatial extent of each soil and vegetation type, and two .csv files providing distribution types and parameter values for each veg/soil type contained in the shapefiles. A minimum slope cutoff value and DEM standard deviation must also be defined before running the tool (Escobar-Wolf et al. 2020). The minimum slope value is defined to prevent any slope stability calculations from being considered for slopes under the threshold value, while the DEM standard deviation is analogous to vertical accuracy. Setting these options to the right value is critical for customizing each trial to match the parameters of the input files. For this study, a minimum slope angle of 10° (assuming all slopes with angle <10° are stable) was selected for each trial, and DEM standard deviations were used based on the vertical accuracy of each DEM. Once all the algorithm's input requirements were met, the final step was to select either the PISA-M or standard ArcMap algorithm for slope calculation. The ArcMap method was chosen for both trials in this study, which utilizes eight-point neighboring pixels for slope calculations. Execution of the TISSA algorithm calculates the FS for each individual pixel within the study area. As a result, the output files from the TISSA tool are raster files (in .tiff format) containing mean FS values for the given area.

For each DEM (12 m and 30 m) there were two trials executed. One of these trials viewed the impacts of unsaturated soil conditions, while the second looked at fully saturated soil conditions. This was achieved by altering the "Hw" (depth of phreatic surface) parameter within the soil properties .csv file. Both trials used a uniform distribution of Hw, with only the maximum height of saturation (minimum depth of phreatic surface) being changed between the two. Unsaturated soil conditions were modeled by ranging Hw from 0.1 to 0.5 of the total soil depth, showing saturation in only the bottom 50% of the soil. On the other hand, saturated conditions were modeled by ranging Hw from 0.1 to 1.0, showing 100% saturation. This allowed for the construction of 4 individual FS maps, two for each DEM (saturated and unsaturated). Each map shows output FS values grouped into 4 categories. The first category (Category A, colored red) shows all FS values that are less than 1.25. The FS value of 1.25 is commonly considered a practical threshold for an unstable slope to account for any uncertainties in the model or calculations. Other categories (Categories B and C) show moderately stable slopes that are considered safe. The final category (Category D) shows all areas with FS > 2, which are considered to be very stable. Finally, the background for each map is a hill-shade constructed using the input DEM. This allows for a useful 3-D display of the mountains terrain surrounding the Enguri Dam Reservoir.

To better understand how the improvement in spatial resolution of input DEMs impacted the quality of output maps, it was important to take a closer look at how the total surface area of each FS category changed for each trial. To do this, the output raster files were converted to ESRI point files, where each pixel could be represented by a single point with a unique FS value. From here, the total number of pixels in each of the four FS categories described above was calculated. Next, a simple calculation multiplying the total number of pixels in each category by the ground area of a single-pixel produced the total area in that category.

$$[\text{Number of Pixels}] \times \left[\text{Pixel Area } \left(\text{km}^2\right)\right] = \text{Total Area } \left(\text{km}^2\right)$$

To account for differences in the total area considered in the slope stability calculations, the ESRI point file resulting from the 12 m DEM had to be edited to ensure it matched that from the 30 m DEM as closely as possible. This was done by deleting all points inside the Enguri Dam Reservoir, as well as points on the outer limits of the 12 m output map that were not included in the 30 m trials. Once the boundaries for the 30 m and 12 m output point files were the same, the total area under consideration was the same, and a meaningful comparison of the input DEM resolution could be achieved. The statistics for each FS category were then calculated and compared for all trials.

15.4 Results

The first DEM utilized in the study was the ASTER 30 m. For Trial 1, unsaturated soil conditions were modeled, resulting in a minimum FS value of 1.22 (Fig. 15.5). To track changes in the spatial extent of the four output FS categories, the surface area of each category was calculated (Table 15.3). The hazardous slopes in Category A were almost undetectable in this trial, with a total area of only 0.01 km^2. The total area modeled in Trial 1 was 16.85 km^2. 96.5% of the calculated area (16.26 km^2) is considered very stable (FS > 2).

For Trial 2, saturated soil conditions were paired with this DEM to produce a minimum FS value of 1.06 (Fig. 15.6). When looking at the output map from this trial, it can quickly be noticed that there is a clear increase in the amount of slope that is unstable when saturated soil conditions are modeled. The statistics now show that the surface area of unstable slope is 0.14 km^2 (Table 15.4). The overall percentage of very stable (FS > 2) slopes has decreased to 94.4%.

The next set of trials utilized the TanDEM-X 12 m DEM. In Trial 3, we once again used unsaturated soil conditions for the model, producing a minimum FS value of 1.18 (Fig. 15.7). Right away, it can be seen that the total area, considered Table 15.5, has now decreased to 16.54 km^2. This minor variation results from the difference in output map resolution (12 m vs. 30 m) causing the two maps to have a slightly different extent. More importantly, it can be seen that the reservoir has

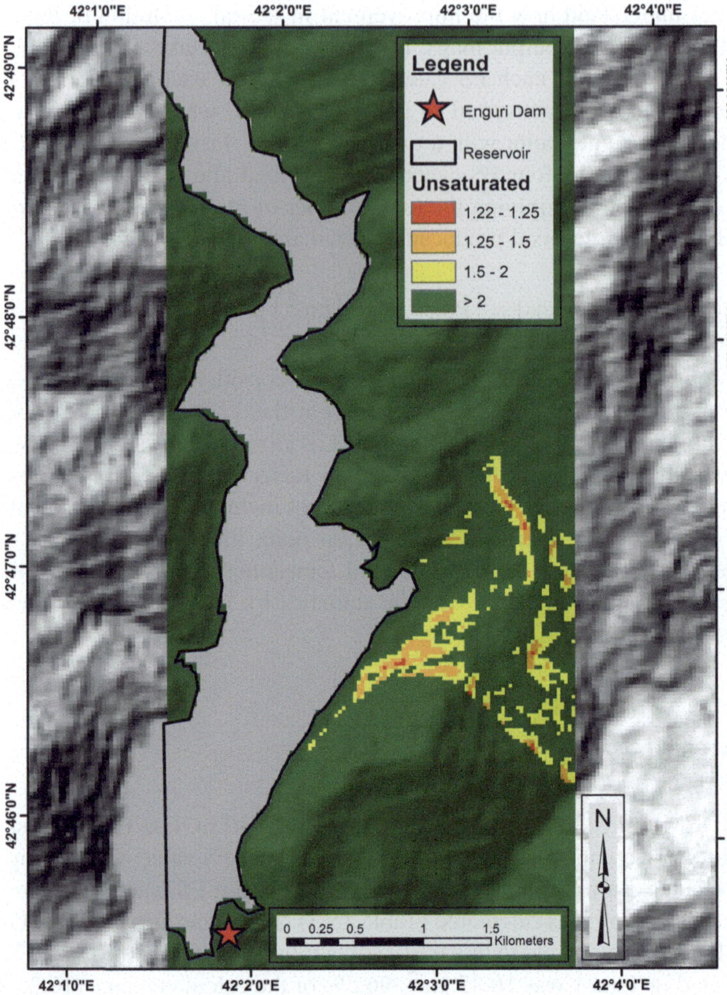

Fig. 15.5 Trial 1 output map. ASTER DEM of 30 m resolution with unsaturated soil conditions. Pixels inside the black reservoir outline were not considered in the analysis

Table 15.3 Area statistics from Trial 1

Category	Factor of safety	# of cells	Cell area (m^2)	TA (m^2)	TA (km^2)
A	Min – 1.25	13	746.437	9703.681	0.01
B	1.25–1.5	209	746.437	156005.333	0.16
C	1.5–2.0	557	746.437	415765.409	0.42
D	> 2.0	21,788	746.437	16263369.36	16.26
				Total:	16.85

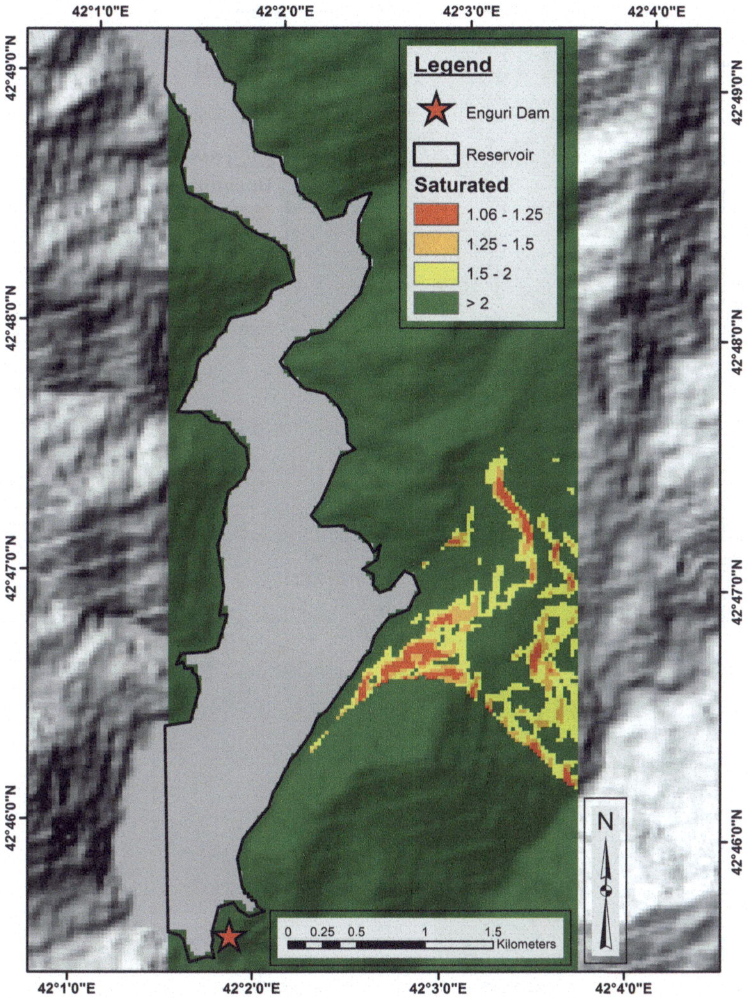

Fig. 15.6 Trial 2 output map. ASTER DEM of 30 m resolution with saturated soil conditions. Pixels inside the black reservoir outline were not considered for the analysis

Table 15.4 Area statistics from Trial 2

Category	Factor of safety	# of cells	Cell Area (m^2)	TA (m^2)	TA (km^2)
A	Min – 1.25	191	746.437	142569.467	0.14
B	1.25–1.5	315	746.437	235127.655	0.24
C	1.5–2.0	752	746.437	561320.624	0.56
D	> 2.0	21,309	746.437	15905826.03	15.91
				Total:	16.85

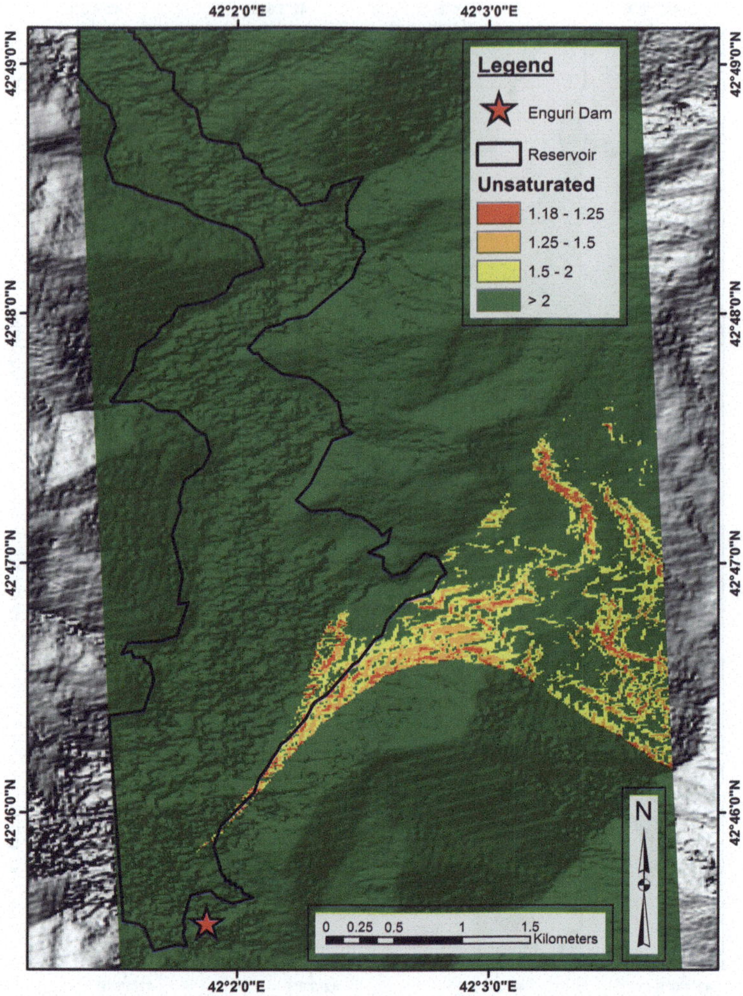

Fig. 15.7 Trial 3 output map. TanDEM-X DEM of 12 m resolution with unsaturated soil conditions. Pixels inside the black reservoir outline were not considered for the analysis

now been considered in the slope stability calculations. Another difference that can be seen in the 12 m resolution output map is the additional slope area calculated at the base of the unstable slope (shown inside the reservoir boundary). The likely reasons for these differences will be discussed further in the next section, but these slope areas inside the reservoir boundary were not considered in the statistical comparison. As for the similar slope areas, we see that the surface area of unstable slopes is now 0.10 km^2 (Table 15.5). The overall percentage of very stable (FS > 2) slopes is 94.4%.

Table 15.5 Area statistics from Trial 3

Category	Factor of safety	# of cells	Cell area (m^2)	TA (m^2)	TA (km^2)
A	Min – 1.25	840	119.421	100313.64	0.10
B	1.25–1.5	2529	119.421	302015.709	0.30
C	1.5–2.0	4474	119.421	534289.554	0.53
D	> 2.0	130,710	119.421	15609518.91	15.61
				Total:	16.54

Table 15.6 Area statistics from Trial 4

Category	Factor of safety	# of cells	Cell area (m^2)	TA (m^2)	TA (km^2)
A	Min – 1.25	3031	119.421	361965.051	0.36
B	1.25–1.5	2976	119.421	355396.896	0.36
C	1.5–2.0	4514	119.421	539066.394	0.54
D	> 2.0	128,021	119.421	15288395.84	15.28
				Total:	16.54

Trial 4 was the final simulation and used saturated soil conditions with the 12 m spatial resolution DEM. The minimum FS value in the output map is 1.05 (Fig. 15.8). Once again, we can see a clear increase in the amount of unstable slopes under saturated soil conditions. The total surface area of unstable slopes has increased to 0.36 km^2 (Table 15.6), while the percentage of very stable (FS > 2) slopes has decreased to 92.3%.

15.5 Discussion

When comparing the output maps from each trial, we can see that there is a clear improvement in map quality when utilizing the 12 m TanDEM – X DEM. In each trial, all values of FS < 2 are limited to the clay soil unit that forms the "hazardous slope". Furthermore, the unstable Category A slope areas are primarily clustered around the Khoko landslide area and the toe of the hazardous clay slope. While it is possible to see these details at both 30 m and 12 m resolution, the maps produced by the 12 m DEM allow for a more detailed view of these locations. The biggest improvement between DEM resolutions can be seen in the area directly adjacent to the dam reservoir, at the toe of the hazardous slope. This area is much more clearly defined at 12 m spatial resolution, allowing for a more accurate delineation of the Enguri reservoir boundary. More land area is exposed in the higher resolution DEM, resulting in additional slope calculations at this location and more area in Category A. This increase in Category A surface area can be observed across all trials. Another major difference between the maps produced from the separate DEM inputs is noticed within the Enguri Reservoir. When the ASTER 30 m DEM was used, there were no FS values calculated within the reservoir. However, inputting

Table 15.7 Changes in the minimum factor of safety values for each trial

	ASTER 30 m DEM	TanDEM-X DEM	Percent change
Minimum FS – Unsaturated	1.22	1.18	−3.3%
Minimum FS – Saturated	1.06	1.05	−0.9%

Table 15.8 Area statistics for Trials 1 and 2, which utilized the ASTER 30 m DEM

Category	Factor of safety	SA (km^2) – Unsaturated	SA (km^2) – Saturated	Percent change
A	Min – 1.25	0.01	0.14	1300%
B	1.25–1.5	0.16	0.24	50%
C	1.5–2.0	0.42	0.56	33.3%
D	> 2.0	16.26	15.91	−2.15%
	Total:	16.85	16.85	

the 12 m DEM resulted in FS values being calculated within the entire reservoir area. But, all resulting FS values fell into the very stable Category D.

Before further discussing the improvements associated with utilizing the higher resolution TanDEM-X DEM, it should first be noted that the resulting minimum FS values from each input DEM resolution were consistent. This is clearly displayed by the change in these values from trials with equivalent soil conditions being less than 4% (Table 15.7). This shows that the physics-based model calculated similar results as the input DEM resolution was altered. However, it is important to realize that there was a decrease in the minimum FS value when inputting the 12 m resolution DEM to the model. This suggests that the details missed by the 30 m DEM were in areas with the steepest slopes where the probability of a landslide is highest, highlighting another key benefit of utilizing high-resolution DEMs in a regional slope stability analysis.

A more detailed analysis of the area statistics associated with each input DEM further highlights how comparable the resulting slope stability calculations are across each trial, while also showing the improvements associated with using a higher spatial resolution. The primary benefit of the output maps produced by the 30 m DEM is that the influence of saturated soil conditions on the slope can clearly be seen. Under unsaturated conditions in Trial 1, there was almost no detection of unstable slopes in the study area. However, under saturated conditions in Trial 2, the surface area of unstable slopes increased to 0.14 km^2 (Table 15.8). The surface area of Categories B and C both increased considerably as well. Most of these weak slopes are centered on the Khoko landslide area, but there are a few unstable areas that are registering at the base of the slope. Unfortunately, the lower 30 m resolution really hampers any attempts to distinguish any details in these areas.

Moving on to the TanDEM-X 12 m DEM, we begin to clearly see the advantages that come with utilizing an input DEM with higher spatial resolution. Even without considering the additional slope area that is observed at the toe of the slope in trials 3 and 4, we can still observe increases in surface area for both Categories A and B within the slope area that was included in both the 30 m and 12 m trials (Table

Table 15.9 Area statistics for Trials 3 and 4, which utilized the TanDEM-X 12 m DEM

Category	Factor of safety	SA (km^2) – Unsaturated	SA (km^2) – Saturated	Percent change
A	Min – 1.25	0.10	0.36	260%
B	1.25–1.5	0.30	0.36	20%
C	1.5–2.0	0.53	0.54	1.89%
D	> 2.0	15.61	15.28	−2.11%
	Total:	16.54	16.54	

15.9). For example, in Category A we see an increase in surface area of 0.22 km^2 under saturated soil conditions when moving from 30 m to 12 m spatial resolution (0.14 km^2 to 0.36 km^2). This has two important implications; first, this shows that there are new unstable slope areas being accounted for at the base of the hazardous slope, adjacent to the reservoir boundary. These values were calculated as being very stable (Category D) when inputting the 30 m resolution DEM. Secondly (and most importantly), it shows that the newly accounted for slope areas are also some of the most unstable in the study area. Not only are these findings backed up by the area statistics (Table 15.9), but the higher resolution DEM also produces output maps of high enough quality to clearly identify the more unstable areas at the base of the slope (Figs. 15.7 and 15.8).

Uncertainties within the slope stability model and calculations originate from limitations in the field data available. As mentioned in the background section, borehole data from the Khoko Landslide area did not extend to bedrock. The minimum depth is known to be 45 m, but accurately assessing the true depth of this clay unit would improve the accuracy of output maps. Also, the clay layer was modeled as a homogeneous layer to simplify the complex interbedded network of calcite and gypsum layers. During the field campaign, jointing and crevasses were observed in the bedrock. These features have been shown to weaken the rock unit and affect slope stability, but no structural measurements were made to quantify these weaknesses. It is important to mention that due to large uncertainties in water table data, a uniform probability distribution of soil water saturation was used for each trial as well. The only difference was the depth of the phreatic surface.

While the scope of this study was focused on comparing the quality of output maps from the GIS-TISSA toolset for different input DEM resolutions to a physics-based slope stability model, it is still important to remember that this is a seismically active area. Saturated soil conditions that are common during the wet season, paired with weak clay slopes, are shown to be the most susceptible to landslides. Seismic inputs only exacerbate these instabilities. As a result, slope areas that are considered unstable should be monitored for any signs of slope movement. Higher-resolution input DEMs produce output maps with higher quality, potentially allowing for the identification of new hazardous slope areas. This is extremely beneficial, as more accurate landslide susceptibility maps will allow for a more focused approach to stabilizing the areas that are shown to be most unstable.

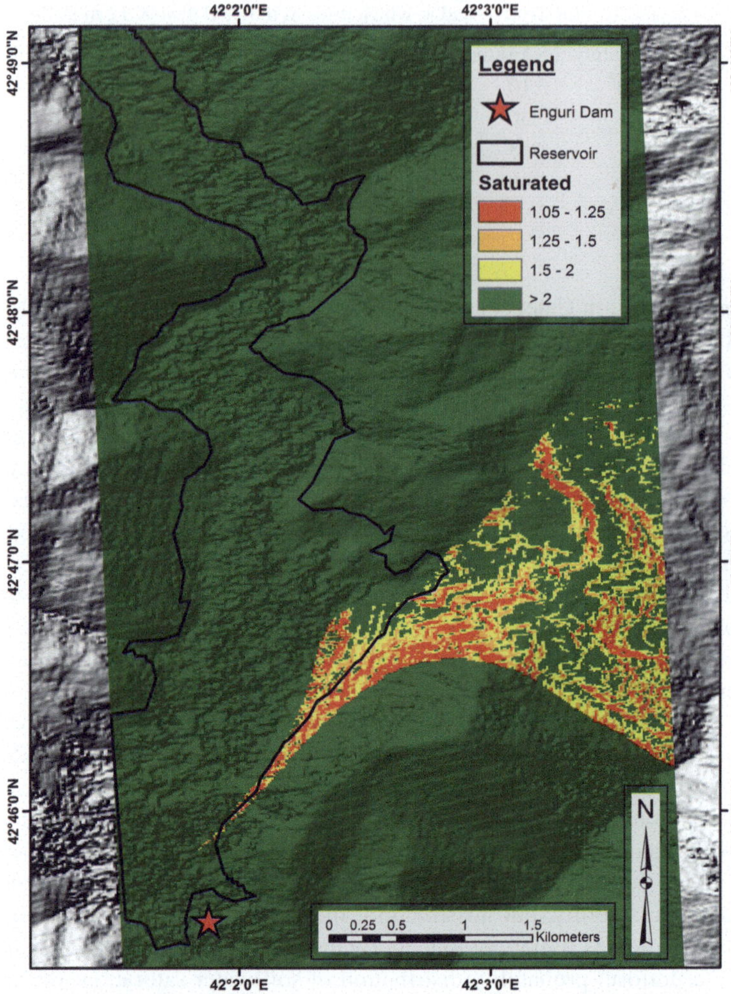

Fig. 15.8 Trial 4 output map. TanDEM-X DEM of 12 m resolution with saturated soil conditions. Pixels inside the black reservoir outline were not considered for the analysis

15.6 Conclusions

The Enguri Hydroelectric Facility is located in an area that makes it very susceptible to landslide related geohazards. The mountains surrounding the area have very steep slopes, are seismically active, and receive high amounts of seasonal rainfall. As a result, there is a high possibility of a landslide impacting the dam and/or reservoir. Such an event could have catastrophic socio-economic impacts for the surrounding communities, and the Republic of Georgia as a whole. The ability to produce accurate regional landslide susceptibility maps is crucial in predicting when and where these landslides are most likely to occur. This study was focused on comparing the output map quality from the GIS – TISSA custom toolset when utilizing input files of different spatial resolution. When comparing the ASTER 30 m DEM to the TanDEM-X 12 m DEM, we were able to identify new unstable slope areas at the higher resolution. These areas were directly adjacent to the dam reservoir, at the toe of the most hazardous slope in the study area. They were completely unaccounted for when inputting the 30 m DEM. The results from this study indicate that higher resolution DEMs greatly increase the resolution and overall quality of output maps, even allowing for the detection of new (and potentially hazardous) slope areas. It was evident from the literature that the use of higher spatial resolution DEMs did not guarantee improved landslide susceptibility analysis in the case of statistical models (Guzzetti et al. 1999; Chang et al. 2019; Claessens et al. 2005; Dietrich et al. 2001; Manzo et al. 2013; Paudel et al. 2016). However, this study demonstrates that in the case of physics-based modeling of landslide susceptibility analysis, a higher spatial resolution DEM can provide improved predictive capability. Such use of improved spatial resolution DEMs will allow for more efficient landslide mitigation practices and continued safe operations of the Enguri Dam hydroelectric facility.

Acknowledgments This study has been conducted in the framework of the NATO project SfP G4934 "Georgia Hydropower Security", of the International Lithosphere Program – Task Force II, of the European Space Agency project no. 32309 "Active tectonics and seismic hazard of southwest Caucasus by remotely sensed and seismological data."

References

Abashidze V, Chelidze T, Jibladze I et al (2010) Installation and utilization of automated data aquisition and transmission system of the tiltmetric network in the Enguri dam. Bull Georg Natl Acad Sci 4(3):58

Aleotti P, Chowdhury R (1999) Landslide hazard assessment: summary review and new perspectives. Bull Eng Geol Environ 58(1):21–44

Bouali EH, Oommen T, Escobar-Wolf R (2016) Interferometric stacking toward geohazard identification and geotechnical asset monitoring. J Infrastruct Syst 22(2):05016001

Brabb EE (1984), Innovative approaches to landslide hazard and risk mapping, 4th International Symposium on Landslides, Toronto. 1:307–323

Chang KT, Merghadi A, Yunus AP et al (2019) Evaluating scale effects of topographic variables in landslide susceptibility models using GIS-based machine learning techniques. Sci Rep 9(1):1–21

Chau KT, Sze YL, Fung MK et al (2004) Landslide hazard analysis for Hong Kong using landslide inventory and GIS. Comput Geosci 30(4):429–443

Chelidze T, Matcharashvili T, Abashidze V (2013) Application of new nonlinear elasticity and nonlinear dynamics tools in real time monitoring of large engineering constructions (case of high arc Enguri dam). Disaster Adv 6(9):84–89

Claessens L, Heuvelink GBM, Schoorl JM et al (2005) DEM resolution effects on shallow landslide hazard and soil redistribution modelling. Earth Surf Process Landf 30(4):461–477

Corominas J, van Westen C, Frattini P et al (2014) Recommendations for the quantitative analysis of landslide risk. Bull Eng Geol Environ 73(2):209–263

Dietrich WE, Bellugi D, Real De Asua R (2001) Validation of shallow landslide model, SHALTAB, for forest management. Water Sci Appl 2:195–227

Escobar-Wolf R, Sanders J, Oommen T et al (2020) A GIS tool for infinite slope stability analysis (in review)

Guzzetti F, Tonelli G (2004) Information system on hydrological and geomorphological catastrophes in Italy (SICI): a tool for managing landslide and flood hazards. Nat Hazards Earth Syst Sci 4(2):212–232

Guzzetti F, Cardinali M, Reichenbach P (1994) The AVI project: a bibliographical and archive inventory of landslides and floods in Italy. Environ Manag 18(4):623–633

Guzzetti F, Carrara A, Cardinali M et al (1999) Landslide hazard evaluation: a review of current techniques and their application in a multi-scale study, Central Italy. Geomorphology 31(1–4):181–216

Guzzetti F, Mondini AC, Carrara A, Cardinali M, Fiorucci F, Santangelo M, Chang KT (2012) Landslide inventory maps: new tools for an old problem. Earth Sci Rev 112(1–2):42–66

Hirano A, Welch R, Lang H (2003) Mapping from ASTER stereo image data: DEM validation and accuracy assessment. ISPRS J Photogramm Remote Sens 57(5–6):356–370

Huang JC, Kao SJ, Hsu ML et al (2006) Stochastic procedure to extract and to integrate landslide susceptibility maps: an example of mountainous watershed in Taiwan. Nat Hazards Earth Syst Sci 6:803–815

Krieger G, Moreira A, Fiedler H et al (2007) TanDEM-X: a satellite formation for high-resolution SAR interferometry. IEEE Trans Geosci Remote Sens 45(11):3317–3341

Manzo G, Tofani V, Segon S et al (2013) GIS techniques for regional-scale landslide susceptibility assessment: the Sicily (Italy) case study. Int J Geogr Inf Sci 27(7):1433–1452

Matcharashvili T, Chelidze T, Abashidze V et al (2010) Changes in dynamics of seismic processes around Enguri high dam reservoir induced by periodic variation of water level. In: Synchronization and triggering: from fracture to earthquake processes. Springer, Berlin, Heidelberg, pp 273–286

Naidu S, Sajinkumar KS, Oommen T et al (2018) Early warning system for shallow landslides using rainfall threshold and slope stability analysis. Geosci Front 9(6):1871–1882

Oommen T, Cobin PF, Gierke JS et al (2018) Significance of variable selection and scaling issues for probabilistic modeling of rainfall-induced landslide susceptibility. Spat Inf Res 26(1):21–31

Paudel U, Oguchi T, Hayakawa Y (2016) Multi-resolution landslide susceptibility analysis using a DEM and random forest. Int J Geosci 7(5):726

Telesca L, Matcharasvili T, Chelidze T et al (2012) Relationship between seismicity and water level in the Enguri high dam area (Georgia) using the singular spectrum analysis. Nat Hazards Earth Syst Sci 12(8):2479–2485

Telesca L, Matcharashvili T, Chelidze T et al (2015) Investigating the dynamical features of the time distribution of the reservoir-induced seismicity in Enguri area (Georgia). Nat Hazards 77(1):117–125

Tibaldi A, Alania V, Bonali FL et al (2017) Active inversion tectonics, simple shear folding and back-thrusting at Rioni Basin, Georgia. J Struct Geol 96:35–53

Tibaldi A, Bonali FL, Russo E et al (2018) Structural development and stress evolution of an arcuate fold-and-thrust system, southwestern Greater Caucasus, Republic of Georgia. J Asian Earth Sci 156:226–245

Tibaldi A, Oppizzi P, Gierke JS et al (2019) Landslides near Enguri dam (Caucasus, Georgia) and possible seismotectonic effects. Nat Hazards Earth Syst Sci 19:71

Van Westen CJ, Castellanos E, Kuriakose SL (2008) Spatial data for landslide susceptibility, hazard, and vulnerability assessment: an overview. Eng Geol 102(3–4):112–131

Varazanashvili O, Tsereteli N, Bonali FL et al (2018) GeoInt: the first macroseismic intensity database for the Republic of Georgia. J Seismol 22(3):625–667

Vishnu CL, Sajinkumar KS, Oommen T et al (2019) Satellite-based assessment of the August 2018 flood in parts of Kerala, India. Geomat Nat Haz Risk 10(1):758–767

Weidner L, Oommen T, Escobar-Wolf R et al (2018) Regional-scale back-analysis using TRIGRS: an approach to advance landslide hazard modeling and prediction in sparse data regions. Landslides 15(12):2343–2356

Chapter 16
Description of a 2-Year, High-Resolution Geodetic Monitoring of the Khoko Landslide, Enguri Reservoir, Georgia

Nazymbek S. Ospanov and Andrey A. Krivchenko

Abstract During a 2-year period, we monitored deformation taking place within the Khoko landslide by means of high-resolution techniques. The mass movement is found on the eastern slope of the Enguri artificial water reservoir. A network of 20 benchmarks was established along the Mestia-Zugdidi road, at the head scarp of the landslide zone, for a total length of 1.8 km. Measurement campaigns were carried out during three different periods, from May 2016 until June 2018. We estimated the displacement velocity and vectors and provided end users with a GIS-based platform for immediate use of the collected data. The whole upper part of the studied slope is affected by continuous deformation with a quite constant displacement rate. During the observation period, maximum average velocity of net displacement amounted to as much as 0.44 mm/day (first period of observation), with an overall average velocity of 0.12 mm/day (entire period of observation). The whole set of benchmarks moved to the west, indicating a coherent displacement of the whole upper landslide area. We suggest that critical rates of displacement were not observed during this period. Anyway, we recommend continuing observations of slope movements using the current geodetic stations, in order to detect any possible alarming changes in deformation rates.

Keywords Khoko · Landslide · Geodetic · Monitoring · Reservoir

16.1 Introduction

An international team of scientists from Italy, Georgia, Azerbaijan, the United Kingdom, US and Kazakhstan worked from November 2015 to September 2018, on an ambitious research program under the aegis of the North Atlantic Treaty

N. S. Ospanov (✉) · A. A. Krivchenko
Kostanay Engineering and Economics University named after M. Dulatov, Kostanay, Republic of Kazakhstan

© Springer Nature B.V. 2021

F. L. Bonali et al. (eds.), *Building Knowledge for Geohazard Assessment and Management in the Caucasus and other Orogenic Regions*, NATO Science for Peace and Security Series C: Environmental Security,
https://doi.org/10.1007/978-94-024-2046-3_16

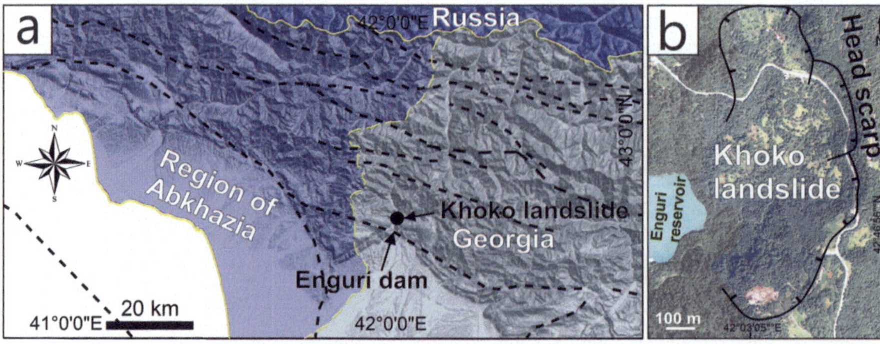

Fig. 16.1 (**a**) Location of the Enguri dam and reservoir in the Greater Caucasus; country borders are from GADM data, © 2018 GADM available at https://gadm.org/, version 3.6 (released on 6 May 2018); black dashed lines represent the main active faults in the region, from the EMME Project (Gulen et al. 2011). Reference system: WGS84/geographic coordinates. (**b**) Sketch map of the Khoko landslide. (After Tibaldi et al. 2019)

Organization (NATO), in the framework of the Science for Peace and Security Programme. In the past, several of the research team members had already taken part in NATO-supported efforts like the one described here (e.g. Kozhurin et al. 2006; Pasquarè et al. 2011; Tibaldi et al. 2015). The 2015–2018 project entitled "Security against geohazards at the major Enguri hydroelectric scheme in Georgia", was aimed at evaluating risks relative to geological hazards at a major, strategic facility of Georgia. The location of the project's study area is shown in Fig. 16.1a. Currently, the Enguri dam is the world's second highest concrete arch dam with a height of 271.5 m (891 ft). It is located north of the town of Jvari and is part of the Enguri hydroelectric power station (HES), partially located in Abkhazia (Tibaldi et al. 2018a). This project has been the first comprehensive research aimed at understanding the geological hazards of this region and proposing possible prevention and remediation measures.

The most threatening geological hazards for the Enguri hydropower plant are represented by earthquakes, landslides and floods, as confirmed by data available in recent research focused on seismic hazard and active tectonics (Tibaldi et al. 2017a, b; 2018b, 2020, 2016; Varazanashvili et al. 2018) and landslide hazard assessment (Tibaldi et al. 2019). These hazards are connected to each other, as earthquakes have the potential of triggering landslides (e.g. Schulz et al. 2012; Tibaldi et al. 2015). The largest, active mass movement in the area, the "Khoko" landslide (11 million to 22 million cubic metres), lies only a few kilometers away from the dam and directly faces the reservoir, posing a threat to the Enguri dam; the main scarp of the landslide is shown in Fig. 16.1b.

The destruction of individual slopes is usually less devastating than other catastrophes, but at the same time such events are more common, and they cause a total damage that is greater than losses due to single destructive natural processes. Furthermore, it is worth noting that most of the damage is due to landslides caused by earthquakes and floods. In recent years, great progress has been made in studying the geomechanical characteristics of soils and their influence on the stability of slopes (e.g. Kim et al. 2004). In fact, ensuring the stability of slopes is the purpose of numerous studies performed by engineers and geologists (e.g. Pasquaré Mariotto and Tibaldi 2016; Tibaldi and Pasquaré 2007; Tibaldi et al. 2015, 2020).

Many hydrotechnical facilities are located in regions with abundant precipitation. As a result, slopes in areas subject to the construction of hydraulic structures, roads, other facilities may slide under the influence of anomalous rainfall, causing the destruction of these facilities, which in turn results in life loss and widespread damage. As underlined by Trejo et al. (2007), the construction of large technical structures such as dams, bridges and multi-storey buildings is essential for the development of any country. However, significant deformation of engineering structures can potentially cause a great number of casualties and severe destruction. Therefore, the preservation and durability of buildings and facilities is an important socio-economic task, which depends not only on the scientifically grounded construction of these structures, but also on the completeness of the investigation of the terrains on which the structures are built.

Consequently, the security of these facilities, especially in the case of dams, requires periodic monitoring, which is conducted through repeated measurements of ongoing deformation and a comprehensive analysis of their structural state, based on a large set of variables that contribute to these deformations. None of the modern large constructions could be built and preserved without systematic observations of slope displacements by means of geodetic techniques. The results of these observations must satisfy the requirements of completeness, timeliness and accuracy (e.g. Erol et al. 2004).

The largest hydroelectric power station in the Caucasus is the Enguri Hydroelectric Power Station, located on the Enguri River, on the way to Upper Svaneti (e.g. Elizbarashvili et al. 2018). The dam impounds the Enguri water reservoir, the volume of which is 1.1 billion cubic meters. The slopes overlooking the reservoir are affected by several mass movements, comprising the huge "Khoko" landslide, situated on the eastern side of the lake (Fig. 16.1b). Due to the size of the landslide, its collapse in the reservoir would represent a major hydrogeological disaster. Furthermore, strong seismicity as a possible trigger for landslides in the area needs to be taken into account: in fact, the area is located at the foot of the Greater Caucasus, which is pervasively cut by active faults, from Georgia to the Caspian Sea (Tsereteli et al. 2016; Varazanashvili et al. 2018). Despite the fact that seismicity can increase slope instability and cause landslides, no dynamic analysis of the stability of the slopes surrounding the Enguri reservoir had ever been carried out.

The work of the Kazakh team has been focused on: *(i)* assessing changes in the state of the massif in the Khoko landslide region by establishing benchmarks of geodetic observation station; *(ii)* estimating displacement speed and values; *(iii)* determining the critical values of deformation preceding the beginning of the active stage of deformation; *(iv)* providing end-users with a GIS-based platform for immediate use of the data collected in the framework of the project.

16.2 Materials and Methods

The main methodology for measuring soil settlements and deformations of engineering structures is geodesy (e.g. Song and Xin 2012), which allows to determine not only the relative displacements of points, but also their absolute magnitude, thanks to practically fixed marks of the geodetic stations. Observations of deformations of structures occupy a significant place in the modern practice of engineering and geodetic works. The volume and complexity of observations and the requirements for their accuracy have increased from year to year and, in connection with this, special methods and measurements have been developed. The choice of the most appropriate methods is based on the constructional features of the structures and on their specific conditions. In order to observe deformations at large engineering structures, such as hydroelectric power stations, it becomes necessary to perform fairly frequent (and sometimes continuous) measurements.

Fundamental (initial) benchmarks are located outside the zone of possible bedrock deformations that occur as a result of the construction of hydraulic facilities and reservoir infilling (Kuznetsov and Krupin 2001). It is preferable to place the benchmarks on bedrock outcrops, located on unflooded marks at points remote from landslide and karst sites, tectonic faults and soils that are prone to expansion or settlement. The benchmarks should be located behind the contour of a landslide or potentially unstable slope. The direction of the alignment is usually perpendicular to the direction of motion of the landslide. The placement of marks at characteristic sites, accessible for observation along the landslide and potentially unstable slope, is determined by means of geometric and trigonometric leveling. Tacheometric surveys (in conditions of difficult visibility or in complex terrains) are carried out, with the aid of intermediate theodolites in order to maintain the alignment of long straight lines. To eliminate the collimation error, the most distant milestone is set twice, with two half-strokes, with the final setting in the middle between the two initial positions. The theodolite, which is installed in place of this milestone, is centered within 1 cm. Intermediate milestones are established by the "self" method.

Horizontal deformation interval for the period between two observations is determined by the formula (16.1):

$$= \frac{\alpha 2 - \alpha 1}{\alpha} \tag{16.1}$$

where α, α1, α2 are the horizontal length of the interval from the initial, previous (16.1) and subsequent (16.2) observations.

Horizontal deformations over the entire observation period, from the initial to the present, are determined by the formula (16.2):

$$\alpha = \frac{\alpha n - \alpha}{\alpha} \tag{16.2}$$

where αn is the horizontal length of the interval from the given observation.

Horizontal deformations corresponding to an increase in the interval are denoted by the sign "+" and are called stretches, and those corresponding to decreasing intervals are negative "−" and are called compressions.

The calculated horizontal distances between the benchmarks (obtained by summing them) determine the distances from the supporting benchmark to each of the profile line benchmarks that fit into the horizontal displacement statement.

When processing observation materials along a profile line, the horizontal displacements E (mm) are determined by the following formula (16.3):

$$E = D_2 - D_1 \tag{16.3}$$

where D_1, D_2 are the distances from the supporting benchmark to the given benchmark, from the initial (or previous) observation and from the subsequent observation.

Horizontal and vertical displacements can also be calculated from differences in the lengths of intervals and exceedances between adjacent benchmarks in the compared series of observations. According to the displacement of the benchmarks, the values of shifts in the area of the rock massif are compiled. The magnitude of the shift γ is defined as the ratio of the difference of the total displacements of nearby benchmarks to the distance between these benchmarks along the normal line to the direction of their displacement (16.4):

$$\gamma = \frac{\Delta bn + 1 - \Delta bn}{m} \tag{16.4}$$

where:

(i) Δb_{n+} is the total displacement of the front benchmark;
(ii) Δb_n is the total displacement of the rear benchmark;
(iii) m is the distance between the benchmarks along the normal line to the direction of displacement of the vectors. The magnitude of the shift is referred to the middle of the interval.

At the beginning of this work, an observation geodetic station (network of benchmarks) was established along the Mestia-Zugdidi road. The length of this station is 1.8 km. Along the road Zugdidi-Mestia, 20 control points were established (Figs. 16.2 and 16.3), so as to measure the dynamics of the landslide movement, due to the danger it poses to the Enguri reservoir.

The work period consisted of 4 initial GPS measurements and 3 periods of estimation of the movement of benchmarks: period *(i)* 20.05.2016–21.11.2016; period 2 |*(ii)* 21.11.2016–22.05.2017; period *(iii)* 22.05.2017–10.06.2018. Control benchmarks are made of metal and are installed at a depth of 1 m into the ground (Fig. 16.2).

Control points are numbered and divided into the following categories: RP, NRP stationary points that are located outside the landslide body, MST-mobile control

Fig. 16.2 Example of benchmark used in the present work

points (inserted in the landslide body), and ST points for leveling the measurements of the total station; their location is shown in Fig. 16.3.

To determine the coordinates in the global coordinate system, we used a high-precision Leica GPS device (accuracy 1.0 mm + 1.5 ppm). The coordinates of points ST1 and SST were measured with great accuracy. The location of such points was determined after a visual inspection of the area and an open terrain was chosen for optimal communication with satellites. Measurements of the frames were made with a high-precision geodetic device with a total station Leica TS11 (accuracy is 1.0 mm + 1.5 ppm).

With the help of high-precision geodetic instruments (total station Leica TS11 and GPS geodetic equipment) two series of instrumental measurements of benchmarks were carried out in the years 2016–2018.

16.3 Results

The comparison between the spatial coordinates of the network, determined at the beginning and at the end of this work, enabled us to determine the total shift vectors-shift trends, as reported in Fig. 16.4. Shift trends can be either natural (caused by gravity or tectonic movements along the boundaries of structural blocks, e.g. Apuani et al. 2013; Tibaldi et al. 2006) or anthropogenic (caused by the redistribution of tensions and deformations in the massif under the influence of man-induced variations in groundwater level and other factors).

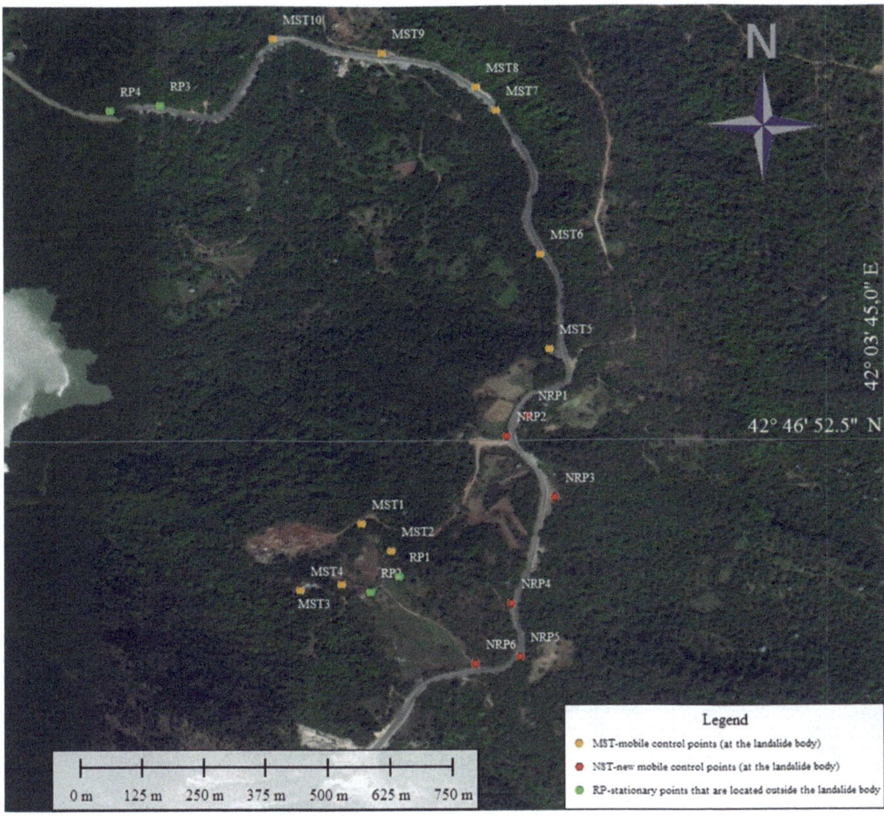

Fig. 16.3 Location of benchmarks, classified in fixed and moving ones. Reference system: WGS84/geographic coordinates

16.3.1 First Observation Period

In the observation period from 20.05.2016 to 21.11.2016 (185 days), the overall benchmarks subsidence was within 1–75 mm. The minimum velocity of net displacement was 0.04 mm/day, the maximum was 0.44 mm/day (Fig. 16.5). According to the graph provided in Fig. 16.6, which displays the amount of net displacements for each benchmarks, the maximum shift was 78.9 mm and the minimum shift was 6.8 mm; the whole set of benchmarks mainly moved to the west. As can be observed in Figs. 16.5 and 16.6, the benchmarks named MST 2, 5, 6, 7, 8, 9 were displaced the most.

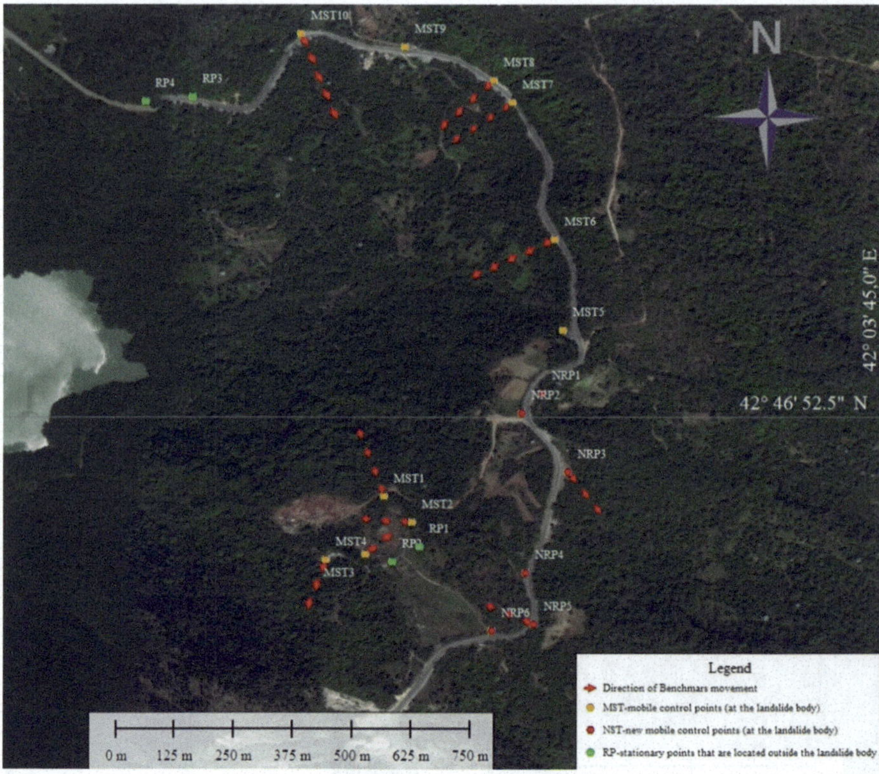

Fig. 16.4 Location of benchmarks, with the horizontal directions of net displacement (red arrows). The length of the red arrows is proportional to the total net displacement in the time-window considered in the present study; values are provided in Fig. 16.12. Reference system: WGS84/geographic coordinates

16.3.2 Second Observation Period

During the observation period from 21.11.2016 to 22.05.2017 (182 days), the amount of vertical displacements (subsidence) was within the range 2–46 mm. The minimum velocity of net displacement was 0 mm/day; the maximum was 0.4 mm/day (Fig. 16.7). According to Fig. 16.8, the maximum shift was 70.6 mm and the minimum shift 2.1 mm. At a general level, the movement was again towards the west.

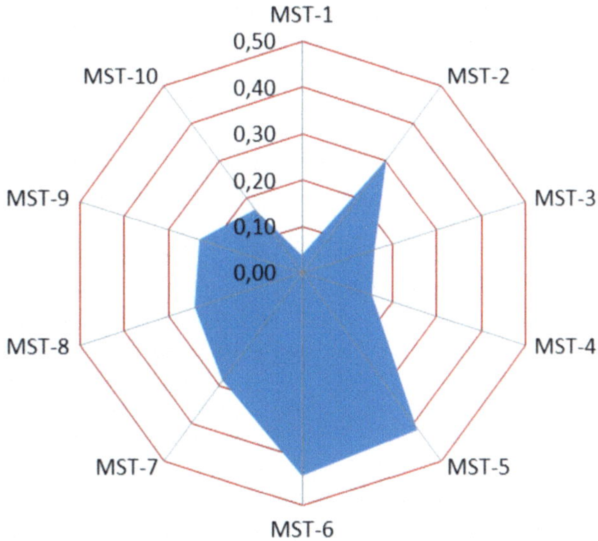

Fig. 16.5 Velocity of net displacements of benchmarks reported in mm/day and measured between 20.05.2016–21.11.2016. Benchmarks names are reported, and their location is shown in Figs. 16.3 and 16.4

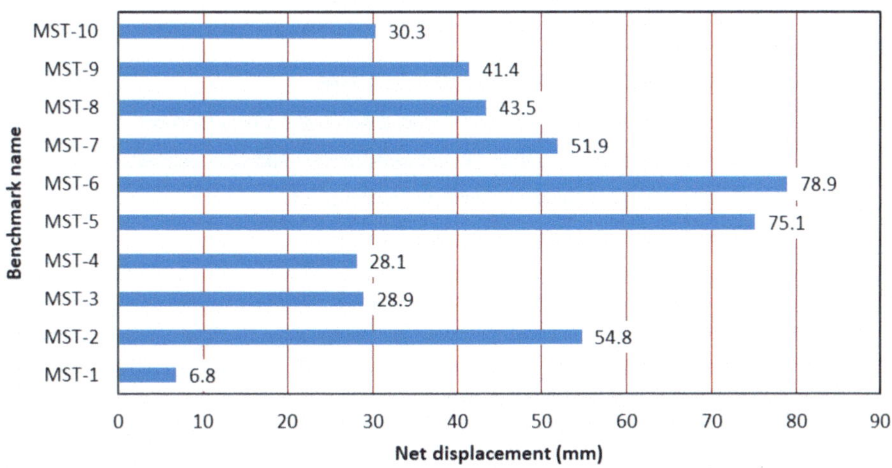

Fig. 16.6 Net displacement of benchmarks in the period 20.05.2016–21.11.2016; their location is shown in Figs. 16.3 and 16.4

16.3.3 Third Observation Period

According to the data collected during the observation period from 22.05.2017 to 10.06.2018 (384 days), the vertical displacement (subsidence) of benchmarks was within 0.3–42 mm. The minimum velocity of the net displacement was 0 mm/day, the maximum 0.2 mm/day (Fig. 16.9). According to Fig. 16.10, which shows net

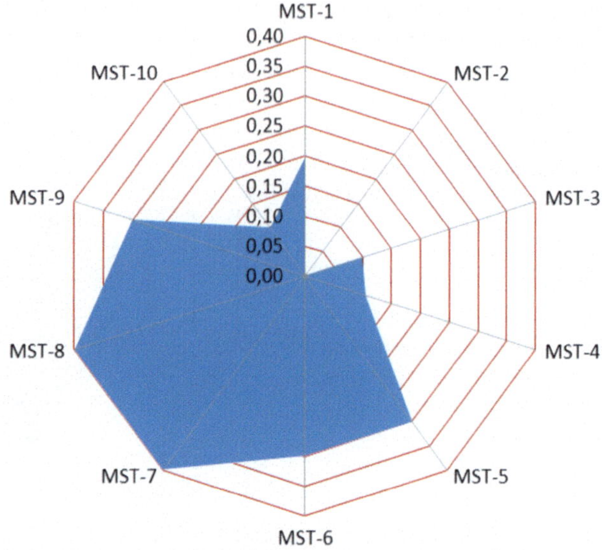

Fig. 16.7 Velocity of net displacements of benchmarks reported in mm/day and measured between 21.11.2016–22.05.2017; their location is shown in Figs. 16.3 and 16.4

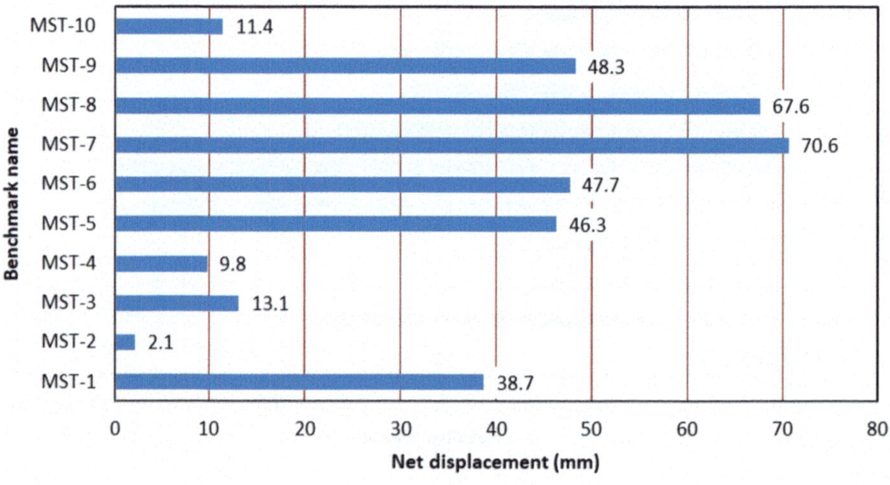

Fig. 16.8 Net displacement of benchmarks in the period 21.11.2016–22.05.2017, their location is shown in Figs. 16.3 and 16.4

displacements at analogous benchmarks, the total maximum shift was 61.7 mm and the minimum was 7.7 mm. By comparing this period with the previous one, a decrease in deformation intensity can be observed. At a general level, the movement was mainly towards the west.

Fig. 16.9 Velocity of net displacement of benchmarks reported in mm/day and measured in the period 22.05.2017–10.06.2018; their location is shown in Figs. 16.3 and 16.4

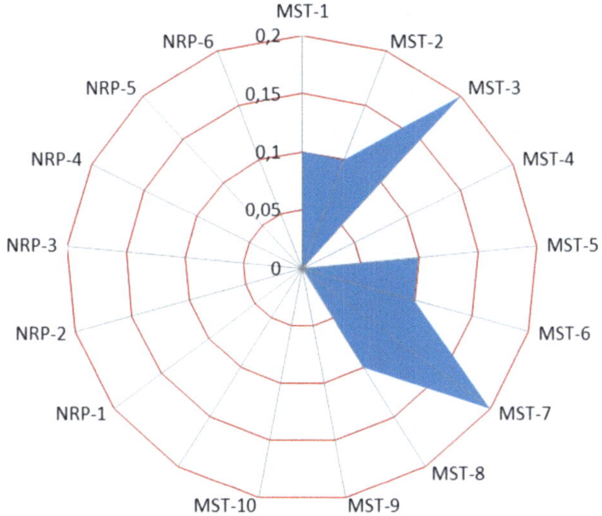

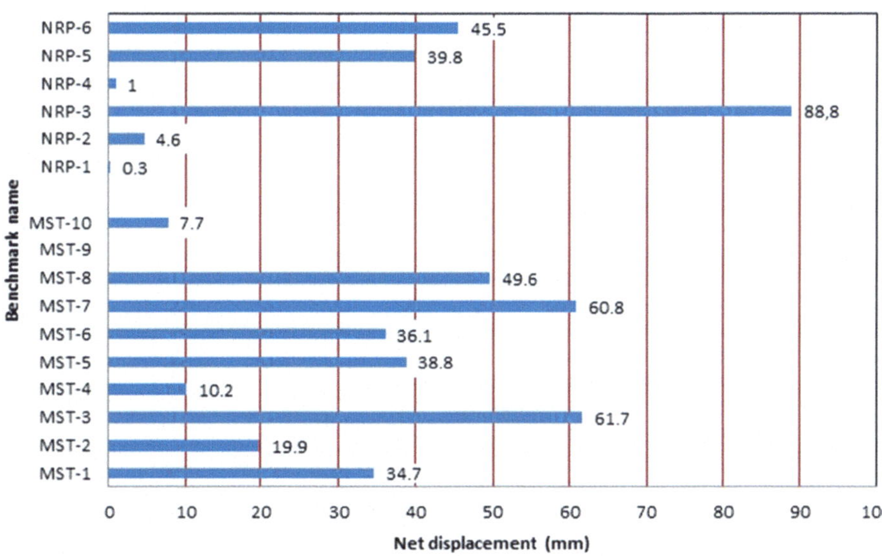

Fig. 16.10 Net displacement of benchmarks in the period 22.05.2017–10.06.2018; their location is shown in Figs. 16.3 and 16.4

16.3.4 Entire Observation Period

The average velocity of net displacement of benchmarks over the entire period was 0.12 mm/day. The average net displacement was 81 mm. The minimum net displacement velocity was 0 mm/day, the maximum 0.25 mm/day (Fig. 16.11). According to the diagram of accumulative net displacements for the 2 years

Fig. 16.11 Average velocity of net displacements of benchmarks for the entire period of study reported in mm/day; their location is shown in Figs. 16.3 and 16.4

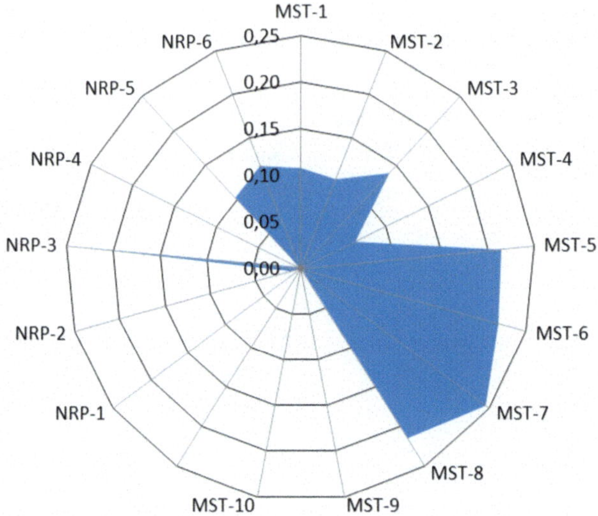

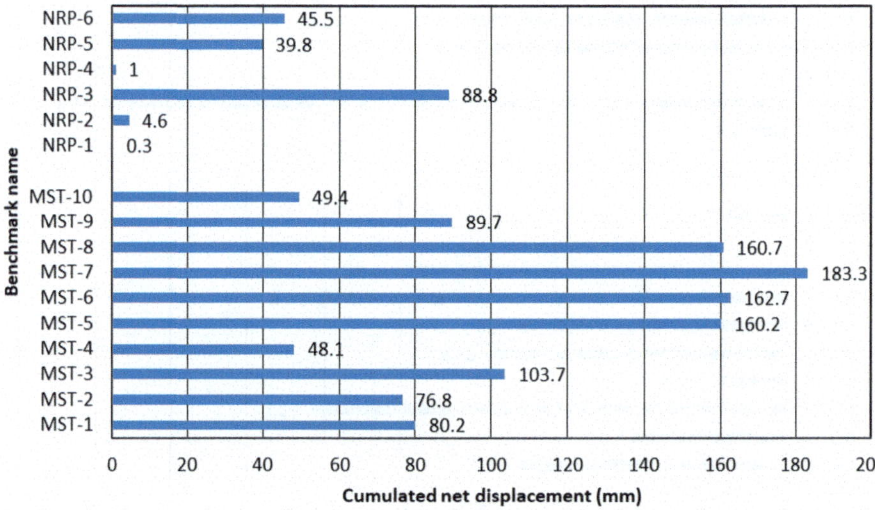

Fig. 16.12 Cumulated net displacement of benchmarks reported in mm for the entire period of study observation; their location is reported in Figs. 16.3 and 16.4

of observations in analogous benchmarks, the total maximum displacement was 183.3 mm and the minimum 48.1 mm (Fig. 16.12). The overall movement is towards the west, as shown in Fig. 16.4.

Figure 16.14 shows the velocity of the movement of benchmarks within the slope mass during the entire period of observations, in comparison with critical values. It has to be pointed out that, over the whole period of observation, the area of the Khoko landslide has shown an average velocity of net displacement that is below a threshold, established at 1 mm/day. This suggests a low landslide hazard for the monitored period.

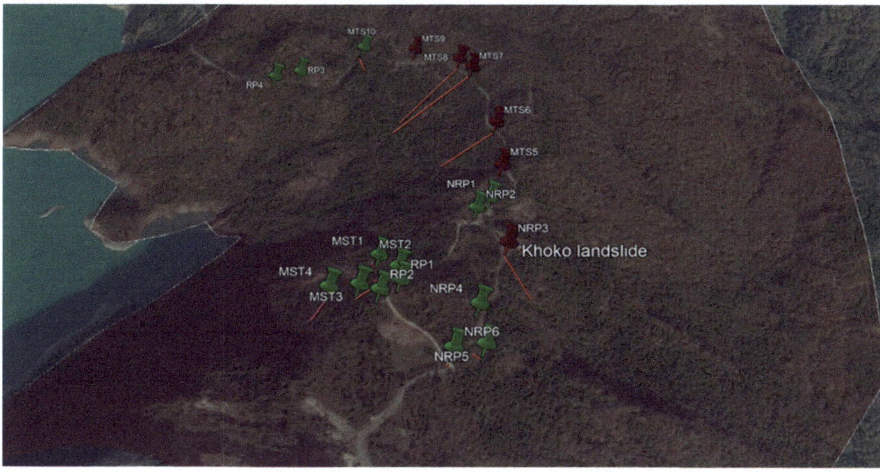

Fig. 16.13 3D, net displacement of the benchmarks; red lines represent the displacement directions of displacement, their length is proportional to the cumulated displacement reported in Fig. 16.12; all benchmarks moved downward. Their location is shown in Figs. 16.3 and 16.4. Google earth V 7.3.2.5776. (November 22, 2015). Enguri Reservoir, Georgia. 42°47′4.47″N, 42° 2′54.12″E. 2020 Maxar Technologies. http://www.earth.google.com [April 11, 2022]

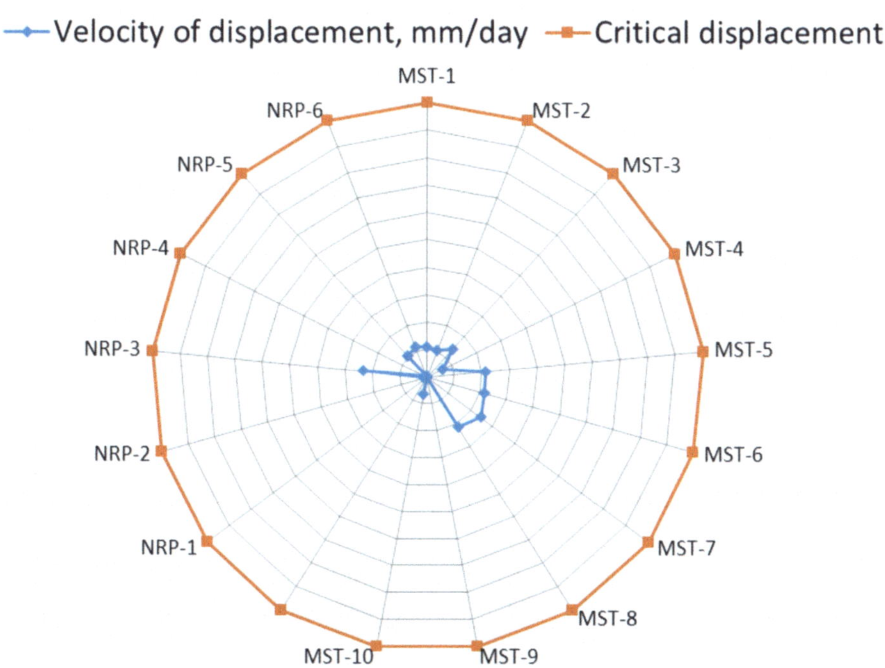

Fig. 16.14 Average velocity of net displacement of benchmarks during the entire period of this study, compared with the critical value of 1 mm/day. Their location is reported in Figs. 16.3 and 16.4

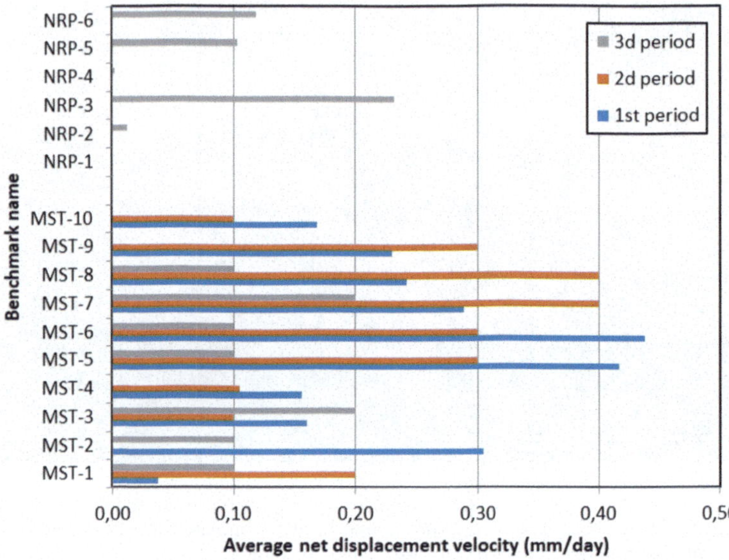

Fig. 16.15 Comparison of the velocity of net displacements (mm/day) of benchmarks for the entire period of study (2015–2018) divided in subperiods. Their location is reported in Figs. 16.3 and 16.4

The diagram in Fig. 16.15 shows the dynamics of net displacement over the entire period of study, subdivided in the three subperiods. It is worth underscoring that there has been a progressive reduction of the intensity of landslide slope deformation.

A detail of the displacement vectors at the Khoko landslide is shown in Fig. 16.16, where it can be noticed that the vectors at sites NRP 5 and NRP 3 point in a direction opposite to all the others. A possible explanation is that these two benchmarks are located near the head scarp of another landslide that affects the mountain slope, dipping to the east: this was already recognized and described as an active landslide by Tibaldi et al. (2019). As mentioned above, the Khoko landslide affects the western side of this mountain ridge.

16.4 Final Remarks

Our study of the evolution of the Khoko landslide from 2016 to 2018, carried out using geodetic methods in the field as well as data elaboration, has enabled us to draw the following main conclusions:

- The scheme of the observation station (network of benchmarks) and the methodology for conducting cost-effective observations, should ensure the completeness and reliability of the resulting information. A more complex system for assessing

Fig. 16.16 Detail of the horizontal displacement vectors of benchmarks located in the southern part of the studied area. Their location is reported in Figs. 16.3 and 16.4; the length of the red arrows is proportional to the values shown in Fig. 16.12

the evolution of slope deformation in the study area is proposed hereunder. In detail, we propose monitoring: *(i)* the development of deformation processes through satellite geodetic technologies; *(ii)* deformation processes by using 3D laser scanning; *(iii)* structural changes in the rock mass by geophysical methods. The proposed system would enable fully monitoring deformation processes and rapidly elaborating remediation measures to ensure the stability of the landslide.

- Our research shows that the study area is presently in an equilibrium stage, characterized by a quite constant rate of displacement. Across the Khoko landslide, movements have not been dangerous during the study period, with an average velocity of movements of 0.12 mm/day. In fact, during the entire period of our work, critical rates of displacement and deformation were not observed.
- Finally, based on the above, we recommend continuously carrying out observations of the slope displacements by way of the current observation stations. The periodicity of the observations is suggested to be once a year. In case critical deformation rates (> 1 mm/day) were recorded, it would be necessary to increase both the frequency of observations and the number of benchmarks along the slope, in order to assess the deformation with as much detail as possible.

References

Apuani T, Corazzato C, Merri A et al (2013) Understanding Etna flank instability through numerical models. J Volcanol Geotherm Res 251:112–126

Elizbarashvili M, Khardziani T, Maisuradze R et al (2018) Mountain environment and population in Georgia: case study of Upper Svaneti. Int J Sust Dev World 25(4):362–370

Erol S, Erol B and Ayan T (2004) A general review of the deformation monitoring techniques and a case study: analysing deformations using GPS/levelling. Presented at the XXth International Society for Photogrammetry and Remote Sensing (ISPRS) congress, Istanbul, 12–23 July 2004

Gulen L, EMME WP2 Team (2011) Active faults and seismic sources of the Middle East region: earthquake model of the Middle East (EMME) project. In: Abstracts of the AGU Fall Meeting, San Francisco, California, 5–9 December 2011

Kim J, Jeong S, Park S et al (2004) Influence of rainfall-induced wetting on the stability of slopes in weathered soils. Eng Geol 75(3–4):251–262

Kozhurin A, Acocella V, Kyle PR et al (2006) Trenching studies of active faults in Kamchatka, eastern Russia: Palaeoseismic, tectonic and hazard implications. Tectonophysics 417(3–4):285–304

Kuznetsov V, Krupin V (2001) Recommendations for carrying out field observations of sediments of ground dams. VNIIG, St. Petersburg

Pasquaré Mariotto F, Tibaldi A (2016) Inversion kinematics at deep-seated gravity slope deformations revealed by trenching techniques. Nat Hazards Earth Syst Sci 16:663–674

Pasquarè F, Tormey D, Vezzoli L et al (2011) Mitigating the consequences of extreme events on strategic facilities: evaluation of volcanic and seismic risk affecting the Caspian oil and gas pipelines in the Republic of Georgia. J Environ Manag 92(7):1774–1782

Schulz WH, Galloway SL, Higgins JD (2012) Evidence for earthquake triggering of large landslides in coastal Oregon, USA. Geomorphology 141:88–98

Song ZF, Xin XP (2012) Review of landslide monitoring. Applied mechanics and materials 166:2821–2827. Trans Tech Publications Ltd.

Tibaldi A, Pasquaré F (2007) Quaternary deformations along the Engadine-Gruf tectonic system, Swiss-Italian Alps. J Quat Sci 23:475–487

Tibaldi A, Bistacchi A, Pasquare F et al (2006) Extensional tectonics and volcano lateral collapses: insights from Ollagüe volcano (Chile-Bolivia) and analogue modelling. Terra Nova 18(4):282–289

Tibaldi A, Corazzato C, Rust D et al (2015) Tectonic and gravity-induced deformation along the active Talas–Fergana Fault, Tien Shan, Kyrgyzstan. Tectonophysics 657:38–62

Tibaldi A, Alania V, Bonali FL et al (2017a) Active inversion tectonics, simple shear folding and back-thrusting at Rioni Basin, Georgia. J Struct Geol 96:35–53

Tibaldi A, Russo E, Bonali FL et al (2017b) 3-D anatomy of an active fault-propagation fold: a multidisciplinary case study from Tsaishi, western Caucasus (Georgia). Tectonophysics 717:253–269

Tibaldi A, Korzhenkov AM, Pasquaré Mariotto F et al (2018a) NATO and earth scientists: an ongoing collaboration to assess geohazards and contribute to societal security in Central Asia and the Caucasus. Episodes 41(3):193–205

Tibaldi A, Bonali FL, Russo E et al (2018b) Structural development and stress evolution of an arcuate fold-and-thrust system, southwestern Greater Caucasus, Republic of Georgia. J Asian Earth Sci 156:226–245

Tibaldi A, Oppizzi P, Gierke JS et al (2019) Landslides near Enguri dam (Caucasus, Georgia) and possible seismotectonic effects. Nat Hazards Earth Syst Sci 19:71

Tibaldi A, Tsereteli N, Varazanashvili O et al (2020) Active stress field and fault kinematics of the Greater Caucasus. J Asian Earth Sci 188:104–108

Trejo D, Moutassem F, Hueste MBD et al (2007) Influence of environmental exposure conditions on mechanical properties of high-strength concrete. ACI Mater J 104(6):643

Tsereteli N, Tibaldi A, Alania V et al (2016) Active tectonics of central-western Caucasus, Georgia. Tectonophysics 691:328–344

Varazanashvili O, Tsereteli N, Bonali FL et al (2018) GeoInt: the first macroseismic intensity database for the Republic of Georgia. J Seismol 22(3):625–667

Chapter 17
Examples of Coastal Hazard Along the Georgian Black Sea Coast

Irakli Gelovani, George Lominadze, George Kavlashvili, and Grigol Russo

Abstract For the last one and a half-century, the Black Sea region of Georgia has been affected by the impact of adverse phenomena, which are the result of the violation of the balance of natural processes along the coastline due to human-induced interventions.

The first of such actions was the building of seaports in Poti and Batumi, during the second half of the nineteenth century. Along the seacoast, dozens of kilometres of shorelines suffered from drastic interference in the deposition balance. In the subsequent decades, because of the withdrawal of abundant material from the beaches and riverbeds, uncontrolled processes of intense erosion along the coast developed for hundreds of kilometres.

The construction of dams in the River Chorokhi basin, since the 1990s, had an extremely negative impact on the stability of the southern section of the Black Sea coast. Because of this, the deposition of material along the coast by the River Chorokhi has almost stopped. A similar process has taken place at the mouth of the River Enguri as well.

It is worth highlighting that it is very likely that an uncontrolled pattern of interference in coastal processes will go on in the future. Therefore, the environmental processes going on along the Black Sea coast need to be further analysed with a systemic approach.

17.1 Introduction

Modern sea coasts are greatly influenced by the variability of climate, and the factors that control it, such as temperature, wind, waves, river discharge, sea-level

---

I. Gelovani · G. Russo
National Environmental Agency of Georgia, Tbilisi, Georgia

G. Lominadze (✉) · G. Kavlashvili
Institute of Geography, Ivane Javakhishvili Tbilisi State University, Tbilisi, Georgia

© Springer Nature B.V. 2021
F. L. Bonali et al. (eds.), *Building Knowledge for Geohazard Assessment and Management in the Caucasus and other Orogenic Regions*, NATO Science for Peace and Security Series C: Environmental Security,
https://doi.org/10.1007/978-94-024-2046-3_17

changes, etc. Other influencing factors include tectonic processes and anthropogenic activities. Out of all the above-mentioned factors, anthropogenic activities are, indeed, those that have the most immediate and accelerated impacts. Along the Georgian, eastern coast of the Black Sea, there are several small and major rivers flowing into it (Fig. 17.1). Storm surges move large volumes of sediment, carried by the rivers, along the coastline, creating accumulative beaches. About 80% of beaches are accumulative, the others being abrasive. The east coast of the Black Sea is arc-shaped. The central, northern and southern districts can be subdivided according to their trend (azimuth) as follows: a central district, trending N170°, a northern district, striking N105°, and a southern district, trending N235°. The strongest sea waves are nearly perpendicular to the central district. These sea disturbances constantly move beach material from the north-western and south-western portion towards the centre of the arc (Kiknadze 1991). Therefore, the existence and sustainability of the beaches on the east coast of the Black Sea entirely depend on the solid runoff of the rivers on the coast. The main reason for the violation of sediment balance is the anthropogenic factor. In the second half of the nineteenth century, in Poti and Batumi, as a consequence of the construction of harbours, the balance of sediment was severely violated over tens of kilometres along the coastline. In the twentieth century, due to the removal of the great quantity of beach-feeding material from the beaches and riverbeds for agricultural and construction purposes, erosion processes on the shore took place over hundreds of kilometres. Along with this, the construction of dams for power plants along the rivers has dramatically reduced, and in some cases completely stopped, the solid discharge from rivers to the coast. In the following chapters, the districts along the east coast of the Black Sea are illustrated, where marked environmental disturbances can be observed, as a result of anthropogenic activities.

For the purpose of our work, we have used comparative-geographic, cartographic methods based on background materials, published scientific articles and online images.

17.2 The Chorokhi Lithodynamic System (Southern Region)

The Adjara coastal zone (55 km of coastline from the Turkish border to the Natanebi River mouth) is completely made of solid sediment from the Chorokhi River (Fig. 17.1). The Chorokhi beach-creating material arrives at the sea estuary and follows two alongshore sediment transport paths. A small alongshore sediment drift is located between the Chorokhi River mouth and the Kvariati village and moves 6.7 km to the south, to the abrasive Cape Kalendere. A strong, dominant main longshore drift influenced by the storms with a western direction, moved 39 km to the north and reached the estuary of the River Natanebi. On the cartographic materials of 1832, the River Chorokhi flowed into the sea with two main tributaries: to the south – in the area of the modern estuary and to the north - at the estuary of the River Mejina. On the 1890 map, the system of the northern branch has been

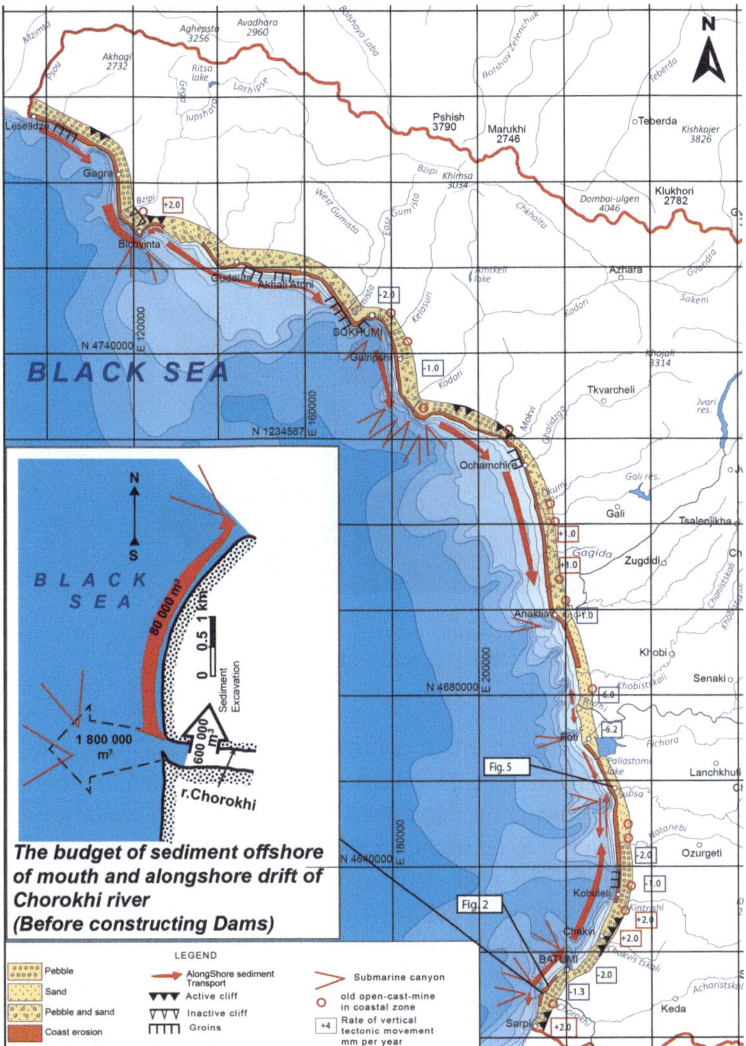

Fig. 17.1 Black Sea Georgian sea coast (Coastal Morphodynamic and Lithodynamic map); Reference system: WGS84 / UTM zone 38N

cancelled. Thus, by the end of the nineteenth century, a new river estuary appeared in the immediate vicinity of the head of the Chorokhi submarine canyon. This was followed by the loss of a large amount of beach-creating material (approximately 90%), brought by the river to the sea at great depths. Under natural conditions, the Chorokhi River was bringing about 5 million m^3 of alluvium annually to the sea, including 2.5 million m^3 of the beach-creating fraction. Of these materials, the amount of the coarser fraction (gravel and pebbles) amounted to about 0.4-0.5

Fig. 17.2 A stretch of the Adlia coastline after the 2013 storm; human-made infrastructures are located in the wave-exposed zone. The location of this site is shown in Fig. 17.1

million m^3 per year (Jaoshvili 1986), while the remaining component was coarse sand. Thinner material (average $D < 0.25$ mm) floated offshore and contributed to the sedimentation process on the seabed. Thus, the concentration of the entire discharge of the Chorokhi River in the southern branch system significantly altered the processes of the coastal zone, from Adlia (southern part of the Batumi city seashore) to the Batumi cape (see Fig. 17.1 for location). The Batumi coastline was formed by a severe shortage of beach-creating sediments.

In Adjara, the sea has washed out about 400 hectares of land, among which more than 40 hectares only from the Adlia-Batumi coastline between 1930 and 1980. Frequently, the buildings constructed for different purposes have been located in the wave-exposed zone (Fig. 17.2).

For the past 150 years, the seacoast has retreated by about 400 m at the estuary of the Mejina River (northern branch of the former Chorokhi River). The washout rate averaged 5.2 m/yr in the 1880-1926 time span.

Further development of the Adjara coast has been greatly influenced by the construction of the Batumi Port in 1878, in the distal part of the Batumi cape; a spur was built to protect the area from storm surges and sanding damage. The construction of the spur has dramatically increased the accumulative cape of Batumi, as shown in Fig. 17.3 (Svishchevski 1939).

In the Adjara region of Georgia, with the purpose of emergency mitigation, in the years 1982 to 1991, from the River Chorokhi mouth area and the Batumi

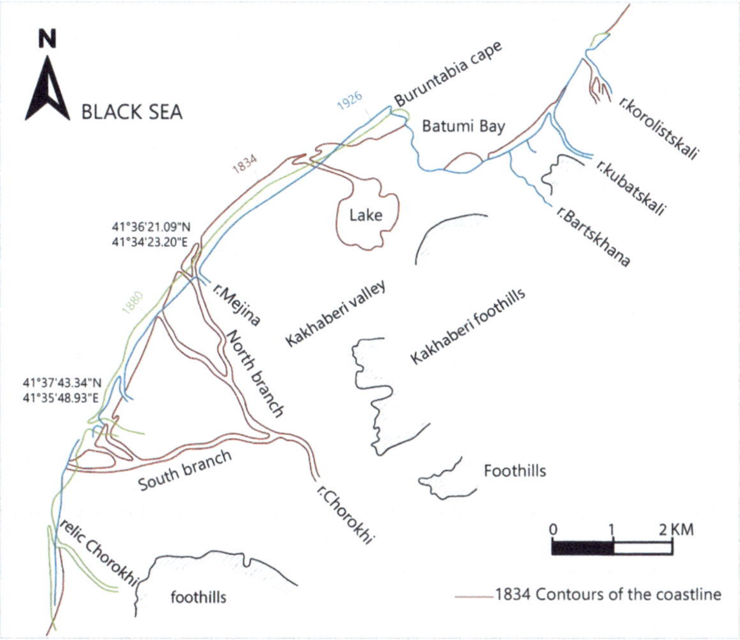

Fig. 17.3 The Kakhaberi plain development stages. (According to Svishchevski 1939). Reference system: WGS84 / geographic coordinates

cape submarine slope, beach-creating materials were extracted and transported to the emergency areas along the coastline (about 6.2 million m^3 of beach-creating sediment). The area of the artificially created coast protecting beaches of Adjara is about 60 hectares (Zenkovich and Schwartz 1987; Meladze 1993; Beridze et al. 2004).

17.3 The Central Coastline

17.3.1 Supsa-Natanebi

The Supsa-Natanebi coastline (southern section of the central coastal area) is particularly interesting, as it has never been washed out entirely and has always been subject to self-recovery processes. The Natanebi and Supsa Rivers played a decisive role in this. The width of 12-kilometer section of the Supsa-Natanebi beach was 60-70 m in 2019. In the early years, the removal of magnetitic sand from the Ureki beach, located south of the Supsa River mouth, (Tavartkiladze et al. 2014), greatly damaged the coast. As a consequence, the beach lost at least 40-50 m in width.

17.3.2 Kolkheti Coastal Zone

Kolkheti coastal zone (Enguri River, Natanebi River) is part of the Poti (Rioni) dynamic system. (Kiknadze 1972). The beaches here are mostly sandy (the exception is a beach made of gravel and pebble in the area of the River Enguri estuary, in the extreme north of the system). In front of the mouths of all big rivers (Enguri, Rioni, Supsa) there are submarine canyons, which cut the shelf area and whose heads are near the edge of the coastline. The solid material flows into the canyons both directly from the Rivers, and by means of alongshore streams. A portion of beach-creating material carried by the Enguri River, as well as almost all the alongshore sediment stream coming from the north (Kodori dynamic system), has been entering the Enguri canyon.

The coastline-forming natural process by the Rioni River continued until the 1960s, before the beginning of Poti harbour construction at the extension of the right bank of the Rioni northern branch. This structure caused the split of the Khobistskali River-Supsa River subsystem into two, morphodynamically different areas. The construction of the Port of Poti shut down the alongshore-movement of sediments northwards, and caused the intensive washout of the coastline along the stretch between the harbour and the Khobistskali River mouth. The opposite was observed towards the south, between the Port of Poti port and the Supsa River estuary. The coast increasing rate grew at the whole site (Kiknadze et al. 1978). In the canyon, gravitational processes started and, as a consequence, the canyon head approached the Rioni north branch. This process posed a risk to the southern mole of the Port of Poti. The situation changed in 1939, when, in order to protect the Port of Poti from flooding, the Rioni riverbed was moved 4 km northward (Fig. 17.4). This led to a dramatic decrease in the solid sediment flow into the submarine canyon; therefore, the distance between the canyon head and the coast increased. At the same time, an intense washout of the Poti-Supsa section started. This process took place especially along the Poti-Maltakva section, where about 300 hectares of coastline was washed out. In Grigoleti, the washout process has continued until the present day, as shown in Fig. 17.5.

An opposite scenario can be observed along the northern section (between Poti and the Khobistskali River) where, as a consequence of 70 years of intense accumulation, about 110-120 million m^3 of solid sediments have accumulated, and the coastline has extended by 2 to 2.5 kilometres (Beridze et al. 2004).

Among other anthropogenic factors, which have influenced the Poti dynamic system, it is worth mentioning the construction of power plants and the associated water reservoirs along the Rioni and Enguri Rivers; this caused a dramatic decrease in the volume of solid sediments carried by these rivers. Along the Rioni River, the average annual volume of solid material diminished from eight million m^3 to 4.4 million m^3, among which 2.1 million m^3 of beach-creating material (Jaoshvili 1986).

Before the construction of the Enguri hydroelectric power plant, the Enguri River carried to the sea 1,500,000 m^3 of material (Jaoshvili 2002), most of which was

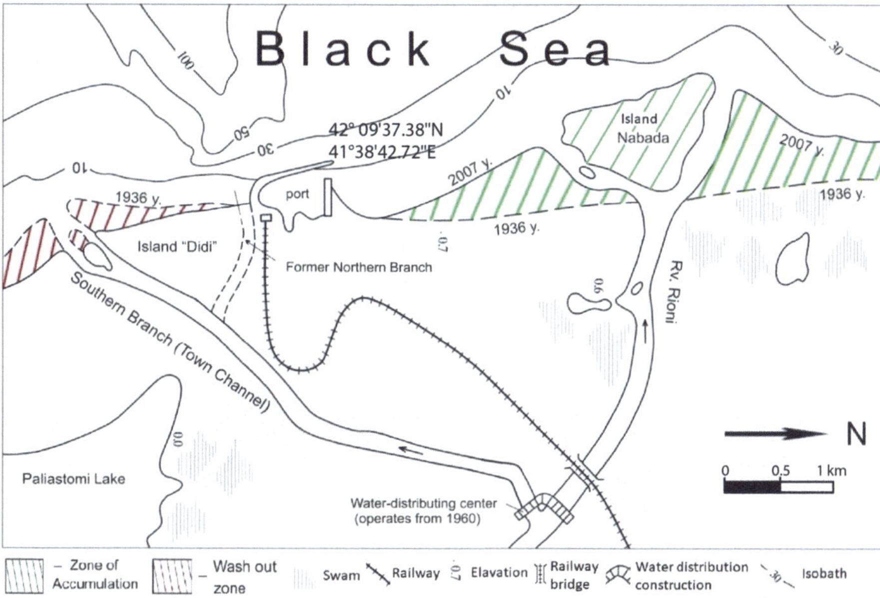

Fig. 17.4 Schematic map of the Rioni River sea mouth changes from 1936 to 2007. (By Russo 2007). Depth is expressed in m below the sea level and is highlighted by the isobaths. Reference system: WGS84 / geographic coordinates

Fig. 17.5 Example of coastline was washed out in Grigoleti in 2014, 10 km south of the Port of Poti (see Fig. 17.1 for the location of Poti)

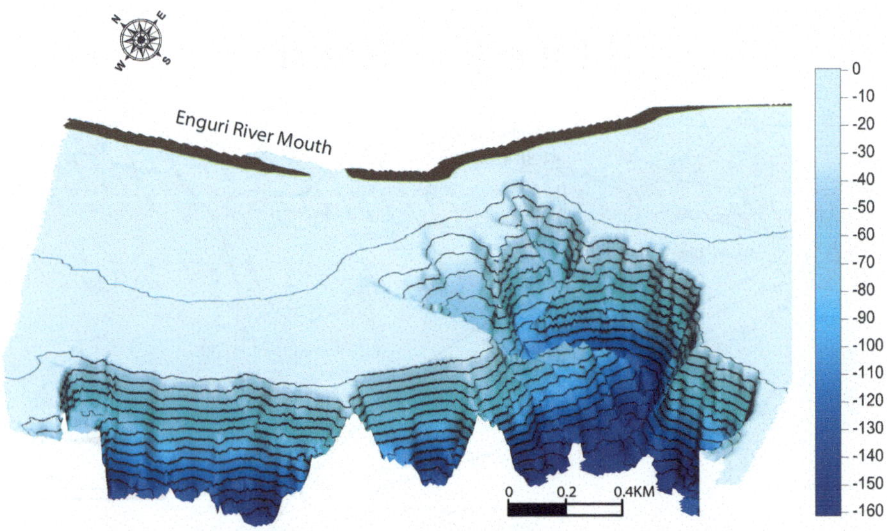

Fig. 17.6 3-D view of the Enguri River Submarine Canyon, the depth ranges from 0 to 160 m, as shown by the column on the left side of the figure, and the isobaths are reported every 10 m)

lost at the canyon, located close to the estuary (Kavlashvili 2018, see Fig. 17.6). In contrast, a portion moved to the south through the Anaklia beaches towards the Khobistskali River.

The HPP built in the 1970s along the Enguri River caused a reduction in the volume of the sediment carried out by the river, of more than 80% (Jaoshvili 1986).

The coastal retreat at Ganmukhuri (2 km north of Anaklia – central part in Fig. 17.1) and Anaklia was at least 200 m. In 1988-1989, the Ministry of Energy made arrangements to deposit 200,000 m^3 of coarse cobblestone, extracted from the Enguri dewatered riverbed, onto the beaches located north of Ganmukhuri. This remediation measure helped preserve the beaches, though the material deposited during this effort has disappeared over time. This event was followed by the first stage of the construction of the Port of Anaklia – with the deepening of the water area and elevation and surfacing of the coast with the extracted material. As a result of this, south of the area under construction, processes of coastal washout were activated.

17.4 Abkhazia (Northern Region)

The Kodori dynamic system starts near the Kodori River mouth and ends 3 km north of the Enguri River mouth; it is fed mainly by the sediments of the Kodori, Dghamisi, Mokva, Ghalizga, and Okumi Rivers; such rivers are reported in Fig. 17.1. In 1936, a port was built in Ochamchire, whose northern mole completely blocked the Kodori River sediments. As a consequence, strong erosive processes developed in the Ochamchire coastal zone. Since, then, Ochamchire has lost a 140-m wide and 4.5-km long coastline. After reinforcement activities (started in 1960) through hydro-technical constructions, washout processes along the Ochamchire coastline stopped; however, the washout rate increased south of the Ghalidzga River (1 km south of Ochamchire), as far as the Enguri River estuary. Before the construction of the Enguri dam, the Enguri River sediments were spread as far north as the Gagida River; after the construction of the dam, this natural process stopped. Washout rates increased in Ganmukhuri (an area near the Enguri River estuary) and in the northern beaches (Beridze et al. 2004).

17.5 Final Remarks

Along the eastern coast of the Black Sea, natural processes have irreversibly violated, negatively affecting the sediment deposition balance. Due to the construction of HPPs, sandy and pebbly material deposited by the two largest rivers on the seashore was entirely removed. The Chorokhi solid runoff averages 5,335,000 m^3 per year, of which the beach-creating materials are 2,500,000 m^3, and pebble volumes are more than 400,000 m^3; the Enguri River solid runoff is 1,500,000 m^3, of which the beach-creating component is about 500,000 m^3. The estuaries of almost all rivers are regulated. Extraction of material for construction purposes from other, much smaller riverbeds, which are important for the coastline balance (the Kintrishi, Supsa, Natanebi Rivers, and others) continues, while material from such river valleys should be extracted solely for shore protection purposes. As this inexpensive material will be depleted in the future, it will be necessary to resort to mining quarries in order to accomplish beach restoration. Ports and shoreline protecting structures disrupted lithodynamic systems and made it impossible to have sediments delivered to beaches from their feeding rivers. Often times, constructions for several purposes are located in the exposure zone of the waves. The east coast of the Black Sea is a good example of how human activities are severely affecting the environment. The adverse impacts of anthropogenic factors can be observed at all locations, with the exception of the Kelasuri dynamic system in Abkhazia (southeast of Sokhumi), as intensive resort construction activities have not begun there yet. In this work, we have pointed out how human interference in natural coastal processes, always leads to catastrophic effects, which can be observed to the present day.

References

Beridze T, Gelovani I, Geladze V, Russo G, Galumova V (2004) The general scheme of regulation and engineering protection of the Black Sea coast of Georgia. Ministry of Urbanization and Construction of Georgia. Research and production center "Napirdatsva" Extract from the decision of the XX international conference "Humanity and the coastal zone of the ocean in the 21st century", 2000, Moscow

Jaoshvili SV (1986). River sediments and beach formation on the Black Sea Coast of Georgia. Tbilisi (in Russian)

Jaoshvili SV (2002). The rivers of the Black Sea. Technical report No 71. European Environment Agency. (http//reports.eea.en.int/technical_report_2002_71)

Kavlashvili G (2018) Map of Enguri River submarine canyon. National Atlas of Georgia. Franz Steiner Verlag, Stuttgart, p 34

Kiknadze A.G. (1972). Zoning of the coastal zone according to the general balance sheet assessment of its dynamic state for the selection of the most optimal measures of coastal protection // Scien.Res L Hydr. and engineer-geol. GPI Problems, Tbilisi: No. 4, p. 14-19

Kiknadze AG (1991) Morphodynamics of the coastal zone and organization of its use. Habilitation for the degree of Geographical Sciences, Tbilisi (in Russian)

Kiknadze AG, Dzidzikashvili N, Amiredjibi R, Jaosvili S, Djanelidze C, Kogoshvili L, Matkava D, Tabidze D (1978) Report "Conditions for the formation and forecast of the development of the Black Sea coast and the limits of the Colchis lowland and recommendations for its stabilization", Tbilisi

Meladze FG (1993) Engineering solutions for the protection of Sea Coasts, Tbilisi, pp 28–52. (in Russian)

Russo GE (2007) Report - The current situation of the coastal zone of Poti and the conditions for its development. Chapter 2, Morphodynamics of the coastal zone. Tbilisi (in Georgian)

Svishchevsky DI (1939) Destruction of the sea coast in Batumi, as a common phenomenon for the East Coast of the Black Sea. Academy of Sciences of USSR:658–689. (in Russian)

Tavartkiladze C., Hantadze A, Elizbarashvili E, Karcivadze N (2014) Ureki-Unique Resort, Tbilisi, Publishing House "Universali" (in Georgian)

Zenkovich VP, Schwartz ML (1987) Protecting the Black Sea – Georgian S.S.R. gravel coast. J Coast Res 3:№2

Part VI
Seismic Microzonation

Chapter 18
Extensive Microzonation as a Tool for Seismic Risk Reduction: Methodological and Political Issues

Dario Albarello and Massimiliano Moscatelli

Abstract Budget limitations require an optimal allocation of resources devoted to seismic risk reduction. In order to allow local Authorities to operate public interventions to strengthen more exposed structures, increase population preparedness and improve emergency planning, a detailed knowledge of seismic hazard at the scale of a municipality (or of the minimal administrative structure in charge for prevention activities) is mandatory. To this purpose, a gap must be filled between seismic hazard assessment at regional scale and seismic response study at the scale of a single building, which is typical anti seismic engineering practice. Seismic Microzonation operates at this intermediate scale, focusing on heterogeneities in the seismic ground motion expected at the scale of a settlement (village or town) as an effect of local subsoil and morphological conditions. Being extensive in nature, its sustainability requires the development of specific methodologies, characterized by cost-effectiveness and, at the same time, able to provide reliable and useful outcomes. In this paper, the seismic microzonation strategy developed in Italy and applied in the last 10 years is briefly outlined. Sharing this experience with other countries prone to damaging earthquakes like Italy, may be of main importance to develop a common strategy to cope with future earthquakes.

Keywords Seismic microzonation · Seismic hazard · Site effects · Guidelines

D. Albarello (✉)
Dipartimento di Scienze Fisiche, della Terra e dell'Ambiente, Università di Siena, Siena, Italy
e-mail: dario.albarello@unisi.it

M. Moscatelli
Consiglio Nazionale delle Ricerche (CNR), Istituto di Geologia Ambientale e Geoingegneri (IGAG), Rome, Italy

© Springer Nature B.V. 2021

F. L. Bonali et al. (eds.), *Building Knowledge for Geohazard Assessment and Management in the Caucasus and other Orogenic Regions*, NATO Science for Peace and Security Series C: Environmental Security,
https://doi.org/10.1007/978-94-024-2046-3_18

18.1 Introduction

The dramatic images (Fig. 18.1) of the effects of a sequence of moderate earthquakes that struck Central Italy in 2016–2017 with maximum Mw 6.5 (e.g., Chiaraluce et al. 2017) showed (once more) the extreme vulnerability of the building heritage in Italy and urgency of long term risk prevention strategies aimed at reducing the impact of future earthquakes.

A basic element for this strategic program is mapping zones more exposed to potentially damaging ground shaking and where major efforts (also in terms of economic resources) are required to improve resilience of local populations and reduce earthquake risk.

Seismic hazard mapping at the regional scale is the first step in this direction (Fig. 18.2). This kind of estimate is provided by only considering large scale properties of the seismogenic process (geometry of seismogenic structures with relevant activity rates and long range radiation pattern) in the framework of a coherent probabilistic approach (e.g., Giardini et al. 2014). Smoothness of the hazard map motion contrasts with the empirical evidence of sharp variations of seismic ground at the scale of hundreds to thousands of meters.

Actually, it is well known that seismic ground motion at vibration periods of engineering interest is strongly controlled by short scale heterogeneity in the litho-technical properties of the shallow subsoil. In particular, sharp seismic impedance contrasts and morphological heterogeneities are responsible for energy trapping and interference phenomena inducing large variations in amplitude and spectral features of seismic ground motion (e.g., Kramer 1996). The dimension of the subsoil volume responsible for these phenomena (the so- called 'local seismic response') depends on the wavelength of seismic waves possibly responsible for

Fig. 18.1 Ruins of the Pescara del Tronto village (42° 45′ 2.3″ N, 13° 16′ 12.54″ E) after the first mainshock (Mw 6.0) of the 2016–2017 seismic sequence. In the village, there were 54 casualties over a population of about 130 inhabitants

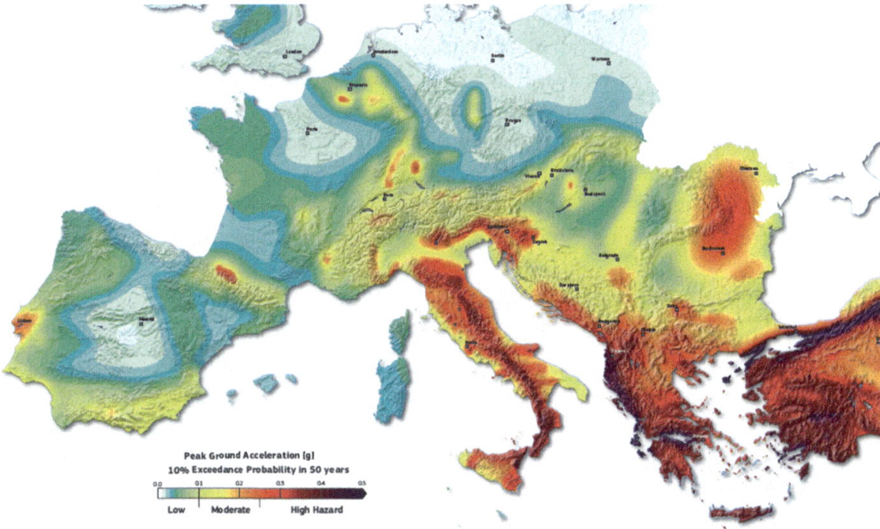

Fig. 18.2 Seismic Hazard map of Central-Western Europe. Colors indicate the Horizontal Peak Ground Acceleration (as a fraction of the gravity acceleration g) of the seismic ground motion characterized by an exceedance probability of 10% for exposure period of 50 years. (Giardini et al. 2014)

the damage of the structure exposed to the seismic event. By assuming that shear waves are the most responsible for damages (since these phases control horizontal ground motion components) and that in the shallow subsoil these are generally characterized by S waves velocities of the order of 200–600 m/s, one can expect that wavelengths of the order of tens to hundreds of meters are of main importance for most common buildings with natural period of the order of 1 s or less (IAEA 2016). These aspects are kept into account by regulations relative to anti seismic design (e.g., BSSC NHERP 2000; Eurocode 8, EN 2004). It appears evident that the seismic characterization of the subsoil at this scale requires detailed and relatively expensive surveys (borehole geophysics, laboratory tests, numerical modeling, etc.). This is feasible when single new buildings are designed or the retrofitting of single important structures is planned, but it becomes rapidly unapproachable when entire settlements are of concern.

Thus, there is a gap between large scale seismic hazard estimates and local seismic response studies, which must be filled to provide local administrations (generally charged for land management and emergency planning) with realistic seismic hazard maps necessary as a support for planning seismic prevention policies. 'Seismic Microzonation' provides this kind of mapping, by operating at an intermediate scale between regional hazard mapping and local seismic response studies. Furthermore, by assuming that all citizens must be protected at the same extent (independently from the economic relevance of their activities), seismic microzonation should be mandatory for all settlements exposed to potentially damaging earthquakes: thus, one can imagine the dimension of the problem, which is particularly dramatic in countries, such as Italy, where a huge amount of distributed settlements exist (about 8100 municipalities), each characterized by small dimensions and strong historical identity. In order to be affordable, extensive seismic microzonation requires the development of specific methodologies, whose main characteristics should be cost effectiveness (to be applied over wide areas) as well as the capability to capture main phenomena and provide outcomes useful for seismic risk reduction, to be effectively applied in city management and emergency planning.

Since the pioneering work by Medvedev (1965) in the former Soviet Union, several attempts have been performed to delineate and standardize procedures for small-scale seismic hazard assessment. In general, each guideline is calibrated for the specific national situation and accounts for the different rules operating in each country. Actually, some of them are guidelines for local seismic response studies finalized to anti-seismic design (e.g., Studer and Ansal 2004; CGS 2008; NZGS 2016) and are based on the application of relatively expensive approaches for the site characterization, which prevents their widespread application over wide areas. In other cases (e.g., TC4 1999; Nath 2011), wide applicability is of main concern. This is the cases of the 'Italian Guidelines for Seismic Microzonation (IGSM), released in 2008 (WGSM 2008) and subsequently updated (WGSMLA 2010; Various Authors 2011), whose main goal is to warrant applicability of seismic microzonation also when available resources are relatively scarce (small settlements, economically depressed areas, etc.). Since these guidelines could be of interest for countries where the latter conditions are widespread, in the following, their main features are shortly outlined. Moreover, after ten years since their early publication, the overall achievements resulting from their application are tentatively delineated.

18.2 The Italian Guidelines for Seismic Microzonation (IGSM)

These guidelines were developed on behalf of public authorities (Italian Conference of Regions and Autonomous Provinces) with the contribution of a large interdisciplinary group, under the coordination of the Civil Protection Department of the Italian Government. On purpose, geologists, seismologists, geotechnical engineers and architects, coming from academy, research institutions and practitioners' organizations worked together by developing a common approach including the most advanced contributions from the different disciplines and professional skills. The efforts for developing a specific common language and a coherent methodological approach were among the main outcomes of this project. A second cornerstone of IGSM is the focus on methodologies, which can be applied, for the most part, by practitioners and public technicians operating locally and trained on purpose. This because one cannot expect that research institutions and academics may be charged with field activities widely distributed on large areas. Furthermore, involving technicians operating locally in strict contact with local communities and administrations may also contribute to improving awareness of citizens with respect to seismic hazard. This aspect should not be underevaluated. In fact, the adoption of anti-seismic policies is costly (both in terms of direct costs and possible of lack of incomes related to different use of land and structures) and requires a long time to be completed. Thus, to be affordable and sustainable, such activities require the consensus of the involved population, political authorities and stakeholders. In this view, a sort of 'bottom-up' approach must be stimulated, where single citizens are made aware of the actual risks, thus prompting local administrations to plan effective actions.

The Seismic Microzonation approach implemented in the IGSM is mainly devoted to providing hazard estimates at the scale of a municipality, to be used for land management and emergency planning. In this regard, the outcomes of microzonation studies developed by following IGMSM should not be used for design. On the other hand, these outcomes represent an important constrain for site response studies developed at the scale of any single building. In this respect, microzonation outcomes complement studies prescribed by Italian seismic rules.

A main aspect of IGSM (shared with the TC4 methodology) is the graduality of the proposed approach, which, to ensure the widest possible applicability, relies on three levels of microzonation studies, involving increasing levels of engagement and economic efforts (e.g., Albarello et al. 2015; Albarello 2017).

At the first level, available information concerning shallow subsoil provided by previous studies eventually available (borehole-data, geological/geomorphological surveys, geotechnical data) is collected and systematized within a codified database (CTMS 2018). This collection, eventually integrated by on-purpose geological/geomorphological surveys (1:5,000–1:10,000) and low-cost geophysical campaigns, are considered to develop a geological model (in the perspective of its seismological use) of the study area. In this context, surface seismic passive measurements are of main concern (e.g., Okada 2003). In particular Horizontal to Vertical Spectral Ratios of ambient vibrations monitored at a single three-directional seismometric station (HVSR method) are used to identify and map the presence of seismic resonance phenomena due to energy trapping, induced by the presence of sharp seismic impedance contrasts in the subsoil (e.g., Lachet and Bard 1994; Bard 1999). In this phase, the area is subdivided in Seismically Homogenous Microzones (SHM), each characterized by similar expected co-seismic phenomena by distinguishing stable microzones, stable microzones with expected ground motion amplification phenomena and unstable microzones (due to surface faulting, landslides, differential soil failures or liquefaction phenomena). All the SHMs are provided with a representative 1D log, where the succession of geotechnical bodies down to the stiff bedrock is reported. Of main importance in the development of the geological model, is the identification and the shaping of geological units potentially representative of the engineering bedrock, which may or may not correspond to the geological bedrock (whatever the definition of this unit would be). Moreover, a range of depth for the engineering bedrock (meters, tens of meters, hundreds of meters) should be assessed for each SHM. Furthermore, Zones have to be identified where important lateral variations and where complex (2D/3D) ground motion amplification effects are expected. Despite of the inherently semi-qualitative character of this propaedeutic level of microzonation, its outcomes are of primary importance for the subsequent levels, since they represent the general frame for planning integrative measurements and interpret relative outcomes. Due to its relative cheapness, level I microzonation can be largely applied also in the cases where resources available at the scale of a single municipality are scarce (small settlements, economically depressed areas, etc.). An example of the first level microzonation map is provided in Fig. 18.3.

In the second level, amplification factors (FA) are determined relative to stable SHM where 1D ground amplification phenomena are expected. An integral representation of spectral amplification is supplied in the form:

$$FA = \frac{\int_{T1}^{T2} Sa(T)\, dT}{\int_{T1}^{T2} Sb(T)\, dT}, \quad (18.1)$$

where T is the building natural period, Sa and Sb are respectively the ordinates of the acceleration response spectra at the outcrop and at the engineering bedrock (reference ground motion). T_1 and T_2 represent the extremes of the considered period interval. In some cases, FV values are also determined where velocity response spectra are considered instead of acceleration response spectra in Eq. 18.1 (Fig. 18.4). In most recent realizations, three values of FA are considered corresponding to three intervals of ground vibration period: 0.1–0.5 s, 0.4–0.8 s and 0.7–1.1 s.

The AF values are determined by the use of specific abacuses (e.g., Peruzzi et al. 2016; Paolucci et al. 2019) supplied by regional authorities. In these abacuses, the relevant FA values are determined by considering the geological context and a pair of dimensions representative of the 1D seismic resonance phenomena: the resonance frequency f_0 and the average V_s value down to a fixed depth h (generally 30 m) or to the engineering bedrock when it is shallower than h. The values to be used are determined for each SHM by surface seismic measurements. As concerns f_0, the HVSR methodology is widely used. The determination of average V_s is generally obtained by surface waves methods (Foti et al. 2011, 2017; Socco et al. 2015). The second level microzonation provides an operative basis for land and emergency planning for most of the study area (Naso et al. 2019) and is expected to represent the main outcome of micozonation studies.

In fact, the third level microzonation only aims at the characterization of complex situations This level of analysis is applied to restricted portions of the study area, where simplified approaches (typical of the first two levels) appear unfeasible. In particular, two kinds of specific situations (both detected at the previous levels of analysis) are of concern: areas where permanent deformations are expected due

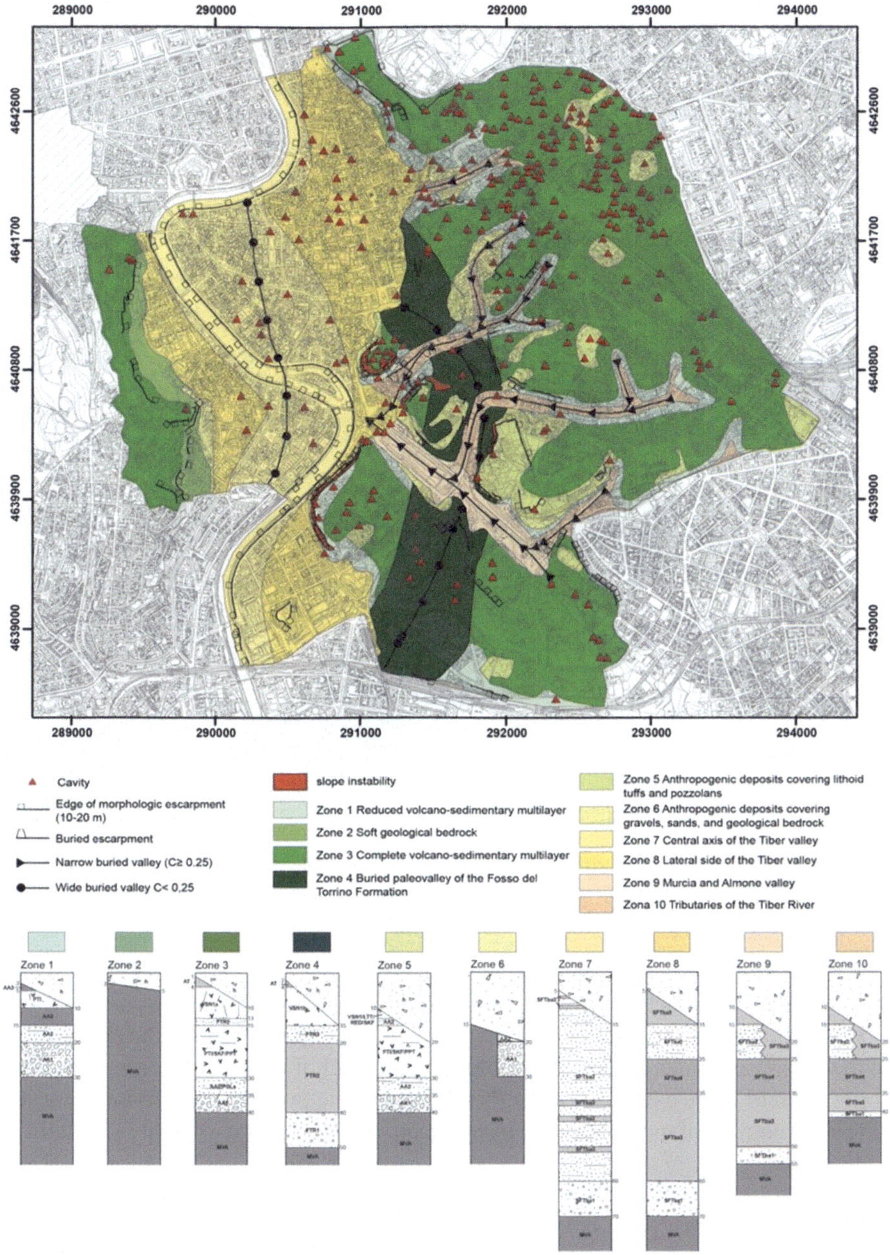

Fig. 18.3 Map of seismic microzonation of level 1 of the historic center of Rome (Italy) with logs (in the lower part of the map) representative of the main lithological features (Moscatelli et al. 2015). Coordinate system: WGS84 UTM33N. The grid at the boundaries of the map is expressed in meters; the values for each grid interval indicate the metric scale for latitude and longitude. The North direction corresponds to highest values in meters at the lateral boundaries of the map

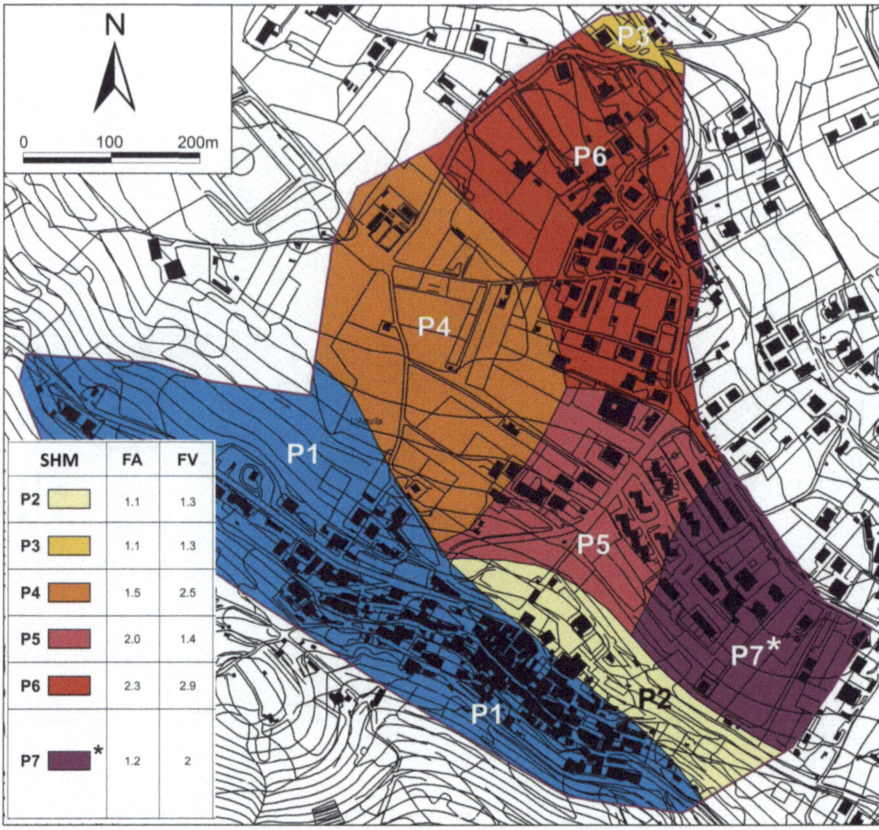

Fig. 18.4 Map of seismic microzonation of level 2 of the village of Pianola (42°19′24.76″N; 13°24′11.51″E) close to the city of L'Aquila (Central Italy). (Modified from the web site: https://www.regione.toscana.it/-/abruzzo-2009-microzonazione)

to landslides, liquefaction, surface faulting etc. (unstable SHMs) and areas where 2D/3D ground motion amplification phenomena are expected due to the complexity of the buried and outcropping morphology. This level of microzonation requires the application of more expensive tools (borehole seismic prospecting, laboratory tests, advanced numerical modeling) and its application is demanded to highly specialized technicians.

18.3 Implementation of Microzonation Studies in Italy

To support microzoning activities, the Italian Central Authority (National Civil Protection) in agreement with Regional Administrations, promoted a multi-year project devoted to the implementation of seismic microzoning in the current practice of city planning. As a whole, a global 7-year plan (2010–2016) was established. With the support of national funds *(about 10^9 € including support for retrofitting)*, 3896 municipalities where preliminary identified as of utmost importance on the basis of the reference hazard map (PGA for a 0% exceedance probability in 50 y ≥ 0.125 g). In order to exploit these funds, local Administrations (regional or local) had to co-fund microzonation activities and warrant their implementation in city and emergency planning. To support these activities from the technical point of view, the Centre for Seismic Microzonation and its Applications (CentroMS; https://www.centromicrozonazionesismica.it/en/) has been established which includes most of the Scientific Institutions operating in Italy in the field seismic hazard assessment: presently, 25 Research Institutions and Universities have been involved, including more than 100 seismologists, geologists, engineers and architects. The main task of the centre is providing guidelines for specific aspects of microzonation activities, training local practitioners and public technicians by assisting and overviewing them during field activities. At the moment, 1653 municipalities have completed microzonation studies and for 2230 municipalities the studies have been completed and under examination for the final validation (Fig. 18.5). Most of these studies are of the first level and only 846 (out of 2883) municipalities have promoted and completed higher level studies.

Data collected in these studies were standardized and implemented in a geographic database. This information is transmitted to the Civil Protection Department where a big public repository of georeferenced geological/geophysical/geotechnical data is established that represents a resource of great importance for future studies (available at www.webms.it).

This system has been also activated after damaging seismic events. As an example, since the beginning of the seismic sequence that affected central Italy starting from August 2016 (Luzi et al. 2017), the CentroMS has been charged with coordinating high level microzonation activities supporting reconstruction in the damaged settlements (Moscatelli et al. 2020). To this purpose, it was requested to complete, within 6 months, third level microzonation studies of 138 municipalities, including a huge amount of small settlements (533), distributed in 4 Administrative Regions (Abruzzo, Lazio, Marche, Umbria) and populated by about 500,000 inhabitants. The budget available for this activity was fixed to 1.8 10^6€. The whole area has been subdivided into 6 Macroareas under the responsibility of a coordinator managing the scientific teams operating in the area (Fig. 18.6). More than 100 researchers have been involved in the task of coordinating, training, and supporting the activity of 114 groups of practitioners operating in the field and providing more complex analyses (borehole and laboratory measurements, 2D numerical modelling). The whole plan was completed on time. Probably, for the first time in Italy (and probably in the world) such an ambitious microzonation study has been carried on in such a short time for an area of this size. The test revealed the effectiveness of the methodological approach.

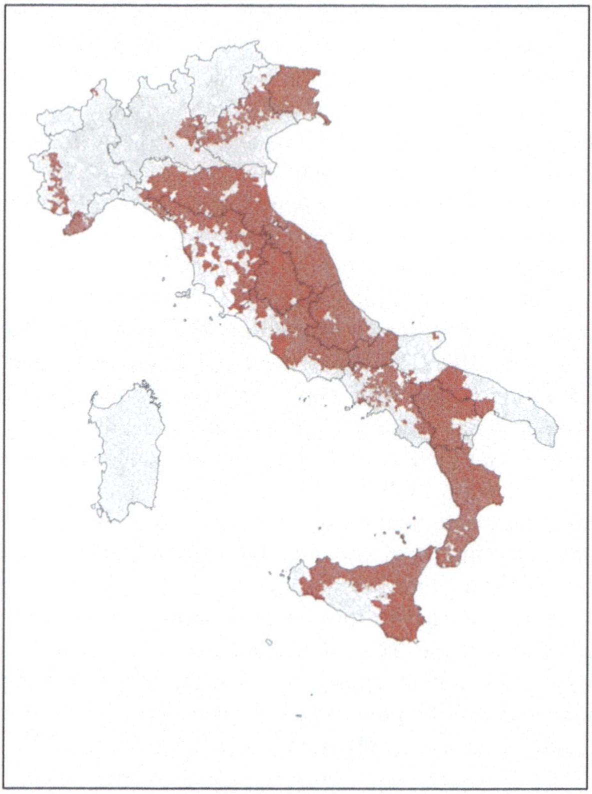

Fig. 18.5 Distribution of the Italian Municipalities where microzonation studies have been completed; regional boundaries are also shown

18.4 Conclusions

The seismic microzonation approach developed in Italy aims at providing local administration with standardized information specifically useful for supporting prevention activities at the scale of a Municipality. Due to this specific aim, microzonation studies complement (do not substitute) site response analysis prescribed by seismic rules for anti-seismic design and fill the gap between the latter ones and regional scale hazard assessments. Beyond the technical importance of this improvement of the present knowledge of seismic characteristics of the Italian territory, three main results have been obtained after seven years of applications of the Italian Guidelines for Seismic Microzonation. First of all, the Italian scientific community was able to cooperate across disciplinary boundaries: geologists, engineers and seismologists coming from different institutions, created technical interdisciplinary teams, whose target-oriented efforts allowed for the development of a common scientific language resulting in guidelines that are progressively improving. Furthermore, a huge amount (several thousands) of homogeneous data has been collected and stored within an homogenous GIS database that will soon

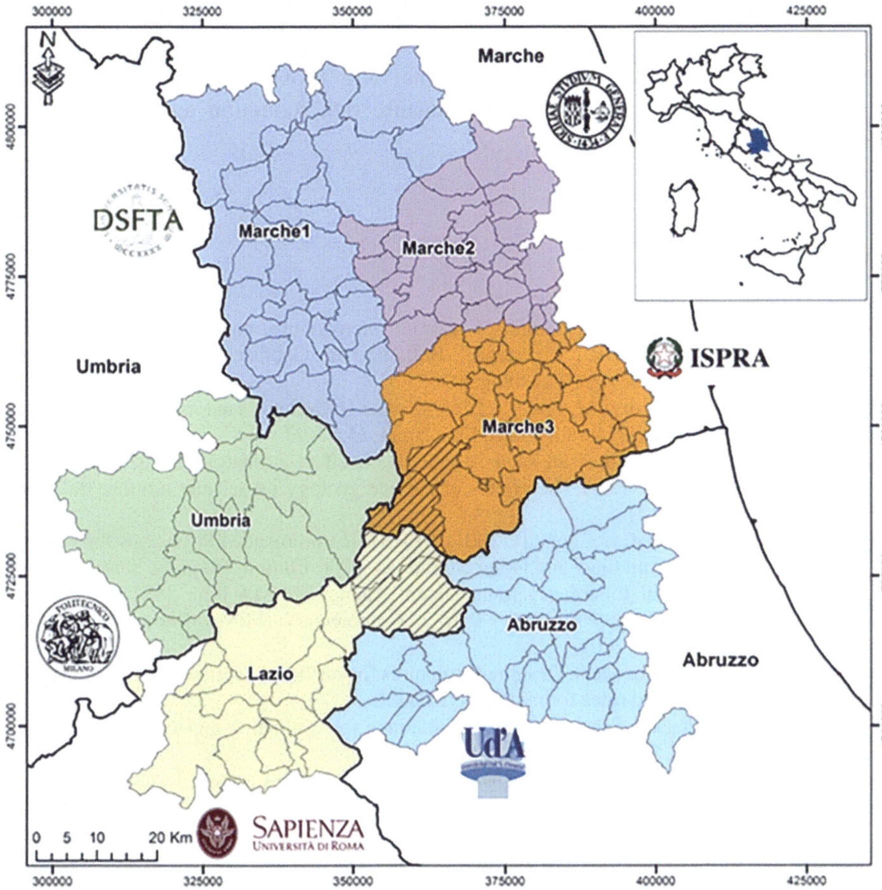

Fig. 18.6 Area of the Central Italy struck by the 2016–2017 seismic sequence; the most damaged four Municipalities hit by the earthquake of 24th August 2016, i.e. Amatrice, Accumoli, Arquata del Tronto, Montegallo, are highlighted in the striped area. The boundaries of the 138 municipalities involved in the seismic microzonation activities are reported. Colors indicate the partition into 6 macroareas, each under the responsibility of a single team, headed by one coordinator of CentroMS: Marche 1 – University of Siena, Marche 2 – University of Catania; Marche 3 – ISPRA; Umbria – University of Milan; Lazio – University of Rome; Abruzzo: University of Chieti. The Institute of Environmental Geology and Geoengineering (IGAG) of the Italian National research Council (CNR) coordinated the project. Coordinate system: WGS84 UTM33N. The grid at the boundaries of the map is expressed in meters; the values for each grid interval indicate the metric scale for latitude and longitude. The North direction corresponds to highest values in meters at the lateral boundaries of the map. (Modified after Moscatelli et al. 2020)

be made available to the scientific community, practitioners and citizens. Finally, hundreds of practitioners (mainly professional geologists) have been trained in microzonation studies and seismic hazard assessment. At the same time, a large number of Regional and Municipal Authorities and public technicians have been involved in microzonation and the respective populations have been made aware of seismic hazard in their villages or towns. This sensitized citizens and technicians represent a well distributed "presidium" that will prove to be of great help in supporting future activities devoted to seismic risk reduction across the whole country.

References

Albarello D (2017) Extensive application of seismic microzoning: methodological and socio-political issues in the Italian experience. Bollettino di Geofisica Teorica ed Applicata 58(4):253–264. https://doi.org/10.4430/bgta0205

Albarello D, Socco VL, Picozzi M, Foti S (2015) Seismic Hazard and land management policies in Italy: the role of seismic investigations. First Break 33:87–93

Bard P-Y (1999) Microtremor measurements: a tool for site effect estimation? In: Irikura K, Kudo K, Okada H, Sasatani T (eds) The effects of surface geology on seismic motion. Balkema, Rotterdam, pp 1251–1127

BSSC (Building Seismic Safety Council) (2001) NEHRP Recommended Provisions for Seismic Regulations for New Buildings and Other Structures. 2000 Edition, Building Seismic Safety Council for the Federal Emergency Management Agency (FEMA Rep. 368, 369), Part 1: Provisions (FEMA Report 368) 374p; Part 2: Commentary (FEMA Report 369) p.444, Washington, D.C., March 2001

CEN-EN1998-1 (2004) Eurocode 8: Design of structures for earthquake resistance. Part 1: General rules, seismic actions and rules for buildings. Brussels

CGS (California Geological Service) (2008) Guidelines for evaluating and mitigating seismic hazards in California. Special Publication 117A:108 pp

Chiaraluce L, Di Stefano R, Tinti E, Scognamiglio L, Michele M, Casarotti E, Cattaneo M, De Gori P, Chiarabba C, Monachesi G, Lombardi A, Valoroso L, Latorre D, Marzorati S (2017) The 2016 Central Italy seismic sequence: a first look at the mainshocks, aftershocks, and source models. Seism Res Lett 88(3):1–15. https://doi.org/10.1785/0220160221

CTMS (Commissione tecnica per la Microzonazione sismica) (2018). Standard di Rappresentazione ed archiviazione informatica, Versione 4.1.1, Dept.of Civil Protection, Rome, novembre 2018 (in Italian). Available on the website: https://www.centromicrozonazionesismica.it/it/download/send/26-standardms-41/71-standardms-4-1

Foti S, Parolai S, Albarello D, Picozzi M (2011) Application of surface wave methods for seismic site characterization. Surv Geophys 32(6):777–825. https://doi.org/10.1007/s10712-011-9134-2

Foti S, Hollender F, Garofalo F, Albarello D, Asten M, Bard P-Y Comina C, Cornou C, Cox B, Di Giulio G, Forbriger T, Hayashi K, Lunedei E, Martin A, Mercerat D, Ohrnberger M, Poggi V, Renalier F, Sicilia D, Socco V (2017) Guidelines for the good practice of surface wave analysis – a product of the INTERPACIFIC project. Bull Eur Earthq Eng, Pub online https://doi.org/10.1007/s10518-017-0206-7

Giardini D, Woessner J, Danciu L (2014) Mapping Europe's seismic Hazard. Eos 95(29):261–262

IAEA (International Atomic Energy Agency) (2016) Seismic hazard assessment in site evaluation for nuclear installations: ground motion prediction equations and site response, IAEA-TECDOC-1796, Wien, 119 pp

Kramer SL (1996) Geotechnical earthquake engineering. Prentice Hall, Upper Saddle River, 672 pp

Lachet C, Bard P-Y (1994) Numerical and theoretical investigations on the possibilities and limitations of Nakamura's technique. J Geophys Earth 42:377–397

Luzi L, Pacor F, Puglia R, Lanzano G, Felicetta C, D'Amico M, Michelini A, Faenza L, Lauciani V, Iervolino I, Baltzopoulos G, Chioccarelli E (2017) The Central Italy seismic sequence between August and December 2016: analysis of strong-motion observations. Seism Res Lett 88(5):1219–1231. https://doi.org/10.1785/0220170037

Medvedev SV (1965) Engineering seismology. Israel Program fo Scientific Translations Ltd., Jerusalem, 260 pp

Moscatelli M, Pagliaroli A, Mancini M, Stigliano F, Marini M, Simionato M, Cavinato GP, Colombi A (2015) Seismic microzonation of level 1 of the historic center of Rome. Rend Online Soc Geol It 33:63–70., 3 figs., 1 tab. https://doi.org/10.3301/ROL.2015.16

Moscatelli M, Albarello D, Scarascia Mugnozza G, Dolce M (2020) The Italian approach to seismic microzonation. Bull Earthq Eng. https://doi.org/10.1007/s10518-020-00856-6

Naso G, Martelli L, Baglione M, Bramerini F, Castenetto S, D'Intinosante V, Ercolessi G (2019) Maps for land management: from geology to seismic hazard. Boll Geofis Teor Appl 60(2):277–294. https://doi.org/10.4430/bgta0263

Nath SK (2011) Seismic microzonation handbook. Geoscience Division, Ministry of Earth Sciences, Government of India, New Delhi, p 518

NZGS (New Zealand Geological Service) (2016) Earthquake geotechnical engineering practice-overview

Okada H (2003) The microtremor survey method, Geophysical monograph series 12. Geophysicists, Society of Exploration, 135 pp

Paolucci E, Tanzini A, Peruzzi G, Albarello D, Tiberi P (2019) Empirical testing of a simplified approach for the estimation of 1D lithostratigraphical amplification factor. Bull Earthq Eng, published online 14 Dec 2019. https://doi.org/10.1007/s10518-019-00772-4

Peruzzi G, Albarello D, Baglione M, D'Intinosante V, Fabbroni P, Pileggi D (2016) Assessing 1D seismic response in microzoning studies in Italy. Bull Earthq Eng 14:373–389. https://doi.org/10.1007/s10518-015-9841-z

Studer J, Ansal A (2004) Seismic microzonation for municipalities, manual. Research Report for Republic of Turkey, Ministry of Public Works and Settlement, General Directorate of Disaster Affairs, World Institute for Disaster Risk Management, Inc, 2004

TC4 (Technical Committee for Earthquake Geotechnical Engineering) (1999) Manual for Zonation on Seismic Geotechnical Hazard. Revised edition, Technical Committee for Earthquake Geotechnical Engineering (TC4) of the International Society of Soil Mechanics and Geotechnical Engineering (ISSMGE), 209 pp

Various Authors (2011) Contributi per l'aggiornamento degli "Indirizzi e criteri per la microzonazione sismica", Ingegneria sismica, 28, 2, Patron editore Bologna, Italy, (in Italian). http://www.protezionecivile.gov.it/resources/cms/documents/aggiornamento_indirizzi_microzonazione_sismi ca.pdf

WGSM (Working Group on Seismic Microzoning) (2008) Indirizzi e criteri per la microzonazione sismica. Conferenza delle Regioni e delle Province autonome – Dipartimento della Protezione Civile, Roma, 3 voll. e DVD, (in Italian). http://www.protezionecivile.gov.it/jcms/it/view_pub.wp?contentId=PUB1137, English version at http://www.protezionecivile.gov.it/http://www.docs/cms/attach_extra/GuidelinesFor-SeismicMicrozonation.pdf?

WGSMLA (Working Group on Seismic Microzoning the L'Aquila Area) (2010) Microzonazione sismica per la ricostruzione dell'area aquilana. Regione Abruzzo, Dipartimento della Protezione Civile, 3 vols and dvd, 796 pp. (in Italian). http://www.protezionecivile.gov.it/jcms/it/view_pub.wp?contentId=PUB25330

Chapter 19
Preliminary Results of Site Effects Assessment in Mtskheta (Georgia)

Nino Tsereteli, Massimiliano Moscatelli, Dario Albarello, Iolanda Gaudiosi, Silvia Giallini, Zurab Gogoladze, Federica Polpetta, Maurizio Simionato, Francesco Stigliano, David Svanadze, Laurentiu Danciu, Otar Varazanashvili, and Giorgi Gaprindashvili

Abstract Seismic hazard and risk assessment is a crucial issue for Georgia, a country located in the Caucasus, which is one of the most seismically active regions in Alpine-Himalayan collision belt. It has long been known that subsoil conditions may have a significant influence on the expression of earthquake motions at the surface, modifying both amplitude as well as the frequency content and duration of the seismic ground motion. In this regard, seismic microzonation studies aimed at identifying and mapping zones, in a given area, characterized by homogeneous seismic behavior, represent a worldwide-accepted tool for the mitigation of seismic risk. In this paper, we provide the preliminary results of the first site effects assessment performed in the urban area of Mtskheta (Georgia). Our main outcomes consist in: (i) a geological map focused on the study area; (ii) a distribution map of resonance frequencies (f_0 and f_1); (iii) an estimation of the average shear-wave velocity of the upper 30 m ($V_{S,30}$) and (iv) the amplification factors from standard acceleration response spectra according to the EC8 classification. The results obtained will represent a first step toward reducing seismic risk in Georgia

N. Tsereteli (✉) · Z. Gogoladze · D. Svanadze · O. Varazanashvili
M. Nodia Institute of Geophysics, I. Javakhisvili Tbilisi State University, Tbilisi, Georgia
e-mail: nino_tsereteli@tsu.ge

M. Moscatelli · I. Gaudiosi · S. Giallini · F. Polpetta · M. Simionato · F. Stigliano
Institute of Environmental Geology and Geoengineering, Italian National Research Council, Rome, Italy

D. Albarello
Dipartimento di Scienze Fisiche, della Terra e dell'Ambiente, Università di Siena, Siena, Italy

L. Danciu
Swiss Seismological Service (SED), ETH, Zürich, Zürich, Switzerland

G. Gaprindashvili
Department of Geology, National Environmental Agency of Georgia, Tbilisi, Georgia

© Springer Nature B.V. 2021

F. L. Bonali et al. (eds.), *Building Knowledge for Geohazard Assessment and Management in the Caucasus and other Orogenic Regions*, NATO Science for Peace and Security Series C: Environmental Security,
https://doi.org/10.1007/978-94-024-2046-3_19

and providing a knowledge base of local seismic hazard, useful for effective seismic risk mitigation strategies.

19.1 Introduction

The assessment of seismic hazard and risk is of great importance for Southern Caucasus and particularly for Georgia, which is the main gateway between Asia and Europe. Seismic hazard is defined as the expected seismic response in a specific area within a given time period. At the regional scale, this assessment is evaluated by applying probabilistic or deterministic (or integrated probabilistic-deterministic) methods to obtain ground motion parameters at a site under conditions of rigid and flat soil represented by soil and topographic classes A and T1, respectively, according to EC8, Eurocode 8-EN 1998-1 (1998).

At the local scale, seismic hazard assessment is made by analyzing the geological, geomorphological, geotechnical and geophysical characteristics of the site, as it is well established that the incoming seismic motion can change in amplitude, frequency and duration due the site-specific local characteristics (site effects). The problem of estimating the influence of site conditions on ground motion is well known since the 1985 Mw 8.1 Mexico City earthquake (Singh et al. 1988). Useful tools for the technicians who are responsible of earthquake-resistant design need now to adequately take into account this influence, thanks to the introduction of seismic response analyses where properly codes (e.g., the European EC8 code) are applied.

The analysis of the influence of local effects on seismic response at the ground surface is also the main issue of seismic microzonation studies, a planning tool applied in Italy since 1980 (Brambati et al. 1980) and devoted to identifying zones in urban areas that are homogenous in terms of seismological and geological characteristics for seismic hazard purposes (e.g., Moscatelli et al. 2014 and references therein), according to the Italian guidelines (Working Group ICMS 2008).

As a rule, the extensive areal coverage of SM studies, made possible by the robust description of geological and geotechnical conditions retrieved at the municipality scale, may allow the definition of synthetic parameters aiming at defining an absolute ranking of seismic hazard (Von Thun et al. 1988). Starting from Seismic Microzonation studies results and by applying a methodology already tested for 138 municipalities in central Italy (Mori et al. 2019), site effects could be accounted for in seismic hazard maps. This latter will be the previous knowledge element for effectively quantifying the damage resulting from the combination of reference ground motion and site effects.

In this paper, we summarize the preliminary results of an ongoing multidisciplinary study in the perspective of the first seismic microzonation of Mtskheta. After a description of the tectonic/geological setting and a brief geological synthesis of the Mtskheta region, we show the dataset of collected measurements, mainly consisting in geophysical acquisitions, both in passive and in active mode (HVSR and MASW, respectively). So far, our outcomes mainly derive from the elaboration and interpretation of the collected geophysical data, that allowed us (i) to achieve a resonance frequencies f_0 and f_1 map (where f_0 is the fundamental frequency and f_1 is a resonance frequency with a frequency value higher than f_0) focused on the Mtskheta study area, (ii) to make an estimation of the average shear-wave velocity of the upper 30 m ($V_{s,30}$), which enables to perform the evaluation of subsoil class affecting the study area (EC8 classification) and to obtain standard acceleration response spectra for each of them. Furthermore, we computed amplification factors from the technical code simplified approach in order to obtain an estimate of the weight of site effects compared to the one of the reference motion. This was also an opportunity to investigate Mtskheta site effects and estimate appropriate elastic design spectra and amplification factors based on different soil classes.

We also point out that it is the first time that these studies are carried out for a Georgian city. The results of this study represent a first step forward in terms of scientific and operational methods for seismic prevention in the Georgian territory.

19.2 Tectonic Setting and Seismicity

The tectonic and geological evolution of Georgia is largely driven by its position in the wide zone of the late Cenozoic collision of the Arabian and Eurasian plates, with the Anatolian micro-plate. This area, in the pre-collisional stage (Late Proterozoic – Early Cenozoic) was composed of a system of island arcs, intra-arc rifts and back-arc basins. During Oligocene-Middle Miocene times, the collision between the Anatolian-Arabian and Eurasian plates caused inversion of the topography, and back arc basins were replaced by the Greater and Lesser Caucasus fold-and-thrust belts, separated by a foreland basin. In the Georgian territory, a complex network of faults delimits distinct tectonic zones, linked to the collisional and post-collisional stages: the Greater Caucasus fold-and-thrust mountain belt, the Adjara-Trialeti Fold-and-Thrust belt (ATFT), the Georgian block, the Rioni and Kura-Kartli Forelands and the Artvin-Bolnisi block. The Dzirula High subdivides the foreland into two basins: the Rioni (R in Fig. 19.1a) foreland to the west and the Kura-Kartli (K in Fig. 19.1a) foreland to the east (Adamia et al. 2010; Tibaldi et al. 2017a, b; Tibaldi et al. 2018; Fig. 19.1). The about N-S directed compressional stress produced by the ongoing continent-continent converging Anatolian-Arabian and Eurasian plates (Tsereteli et al. 2016), led to the delineation of three distinctive principal trends of active faults: (i) WNW-ESE trending compressional structures (reverse faults, thrusts and strongly deformed fault-propagation folds with a lateral-dextral-component of motion), also called "Caucasian"; (ii) NE-SW striking, mainly transversal faults, mainly compressional structures with sometime considerable strike-slip components; (iii) tensional submeridional faults associated with the intensive Neogene-Quaternary volcanism in the Transcaucasian region (e.g., Pasquarè et al. 2011).

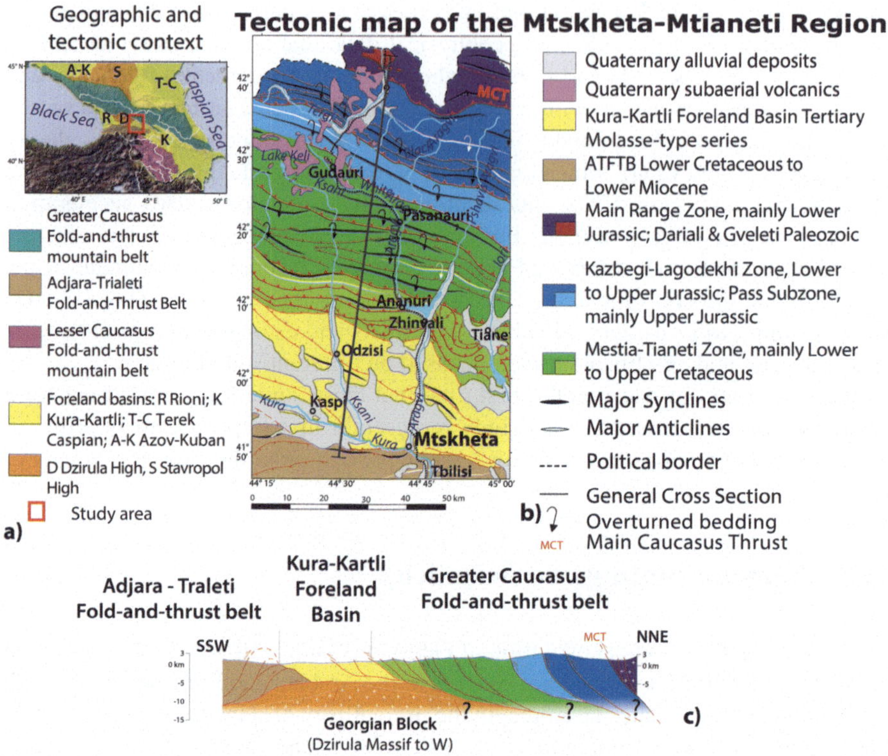

Fig. 19.1 (**a**) Geographic and tectonic setting of the Georgian territory; (**b**) Tectonic map of the Mtskheta-Mtianeti Region; on the right-hand side of the picture is provided the color legend of geological units and geological elements. (**c**) General cross section showing the main tectonic setting. (Modified after Mauvilly et al. 2015. Reference system: WGS84/geographic coordinates)

The territory of Georgia, being located in the Caucasus Region, is considered a highly seismically active zone, as confirmed by manifestation of both historical and instrumental strong earthquakes with epicenters located along the fault systems of the Greater and Lesser Caucasus and the intermountain depressions (Fig. 19.2), as well as by fault data analysis.

In detail, the city of Mtskheta (hereinafter referred as Mtskheta) is placed in the Kura-Kartli Foreland Basin, at the boundary with the northern margin of the ATFT. In Fig. 19.3 it is possible to see a schematic representation of the main faults affecting the study area, with a set of selected earthquakes belonging to the Georgian catalogue (http://www.enguriproject.unimib.it/index14c4.html?page_id=1874), each one associated with instrumental (in red) or historical (in blue) related magnitude information. In particular, the Surami Fault (SF in Fig. 19.3) represents the northern tectonic border of the ATFT. It is a southward dipping overthrust running ESE-WSW through the territory of Mtskheta (Fig. 19.2). Further active faults belonging to the ATNF affect the study area: the left-lateral, strike slip Bakuriani Fault (BF in Fig. 19.3) and the right-lateral, strike slip Tbilisi Fault (TF in Fig. 19.3). Finally, the faults belonging to the Kura-Kartli Foreland Basin also

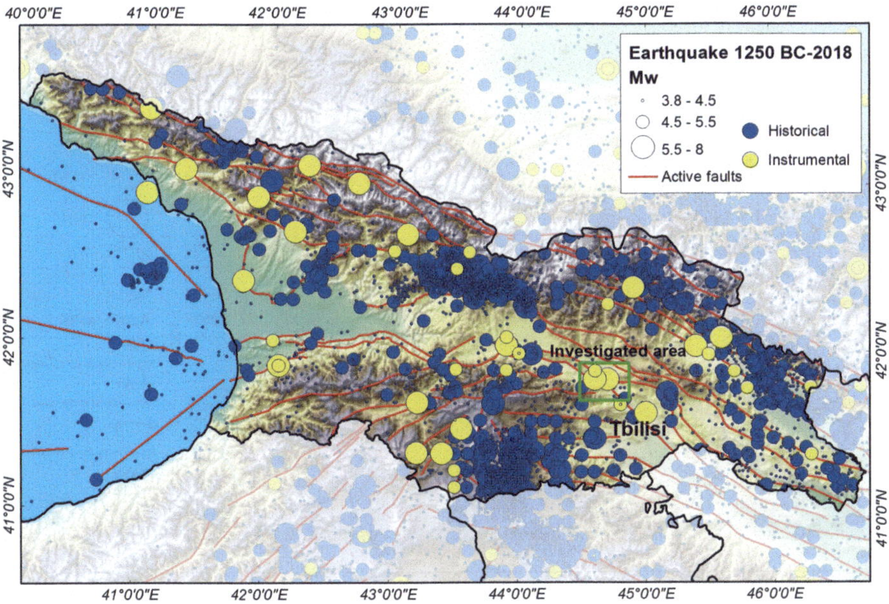

Fig. 19.2 Map of active faults (red line) in Georgia and the epicenters of historical and instrumental earthquakes until 2019. The green box represents the study area. Reference system: WGS84/geographic coordinates

contribute to the overall seismicity of the Mtskheta Region. In particular, the Eldari Fault (right-lateral strike slip with a reverse component; EF in Fig. 19.3) and the Kaspi Fault (northward dipping overthrust; KF in Fig. 19.3) (Adamia et al. 2008) are located roughly 10 km, respectively, north-east and north-west of the study area. It should be noted that all these active faults run within 20–30 km distance of Mtkheta. The western and eastern segments of the Surami Fault, with a length of 68 km and 41 km and a potential, associated magnitude of Mw 7.2 and 6.9, respectively (Wells and Coppersmith 1994), represent the major seismogenic fault segments of the Region. Seismogenic zones (SSzs) lying in the study area are represented in Fig. 19.3, according to which seismicity may be distributed in 2 main zones, each associated with a Max magnitude (Mmax) value. In detail, through geological and seismological methods, we estimated a Mmax = 6 and a Mmax = 7, respectively for the northern and southern SSz, corresponding to the light grey and dark grey colors in Fig. 19.3.

Adamia et al. (2008) identified the Surami Fault as the one associated with the 1275 Mtskheta earthquake, the strongest seismic event ever occurred in the study area. Written sources enable to assign to this event an Intensity (Io) 9 (MSK64 scale) and a Magnitude (M_W) 6.5, and to locate its epicentre in Mtskheta. This seems to be confirmed by two coseismic dislocation structures documented in the Natskhori valley, north of Mtskheta, by Kogoshvili (1970), who, on the basis of geological-sedimentological evidence, attributed them to an impulsive (like earthquake) event, such as the 1275 Mtskheta earthquake; in fact, the discovered structures were about

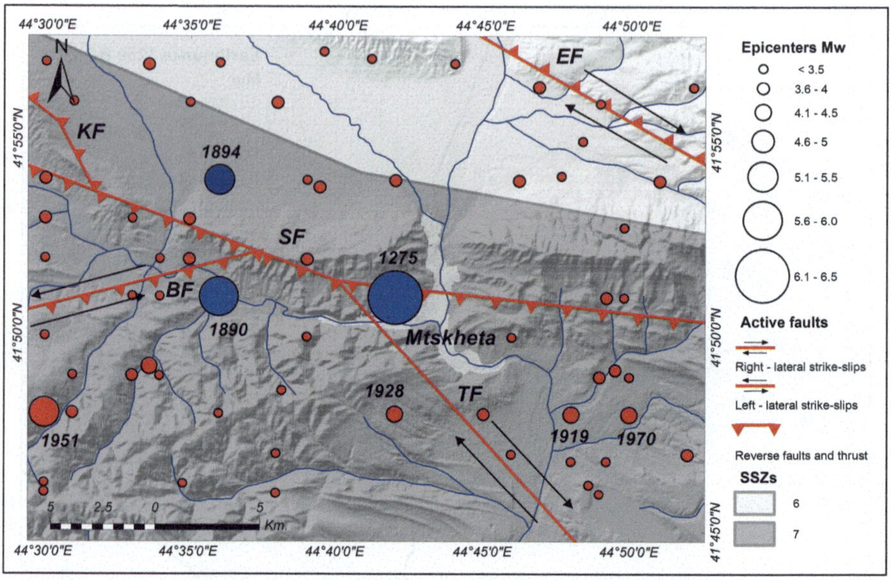

Fig. 19.3 Simplified active tectonic scheme and seismicity of the Mtskheta Region, representing an enlargement of the green box in Fig. 19.2. *KF* Kaspi Fault, *SF* Surami Fault, *BF* Bakuriani Fault, *TF* Tbilisi Fault, *EF* Eldari – Fault. Reference system: WGS84/geographic coordinates

the same age as this earthquake as well as a large seismogenic landslide, according to Varazanashvili et al. (2011) and (2018).

Historical data (Koçyiğit et al. 2001; Adamia and Gujabidze 2004; Adamia et al. 2008; Varazanashvili et al. 2011, 2018; Earthquake Catalogue of Georgia and Adjacent Areas 2012; Zare et al. 2014) provide information about two further strong earthquakes that struck the Mtskheta region before 1900: the 1890 ($M_W = 5.6$; I0 = 7) and the 1894 ($M_W = 5.4$; $I_0 = 6$) earthquakes. In regard to the instrumental period, major seismic events that took place in the study area are referred to the 1951 ($M_W = 5.2$), 1928 ($M_W = 4.4$), 1919 ($M_W = 4.2$), and 1970 ($M_W = 4.1$) seismic events, whose epicenters are shown in Fig. 19.3.

Based on the above, we conclude that Mtskheta is located in a moderately seismically active zone, whose seismicity can be also expressed taking into account the Gutenberg-Richter law. Indeed, we calculated that this area is characterized by a seismic productivity of 2.7 and b-value = 0.82.

19.3 Geological Setting

Mtskheta is located at the confluence of the Kura and Aragvi rivers (Fig. 19.4). Here, the compressional stress affecting the study area is well represented by the Mtskheta anticline, belonging to the ATFTB tectonic unit. In the study area, the

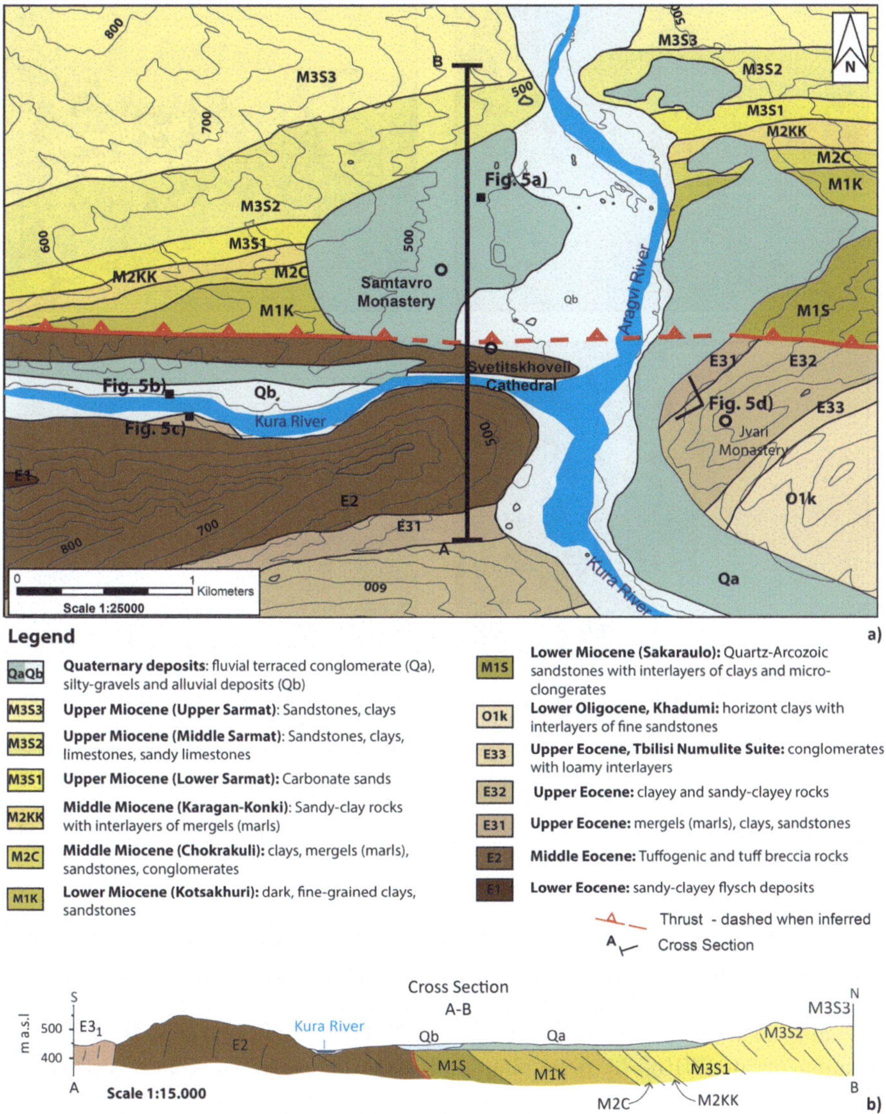

Fig. 19.4 (**a**) Geological map of the Mtskheta Region (approximately 20 km^2) with reported location of some important heritage sites of Mtskheta (black circles); (**b**) geological cross section traced from the south to the north. Data derived from Project Mtskheta Land Use Master Plan (LUMP) and provided by the Mtskheta City Hall. Reference system: WGS84/geographic coordinates

fold axis traces the latitudinal direction of the Kura river. The northern limb of the Mtskheta fold, complicated by additional folding (Gamkrelidze et al. 2019), consists of a Middle Eocene volcanic formation and the Upper Eocene clayey-sandy rocks (Fig. 19.4), the latter cropping out in the southernmost part of the study area (Fig. 19.5c).

Fig. 19.5 Field photographs see Fig. 19.4 for their location in map. Fluvial terraced conglomerates cropping out in Mtskheta, on the right bank of the Aragvi River (**a**) and on the left bank of the Kura River (**b**). Arrows indicate flow direction as deduced by tractional a(t)–b(i) imbrication of clasts (sensu Collinson et al. 2006). Photos taken respectively from the East and the Southwest (**c**) Late Eocene Sandstones cropping out on the right bank of the Kura River (photo taken from North); (**d**) Panoramic view of Mtskheta from East, showing the confluence of the Kura and Aragvi Rivers. On the right is indicated the location of the Svetiskhoveli Cathedral

In detail, the city is located within the intramontane depression of the Kura-Kartli Basin. Here, the sedimentary rocks consist of Early Cenozoic, mainly shallow water formations, and of Upper Cenozoic marine and continental molasses, mainly consisting of marls, sandstones-siltstones and clastic limestone flysch (Fig. 19.4) (Adamia et al. 2010). In particular, molasses represent the upper part of the stratigraphic column of the Basin and here are overthrusted by the Achara Trialeti fold and thrust belt units (Gamkrelidze 1964). Beyond the Mtskheta overthrust, on the right bank of the Aragvi River, it is possible to observe the Skhaltba ridge, consisting of a north-dipping monocline with the following exposed formations: Quartz-Arcozoic sandstones of the Sakaraulo Formation (Lower Miocene-Middle Miocene), sandstone and clays of the Kotsakhuri Formation (Lower Miocene); clays, marls, sandstones and conglomerates of the Chokrakuli Formation (Middle Miocene) Sandy-clay rocks with interlayers of marls of the Karagan-Konki Formation (Middle Miocene) carbonate sand and sandstone of the Lower and Middle

Sarmatian, continental sandstones and clays of the Upper Sarmatian. The Flysch and tertiary molasses represent the geological bedrock in the study area, partially covered by Quaternary alluvium deposits, mainly consisting of fluvial terraces conglomerates and silty-gravels/loams deposits (Figs. 19.5a–b). In Fig. 19.4 is provided the geological map of the Mtskheta Region, with the location of some representative pictures.

19.4 Dataset of Geophysical Measurements

In the study area, a considerable amount of geophysical investigations has been performed in the framework of the ongoing seismic microzonation activities.

In particular, we acquired ambient vibrations measurements within an area of approximately 3 km^2 which extends across the whole urban area of Mtskheta. Their relative location is shown in Fig. 19.6. The eighty recordings of about 30 minutes of ambient vibrations were collected by using a Tromino seismograph. At each point, we computed HVSR (Horizontal to Vertical Noise Spectral Ratio) curves, evaluating the ratio between the horizontal and the vertical component of ambient vibration measured at the same station (e.g., Nakamura 1989; Lunedei and Malischewsky 2015; Sánchez-Sesma 2017). HVSR curves can provide a reliable estimate of the resonant frequencies due to soft sediments (e.g., Bonnefoy-Claudet et al. 2006; Lunedei and Albarello 2010; Imposa et al. 2017; Panzera et al. 2018).

We computed HVSR as the geometric average of the horizontal to vertical spectral ratios for time windows selected from the records with duration of 25 s and no overlap between consecutive windows. We used the same parameters to compute HVSR for all our measurements, in order to guarantee consistent results throughout the dataset. Fourier amplitude spectra were smoothed before computing the spectral ratios using a triangular smoothing function f with a width equal to 5% the central frequency.

The geophysical dataset consists also of eight linear arrays of vertical 4.5 Hz geophones recording active data (Multichannel Analysis of Surface Waves, MASW). Their location is plotted in Fig. 19.6 with a blue line. The linear arrays were composed of 24 geophones with a regular spacing of 2 m. We used a hummer as active source performing two shots, respectively at 2-m offset from the first and last geophone. Surface-waves analyses was applied to the vertical signals recorded by the arrays. We applied frequency-wavenumber analysis (Park et al. 1999) through the Geopsy analysis tool (www.geopsy.org). These analyses enabled obtaining shear-wave velocity (Vs) profiles. Inversion procedure was performed using HV-Inv software (https://w3.ual.es/GruposInv/hv-inv/, Piña-Flores et al. 2017).

Fig. 19.6 Location of the acquired geophysical data. Location and label of the MASW measurement sites are indicated with a blue line, while ambient vibrations measurements sites are plotted with a red triangle. Reference system: WGS84/geographic coordinates

19.5 Results

From the HVSR curves, we obtained the identification of resonance frequencies (f_0 and f_1 hereafter) of the deposits characterizing the subsoil of the area (where f_0 is the fundamental frequency and f_1 is a resonance frequency with a frequency value higher than f_0). The presence of broadband peaks that would lead to classify f_0 and f_1 as unclear peaks in some cases according to SESAME criteria (Bard 2002) obliged us to manually pick the HVSR curves assuming the following criteria: the peak is recognized only when HVSR curves present values higher than 2. In Fig. 19.7, the manually identified f_0 and f_1 frequencies observed in HVSR curves are plotted. In particular, the stations along the river show high frequency effect, probably due to the presence of very shallow layers. f_0 frequency of values lower than 3.5 Hz are distinguishable on the eastern portion of the investigated area, around the historic Svetitskhoveli Cathedral, while lower peaks amplitude and no peak can be recognized towards the N.

The peak frequencies were analysed also in terms of recurrence, as shown in the histogram of Fig. 19.8: the peak frequencies are mostly in the 1.5–6 Hz interval, while f_1 frequencies have been observed only at seven sites.

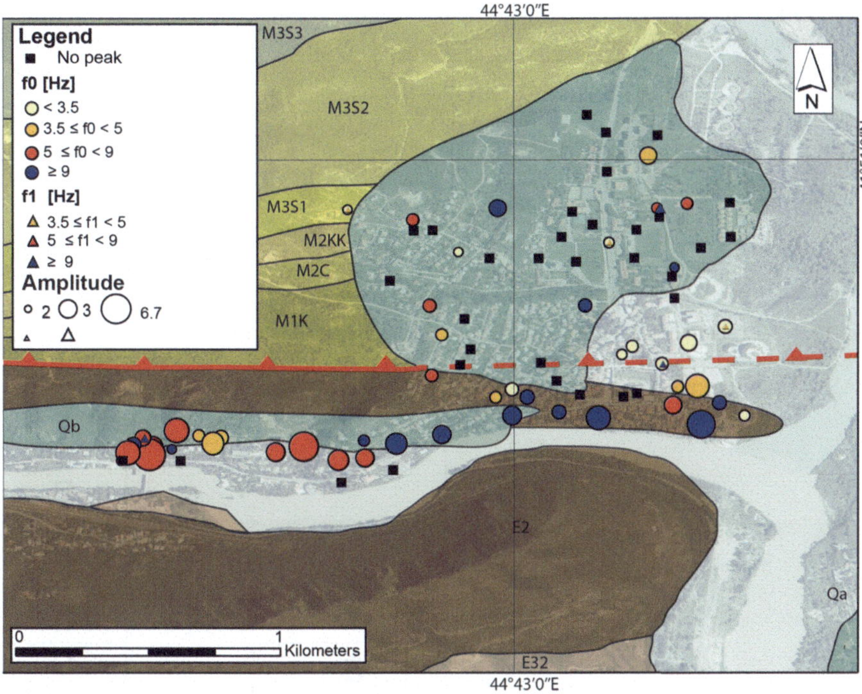

Fig. 19.7 f_0 and f_1 natural frequencies map of the area. Circles and squares size is proportional to amplitude between 2 and 6.7 and 2 and 3, respectively. Codes of geological units refer to the geological legend reported in Fig. 19.4. The orthophoto of Mtskheta has been used as background. Reference system: WGS84/geographic coordinates

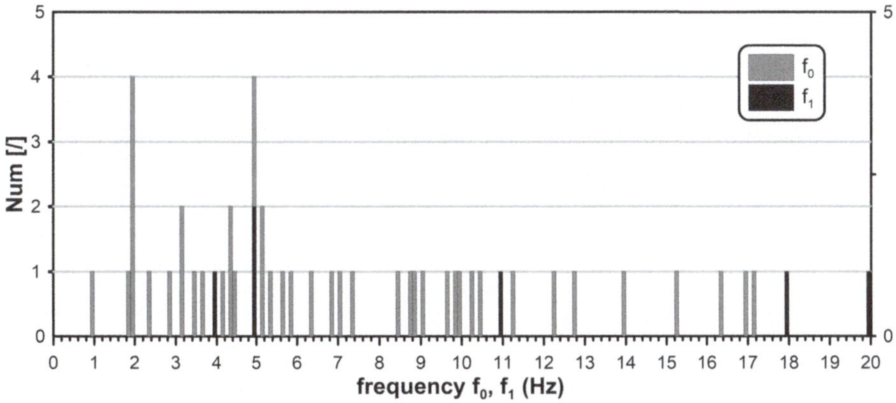

Fig. 19.8 Distribution of f_0 and f_1 values on the 47 noise measurements. Thirty-three noise measurements which showed no peaks are excluded from this plot

Table 19.1 Estimated value of $V_{S,30}$, the resonance depth and the corresponding EC8 classes

Name	$V_{s,30}$ (m/s)	depth to bedrock *H* (m)	EC8 class
MASW_A	360	19	E
MASW_B	404	22	B (?)
MASW_C	290	25	E (?)
MASW_D	391	21	B (?)
MASW_F	767	3	A
MASW_G	420	20	B (?)
MASW_H	410	18	B (?)
MASW_I	498	19	B(?)

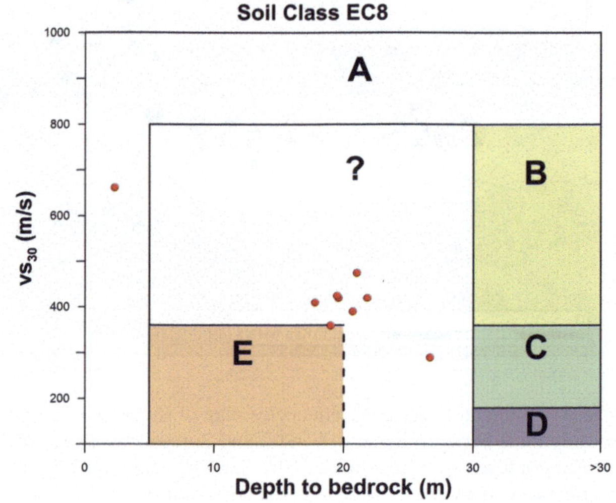

Fig. 19.9 Distribution map of EC8 classes based on the MASW measurements

Estimated $V_{S,30}$ obtained from MASW measurements varies from 290 to 767 m/s, while depth to bedrock (*H*, hereafter) varies from 3 to 25 m, highlighting the presence of very shallow deposits in the area. H progressively increases going from W to E and shows a minimum in correspondence with the MASW_F (Table 19.1).

We classified the V_s profiles according to the EC8 regulation (Table 19.1).

According to Bouckovalas et al. (2006), the definition of EC8 ground types is not complete, as sites with $V_{S,30} > 360$ m/s and $H = 5 - 30$ m, as well as sites with $V_{S,30} < 360$ m/s and $H = 20 - 30$ m cannot be classified as any ground type. In fact, most of the investigated sites belong to the gap zone of EC8 classification (Fig. 19.9).

Recently, a new classification approach has been adopted by the Italian regulation (NTC18 2018). The NTC2018 introduced the equivalent shear wave velocity $V_{S,H}$. This latter is computed between 0 (ground surface) and H (depth of the seismic bedrock); if H > 30 m the procedure is limited to the upper 30 m (i.e. H = 30 m) thus obtaining Vs,30. For the sake of completeness, this reclassification is shown in Table 19.2.

Table 19.2 $V_{S,H}$ and corresponding soil class according to the NTC18 technical code classification

Name	$V_{S,H}$ (m/s)	NTC18 class
MASW_A	267	E
MASW_B	343	E
MASW_C	350	E
MASW_D	391	B
MASW_F	372	A
MASW_G	322	E
MASW_H	364	B
MASW_I	409	B

Both regulation classifications showed that investigated sites may belong to A, B or E soil classes.

These results represent a valuable starting point in the Seismic Microzonation (hereafter, SM, Working Group ICMS 2008) perspective. In this regard, it is worth underscoring that SM studies are a fundamental base in the forecasting of the ground modifications induced by local conditions. SM studies make it possible to rationalize the knowledge of modification of ground motion, by recognizing and characterizing, at sufficiently large scale (i.e. municipality or sub-municipalities) stable areas, stable areas susceptible to local amplification and areas subject to instability (e.g. landslides, surface fractures, soil liquefaction).

As a consequence, they can be considered a very relevant "tool" able to accomplish the purpose of the site-specific hazard assessment.

In particular, according to the Guidelines for SM (Working Group ICMS2008), SM studies are divided into three levels (1, 2 and 3) of in-depth analysis. Level 1 summarizes and represents on thematic maps (hereafter, SM maps) the areas defined as Seismically Homogeneous Microzones (SHM), which are alternatively:

- stable zones where the seismic ground motion is not modified with respect to the expected one in ideal conditions of flat and rigid rock (not affected by fractures and by significant alteration);
- stable zones prone to amplifications, where the seismic motion is modified due to the geological/ geophysical/geotechnical and morphological characteristics of the territory;
- zones prone to instability, where earthquake-related phenomena (i.e., landslides, soil liquefaction, superficial faulting and differential ground subsidence or permanent instabilities) can be triggered by an earthquake and create morphological steps.

Levels 2 and 3 allow to associate to the stable zones prone to amplification defined in the SHM maps: (1) the values of the amplification factor (AF) and (2) response spectra (only for the level (3). Through the acquisition of new data, level 2/3 may revise the SHM maps of the level 1, eventually characterizing permanent instabilities through specific parameters.

Several outputs can be obtained in the aforementioned level 3, such as:

- maps of Amplification Factors (AF) in chosen vibration period ranges of engineering interest;
- transfer functions from 1D and 2D numerical modelling;
- acceleration response spectra.

AFs are the integrals of the ratio between the elastic response spectrum at the study site and the integrals of the acceleration response spectra at the reference site in selected period ranges, whose choice has to be made taking into account the typical fundamental periods of buildings in studied site.

In this study, we considered three different ranges of period (T_n), following those proposed in Pergalani et al. (2019): $0.1 \leq T_1 \leq 0.5$ s; $0.4 \leq T_2 \leq 0.8$ s and $0.7 \leq T_3 \leq 1.1$ s. These intervals are closely linked to the building characteristics, essentially the number of floors (Pergalani et al. 2019): about one to three floors for the first interval; three to six floors for the second interval and six to nine floors for the third interval.

Preliminarily and before performing advanced site response analyses by means of numerical modelling (which would require more refined data, e.g., boreholes, laboratory tests), a simplified approach based on the recognition of reference ground type conditions may be used. In fact, Derras et al. (2017) showed that Vs_{30} is an adequate proxy where resonance frequencies at the sites are higher than 1.67 Hz (periods lower than 0.6 s). According to analysis of Fig. 19.8, this is true for the majority of the investigated sites.

In order to have a preliminary estimate of the amplification expected at the sites, we computed AFs for the B and E EC8 soil class spectra in the following section.

19.6 Response Spectra and Amplification Factors with the Conventional $V_{s,30}$ Site Proxy

We computed the reference spectrum (or target spectrum), referring to the EC8 code and, in detail, we used the anchoring value of 0.41 g which is related to the PGA with a 10% exceedance probability in 50 years on a rock site according to the procedure shown hereafter.

At first, a Probabilistic Seismic Hazard Assessment (PSHA) for Mtskheta was evaluated. It followed the international standards developed in the GEM -EMME (Global Earthquake Model - Earthquake Model for Middle East Region) Project, aimed at developing a homogeneous seismic hazard model across the Middle East and Caucasus Regions. Briefly, by using the Open Quake software (Pagani et al. 2014), we calculated the exceedance probabilities (POEs) of the specific ground motion levels for PGA and pseudo-accelerations (SA) spectra with 5% critical damping within 50 years in Mtskheta for the reference rock only ($Vs_{30} = 800$ m/s). Seismogenic sources were presented by area source and active faults combined

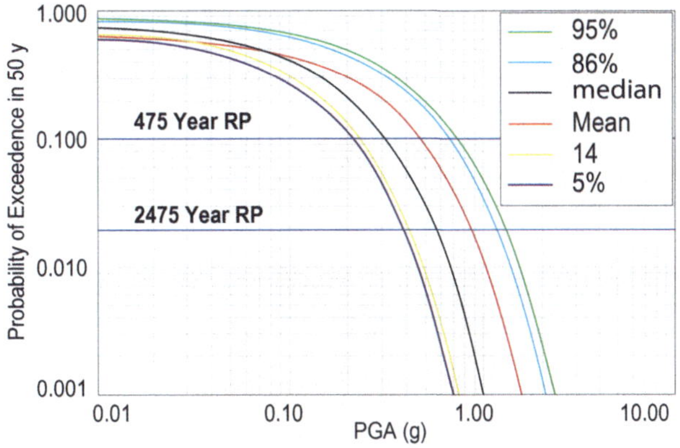

Fig. 19.10 Mean and 5, 14, 50, 84 and 95% quantile PGA hazard curves for Mtskheta

with background seismicity. Area sources are presented as active shallow crust sources, volcanic sources for the highland of Javakheti and subduction sources in slab (northeastern Caucasus, Mumladze et al. 2015). For ground motion prediction equation (GMPE), the following models were selected: Akkar et al. (2014) and Kotha et al. (2016) as pan-regional models, Chiou and Youngs (2014) an NGA West2 model, Cauzzi et al. (2014), a global model which includes Japanese data too. Kotha et al. (2016) and Chiou and Youngs (2014) are the lower and the upper bound respectively and a 0.15 weight is assigned to them; the central part of the ground motion distribution is captured by Akkar et al. (2014) and Cauzzi et al. (2014), with equal weights of 0.35. Two ground motion models, i.e. Bradley (2013) and Faccioli et al. (2010), are selected and used for volcanic sources with equal weights of 0.50. For deep seismicity sources we used Lin and Lee (2008), Abrahamson et al. (2016), Montalva et al. (2017). PSHA was obtained from a set of realizations, in accordance with the four main levels of logic tree branches: seismic sources models, recurrence parameter - b value; Mmax and GMPEs (see also Tsereteli et al. 2020 "The 2020 National Seismic Hazard Model for Georgia Sakartvelo", this volume).

Seismic hazard estimates are shown for the central part of Mtskheta. The value of the mean PGA is 0.41 g with a 10% of exceedance in 50 years. This high value is mainly linked to the SF active fault, being characterized by high slip rates. Figure 19.10 presents the hazard curves in terms of PGA [g] corresponding to mean and 5, 14, 50, 84 and 95% quantile computations for Mtskheta. Such estimations provide information about the uncertainty ranges associated with the ground motion estimations at all return periods.

In order to identify the main sources contributing to the hazard at the study site we performed a disaggregation probabilistic hazard analysis.

This involves the determination of potential earthquake variables, principally magnitude, distance and epsilon (number of standard deviations from which the recorded seismic intensity logarithm deviates from the expected average value according to a specific propagation model or GMPE) for each end branch of the logic tree structure. Figure 19.11 shows the results of seismic hazard disaggregation

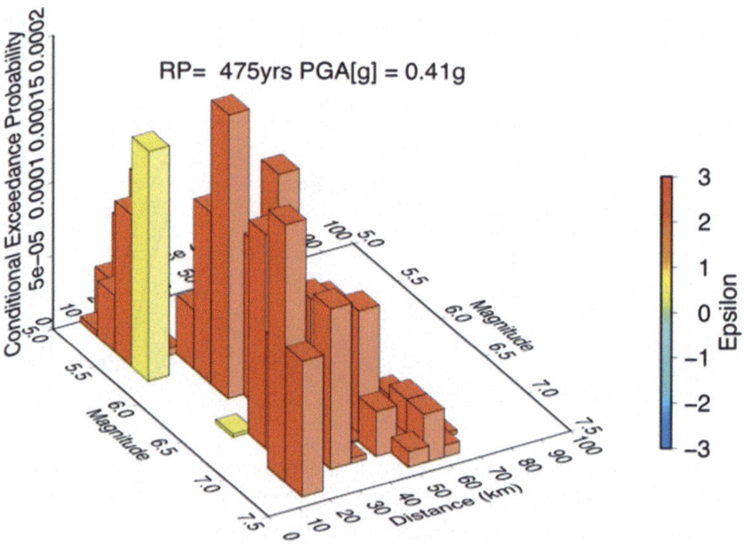

Fig. 19.11 Disaggregation plots for the 475-year Return Period (RP) obtained for Mtskheta

computed for the study site by considering a 475-year return period (RP) and thus, defines seismic events that contribute to the selected seismic hazard level (McGuire 1995; Bazzurro and Cornell 1999).

We observed that for sources at relatively close distances (0–30 km), the level of PGA (0.41 g) for the 475-year RP dominates, while it seems that a minor contribution is given by seismic sources located at medium distances (up to 70 km).

We then defined also the 475-year RP EC8 response spectra (type I for surface wave magnitude, Ms > 5.5) from the simplified approach for soil Classes A, B and E (the comparison among them is shown in Fig. 19.12).

The AF values AFs, as defined in the previous section, are analysed. They are obtained as the ratio between the integral of the acceleration response spectra derived from EC8 for the two soil classes B and E and for a soil class A with reference to the three selected ranges of periods. Topographic amplification factors have been not considered because slopes above 15° do not fall within the area of investigation. It may be underlined that AFs values computed starting from NTC18 spectra should not significantly differ from the AFs obtained from EC8 provisions, being NTC18 based on the same type 1 EC8 spectra, if computed for similar anchoring value (i.e., ag or peak ground acceleration, PGA, on stiff soil equal to 0.41 g). Moreover, NTC2018 provisions delete the ambiguity about the soil classes.

For the two above reasons, once solved the ambiguity for the classes which do not belong to B or E EC8 soil classes, a preliminary distribution of AFs can be generated (Fig. 19.13).

The behaviour of the AFs is shown in Fig. 19.13. AF values decrease from the T_1 interval (where AF values is equal to 1.2 for B/A and to 1.4 E/A, respectively)

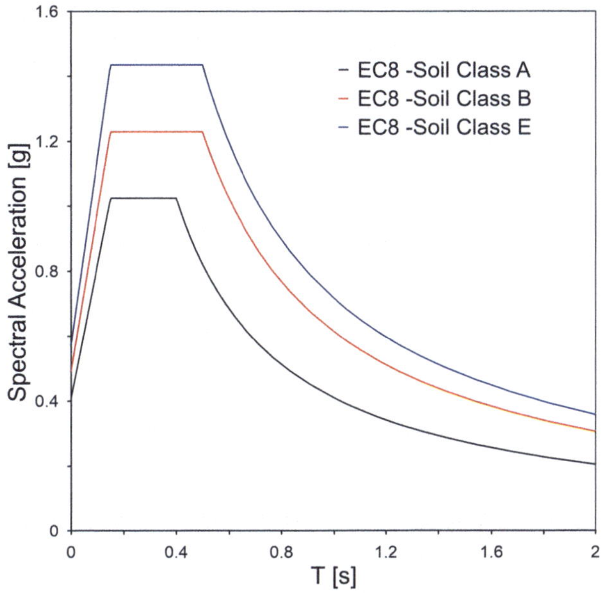

Fig. 19.12 Comparison among 475 years RPEC8 response spectra (type 1, Ms > 5.5) from simplified approach for soil Classes A, B and E

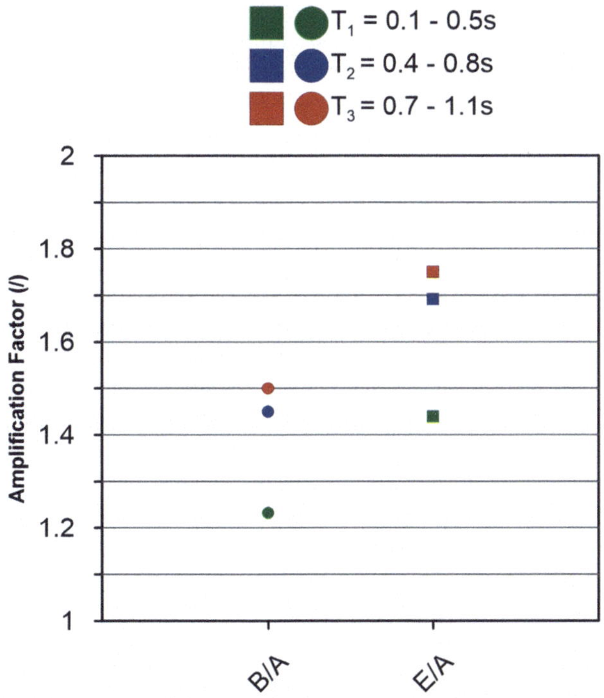

Fig. 19.13 AF distribution derived from a simplified approach for a 475-years return period EC8 standard spectra for soil categories B/A and E/A

to the T2 and T3 interval (AF values equal to 1.45 and 1.5 respectively for B/A and to 1.7 and 1.75 respectively for E/A). In light of the obtained AFs, it is clear that buildings constructed on sites belonging to an E soil class may expect to experience greater ground motions if compared to the same buildings constructed on B sites.

19.7 Conclusions

We performed a first screening of the resonance frequencies f_0 read in the HVSR curves. It showed that the most significant HVSR variability is located around the Svetitskhoveli Cathedral (in the eastern part of the investigated area) as well as the presence of geological heterogeneities in the subsoil. Most noise measurements recorded no peak in the HVSR curves, while the distribution of the curves characterized by at least one peak, showed that f_0 and f_1 are concentrated at frequencies below 6 Hz. Despite amplitudes in these cases are not significant, they enable us to conclude that, in case of a coexistence between the occurrence of high resonance frequencies (3–6 Hz) and a peculiar building typology, mainly represented by old, highly vulnerable, two- to four-storey high masonry buildings, the expected damages might be particularly enhanced for those sites.

The lowest f_0 peaks highlight the part of the city where the deposits are expected to be thicker, in agreement with the geological features described in the geological map. This trend is confirmed by the MASW measurements results.

A simplified approach based on the $V_{S,30}$ was applied in this manuscript. Although it has to be considered only preliminary, as it is not highly accurate in catching the most important features of the expected ground motion at the site (Gallipoli and Mucciarelli 2009; Luzi et al. 2011), it has allowed to preliminary estimate AFs. These values, computing amplification factors in determined period intervals (which have a correlation with buildings characteristics), confirmed that ad hoc site response analyses have to be strongly encouraged to define the seismic action. Moreover, these studies are even more relevant in cities like Mtskheta, which is the former capital of Georgia and hosts approximately 7700 residents, with increasing level of urbanization, population density and infrastructures.

Mtskheta is also known as the Georgian City-Museum. Its historic monuments have been listed among UNESCO World Heritage Sites since 1994 and are excellent examples of medieval religious architecture in the Caucasus, representing the high artistic and cultural level attained by this Region. The presence of such an outstanding cultural heritage highlights the need to foster further studies aimed at reducing seismic risk at Mtskheta.

Acknowledgments Authors are grateful to the Shota Rustaveli National Science Foundation, for making the geophysical investigation possible. All the activities were carried out in the framework of the National Project 216758: "Evaluation of seismic hazard for Georgia and seismic risk for city Mtskheta with modern approaches".

We are extremely thankful to the Guest Editors and the two anonymous reviewers for their precious suggestions, which have substantially improved the quality of the present paper.

References

Abrahamson N, Nicholas G, Kofi A (2016) BC hydro ground motion prediction equations for subduction earthquakes. Earthquake Spectra 32(1):23–44

Adamia S, Gujabidze G (2004) Geological Map of Georgia 1:500,000 (on the basis of 1:200,000 and 1:50,000 scale State Geological Maps of Georgia). Tbilisi, Georgia, Department of Geology, Noida Institute of Geophysics.. http://www.ig-geophysics.ge/Georgia_Geology_1.jpg. Accessed 12 Jan 2017

Adamia S, Mumladze N, Sadradze N et al (2008) Late Cenozoic tectonics and geodynamics of Georgia (SW Caucasus). Georgian Int J Sci Technol 1:77–107

Adamia Sh, Alania V, Chabukiani A et al (2010) Evolution of the Late Cenozoic basins of Georgia (SW Caucasus): a review. Geological Society, London, Special Publications, 340, 239–259, 1 January 2010., https://doi.org/10.1144/SP340.11

Akkar S, Sandikkaya MA, Bommer JJ (2014) Empirical ground-motion models for point- and extended-source crustal earthquake scenarios in Europe and the Middle East. Bull Earthq Eng 12(1):359–387

Bard P-Y (2002) Extracting information from ambient seismic noise: the SESAME project (Site EffectS assessment using AMbient excitations). European project EVG1-CT-2000-00026 SESAME

Bazzurro P, Cornell CA (1999) Deaggregation of seismic hazard. Bull Seismol Soc Am 89:501–520

Bonnefoy-Claudet S, Cornou C, Bard P-Y et al (2006) H/V ratio: a tool for site effects evaluation. Results from 1-D noise simulations. Geophys J Int 167:827–837. https://doi.org/10.1111/j.1365-246X.2006.03154.x

Bouckovalas GD, Papadimitriou AG, Karamitros D (2006) Compatibility of EC-8 GroundTypes and site effects with 1-D wave propagation theory. In: Bouckovalas (ed) Proceedings of ISSMGE ETC-12 workshop, NTUAAthens, Greece

Bradley BA (2013) A summary of the observed ground motions from the 21 July 2013 Mw 6.6 earthquake in Cook Strait, off the coast of Seddon. https://sites.google.com/site/brendonabradley/

Brambati A, Faccioli E, Carulli GB et al (1980) Studio di microzonizzazione sismica dell'area di Tarcento (Friuli).. Regione Autonoma Friuli Venezia-Giulia

Cauzzi C, Faccioli E, Vanini M et al (2014) Updated predictive equations for broadband (0.01-10 s) horizontal response spectra and peak ground motions, based on a global dataset of digital acceleration records. Bull Earthq Eng 13(6):1587–1612

Chiou BSJ, Youngs RR (2014) Update of the Chiou and Youngs NGA model for the average horizontal component of peak ground motion and response spectra. Earthquake Spectra 30(3):1117–1153

Collinson J, Mountney N, Thompson D (2006) Sedimentary structures, 3rd edn. Terra Publishing, Hertfordshire, England, p 292

Derras B, Bard PY, Cotton F (2017) VS30, slope, H800 and f0: performance of various site-condition proxies in reducing ground-motion aleatory variability and predicting nonlinear site response. Earth Planets Space 69:133. https://doi.org/10.1186/s4062-3-017-0718-z

Eurocode 8-EN 1998-1 (1998) General rules, seismic actions and rules for buildings. CEN European Committee for Standardization, Bruxelles

Faccioli E, Bianchini A, Villani M (2010) New ground motion prediction equations for t>1 s and their influence on seismic hazard assessment. In: Proceedings of the University of Tokyo

Symposium on Long-Period Ground Motion and Urban Disaster Mitigation, Tokyo, Japan, 17–18 March 2010
Gallipoli MR, Mucciarelli M (2009) Comparison of site classification from VS30, VS10, and HVSR in Italy. Bull Seismol Soc Am 99:340–351
Gamkrelidze P (1964) Geology of the USSR, vol. X, Georgian SSR. Nedra, Moscow, pp 1–655. (In Russian)
Gamkrelidze P, Okrostsvaridze A, Maisadze F et al (2019) Main features of geological structure and geotourism potential of Georgia, the Caucasus. Modern Environ Sci Eng 5(5):422–442
Imposa S, Panzera F, Grassi S et al (2017) Geophysical and Geologic surveys of the areas struck by the August 26th 2016 Central Italy earthquake: the study case of Pretare and Piedilama. J Appl Geophys 145:17–27. https://doi.org/10.1016/j.jappgeo.2017.07.016
Koçyiğit A, Yilmaz A, Adamia S et al (2001) Neotectonics of East Anatolian plateau (Turkey) and Lesser Caucasus: implication for transition from thrusting to strike-slip faulting. Geodynamica Acta 14:177–195. https://doi.org/10.1080/09853111.2001.11432443
Kogoshvili LV (1970) Live tectonics of Georgia and its impact on the relief. Metsniereba Publ House, Tbilisi, p 220. (in Russian)
Kotha SR, Bindi D, Cotton F (2016) Partially non-ergodic region specific GMPE for Europe and Middle-East. Bull Earthq Eng 4(4):1245–1263
Lin P-S, Lee C-T (2008) Ground-motion attenuation relationships for subduction-zone earthquakes in northeastern Taiwan. Bull Seismol Soc Am 98(1):220–240. https://doi.org/10.1785/0120060002
Lunedei E, Albarello D (2010) Theoretical HVSR curves from full wavefield modelling of ambient vibrations in a weakly dissipative layered Earth. Geophys J Int 181:1093–1108
Lunedei E, Malischewsky P (2015) A review and some new issues on the theory of the H/V technique for ambient vibrations. In: Perspectives on European Earthquake Engineering and Seismology, vol 2, pp 371–394. https://doi.org/10.1007/978-3-319-16964-4_15
Luzi L, Puglia R, Pacor F et al (2011) Proposal for a soil classification based on parameters alternative or complementary to Vs, 30. Bull Earthq Eng 9:1877–1898
Mauvilly J, Koiava K, Gamkrelidze I et al (2015) Tectonics in the Greater Caucasus: a north-south section along the Georgian Military Road. Presented at the 13th Swiss Geoscience Meeting, Basel, Switzerland
McGuire RK (1995) Computations of seismic hazard. Ann Geofis 36:181–200
Montalva GA, Bastías N, Rodriguez-Marek A (2017) Ground-motion prediction equation for the Chilean subduction zone. Bull Seismol Soc Am 107(2):901–911. https://doi.org/10.1785/0120160221
Mori F, Gaudiosi I, Tarquini E et al (2019) HSM: a synthetic damage-constrained seismic hazard parameter. Bull Earthq Eng 1-24 (on line-first)
Moscatelli M, Pagliaroli A, Cavinato GP et al (2014) Seismic microzonation of Palatine hill, Roman Forum and Coliseum Archeological Area. Bull Earthq Eng 12:1269–1275
Mumladze T, Forte AM, Cowgill ES et al (2015) Subducted, detached, and torn slabs beneath the Greater Caucasus. GeoResJ 5:36–46. https://doi.org/10.1016/j.grj.2014.09.004
Nakamura Y (1989) A method for dynamic characteristics estimation of subsurface using microtremor on the ground surface. QR Railway Tech Res Inst 30:25–33
NTC (2018) Norme Tecniche per le Costruzioni. Decreto Ministero Infrastrutture, Italy. GU Serie Generale n 42 del 20-02-2018 - Suppl. Ordinario n 8
Pagani M, Monelli D, Weatherill G et al (2014) OpenQuake-engine: an openhazard (and risk) software for the global earthquake model. Seismol Res Lett 85:692–702. https://doi.org/10.1785/0220130087
Panzera F, Lombardo G, Imposa S et al (2018) Correlation between earthquake damage and seismic site effects: the study case of Lentini and Carlentini, Italy. Eng Geol 240:149–162. https://doi.org/10.1016/j.enggeo.2018.04.014
Park CB, Miller RD, Xia J (1999) Multichannel analysis of surface waves. Geophysics 64:800–808

Pasquarè F, Tormey D, Vezzoli L et al (2011) Mitigating the consequences of extreme events on strategic facilities: evaluation of volcanic and seismic risk affecting the Caspian oil and gas pipelines in the Republic of Georgia. J Environ Manag 92(7):1774–1782

Pergalani F, Pagliaroli A, Bourdeau C et al (2019) Seismic microzoning map: approaches, results and applications after the 2016-2017 Central Italy seismic sequence. Bull Earthq Eng. https://doi.org/10.1007/s10518-019-00640-1

Piña-Flores J, Perton M, García-Jerez A et al (2017) The inversion of spectral ratio H/V in a layered system using the diffuse field assumption (DFA). Geophys J Int 208(1):577–588. https://doi.org/10.1093/gji/ggw416

Sánchez-Sesma FJ (2017) Modeling and inversion of the microtremor H/V spectral ratio: physical basis behind the diffuse field approach. Earth Planets Space 69(1):1–9. https://doi.org/10.1186/s40623-017-0667-6

Singh SK, Lermo J, Dominguez T et al (1988) The Mexico earthquake of September 19, 1985–a study of amplification of seismic waves in the Valley of Mexico with respect to a hill zone site. Earthquake Spectra 4:653–673

Tibaldi A, Russo E, Bonali FL et al (2017a) 3-D anatomy of an active fault-propagation fold: a multidisciplinary case study from Tsaishi, western Caucasus (Georgia). Tectonophysics 717:253–269

Tibaldi A, Alania V, Bonali FL et al (2017b) Active inversion tectonics, simple shear folding and back-thrusting at Rioni Basin, Georgia. J Struct Geol 96:35–53

Tibaldi A, Bonali FL, Russo E et al (2018) Structural development and stress evolution of an arcuate fold-and-thrust system, southwestern Greater Caucasus, Republic of Georgia. J Asian Earth Sci 156:226–245

Tsereteli N, Tibaldi A, Alania V et al (2016) Active tectonics of central-western Caucasus, Georgia. Tectonophysics 691:328–344

Tsereteli N, Danciu L, Varazanashvili O et al (2020) The 2020 National Seismic Hazard Model for Georgia (Sakartvelo). F. L. Bonali et al. (eds.), *Building Knowledge for Geohazard Assessment and Management in the Caucasus and other Orogenic Regions*, NATO Science for Peace and Security Series C: Environmental Security, Chapter 8. https://doi.org/10.1007/978-94-024-2046-3_8

Varazanashvili O, Tsereteli N, Tsereteli E (2011) Historical earthquakes in Georgia (up to 1900): source analysis and catalogue compilation, Monograph. Pub Hause MVP, Tbilisi, p 81

Varazanashvili O, Tsereteli N, Bonali FL et al (2018) GeoInt: the first macroseismic intensity database for the Republic of Georgia. J Seismol 22(3):625–667

Von Thun JL, Rochim LH, Scott GA et al (1988) Earthquake ground motion for design and analysis of dams. In: Earthquake engineering and soil dynamics II – recent advance in ground-motion evaluation, geotechnical special publication 20. ASCE, New York, p 463

Wells DL, Coppersmith KJ (1994) New empirical relationships among magnitude, rupture length, rupture width, rupture area, and surface displacement. Bull Seismol Soc Am 84(4):974–1002

Working Group ICMS (2008) Guidelines for seismic microzonation. Conference of regions and autonomous provinces of Italy – Civil Protection Department. English version available at http://www.protezionecivile.gov.it/httpdocs/cms/attach_extra/GuidelinesForSeismicMicrozonation.pdf?

Zare M, Amini H, Yazdi P et al (2014) Recent developments of the Middle East catalogue. J Seismol 18(4):749–772. https://doi.org/10.1007/s10950-014-9444-1

Chapter 20
Rheological Properties of Soils in Assessing the Seismic Hazard of the South Ukrainian Nuclear Power Plant

Oleksandr Kendzera, Jeremiah Rushchitsky, and Yuliia Semenova

Abstract Rheological properties of soils influence the surface response of strata to seismic ground motion. This paper presents modelling of site amplification in the South Ukrainian nuclear power plant. The territory of the South Ukrainian nuclear power plant was divided into 3 zones according to the results of seismic microzonation. For each zone, the resulting frequency response and a set of design accelerograms were separately calculated. The effects of the rheological properties of the soil on site amplification and resonance frequency are analyzed for each zone. In the conditions of the site of the South Ukrainian nuclear power plant, the roof of the bedrock changes quite sharply. This leads to a shift of the resonance maxima in a wide range of frequencies. The study of the variability of the parameters of ground motion during earthquakes depending on the rheological properties of the rocks and the structure of the surface part of the section of the site is an essential element in earthquake-resistant design, reconstruction and extension of the life of buildings and structures.

Keywords Site amplification · Frequency response · Resonance frequency · Ground response analysis · Design accelerogram

20.1 Introduction

Seismic hazard parameters at a site not only depend on the earthquake's magnitude and the distance from the origin, but also on the topography and subsurface geology.

O. Kendzera (✉) · Y. Semenova
Subbotin Institute of Geophysics, Kyiv, Ukraine
e-mail: skendzera@gmail.com

J. Rushchitsky
SP Timoshenko Institute of Mechanics, Kyiv, Ukraine

© Springer Nature B.V. 2021
F. L. Bonali et al. (eds.), *Building Knowledge for Geohazard Assessment and Management in the Caucasus and other Orogenic Regions*, NATO Science for Peace and Security Series C: Environmental Security,
https://doi.org/10.1007/978-94-024-2046-3_20

Site effect studies are extremely important, for which the geodynamical characterization of the subsite's layers is required. Damage due to earthquakes generally decreases as a function of distance from the epicenter. However, as a consequence of destructive earthquakes it has been observed that damage is often more severe on unconsolidated deposits than rocks.

The 1985 Mexico City earthquake was triggered about 400 km away from the city, but ground motion was greatly amplified by the underlying soft sediments. In the city itself, seismic motion was amplified up to a factor of 60 as compared to the bedrock, because of the soft clay deposit underneath the urban center. Thus, surface geology is one of the dominant factors that affect the amplitude, duration, and frequency content of strong ground-motion. The earthquake engineering community has generally accepted that soft soils can play a key role in ground motion, and therefore need to be included in seismic zoning.

All buildings have their own, natural resonance frequencies, and the subsoil has different resonance frequencies as well.

The generally hard bedrock has higher eigen frequencies then softer sediments. If the period of ground motion matches the eigen resonance of a building, this will undergo the possible largest oscillations and can suffer the greatest damage.

The frequencies at which soil oscillations can be amplified are determined by the frequency response of soils. Analytically, the frequency response is the ratio of the spectrum of oscillations on the surface to the spectrum of oscillations on bedrock. But this problem is incorrect according to Hadamard.

The registration of strong earthquakes at a construction site is the most accurate method for determining the frequency response. But in conditions of weak seismicity of the territory, this method cannot be used. Other methods that recommend building codes in Ukraine do not take non-linear effects into account.

During seismic loading, nonlinear dynamic soil behavior can occur. This nonlinear soil behavior has to be incorporated in site response analysis.

20.2 Ground Response Analysis

There are three basic approaches to carrying out ground response analysis (Kramer 1996). One is known as linear ground response analysis; another is called non-linear ground response analysis depending on what type of behavior of the soil material we are considering for an earthquake load. When we are considering the shear stress versus shear strain relationship of the linear material, we call it linear ground response analysis. When we are considering the relationship as non-linear, we call it non-linear ground response analysis.

Within the non-linear analysis, two cases may apply: one is the equivalent linear, the other is the absolutely non-linear.

In an equivalent linear approach, soil behavior is approximated as a Kelvin-Voigt solid. The shear stress-shear strain relationship is expressed as:

$$\tau = G\gamma + \eta \frac{\partial \gamma}{\partial t},$$

where τ – shear stress, G – shear modulus, γ – shear strain and η – viscosity, t – time.

Then the equation of motion for a viscoelastic medium has the next form:

$$\rho \frac{\partial^2 u}{\partial t^2} = G \frac{\partial^2 u}{\partial z^2} + \eta \frac{\partial^3 u}{\partial z^2 \partial t},$$

where ρ – density, u = u (t, z) – horizontal displacement, z – depth below the ground surface.

The algorithm in the original program SHAKE (Schnabel et al. 1972) is based on the continuous solution to the wave equation (Kanai 1951; Lysmer et al. 1971), which was adapted for transient motions using the Fast Fourier Transform techniques. To date, this algorithm is used by the softwares Proshake (EduPro Civil Systems 1998), EERA (Bardet et al. 2000), DEEPSOIL (Hashash et al. 2017), etc. An equivalent linear procedure is used to account for the nonlinearity of the soil using an iterative procedure to obtain values for shear modulus and damping ratio that are compatible with the strain induced in each sublayer.

In non-linear analysis, the following dynamic equation of motion is solved:

$$[M]\{\ddot{u}\} + [C]\{\dot{u}\} + [K]\{u\} = -[M]\{I\}\,\ddot{u}_g(t),$$

where [M] – matrix of mass, [C] – matrix of damping, [K] – matrix of stiffness, $\{\ddot{u}\}$ – acceleration vector, $\{\dot{u}\}$- velocity vector, $\{u\}$ — displacement vector, {I} – unit vector, $\ddot{u}_g(t)$ – input accelerogram.

The equation is solved numerically at each time step using a time integration method. The mass, stiffness and damping matrices are updated at each time increment to incorporate non-linearity of the soil.

The most currently available, nonlinear ground response analysis computer programs characterize the stress-strain behavior of the soil by the following cyclic stress-strain models:

- Ramberg-Osgood model – CHARSOIL (Streeter et al. 1974);
- Iwan model –N0NLI3 (Joyner and Chen 1975);
- Martin-Davidenkov model – MASH (Martin and Seed 1978);
- Hyperbolic model DESRA (Lee and Finn 1978), TARA (Joyner and Chen 1975);
- Hardin-Drnevich-Cundall-Pyke (HDCP) – TESS1 (Pyke 1985);
- Iwan – Mroz (IM – model) – NERA (Bardet and Tobita 2001);
- Modification on the hyperbolic model – DeepSoil (Hashash and Park 2001).

When constructing a computational seismic geological model for mathematical modeling of the effect of the soil strata on seismic effects, data on tectonics, lithology, boundary geometry, and physical-mechanical and non-linear soil properties of the site should be taken into account. On the other hand, calculations are limited by the capabilities of computational algorithms. In each case, it is necessary to find the optimal relationship between the complexity of the computational model of the geological environment and the capabilities of computational algorithms for calculating the movement of soil particles in complex models of the environment during earthquakes.

20.3 Site Amplification in the South Ukrainian Nuclear Power Plant

The territory of the South Ukrainian nuclear power plant was divided into 3 zones according to the results of seismic microzonation. For each zone, the resulting frequency response and a set of design accelerograms were separately calculated.

To assess the influence of subsurface geology, we used one-dimensional (horizontally layered models of the medium), which is quite acceptable under conditions of close horizontal bedding of the main geological elements of the sedimentary sequence.

The geometric parameters and density properties of the models were estimated based on a priori engineering and geological information. The velocity properties of the models were determined on the basis of a priori geophysical data and field studies.

Properties of soil profiles of zones 1–3 of the territory of the South Ukrainian nuclear power plant are listed in Tables 20.1, 20.2 and 20.3.

From Tables 20.1, 20.2 and 20.3 it follows that Mean shear wave velocity of sedimentary layers varying from 200 to 350 m/s in all zones. Shear wave velocity of the bedrock equal to 2000 m/s.

To take into account the features of the nonlinear behavior of soils in this work, we used the equivalent linear model. In equivalent linear method, strain-dependent secant shear modulus and damping ratios are used, and the response of each layer is obtained through an iterative procedure until the assumed properties are

Table 20.1 Properties of soil profile of zone 1 of the territory of the South Ukrainian nuclear power plant

Soil type	Shear-wave velocity (Vs), m/s	Mass density (ρ), g/cm^3	Thickness, m
Bulk soil and soil-vegetable layer	200	1.69	2.5
Loamy loam	230	1.83	1.0
Clay	260	1.88	2.0
Indigenous clays	260	1.88	1.7
Indigenous sands	300	1.90	3.0
Kaolins (loam)	300	1.90	2.3
Woody soil and woody sand	350	2.00	1.5
Rock	2000	2.60	–

Table 20.2 Properties of soil profile of zone 2 of the territory of the South Ukrainian nuclear power plant

Soil type	Shear-wave velocity (Vs), m/s	Mass density (ρ), g/cm^3	Thickness, m
Bulk soil and soil-vegetable layer	200	1.69	1.5
Loamy loam	230	1.83	1.0
Clay	260	1.88	2.0
Indigenous clays	260	1.88	2.5
Woody soil and woody sand	350	2.00	1.5
Rock	2000	2.60	–

Table 20.3 Properties of soil profile of zone 3 of the territory of the South Ukrainian nuclear power plant

Soil type	Shear-wave velocity (Vs), m/s	Mass density (ρ), g/cm^3	Thickness, m
Soil-vegetable layer	200	1.69	4.0
Loamy loam	230	1.83	1.3
Indigenous clays	300	1.90	1.7
Indigenous sands	300	1.90	1.0
Kaolins (loam)	300	1.90	7.0
Woody soil and woody sand	350	2.00	1.5
Rock	2000	2.60	–

compatible with computed strain levels in all layers. Equivalent linear site response analyses were performed using Proshake computer program. Shear modulus (G) and damping ratio (D) are efficient and important parameters in estimation of seismic response of a site. The strain-dependent shear modulus and damping ratio for each layers composing the section were selected on the basis of models obtained for different soil (Kendzera and Semenova 2017).

Figure 20.1 compares calculated frequency characteristics of the soils of the three zones identified by the results of seismic micro-zoning within the South Ukrainian nuclear power plant.

According to the results presented in Fig. 20.1, it can be seen that the frequency response of the soil of zone 1 of the South Ukrainian nuclear power plant is characterized by a frequency range of resonant amplification of seismic oscillations from 2.06 to 4.0 Hz with a maximum gain of 4.93. In the indicated frequency range, one clear maximum is observed at a frequency of 3.26 Hz. The frequency response of the soil of zone 2 is characterized by the frequency range of resonant amplification of seismic oscillations from 6.14 to 8.14 Hz with a maximum gain of 9.76. In the specified frequency range, there is one clear maximum at a frequency of 7.26 Hz. The frequency response of the soil of zone 3 is characterized by the frequency range of resonant amplification of seismic oscillations from 2.82 to 4.3 Hz with a maximum gain of 7.79. In the indicated frequency range there is one clear maximum at a frequency of 3.68 Hz. Comparing the frequency characteristics of the

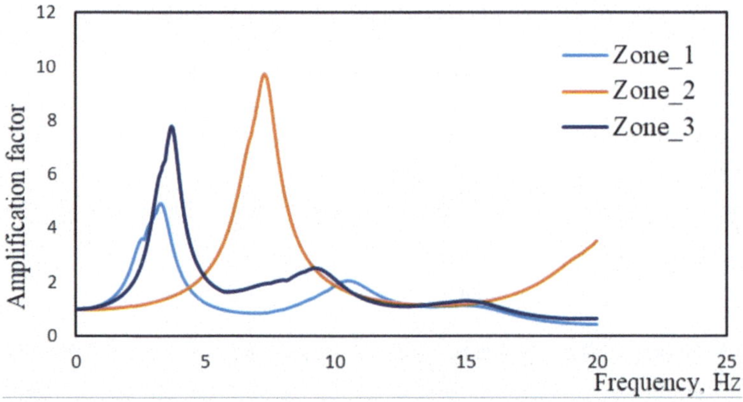

Fig. 20.1 Frequency characteristics of the soils of the three zones identified by the results of seismic micro-zoning within the South Ukrainian nuclear power plant

soil of the three zones visualized within the territory of the South Ukrainian nuclear power plant, it can be seen that zone 1 is characterized by lower frequency natural vibrations compared to zones 2 and 3.

Analyzing the calculated frequency characteristics, it can be seen that they are different depending on the ground conditions of the zone of site. Soil amplification of seismic waves in zone 2 is greater than in zone 1 and zone 2. This phenomenon can be explained by the fact that the thickness of sedimentary deposits in the context of zone 2 is the smallest. The thickness of sediment in zone 2 is 8.5 m, while in zone 1 it is 14 m, and in zone 3 it is 16.5 m. In the conditions of the site of the South Ukrainian nuclear power plant, the roof of the bedrock changes quite sharply. This leads to a shift of the resonance maxima in a wide range of frequencies. Zone 1 differs from zone 3 in lower values of Vs, which explains the greater absorption of seismic energy and a smaller gain. However, it is seen that more loose, water-saturated soils have lower frequency oscillations than dense clay or coarse-grained soils.

The obtained frequency characteristics were used to generate design accelerograms simulating projective earthquakes from the Vrancea zone (see Fig. 20.2) and from local origin zones (see Fig. 20.3).

Design accelerograms should be generated only for specific sites of the proposed seismic resistant construction. Their use will ensure an acceptable level of seismic resistance while reducing the cost of the object. The use of accelerograms, which do not take into account the resonant and nonlinear properties of soils at a specific site under study, can lead to incorrect engineering calculations of the seismic resistance of buildings.

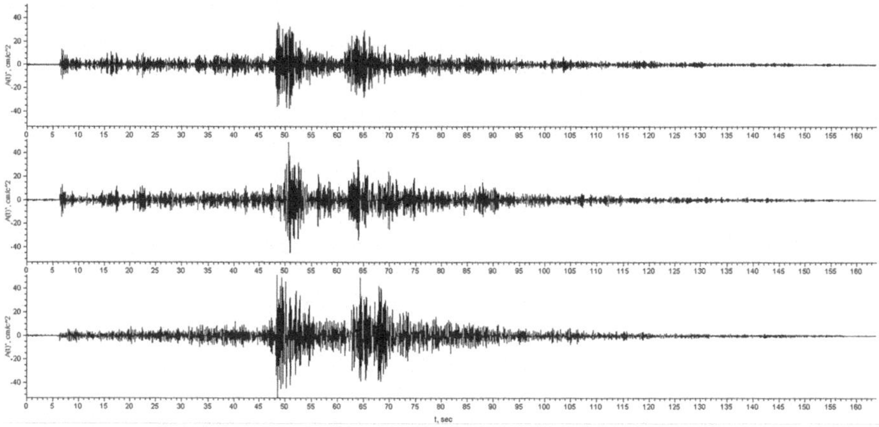

Fig. 20.2 An example of a three-component design accelerogram for modeling a project earthquake from the Vrancea zone in the Zone-1 section on the free surface of the ground at the site of the South Ukrainian nuclear power plant

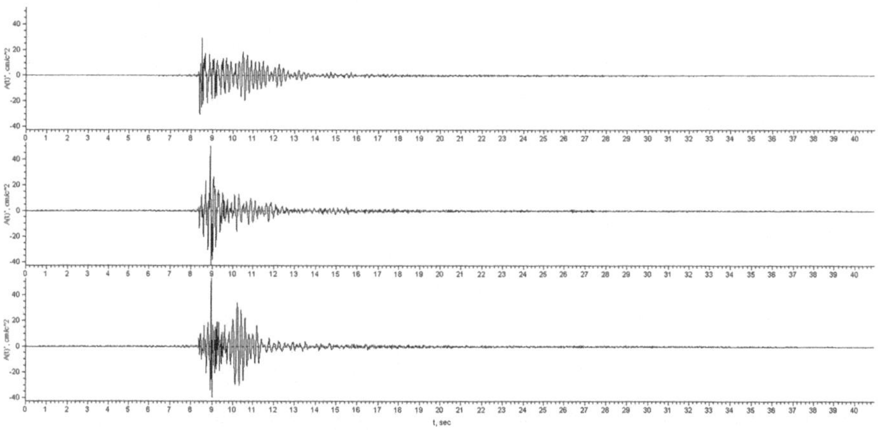

Fig. 20.3 An example of a three-component design accelerogram for modeling a project earthquake from the local origin zone in the Zone-1 section on the free surface of the ground at the site of the South Ukrainian nuclear power plant

20.4 Main Conclusion

The destructions and damages of earthquake-resistant structures during earthquakes are associated not only with poor construction quality and unfavorable ground conditions of their sites, but also with the coincidence of the eigen frequencies of structures with maximal spectral density of frequency characteristics of the soil strata (resonant effects); liquefaction of soils or partial loss of their bearing capacity (nonlinear effects). Rheological properties of soils influence the surface response of strata to seismic ground motion. An analysis of the site effects associated with the rheological properties of soils is very important since the amplification of seismic motion in some specific areas can be very strong.

This paper presents modelling of site amplification in the South Ukrainian nuclear power plant. The territory of the South Ukrainian nuclear power plant was divided into 3 zones according to the results of seismic microzonation. For each zone, the resulting frequency response and a set of design accelerograms were separately calculated. The effects of the rheological properties of soil on site amplification and resonance frequency are analyzed for each zone. In the conditions of the site of the South Ukrainian nuclear power plant, the roof of the bedrock changes quite sharply. This leads to a shift of the resonance maxima in a wide range of frequencies.

Comparing the frequency characteristics of the soil of the three zones visualized on the territory of the South Ukrainian nuclear power plant, it was found that zone 1 is characterized by lower natural oscillations frequencies compared to zones 2 and 3.

Accounting for seismic vibrations in the low frequency range is especially necessary for the seismic resistance of objects. Since objects located on the territory of Ukraine can be damaged by strong earthquakes from the Vrancea zone. Seismic effects from these earthquakes are characterized by low-frequency long-term oscillations and spread over long distances without significant attenuation, which can cause dangerous resonance phenomena.

In general, all three zones on the territory of the South Ukrainian nuclear power plant will have a different soil response to seismic effects. Although the zones are located very close to each other.

Therefore, in each case, it is necessary to calculate the own frequency response of the soil and a set of design accelerograms.

The study of the variability of the parameters of soil movements during earthquakes depending on the rheological properties of the rocks and the structure of the surface part of the section of the site is an essential element in earthquake-resistant design, reconstruction and extension of the life of buildings and structures.

Amplification of seismic waves in surface geological layers should be taken into considered for the already existing residential areas as well as for the new sites to be opened to construction with respect to decreasing possible damages and collapses in earthquake.

References

Bardet JP, Tobita T (2001) NERA. A computer program for nonlinear earthquake site response analyses of layered soil deposits. University of Southern California, Los Angeles

Bardet J, Ichii K, Lin C (2000) Manual of EERA: a computer program for equivalent-linear earthquake site response analysis of layered soil deposits. University of Southern California, Los Angeles

EduPro Civil Systems Inc. (1998) ProShake – ground response analysis program. Redmond, Washington. User's manual

Hashash YMA, Park D (2001) Non-linear one-dimensional seismic ground motion propagation in the Mississippi embayment. Eng Geol 62(1–3):185–206

Hashash YMA, Musgrove MI, Harmon JA et al (2017) DEEPSOIL 7.0, user manual. University of Illinois at Urbana-Champaign

Joyner WB, Chen ATF (1975) Calculation of nonlinear ground response in earthquakes. Bull Seismol Soc Am 65(5):1315–1336

Kanai K (1951) Relation between the nature of surface layer and the amplitude of earthquake motions. Bulletin Tokyo Earthquake Research Institute

Kendzera AV, Semenova YV (2017) Deformation characteristics of computational model of soil strata. Geology 78:17–29

Kramer SL (1996) Geotechnical earthquake engineering. Prentice-Hall, New Jersey, p 273

Lee MKW, Finn WDL (1978) DESRA-2, dynamic effective stress response analysis of soil deposits with energy transmitting boundary including assessment of liquefaction potential. In: Soil mech. Series, 38, Department of Civil Engineering. University of British Columbia, Vancouver

Lysmer J, Seed HB, Schnable PB (1971) Influence of base rock characteristics on ground response. Bull Seismol Soc Am 61:1213–1232

Martin PP, Seed HB (1978) MASH: a computer program for the non-linear analysis of vertically propagating shear waves in horizontally layered deposits. In: Report no. EERC. University of California, Earthquake Engineering Research Center, Berkley, California, pp 78–23

Pyke RM (1985) TESS1: a computer program or nonlinear ground response analysis, California. User manual, TAGA engineering software services, Lafayette

Schnabel PB, Lysmer PB, Seed HB (1972) SHAKE: a computer program for earthquake response analysis of horizontally layered sites. In: Report EERC 72–12, earthquake engineering research center. University of California, Berkeley

Streeter VL, Wylie EB, Richart FE (1974) CHARSOIL: soil motion computations by characteristics method. J Geotech Eng Div 100(3):247–263

Chapter 21
On the Seismic Waves Propagating in the Layered Earth Stratum

Jeremiah Rushchitsky, Oleksandr Kendzera, and Yuliia Semenova

Abstract The direct method of mechanics is utilized in studying the passing and reflecting of *SH*-wave while it propagates through the finite number of ground layers with differing thicknesses and rheological properties. The mechanical properties of the ground layers are described by the Hookean and standard rheological models. The explicit formulas for evaluation of the wave number and attenuation of the harmonic *SH*-wave are obtained. The wave attenuation is studied by two variants – by the spatial coordinate and in time. The obtained formulas can be treated as some continuation of the classical works of Prof. Savarensky (Russia) in the area of elastic seismic waves and Prof. Stepanishen (USA) in the area of viscoelastic seismic waves.

Keywords Multi-layered stratum · Seismic *SH*-wave · Elastic model · Visco-elastic model · Wave attenuation

21.1 Introduction

Materials science distinguishes seismic waves into a separate section, based on three main features of these waves: (1) the medium in which seismic waves propagate is the Earth; (2) seismic waves are waves with the low-frequency values; (3) sources for seismic waves are earthquakes, volcanic eruptions, the movement of magma through the Earth, large natural landslides, or large anthropogenic (human-caused) explosions. The seismic wave propagation medium is generally considered to be deformable (material) and various deformation models of materials are used to

J. Rushchitsky (✉)
SP Timoshenko Institute of Mechanics, Kyiv, Ukraine
e-mail: rushch@inmech.kiev.ua

O. Kendzera · Y. Semenova
Subbotin Institute of Geophysics, Kyiv, Ukraine

© Springer Nature B.V. 2021

F. L. Bonali et al. (eds.), *Building Knowledge for Geohazard Assessment and Management in the Caucasus and other Orogenic Regions*, NATO Science for Peace and Security Series C: Environmental Security,
https://doi.org/10.1007/978-94-024-2046-3_21

model it, most often choosing the elastic deformation model. Therefore, in the analysis of seismic waves, it is important to know the physical characteristics of the various components of the Earth's layers. In the elastic deformation model, this entails the knowledge of density and elastic constants. Historically, seismic waves have been analyzed in the framework of the theory of elasticity, which operates with such types of waves as bulk waves, plane waves, surface waves. It is often accepted that the source of seismic waves is localized and generates spherical waves. However, when the object of the study of seismic waves is far enough from the source, the waves are considered the plane ones. Therefore, plane waves are dealt with by most of seismic wave science.

This presentation is devoted to the theoretical analysis of the propagation of a plane, horizontally-polarized harmonic wave (*SH*-wave) from rocky soil through soil layers to the Earth's surface. Such an analysis results in formulas for calculating the oscillations of the Earth's surface and the required coefficients of reflection and propagation of waves across the boundaries between the elements of the system "rocky massif – soil layers – atmosphere". Knowledge of the oscillation of the Earth's surface due to the arrival of the seismic *SH*-wave is a necessary component of seismic micro-zoning of building sites. Therefore, the importance of theoretical justification for data on the vibration of the Earth's surface is obvious (Savarenskii 1972; Stepanishen and Strozeski 1972; Kendzera et al. 1979; Slawinski 2003; Shearer 2009; Rushchitsky 2012, 2019; Kendzera et al. 2020).

The first part of this work contains a certain extension of the classical approach and the final results of Savarensky in the field of elastic seismic waves (Rushchitsky 2012). Five variants of the analysis of elastic classical and seismic linear harmonic *SH*-wave are considered here: two-component elastic system "upper half-space – lower half-space" with the plane boundary of half-space; three-component elastic system "upper half-space – layer – lower half-space"; three-component system "atmosphere – elastic soil layer – elastic rock massif"; $(N + 2)$-component elastic system "upper half-space – N layers – lower half-space"; $(N + 2)$-component system "atmosphere – elastic soil layers – elastic rock massif".

The second part contains some extension of the classical approach and the final results of Stepanishen in the field of viscoelastic seismic waves (Stepanishen and Strozeski 1972). It used the rheological standard (viscoelastic three-element) soil deformation model and analyzed the seismic harmonic *SH*-wave in the three-component system "atmosphere – viscoelastic soil layer – elastic rock massif". and in the $(N + 2)$ component system "atmosphere – N viscoelastic soil layers – elastic rock mass".

21.2 To the Theory of Elastic Seismic Harmonic SH-Waves

While the seismic *SH*-wave being analyzed within the framework of the simplest classical problem of mechanics, it is usually considered that the wave propagates in a two-component elastic system "upper half-space $z > 0$ – lower half-space $z < 0$" with an interface $z = 0$ in the form of a plane Oxy. Such a three-dimensional problem is reduced to a two-dimensional one, assuming the wave vector has the classical form $\overrightarrow{k} = (k_1, 0, k_3)$ or $\overrightarrow{k} = (k \sin\theta, 0, k \cos\theta)$, if a notion of the angle θ of incidence of the wave at the interface is introduced as the angle between the wave direction and the vertical coordinate z. Then the dependence on the coordinate y can be neglected and the problem considered a plane one. In the plane, all parameters relating to the upper half-plane $\infty \leq z \leq 0$ are denoted by index 1 and those parameters related to the lower half-plane $-\infty \leq z \leq 0$ are denoted by index 2. It is assumed further that each half-plane has its physical properties of linear elastic isotropic material (in this case – the shear modulus μ_α and density ρ_α $(\alpha = 1; 2)$) and the wave propagates from the bottom up to the top – from the lower half-plane to the upper one and from left to right. In this case, the incident on interface wave may be represented in the form

$$u_{2(inc)}(x, z, \theta_2, t) = A_2 e^{i[(k_2 \sin\theta_2)x + (k_2 \cos\theta_2)z - \omega t]}, \tag{21.1}$$

where all parameters are assumed to be known including the incident wave angle θ_2.

When passing through the interface, the wave (21.1) is divided into two waves – one is symmetrically reflected from the interface (that is, at a known angle $(-\theta_2)$) and propagates with an unknown amplitude and the other continues to propagate in the upper half-plane at an unknown angle (is refracting) and with unknown amplitude.

Usually, these two waves are represented as follows:

incident wave

$$u_{2(inc)}(x, z, \theta_2, t) = V A_2 e^{i[(k_2 \sin\theta_2)x - (k_2 \cos\theta_2)z - \omega t]}, \tag{21.2}$$

reflected wave

$$u_{1(inc)}(x, z, \theta_1, t) = W A_2 e^{i[(k_1 \sin\theta_1)x + (k_1 \cos\theta_1)z - \omega t]}. \tag{21.3}$$

Here, the standard and at this step unknown constant coefficients before amplitudes are introduced. They characterize the interface – the reflection coefficient V as the ratio of amplitudes of reflected VA_2 and incident A_2 waves and the transmission coefficient W as the ratio of amplitudes of transmitted WA_2 and incident A_2 waves.

To determine the unknown quantities θ_1, V, W, the boundary conditions of full mechanical contact at the interface are used – equality of displacements and shear stresses (in conditions of the displacement field (21.1) the shear stress $\sigma_{yz} = \mu(\partial u/\partial z)$ only is nonzero one

$$u_{2(inc)}(x, 0, t) + u_{2(ref)}(x, 0, t) = u_{1(tra)}(x, 0, t), \tag{21.4}$$

$$\sigma_{yz(2inc)}(x,0,t) + \sigma_{yz(2ref)}(x,0,t) = \sigma_{yz(1tra)}(x,0,t), \tag{21.5}$$

or

condition (21.4)

$$\begin{aligned} & A_2 e^{i[(k_2\sin\theta_2)x + (k_2\cos\theta_2)(z=0) - \omega t]} + V A_2 e^{i[(k_2\sin\theta_2)x - (k_2\cos\theta_2)(z=0) - \omega t]} \\ & = W A_2 e^{i[(k_1\sin\theta_1)x + (k_1\cos\theta_1)(z=0) - \omega t]}, \\ & (1+V)\, e^{i(k_2\sin\theta_2)x} = W e^{i(k_1\sin\theta_1)x}, \end{aligned} \tag{21.6}$$

condition (21.5)

$$\begin{aligned} & \sigma_{yz} = \mu \partial u/\partial z = \mu\,(\partial/\partial z)\, A e^{i[(k\sin\theta)x + (k\cos\theta)z - \omega t]} = i\mu k\cos\theta \cdot u, \\ & (\mu_2 k_2\cos\theta_2)\,(1-V)\, e^{i(k_2\sin\theta_2)x} = (\mu_1 k_1\cos\theta_1)\, W e^{i(k_1\sin\theta_1)x}. \end{aligned} \tag{21.7}$$

The condition (21.6) is fulfilled when the Snell's refraction law

$$k_2\sin\theta_2 = k_1\sin\theta_1 \tag{21.8}$$

and equality $1 + V = W$ are fulfilled. It is considered that formulas

$$V = \frac{\mu_2 k_2\cos\theta_2 - \mu_1 k_1\cos\theta_1}{\mu_2 k_2\cos\theta_2 + \mu_1 k_1\cos\theta_1},\ W = \frac{2\mu_2 k_2\cos\theta_2}{\mu_2 k_2\cos\theta_2 + \mu_1 k_1\cos\theta_1} \tag{21.9}$$

finish the analysis of the problem on propagation of harmonic *SH*-wave in the two-component elastic system “upper half-plane – lower half-plane”.

An analysis of the three-component elastic system “upper half-space – layer – lower half-space” is carried out; three-component system “atmosphere – elastic soil layer – elastic rock massif”; $(N + 2)$-component elastic system “upper half-space – N layers – lower half-space” form the basis for solving the problem of the $(N + 2)$-component system “atmosphere – N elastic soil layers – elastic rock massif”.

It should be noted that modeling the soil thickness over a rock massif with one layer is a significant simplification of the real situation when there are many layers and they differ in thickness and lithological composition and, accordingly, in physical properties. Therefore, mainly the soil thickness is modeled by a certain (for different localities different) finite number of layers with different thicknesses (capacities), density and rheological properties.

It should also be recalled that the problem of seismic waves takes into account the distinction from the classical problem, both in terms of terminology and the statement of the problem precisely through the consideration of the seismic wave. In particular, the upper component of the system is the atmosphere that is characteristic of seismic waves in the soils thickness, which are detected on the Earth's surface by seismic activity (earthquakes, etc.). It is considered that the atmosphere reflects the wave back into the soil layer (part of the wave energy goes into the vibrations of the layer roof). Also, the upper part of the soil layer is called the roof and the lower – the bottom, the lower half-space is called the underlying rock massif.

A statement of the problem of the $(N + 2)$-component system is as follows: the wave propagates from the underlying rock massif at a given angle to the plane of separation (interface) between the massif and the first layer. Denote all the parameters of the massif by index $(N + 2)$, the parameters of the layers – by indices from $(N + 1)$ up to 2, the atmosphere – by index 1. It is necessary to find the parameters of the wave that will come to the Earth's surface.

Here, a new situation arises when in the upper layer each wave reflected from the roof into the bottom forms a new wave in the lower layer, which is reflected by the infinite number of times from the bottom of this layer to its roof and back. But at each arrival to the boundary with the massif a new transmitted wave is formed in the massif and at each arrival to the boundary with the upper layer a new wave is transmitted into the upper layer. This process is endless and at each step some part of the energy is trans mitted back from the top layer to the rock massif.

The above procedure of the direct approach of mechanics gives the following result: when interacting with N boundaries $z = h_{N-1}, \ldots, z = h_2$ without taking into account the multiple reflection of waves in the ground layers, a wave falling from a rock massif

$$u_{(N+2)(inc)}\left(x, z, \theta_{N+2}, t\right) = A_{N+2}e^{i\left[\left(k_{N+2}\sin\theta_{N+2}\right)x+\left(k_{N+2}\cos\theta_{N+2}\right)z-\omega t\right]}$$

is transformed on the Earth's surface into vibrations

$$u\left(x, z, t\right) = 2W_2 \cdots W_{N+1}A_{N+2}\cos\left[\left(k_2\sin\theta_2\right)x - \omega t\right]\cos\left(k_2 h\cos\theta_2\right). \tag{21.10}$$

The formula for finding the displacement field in a rock massif is as follows

$$\begin{aligned} u_{N+2}\left(x, z, \theta_{N+2}, t\right) &= A_{N+2}e^{i\left[\left(k_{N+2}\sin\theta_{N+2}\right)x-\omega t\right]} \\ &\times\left(e^{i\left(k_{N+2}\cos\theta_{N+2}\right)z} + V_{N+1}e^{-i\left(k_{N+2}\cos\theta_{N+2}\right)z}\right). \end{aligned} \tag{21.11}$$

It should be noted that the last two formulas represent the main characteristics that you need to know as a result of solving the problem of seismic harmonic *SH*-wave in the $(N + 2)$ component system "atmosphere – N elastic ground layers – an elastic rock massif.

It is also worth noting that the assumption of elastic deformation of the layers is some simplification of the real situation: when the layers being deformed, they behave as the rheological bodies. Therefore, we will consider in following a generalization of the formulas obtained in the case when the deformations of rock massif and soil layers occur according to rheological laws.

21.3 To the Theory of Viscoelastic Seismic Harmonic *SH*-Waves

Further, an analysis is limited to a standard rheological model. But it should be remembered that the problem of propagation of the *SH*-wave within the linear theory of viscoelasticity is a classic one. Since any one-dimensional harmonic wave of displacement has the form $u(x,t) = u^o e^{i(kx - \omega t)}$, that is, oscillations over time occur with frequency ω, the generally accepted method of obtaining the equation of propagation of such a wave in the theory of viscoelasticity uses the so-called complex moduli. They are introduced on the basis of a representation of the equation of motion, which in the theory of elasticity is obtained using an integral visco-elastic operators by replacing in the equation of elastic theory of elastic moduli for these operators.

But for the simple models to which the standard model is related, a direct approach can be applied.

On the first step of this approach, the fact is used that any one-dimensional motion for any dependence of stress on deformation (specifically for both elastic deformation law and viscoelastic law) is described by the equation

$$\rho\left(\partial^2 u/\partial t^2\right) = (\partial\sigma/\partial x).$$

Further, the direct approach consists in substitution of the constitutional law for the standard model into this equation. Such an approach makes it possible to directly calculate the parameters of a plane harmonic *SH*-wave of displacement, which is just the subject of analysis.

The wave representation in this case uses the notions of the constant amplitude A, real wavenumber k and relaxation coefficient $\gamma > 0$

$$u(x,t) = Ae^{-\gamma x - i(kx - \omega t)}. \tag{21.12}$$

In the standard model, the stress-strain relationship equation (constitutive equation) has a form of a regular first order differential equation with constant coefficients

$$\frac{\eta_K}{\mu_K + \mu_S}\dot{\sigma}(t) + \sigma(t) = \frac{\mu_S \eta_K}{\mu_K + \mu_S}\dot{\varepsilon}(t) + \frac{\mu_S \mu_K}{\mu_K + \mu_S}\varepsilon(t),$$

which is usually written as

$$n\dot{\sigma}(t) + \sigma(t) = En\dot{\varepsilon}(t) + H\varepsilon(t), \tag{21.13}$$

where $E = \mu_S$ is the short-term (instantaneous) modulus of elasticity, $H = \mu_S\mu_K/(\mu_K + \mu_S) < E$ is the long-term modulus of elasticity, $n = \eta_K/(\mu_K + \mu_S)$ is the relaxation time.

If to introduce the shear modulus μ ($\mu_0 = E, \mu_\infty = H$) into (21.13), then the equations

$$n\frac{\partial \sigma}{\partial t} + \sigma = \mu_0 n \frac{\partial^2 u}{\partial x \partial t} + \mu_\infty \frac{\partial u}{\partial x} \text{ or } \rho\frac{\partial^2 u}{\partial t^2} + n\rho\frac{\partial^3 u}{\partial t^3} = \mu_0 n \frac{\partial^3 u}{\partial x^2 \partial t} + \mu_\infty \frac{\partial^2 u}{\partial x^2} \tag{21.14}$$

can be obtained.

Note that this is not the simplest wave equation as compared with the classical wave equation that describes the propagation of elastic *SH*-wave – Eq. (21.14) contains an additional two third-order terms.

In the study of harmonic waves, which are described by rheological Eq. (21.14), two cases are possible. It is considered in the first case that the wave attenuates in a spatial coordinate and has the form (21.12) $u(x, t) = Ae^{-\gamma x - i(kx - \omega t)}$. The second case considers the damping in time and then the wave has a different form

$$u(x, t) = Ae^{i[kx-(\omega+i\alpha)t]}. \tag{21.15}$$

Let's start with case 1 and check that Eq. (21.14) has a wave-like solution (21.12). To do this, substitute (21.12) in (21.14) and obtain the equation of link between the wavenumber and the damping coefficient

$$\begin{gathered}\left[\left(v_T^o\right)^{-2}\omega^2 + \left(\mu_\infty/\mu_0\right)\left(\gamma^2 - k^2\right) - 2n\omega\gamma k\right] + \\ + i\left[n\left(v_T^o\right)^{-2}\omega^3 + n\omega\left(\gamma^2 - k^2\right) + 2\left(\mu_\infty/\mu_0\right)\gamma k\right] = 0 \quad v_T^o = \sqrt{\mu_0/\rho}.\end{gathered} \tag{21.16}$$

From the equality of the imaginary part of this equation to zero the expression γ through k with taking into account the denotation $(\mu_0/\mu_\infty) = \left(v_T^{o2}\right) / \left(v_T^{\infty 2}\right) = m$

$$\gamma = -\frac{1}{n v_T m}\left[1 \mp \sqrt{1 + (n\omega)^2 m^2 \left(1 - \frac{v_T^2}{v_T^{o2}}\right)}\right] \tag{21.17}$$

can be obtained.

It is solved by the standard method and all coefficients in (21.17) are the real quantities and are expressed by known rheological model parameters and wave frequency.

With the known attenuation coefficient, the velocity of the wave is determined by the formula

$$v_T = v_T^o\sqrt{n\omega^2 / \left(n\omega^2 + 2\alpha + 3n\alpha^2\right)}. \tag{21.18}$$

An advantage of using the standard rheological model to the analysis of seismic waves can be considered to be the ability to analytically calculate the wave characteristics. For each layer, three constants of the standard model (short-term modulus, long-term modulus, relaxation time) and the wave parameters at the inlet of the layer (amplitude, incidence angle, frequency, attenuation coefficient) must be known. The constants of the standard model are determined from the standard (regulated) experiments on the creep of the samples at shear.

Further, the seismic harmonic *SH*-wave in the three-component system "atmosphere – viscoelastic soil layer – elastic rock massif", which is the basic for the next more complex problem. Let us choose the damping by the spatial coordinate and suppose that the wave that falls at a known angle from the bottom up and from left to right from the elastic rock mass to the bottom of the viscoelastic soil layer $z = 0$ has the form

$$u_{2(inc)}\left(x, z, \theta_2, t\right) = A_2 e^{i[(k_2 \sin\theta_2)x + (k_2 \cos\theta_2)z - \omega t]}, \tag{21.19}$$

where all the parameters are considered known.

When being crossing the bottom, this wave is divided into the reflected elastic wave

$$u_{2(ref)}\left(x, z, \theta_2, t\right) = V_1 A_2 e^{i[(k_2 \sin\theta_2)x - (k_2 \cos\theta_2)z - \omega t]}$$

and the refracted viscoelastic wave

$$u_{1(tra)}\left(x, z, \theta_1, t\right) = W_1 A_2 e^{i[(k_1 \sin\theta_1)x + ((k_1 + i\gamma)\cos\theta_1)z - \omega t]}.$$

Further, two waves should be considered that are already generated in a layer and are already viscoelastic: the refracted wave (21.20) and the reflected from the atmosphere wave

$$u_{1(ref)}\left(x, z, \theta_1, t\right) = V_0 W_1 A_2 e^{i[(k_1 \sin\theta_1)x - ((k_1 + i\gamma)\cos\theta_1)z - \omega t]}. \tag{21.20}$$

These waves create a displacement field that is at the boundary of the atmosphere as follows

$$\begin{aligned} u_1\left(x, h, \theta_1, t\right) = {} & W_1 A_2 e^{i[(k_1 \sin\theta_1)x - \omega t]} \\ & \times \left(e^{-\gamma_1 h \cos\theta_1} e^{ihk_1 \cos\theta_1} + V_0 e^{\gamma_1 h \cos\theta_1} e^{-ihk_1 \cos\theta_1}\right), \end{aligned} \tag{21.21}$$

For finding the unknown quantities θ_1, V_1, W_1, V_0, the boundary conditions of full mechanical contact on the interfaces are used – equality of displacements and shear stresses at $z = 0$ and equality of zero shear stresses at $z = h$.

As a result, the expressions for the coefficients V_o, V_1, W_1 can be obtained

$$V_0 = e^{2ih(k_1 + i\gamma_1)\cos\theta_1} \text{ or } V_0 = e^{-2\gamma_1 \cos\theta_1 h} \cos\left(2k_1 h \cos\theta_1\right). \tag{21.22}$$

$$W_1 = \frac{2\left(\mu_2 k_2 \cos\theta_2\right)}{\left[\begin{array}{l}\left(\mu_2 k_2 \cos\theta_2\right)\left(1 + e^{2k_1 \cos\theta_1\, h}\ \cos\left(2\gamma_1 \cos\theta_1\ h\right)\right) + \\ + \left(e^{i\varphi_1} M_{z1} \mu_1^{\infty} \cos\theta_1\right)\left(1 - e^{2k_1 \cos\theta_1\, h}\ \cos\left(2\gamma_1 \cos\theta_1\ h\right)\right)\end{array}\right]}, \tag{21.23}$$

$$V_1 = \frac{\left[\begin{array}{l}\left(\mu_2 k_2 \cos\theta_2\right)\left(1 + e^{2k_1 \cos\theta_1\, h}\ \cos\left(2\gamma_1 \cos\theta_1\ h\right)\right) - \\ - \left(e^{i\varphi_1} M_{z1} \mu_1^{\infty} \cos\theta_1\right)\left(1 - e^{2k_1 \cos\theta_1\, h}\ \cos\left(2\gamma_1 \cos\theta_1\ h\right)\right)\end{array}\right]}{\left[\begin{array}{l}\left(\mu_2 k_2 \cos\theta_2\right)\left(1 + e^{2k_1 \cos\theta_1\, h}\ \cos\left(2\gamma_1 \cos\theta_1\ h\right)\right) + \\ + \left(e^{i\varphi_1} M_{z1} \mu_1^{\infty} \cos\theta_1\right)\left(1 - e^{2k_1 \cos\theta_1\, h}\ \cos\left(2\gamma_1 \cos\theta_1\ h\right)\right)\end{array}\right]}. \tag{21.24}$$

Note that the refraction angle in the layer is changed with frequency, since the wave number in the viscoelastic layer depends on the frequency of the wave.

Thus, formulas (21.21), (21.22), (21.23) and (21.24) give the solution of the problem on propagation of the harmonic *SH*-wave in the three-component elastic system "atmosphere – viscoelastic soil layer – elastic rock massif", in the condition that the deformation of viscoelastic layer is described by the standard model and the wave attenuation is taken into account by the variant 1.

Therefore, the viscoelasticity of deformation of the soil layer is manifested in the change in the amplitude of oscillations, which already depends on the form of wave attenuation, phase shift, parameters of the standard rheological model (short-term and long-term moduli, and relaxation time), as well as the angles of incidence of the wave and thickness of layer.

The obtained formulas show that the corresponding formulas for the case of maximum consideration of the reflection of waves in a problem where the layer is viscoelastic may be written for reasons of analogy:

the elastic massif with the elastic layer from above

$$u_2\left(x, z, \theta_2, t\right) = \left[e^{i\left(k_2 \cos\theta_2\right)z} + \left(V_1 + \left(\left(W_1\right)^2 / \left(1 - V_1\right)\right)\right) e^{-i\left(k_1 \cos\theta_1\right)z}\right] A_2 e^{i\left[\left(k_2 \sin\theta_2\right)x - \omega t\right]},$$

the surface of the viscoelastic layer

$$u_1\left(x, h, \theta_1, t\right) = \frac{W_1}{1 - V_0 V_1} A_2 e^{i\left[\left(k_1 \sin\theta_1\right)x - \omega t\right]}\left(e^{-\gamma_1 h \cos\theta_1} e^{ihk_1 \cos\theta_1} + V_0 e^{\gamma_1 h \cos\theta_1} e^{-ihk_1 \cos\theta_1}\right).$$

It should be noted that in the system "atmosphere – elastic soil layer – elastic rock massif" a mechanism of returning the waves from both the atmosphere and the massif is absent. Therefore, a repeated reflection of the waves in the viscoelastic ground layer can be fully taken into account.

The seismic harmonic *SH*-wave was then considered in the $(N + 2)$-component system "atmosphere – N viscoelastic soil layers – elastic rock massif". The repetition of the procedure tested in the previous problem gives the following results

$$V_N = e^{i[((k_{(N+1)}\cos\theta_{(N+1)} - k_N\cos\theta_N)) - (\gamma_{(N+1)}\cos\theta_{(N+1)} - \gamma_N\cos\theta_N)]h_{N+1} + i(\varphi_{N+1} - \varphi_N)} \times$$
$$\times \frac{(e^{i\varphi_N} M_{zN}\mu_N^\infty \cos\theta_N) - \left(e^{i\varphi_{N+1}} M_{z(N+1)}\mu_{N+1}^\infty \cos\theta_{N+1}\right) e^{2[((ik_N - \gamma_N)\cos\theta_{(N)})h_{N+1} + i\varphi_N]}}{(e^{i\varphi_N} M_{zN}\mu_N^\infty \cos\theta_N) + \left(e^{i\varphi_{N+1}} M_{z(N+1)}\mu_{N+1}^\infty \cos\theta_{N+1}\right) e^{2[((ik_N - \gamma_N)\cos\theta_N)h_{N+1} + i\varphi_N]}} \tag{21.25}$$

$$W_N = e^{i[((k_{(N+1)}\cos\theta_{(N+1)} - k_N\cos\theta_N)) - (\gamma_{(N+1)}\cos\theta_{(N+1)} - \gamma_N\cos\theta_N)]h_{N+1} + i(\varphi_{N+1} - \varphi_N)} \times$$
$$\times \frac{2\left(e^{i\varphi_{N+1}} M_{z(N+1)}\mu_{N+1}^\infty \cos\theta_{N+1}\right)}{(e^{i\varphi_N} M_{zN}\mu_N^\infty \cos\theta_N) + \left(e^{i\varphi_{N+1}} M_{z(N+1)}\mu_{N+1}^\infty \cos\theta_{N+1}\right) e^{2[((ik_N - \gamma_N)\cos\theta_N)h_{N+1} + i\varphi_N]}}. \tag{21.26}$$

$$V_3 = e^{i[((k_4\cos\theta_4 - k_3\cos\theta_3)) - (\gamma_4\cos\theta_4 - \gamma_3\cos\theta_3)]h_4 + i(\varphi_4 - \varphi_3)} \times$$
$$\times \frac{(e^{i\varphi_3} M_{z3}\mu_3^\infty \cos\theta_3) - (e^{i\varphi_4} M_{z4}\mu_4^\infty \cos\theta_4) e^{2[((ik_3 - \gamma_3)\cos\theta_3)h_4 + i\varphi_3]}}{(e^{i\varphi_3} M_{z3}\mu_3^\infty \cos\theta_3) + (e^{i\varphi_4} M_{z4}\mu_4^\infty \cos\theta_4) e^{2[((ik_3 - \gamma_3)\cos\theta_3)h_4 + i\varphi_3]}}, \tag{21.27}$$

$$W_3 = e^{i[((k_4\cos\theta_4 - k_3\cos\theta_3)) - (\gamma_4\cos\theta_4 - \gamma_3\cos\theta_3)]h_4 + i(\varphi_4 - \varphi_3)} \times$$
$$\times \frac{2(e^{i\varphi_4} M_{z4}\mu_4^\infty \cos\theta_4)}{(e^{i\varphi_3} M_{z3}\mu_3^\infty \cos\theta_3) + (e^{i\varphi_4} M_{z4}\mu_4^\infty \cos\theta_4) e^{2[((ik_3 - \gamma_3)\cos\theta_3)h_4 + i\varphi_3]}}. \tag{21.28}$$

The upper atmosphere boundary layer requires the analysis of four viscoelastic waves, which is typical of all seismic *SH* waves considered in condition when the upper component of the system is the atmosphere. The final formulas for the top layer of thickness h_2 are as follows

$$W_2 = \frac{2\left(e^{i\varphi_3} M_{z3}\mu_3^\infty \cos\theta_3\right) e^{i[((k_3\cos\theta_3 - k_2\cos\theta_2)) - (\gamma_3\cos\theta_3 - \gamma_2\cos\theta_2)]h_3 + i(\varphi_3 - \varphi_2)}}{\left(e^{i\varphi_2} M_{z2}\mu_2^\infty \cos\theta_2\right) + \left(e^{i\varphi_3} M_{z3}\mu_3^\infty \cos\theta_3\right) e^{2[((ik_2 - \gamma_2)\cos\theta_2)h_3 + i\varphi_2]}}, \tag{21.29}$$

$$V_2 = e^{i[((k_3\cos\theta_3 - k_2\cos\theta_2)) - (\gamma_3\cos\theta_3 - \gamma_2\cos\theta_2)]h_3 + i(\varphi_3 - \varphi_2)} \times$$
$$\times \frac{(e^{i\varphi_2} M_{z2}\mu_2^\infty \cos\theta_2) - (e^{i\varphi_3} M_{z3}\mu_3^\infty \cos\theta_3) e^{2[((ik_2 - \gamma_2)\cos\theta_2)h_3 + i\varphi_2]}}{(e^{i\varphi_2} M_{z2}\mu_2^\infty \cos\theta_2) + (e^{i\varphi_3} M_{z3}\mu_3^\infty \cos\theta_3) e^{2[((ik_2 - \gamma_2)\cos\theta_2)h_3 + i\varphi_2]}}. \tag{21.30}$$

The basic formula – the formula for vibrations on the Earth's surface – is as follows

$$u_2(x, h_2, t) = 2W_2 \cdots W_{N+1} A_{N+2} e^{-\gamma_2 h_2 \cos\theta_2} \cos(h_2 k_2 \cos\theta_2) \cos[(k_2 \sin\theta_2)x - \omega t]. \quad (21.31)$$

Thus, the presented above formulas solve the problem of harmonic *SH*-waves in the $(N + 2)$-component system "atmosphere – viscoelastic soil layers – elastic rock massif" when the deformation of layers is described by a standard rheological model with different rheological constants and the repeated reflections of waves are not taken into account.

Two variants of wave attenuation in a layer are considered. The formulas for each variant can be divided into three groups. The first group contains formulas for calculating the velocity of a wave and its attenuation coefficient. The variant 1 gives here the somewhat simpler formulas. The second group concerns the reflection and transparency coefficients. Here already the variant 2 is described by some simple formulas. The third group includes formulas for waves in a rock massif and vibrations on the Earth's surface. These formulas are actually identical for both variants.

It is worth pointing out at the end that the formulation of the problem as a problem of mechanics implies that the half-spaces and layers are deformed according to certain laws and that the geometric picture is such that all the boundaries between the layers are the planes. It is also assumed that the conditions of full mechanical contact (equality of displacements and stresses) are fulfilled at the boundaries.

The wave in this work is called seismic due to two circumstances – the wave propagates in the lower half-space at a given angle to the interface of this half-space with the first bottom soil layer (which corresponds to the real situation regarding a seismic wave in a rocky massif) and the upper half-space is considered to be a medium in which the wave does not propagate (which corresponds to the real situation when the wave comes from the last soil (upper) layer to the Earth's surface and the upper half-space is atmosphere).

21.4 Main Conclusions

The direct method of mechanics is utilized in studying the passing and reflecting of SH-wave while it propagates through the finite number of ground layers with differing thicknesses and rheological proper ties. The mechanical properties of the ground layers are described by the Hookean and standard rheological models. The explicit formulas for evaluation of the wave number and attenuation of the harmonic SH-wave are obtained. The wave attenuation is studied by two variants – by the spatial coordinate and in time. The obtained formulas can be treated as some continuation of classical works of Prof. Savarensky (Russia) in the area of elastic seismic waves and Prof. Stepanishen (USA) in the area of viscoelastic seismic waves.

The overall conclusion of this study is that the solving procedure is not formalized and at all steps exhibits physical content and mathematical transparency. The algebraic formulas obtained for oscillations on the Earth's surface and the aggregate wave in a rock solid are simple and suitable for computer simulation. An applicability of the standard rheological model of deformation of layers to the simple mathematical description of the attenuation of the amplitude wave, which actually exists when the wave propagates in the soil layers.

The perspective of the presented approach can be revealed when it will be applied in cases of using other rheological models, which are described by differential relations, since, in these cases, the explicit algebraic formulas will be also obtained for the required parameters of seismic waves.

References

Kendzera OV, Kosarev GL, Savarenskii EF (1979) Determination of the true velocity value of soil vibrations from the seismogram. Geofizicheskii zhurnal 1(1):56–62

Kendzera OV, Rushchitsky JJ, Semenova YV (2020) Allowance for the resonant properties of soils in seismic microzoning. In: Naukova Dumka (in press)

Rushchitsky JJ (2012) Theory of waves in materials. Ventus Publishing ApS, Copenhagen

Rushchitsky JJ (2019) Cylindrical surface seismic wave: revisiting the classical Biot's problem. In: Kanao M, Toyo-kuni G (eds) Seismic waves – probing earth system, pp 43–63

Savarenskii EF (1972) Seismic waves. Sewa, Nedra, Moscow

Shearer PM (2009) Introduction to seismology. Cambridge University Press, Cambridge

Slawinski MA (2003) Seismic waves and rays in elastic media. Elsevier, London

Stepanishen PR, Strozeski B (1972) Reflection and transmission of acoustic wide band plane waves by layered viscoelastic media. J Acoust Soc Am 71(1):9–21

Part VII
Innovative Approaches for Geohazard and Risk Assessment

Chapter 22
Commercial-UAV-Based Structure from Motion for Geological and Geohazard Studies

Fabio L. Bonali, Noemi Corti, Elena Russo, Fabio Marchese, Luca Fallati, Federico Pasquaré Mariotto, and Alessandro Tibaldi

Abstract In the present work we applied the use of the UAV-based Structure from Motion technique (SfM) to geological and geohazard studies, with emphasis placed on active tectonics and volcano-tectonics cases. Our aim is to obtain high-resolution orthomosaics and Digital Surface Models (DSMs) in two study areas: the Theistareykir Fissure Swarm within the Northern Volcanic Zone (NVZ) of Iceland and the active Khoko landslide, Enguri reservoir, in the Greater Caucasus, Georgia. The first is affected by seismic and volcanic hazard, the second by landslide and hydrogeological hazard. Regarding the NVZ, by analysing the resulting Orthomosaics and DSMs we collected a total of 453 quantitative measurements of the amount of opening and opening direction of Holocene extension fractures and 36 measurements of the height of fault scarps. These data allowed us to assess an overall spreading direction of N106.4° during Holocene times within the studied rift zone, which has been compared with geodetic motion vectors, and a stretching ratio of 1.013–1.017 for 8–10 ka old lava units. We conclude that deformation in the area is related to both dyke intrusions and extensional tectonics. In the Greater Caucasus, we applied the method to identify the main geomorphological features related to the Khoko landslide and to measure the scarp height of the principal slip surfaces, in order to improve geomorphological knowledge of the landslide, and contribute to the assessment of the hydrogeological hazard of the area. At a general

F. L. Bonali · E. Russo · A. Tibaldi
Department of Earth and Environmental Sciences, University of Milan-Bicocca, Milan, Italy

CRUST- Interuniversity Center for 3D Seismotectonics with Territorial Applications, Chieti Scalo, Italy
e-mail: n.corti3@campus.unimib.it

N. Corti · F. Marchese · L. Fallati
Department of Earth and Environmental Sciences, University of Milan-Bicocca, Milan, Italy
e-mail: n.corti3@campus.unimib.it

F. Pasquaré Mariotto
Department of Theoretical and Applied Sciences, Insubria University, Varese, Italy

© Springer Nature B.V. 2021

F. L. Bonali et al. (eds.), *Building Knowledge for Geohazard Assessment and Management in the Caucasus and other Orogenic Regions*, NATO Science for Peace and Security Series C: Environmental Security,
https://doi.org/10.1007/978-94-024-2046-3_22

level, our results suggest that the use of UAV-based SfM is a convenient and efficient way to collect plenty of data aimed at better assessing geohazards in areas prone to catastrophic natural phenomena like earthquakes, volcanic eruptions and landslides.

Keywords UAV or drone · Structure from motion · Caucasus · Iceland · Tectonics · Geohazard

22.1 Introduction

Hereunder, we introduce and illustrate the Structure from Motion techniques applied to geological and geohazard studies: indeed, this represents a powerful technique to collect a wealth of data aimed at geohazard assessment and may be regarded as a crucial advancement with respect to research based only on classical fieldwork. Particularly, we focused our attention on two different key sites and topics: (i) extension fractures and normal faults in the Theistareykir Fissure Swarm (Northern Volcanic Zone, Iceland); (ii) the active Khoko landslide near the Enguri dam (Greater Caucasus, Georgia).

Data collection and field observations are essential, especially when dealing with geohazards assessment but, unfortunately, very often the sites are characterized by difficult logistic conditions, and thus they are not easily reachable. In order to overcome these limitations, remote sensing-derived images have been increasingly used in the scientific community over the last decade, allowing to investigate and collect geological data in inaccessible and wider areas.

In structural geology, studies aimed at seismic hazard assessment and field or remote-based studies have been focused on joints, faults, extension fractures and folds (e.g. Gudmundsson 1987; Cembrano and Lara 2009; Tadini et al. 2014; Tibaldi et al. 2017a, b). Joints are usually a few meters to some tens of meters long, extension fractures can be hundred meters up to kilometres long, whereas folds and faults can reach tens of kilometres in length (e.g. Tibaldi et al. 2015; Tibaldi and Bonali 2018). Mapping the activity of these types of structures has been based on satellite images interpretation since the 1980s (e.g. Wise et al. 1985; Marino and Tibaldi 1988; Tibaldi and Ferrari 1991, 1992; Ramli et al. 2010), and since the 1990s, these techniques have been applied to locate landslides and thus to assess slope instability and the related geohazards (e.g. Soeters and Cornelis's 1996; Nichol and Wong 2005; Nichol et al. 2006), as well as to investigate land surface changes with multi-temporal satellite images (e.g. Cheng et al. 2004). Space based remote studies offer some revolutionary features: (i) global coverage; (ii) time repetition; (iii) synoptic plan view; (iv) wide spectrum of sensors, with different wavelength and resolution. On the other hand, satellite images present also some limitations, that can be overcome by using Unmanned Aerial Vehicles (UAV) (Fig. 22.1): (i) they do not allow to survey a defined area on a specific date; (ii) resolution is high enough to allow a detailed structural mapping, but data collection in some cases, such as the opening direction along extension fractures in Iceland, cannot be

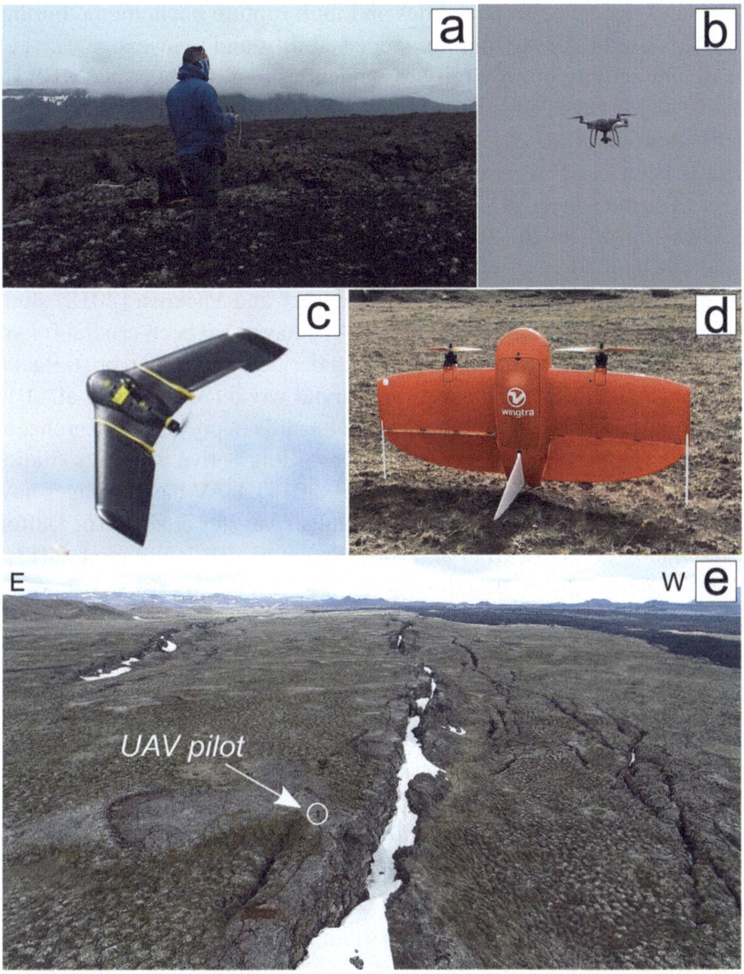

Fig. 22.1 (**a**) UAV pilot at work, (**b**) Commercial UAV quadcopter in action, (**c**) Fixed wings UAV type, (**d**) VTOL (vertical take-off and landing) UAV type – this picture is courtesy of Joel Ruch -, (**e**) Drone-captured view of a fracture field, a typical feature within the active rifts of the Northern Volcanic Zone of Iceland, where it is possible to observe both extension fractures and major dip-slip faults. UAV pilot shown for scale

properly provided (e.g. Bonali et al. 2019); (iii) they can be expensive and therefore, sometimes, they are not used for geological studies. As a consequence of that, UAVs have been increasingly used to enhance knowledge on different types of geological hazards over the past decade. Regarding seismic hazard, Baiocchi et al. (2013) and Freeman et al. (2019) monitored urban areas affected by earthquakes, Pesci et al. (2018) and Yao et al. (2019) evaluated soil deformation due to liquefaction and Zekkos et al. (2018) created 3D models in order to analyse post-earthquake effects in New Zealand, focusing on landslides and fault rupture phenomena. Furthermore, several authors used UAVs to monitor slope stability, and to investigate slope failure mechanism and activity in proximity of landslides (e.g. Gong et al. 2010; Hu et al. 2012; Rathje and Franke 2016; Brook and Merkle 2019; Blistan et al. 2019; Cignetti et al. 2019; Warrick et al. 2019). Regarding volcanic hazard, UAV-acquired aerial images have been used to assess deformation during volcanic eruptions (Müller et al. 2017; Darmawan et al. 2018; De Beni et al. 2019) and to study features of lava flows (Favalli et al. 2018). In regard to flood hazard, Govedarica et al. (2018) and Hashemi-Beni et al. (2018) created Digital Elevation Models (DEMs) aimed at assessing flood risk, while Langhammer and Vackova (2018) studied the morphological consequences of flooding. UAVs have also been crucial for studying geysers and hot springs (Walter et al. 2018; Federico et al. 2019), post-glacial uplift (De Quay et al. 2019) and for quantifying fluvial wood (Sanhueza et al. 2019), and have been employed also to assess geological hazards in proximity of archaeological sites (Themistocleous and Danezis 2018). Regarding active tectonics studies, after the first attempts with balloons (Johnson et al. 2014), UAV-based images have been also recently used to map and study active faults (Angster et al. 2016; Deffontaines et al. 2016; Jiao et al. 2016; Bi et al. 2017; Gao et al. 2017; Wu et al. 2017; Cheng et al. 2018; Rao et al. 2018; Bonali et al. 2019; Trippanera et al. 2019; Weismüller et al. 2019).

UAVs became this popular for scientific purposes, thanks to a number of characteristics that make them more competitive and useful than satellites: (i) appropriate data acquisition and time can be chosen; (ii) flight height can be adjusted to obtain the required (very high) spatial resolution; (iii) flights can be repeated daily or several times per day; (iv) three-dimensional models of terrains can be easily produced from the acquired RGB camera; (v) it is possible to survey vertical rock cliffs; (vi) the cost of data acquisition is substantially lower than that of high-resolution satellite imagery; (vii) UAVs can carry various types of sensors designed for specific purposes.

In view of the above, we aim to contribute to the use of the UAV-based SfM approach for geological and geohazard studies, focusing on structural geology and geomorphology in two different geological environments, both characterized by difficult logistic conditions and impending geohazards. The first is the Theistareykir Fissure Swarm, belonging to the Northern Volcanic Zone, in Iceland (Fig. 22.2), affected by volcanic and seismic hazards. In this case, we surveyed normal faults and extension fractures (Fig. 22.3), collecting the azimuth and scarp height for faults and the azimuth, opening direction and amount of opening for extension fractures. As a result, we reconstructed the Holocene strain field of a wide area within the northern Iceland rift and compared our results with other geophysical data, such as geodetic measurements. The second key area is the Khoko landslide, located near the Enguri Dam in the southwestern sector of the Greater Caucasus (Fig. 22.4); here, we used the UAV-based SfM technique in order to recognize the main geomorphological features related to the presently, active landslide. These surveys contribute to providing new quantitative data regarding tectonic and magmatic-induced superficial deformation that can be used for better assessing seismic and

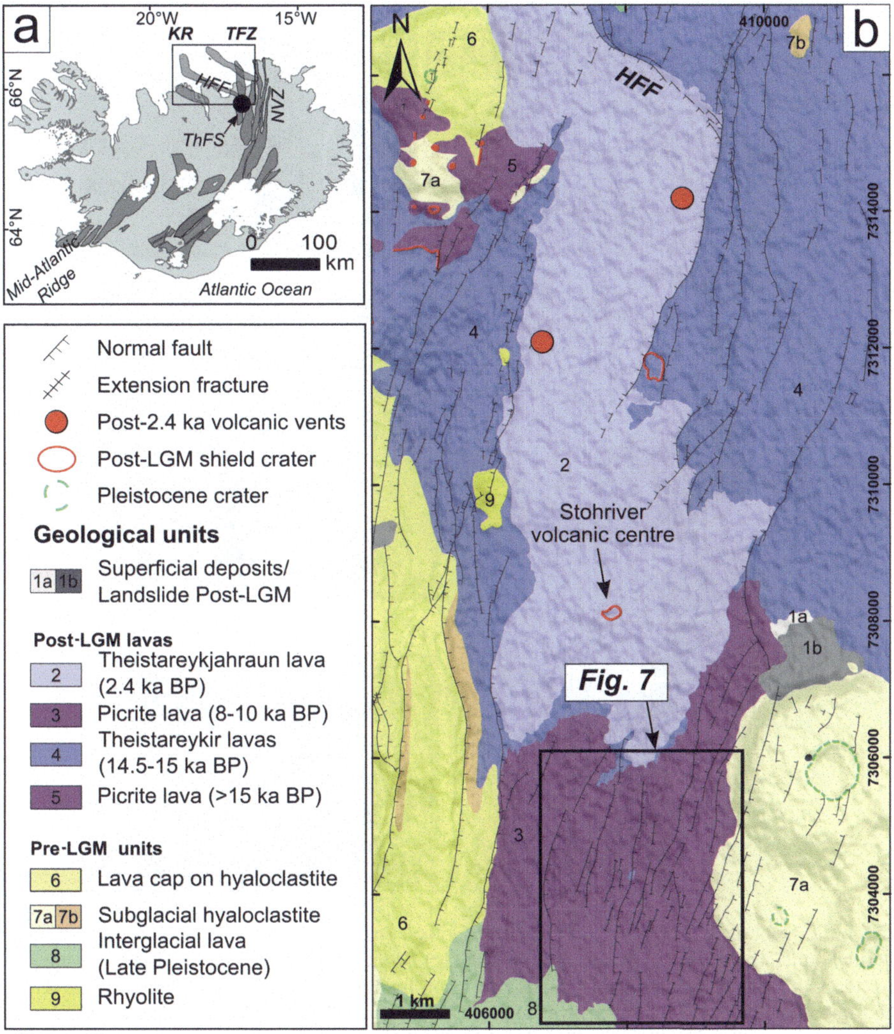

Fig. 22.2 (**a**) Location of the study area (black dot), Tjornes Fracture Zone (black box) and volcano-tectonic rift zones of Iceland (after Hjartardóttir et al. 2016), reference system: WGS84/geographic coordinates. *NVZ* Northern Volcanic Zone, *TFZ* Tjornes Fracture Zone (black box), *HFF* Husavik-Flatey transform fault, *ThFS* Theistareykir Fissure Swarm, *KR* Kolbeinsey Ridge (**b**) Geological map of the study area, indicated by the black rectangle, based on Saemundsson et al. (2012), reference system: WGS84/UTM zone 28 N. Figure 22.7 is located. *LGM* Last Glacial Maximum. (Modified after Bonali et al. 2019)

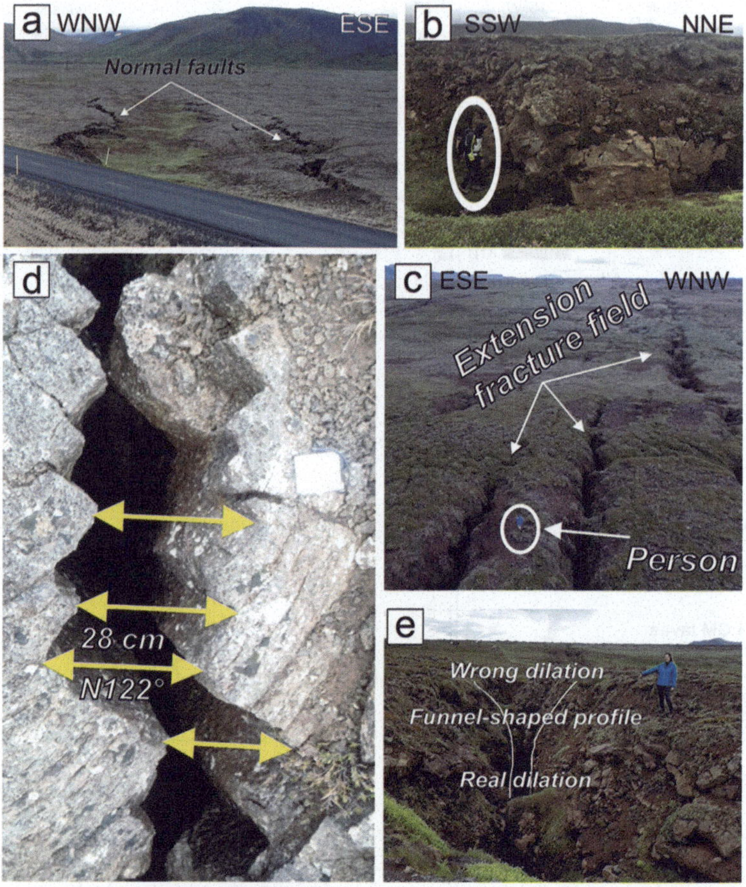

Fig. 22.3 Geological objects of interest. (**a**) Graben structures, with the two opposite-dipping normal faults, (**b**) Normal fault, person for scale, (**c**) Drone-captured picture of a field of extension fractures in Iceland, person for scale, (**d**) Extension fracture with indication of opening direction and amount of opening. Yellow arrows represent piercing points, (**e**) Detail of an extension fracture, with real and wrong dilation indicated, the latter being caused by the funnel-shaped profile

volcanic hazard in the first case; regarding the second case, new data on landslide scarps can be used for further landslide and hydrogeological hazard assessment in Enguri reservoir area.

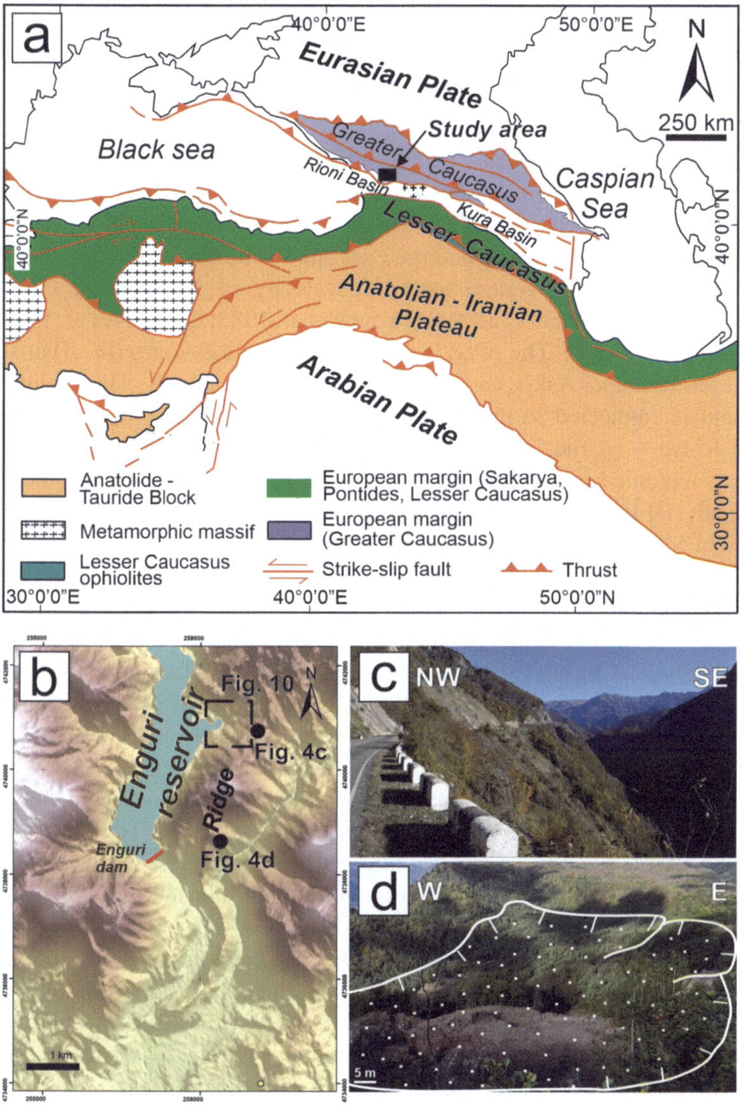

Fig. 22.4 (**a**) Tectonic map of the Arabia – Eurasia collision zone (from Tibaldi et al. 2019), where a black box shows the location of the study area, reference system: WGS84/geographic coordinates; (**b**) DEM of the Enguri Dam area, reference system: WGS84/UTM 38 N. Box locates Fig. 22.10; black dots locate Fig. 22.4c, d, (**c**) Segment of the Jvari-Khaishi-Mestia road crossing the eastern side of the slope, on the left side fresh minor landslides are present; (**d**) Oblique view (looking north) of the area, facing the Enguri water reservoir, affected by active slope deformation, indicated with a dotted white area. The scale is referred only to the forefront zone

22.2 Geological Background of the Selected Key Areas

22.2.1 *Theistareykir Fissure Swarm, NE Iceland*

The key site we selected to study extension fractures and normal faults using UAV-based SfM techniques, is the southern sector of the active Theistareykir Fissure Swarm (ThFS), that represents the westernmost active rift of the Northern Volcanic Zone (NVZ), the northernmost emerged section of the Mid-Atlantic Ridge, active since 8–9 Ma with an average spreading velocity of 19 mm/year (Fig. 22.2a; Saemundsson 1974; Jancin et al. 1985; Young et al. 1985; De Mets et al. 1994; Bergerat and Angelier 2008). The NVZ is composed of five rift zones (the Theistareykir, Krafla, Fremrinámar, Askja and Kverkfjöll volcanic systems, Hjartardòttir et al. 2016), and is connected to the Kolbeinsey Ridge (KR) – expression of the Mid-Atlantic Ridge – by the Tjornes Fracture Zone (TFZ); the latter is composed of three main tectonic lineaments reported in the box in Fig. 22.2a, where the Husavik Flatey Fault (HFF) is the central one.

The ThFS is a 10 km-wide elongated swarm of structures composed of: (i) normal faults, characterized by metric to decametric vertical offsets (Figs. 22.1e and 22.3a–b); (ii) extension fractures, which are characterized by dilation up to several metres (Figs. 22.1e and 22.3c–e); and (iii) eruptive fissures. These structures strike predominantly N-S to NNE-SSW and show a change in their azimuth only in the region where the HFF, a dextral strike-slip fault belonging to the TFZ, intersects the ThFS. In this area, the structures tend to rotate in an anticlockwise manner with respect to the overall orientation of the rift, with azimuth between NNW-SSE and N-S (Gudmundsson et al. 1993; Fjäder et al. 1994; Magnusdottir and Brandsdottir 2011; Pasquaré Mariotto et al. 2015; Hjartardóttir et al. 2016; Tibaldi et al. 2016a, b). From a lithological point of view, the ThFS is characterized by a wide range of depositional units, consisting of sedimentary and volcanic units of Miocene to Pleistocene age, and lava flows of Pliocene, Pleistocene and Holocene age (Fig. 22.2b; Saemundsson 1974; Garcia et al. 2002; Saemundsson et al. 2012). During historical times, the ThFS has been affected by the following rifting and volcanic unrest episodes: (i) between 1867 and 1868, an M 6 earthquake hit in the northern part of the ThFS, accompanied by volcanic eruptions within the Mánáreyjar volcanic system, situated offshore (Thoroddsen 1925; Halldórsson et al. 1984; Magnúsdóttir and Brandsdóttir 2011); (ii) between 1884 and 1885, an M 6.3 earthquake hit in near the coast, representing the apex of an earthquake swarm along the rift (Halldórsson et al. 1984); (iii) between 2007 and 2009 an 8-cm uplift in correspondence of the Theistareykir central volcano was observed, due to the inflation of the underlying magma chamber (Metzger et al. 2011; Metzger and Jónsson 2014). The area we chose is located south of the intersection with the HFF and of the main Theistareykir volcano (Fig. 22.2b). From a lithological point of view, only the "Borgarhraun" picritic lava (dated 8–10 ka BP) crops out in the study area (Saemundsson et al. 2012). Our UAV surveys are focused on recent lavas, with the purpose of analysing deformation caused by tectonic and volcano-tectonic activity without the influence of glaciers that dominated the area during the Late Glacial Maximum: thus, we did not include in our study the Late Quaternary interglacial lava located to the SW and the Weichselian subglacial hyaloclastite to the E. In this area, the analysis of the Orthomosaics and Digital Surface/Terrain Models (DSMs/DTMs) obtained through UAV-based SfM approach enabled us to measure the opening direction and amount of opening along the extension fractures, in order to describe the Holocene strain field, and the height of active fault scarps. The newly surveyed areas have been merged with those from Bonali et al. (2019) in order to present a geological study focused on a larger area and, consequently, a more complete analysis and evaluation of the ongoing volcano-tectonic processes.

22.2.2 The Active Khoko Landslide in the Greater Caucasus

The second key area is located near the Enguri Dam, in the southwestern sector of the Greater Caucasus, Georgia (Fig. 22.4a–b). The Greater Caucasus is a fold-and-thrust belt, separated from the Lesser Caucasus by the Black Sea – Rioni Basin and the Caspian – Kura intermontane depression. (Adamia et al. 1977, 2010; Banks et al. 1997; Mosar et al. 2010; Sosson et al. 2010). These regions developed as foreland basins (Oligocene – Early Miocene), and were later included in the fold-and-thrust belts (Forte et al. 2010; Adamia et al. 2011; Alania et al. 2016). These, in turn, were affected by intensive deformation over the last 14–15 Ma, represented mainly by fault slip along thrust systems, whose rate reached a peak at the end of the Miocene (Vincent et al. 2007, 2011; Avdeev and Niemi 2011; Alania et al. 2016; Adamia et al. 2017).

Deformation has been active up to the Quaternary, and is still affecting the area; in detail, historical and instrumental seismic catalogues, earthquake focal mechanism solutions, seismic reflection sections and geological-structural field surveys show that an active compressional tectonics is still present both at the core of the Greater Caucasus and along its southern border and at the fold-and-thrust belts (Tsereteli et al. 2016; Tibaldi et al. 2017a, b). According to these authors, at the core of the Greater Caucasus there are a number of thrusts striking WNW-ESE, dipping towards NNE and marked by pure dip-slip movements. Furthermore, along the southern border of the Greater Caucasus, where our study area is located, thrust planes dipping towards NNE can be observed (Fig. 22.4a; Tibaldi et al. 2017a).

From a stratigraphic point of view, the axial zone of the Greater Caucasus is characterized by the presence of the pre-Mesozoic basement and Jurassic sedimentary rocks, while in the outermost zones, Cretaceous and Cenozoic rocks crop out (Mosar et al. 2010; Adamia et al. 2011). The study area is dominated by Jurassic volcanic and terrigenous rocks and Cretaceous carbonatic deposits, dipping mostly towards S (Tibaldi et al. 2019). The same authors observed that in the Enguri Dam area, Cretaceous rocks crop out, striking E-W and dipping 60–70° towards the S, becoming steeper going southward and subhorizontal in the northern part of the Enguri reservoir. Below the Cretaceous rocks, Jurassic deposits made of sandstones, gypsum, tuffs and tuff-breccia are present, and they locally crop out along the southeastern side of the reservoir, dipping to the S. From a geomorphological point of view, there are two parallel river valleys in the area with a nearly N-S trend, and the Enguri reservoir is located in the westernmost one (Fig. 22.4b). Between the two valleys there is a mountain with steep slopes, locally reaching 70°. This leads to several processes of landsliding, which affect the opposite sides of the mountain ridge, and also the Jvari-Khaishi-Mestia main road (Fig. 22.4c), that runs along the eastern side of Enguri reservoir (Fig. 22.4b). The western slope of the mountain ridge is characterized by different scarps facing westward, directly over the Enguri reservoir. Most of the scarps are perpendicular to the slope dip, and the head scarps are affected by several fractures and fissures, especially where the Jvari-Khaishi-Mestia road intersects the head scarp. Tibaldi et al. (2019) monitored this displacement between November 2015 and May 2017, obtaining a vertical component of 14 cm, and an average slip rate of 9.3 cm/yr at the northernmost structure. Monitoring activity has been carried by establishing benchmarks of geodesic observation station and Tacheometric surveys.

The western side of the mountain faces directly the Enguri reservoir (Fig. 22.4d), which represents the major source of hydroelectric energy in the country thanks to the Enguri dam, the sixth highest dam in the world. For this reason, we decided to choose the western landslide, locally known as Khoko, as a key site for this work, and we used UAV-SfM derived models to recognize the main geomorphological features present in the area and to measure the scarp height of the principal slip surfaces; this was done with the goal of improving geomorphological knowledge of the landslide, and contributing to the assessment of the hydrogeological hazard in the area. The UAV survey has been made possible in the framework of NATO Project G4934 titled: "*Security against geohazards at the major Enguri hydroelectric scheme in Georgia*" (http://www.enguriproject.unimib.it/).

22.3 UAV-Based Methodology

22.3.1 UAV Selection

Different types of UAVs can be used for geological studies (Fig. 22.1): balloons, multi-rotor (Fig. 22.1b), fixed-wing (Fig. 22.1c) and hybrid (Fig. 22.1d). While balloons do not need fuel or a battery, on the other hand they cannot be remotely controlled. Hybrid types allow to switch between flying like a fixed-wing aircraft and hovering like a multi-rotor one: an example of Tailsitter VTOL (Vertical Take-off and Landing) is shown in Fig. 22.1d (the picture is courtesy of Joël Ruch). We decided to work with a commercial quadcopter because it can be easily transported during field work, with several batteries, it can fly at very low heights attaining a great field resolution, and take-off and landing operations are easy; all these points are crucial especially in difficult logistic terrains (e.g. outcropping lavas or remote beach areas, Bonali et al. 2019; Fallati et al. 2019). Considering all the above, we decided to use multi-rotor UAVs (Fig. 22.1b) for the present work, in particular the DJI Phantom 4 and the DJI Phantom 4 PRO quadcopters, for the 2017 and 2018 survey respectively, which are among the most frequently used types of UAV.

The DJI Phantom 4 is equipped with a 12.4 Megapixels camera, including EXIF (Exchangeable Image file Format) information and GPS coordinates, provided by the integrated Satellite Positioning Systems GPS/GLONASS (referred to the WGS84 datum), and it is characterised by a long flight time ($\approx$28 min). On the other hand, the DJI Phantom 4 PRO, which represents an upgrade with respect to the DJI Phantom 4, is equipped with a 20 Megapixels camera, and its flight time is extended up to 30 min approximately. The degree of resolution provided by both instruments and the set of attached information are useful for SfM processing aimed at research purposes.

22.3.2 UAV-Based Structure from Motion Workflow: Data Collection (Part I)

Our workflow aimed at the reconstruction of DSMs and Orthomosaics and it can be separated in two parts (Fig. 22.5a): Part I is devoted to data collection (photos and Ground Control Points – GCPs) whereas Part II is aimed at data processing and model reconstruction.

Regarding Part I, the first step is to define the area to be surveyed: UAV flight missions, including paths orientations, are planned and managed using the DJI Ground Station Pro software (https://www.dji.com/ground-station-pro). In doing this, the pilot needs also to take into account the wind direction that can affect UAV flight performance and therefore the quality of the captured images: it is much better to have the wind parallel to the main and longer paths (e.g. Fig. 22.5b).

The photo overlap has also to be defined, as suggested by Gerloni et al. (2018), Bonali et al. (2019), Krokos et al. (2019) and Antoniou et al. (2019). In fact, it is recommended that UAV-captured photos should have an overlap of 90% along the path and 80% in lateral direction, in order to obtain a greater alignment of images and to reduce the distortions on the resulting Orthomosaics. In the present work, we captured the photos in a range of 90–85% overlap along the path and 80–75% in lateral direction, thus allowing a larger areal coverage. The photos have been captured every 2 s (equal time interval mode), with a constant velocity of 3 m/s (low speed) and in optimal light conditions suitable for the camera ISO range (100–1600), to minimize the motion blur, to avoid the rolling shutter effect, as well as to achieve well-balanced camera settings (exposure time, ISO, aperture) and ensure sharp and correctly exposed images with low noise (e.g. Vollgger and Cruden 2016). Finally, in order to reduce shadows around elevated features, we flew with the sun to the zenith in both areas.

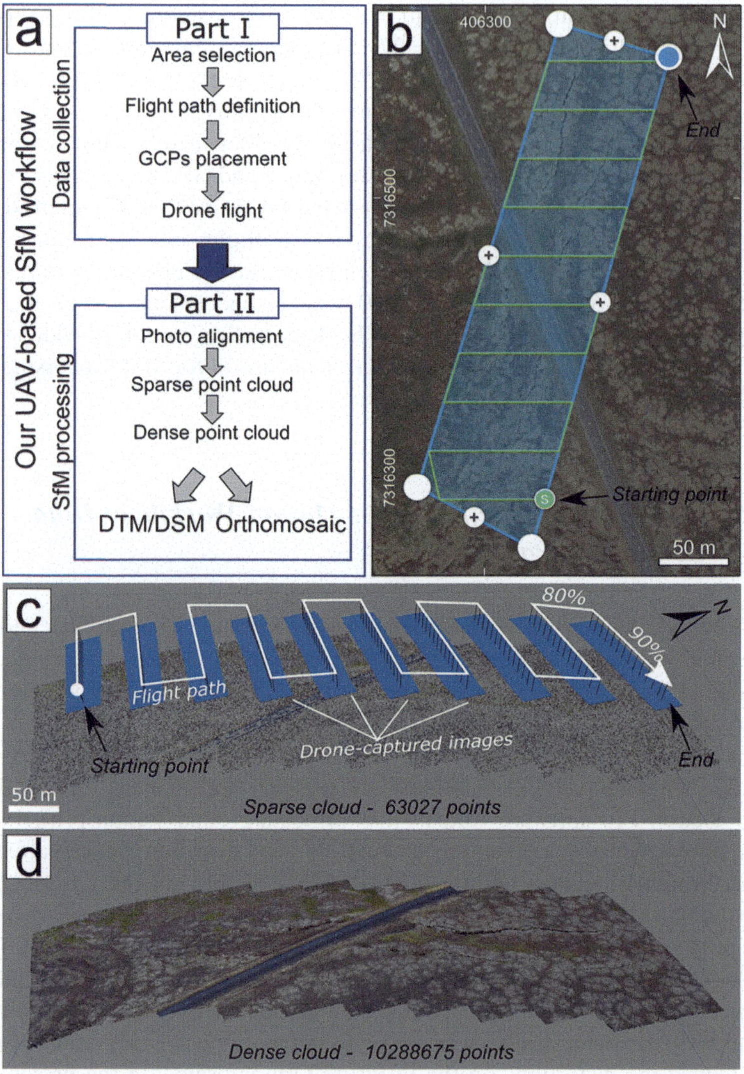

Fig. 22.5 (**a**) Overall workflow used in the present research to generate DTMs/DSMs and Orthomosaics using SfM software, (**b**) UAV flight path (green line) at 50 m flight height. Wind conditions have also been taken into account. Blue square indicates the surveyed area, (**c**) Sparse and (**d**) Dense cloud generated by SfM software. The highest suggested overlap among pictures is also reported. Computed camera positions are represented as blue rectangles, black lines show the pitch angle and camera orientation

In order to allow for the co-registration of datasets and the calibration of models resulting from SfM processing (e.g. Orthomosaics and DSMs), World Geodetic System (WGS84) coordinates of, at least, four artificial Ground Control Points (GCPs) were established near each corner within each surveyed area and one in the central part (e.g. Fig. 22.6a) (e.g. James and Robson 2012; Turner et al. 2012; Westoby et al. 2012; James et al. 2017) for reducing the 'doming' effect resulting from SfM processing, for a total of 5 GCPs in each surveyed area. For surveying GCPs, the best way is to use well visible artificial markers (printed coded targets). By the way, in critical logistic condition, we have also used natural and ecological targets (e.g. Fig. 22.6b–d) in order to save time during GCPs survey. In general, we collected the GCPs only over objects expected to be clearly visible from the orthophotos and with a shape or structure that allowed us to be as accurate as possible, with 5–10 cm size. During GPS points collections, we took ground photos from different perspectives to facilitate the precise placement of the points on the orthophoto. Natural targets can be represented by very well visible stones, lava flow borders (e.g. Fig. 22.6b), or piercing points along fractures; as ecological target, we used five slices of bread forming an X on the ground, pointing the central part of the X (Fig. 22.6d), which is 10 cm size. Even though this last approach can introduce very few centimetres of accuracy, GCPs survey process can be considerably speeded up; velocity in GCPs and drone survey is a fundamental issue in place like Iceland where the weather conditions can change rapidly and the target area can be very far away from roads and tracks. GCPs were surveyed with the Emlid Reach RS©, low-cost single frequency receivers (Rover and Base) in RTK configuration (with centimeter-level accuracy). For the surveys in the ThFS, the Rover was linked in real time by NTRIP (Networked Transport of RTCM via Internet Protocol) to the Icelandic CORS network (ICECORS – https://www.lmi.is/en/icecors-network/), used as reference station. Emlid Reach RS© runs an open-source RTK processing software called RTKLIB by Takasu (2009), that is available directly on receiver and does not require any computer to run, if the base stations are at 15 km maximum distance from the rover. In our case, the medium distance from our rovers to the ICECORS network (Myvatn base station) was 35 km so we performed an additional post-processing correction using the RTKLIB software for computers (http://www.rtklib.com/). We applied a Post Processed Kinematic (PPK) correction using the raw RINEX data, registered by the rover, and the RINEX data of the base (downloadable from https://www.lmi.is/en/icecors-network/). With this process, we reached an RTK fix status that, considering the distance from the base, it is in the order of few centimetres.

Instead, in regard to the survey on Khoko landslide, due to the absence of an available NTRIP network, we used the Long-Range Radio (LoRa 868/915 MHz) connection mode. The Base was set on a fix position and sent a real-time correction to the Rover (Hill et al. 2019). The LoRa mode can be advantageous in the absence of international GNSS service or CORS network. All the z (altitude above the sea level) values of the GCPs were corrected using the regional geoid model to obtain the orthometric height for the models.

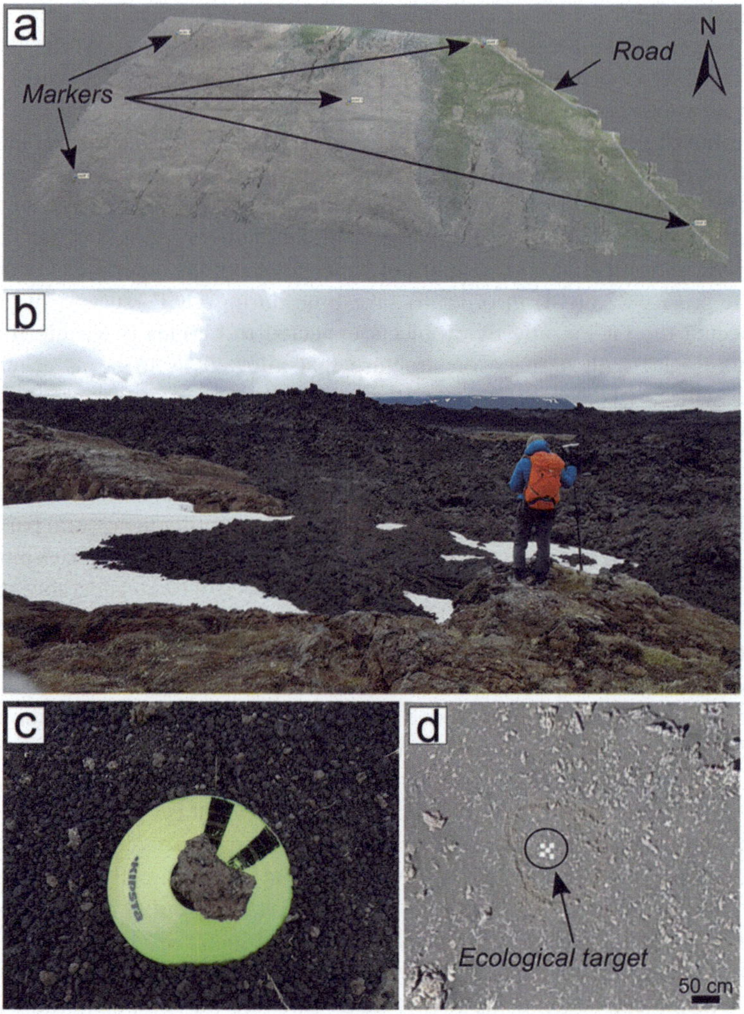

Fig. 22.6 (**a**) Dense cloud related to Area 09 (located in Fig. 22.7) with reported Ground Control Points used to georeference the model, (**b**) Operator collecting Ground Control Points on a natural target, (**c**) artificial target and (**d**) ecological target used in the present work

22.3.3 UAV-Based Structure from Motion Workflow: Data Processing and Model Reconstruction (Part II)

Part II of our workflow focus on data processing aimed at model reconstruction, in terms of DSMs and Orthomosaics. Photos have been processed with the use of Agisoft Metashape (http://www.agisoft.com/), a commercial Structure from Motion software (SfM). This software is becoming increasingly widely used in the geological community for both UAV and field based SfM studies for its user-friendly interface, intuitive workflow and high quality of points clouds (Benassi et al. 2017; Burns and Delparte 2017; Cook 2017). The SfM technique allowed to identify matching features in different photos, collected along a defined flight path (Fig. 22.5b), and combined them to create a sparse and dense cloud (Fig. 22.5c–d), eventually a mesh, an Orthomosaic, and a DTM/DSM as final products (further details in Stal et al. 2012; Westoby et al. 2012). The overall workflow for model reconstruction is shown in Fig. 22.5a; further details are provided hereunder. The first step was to obtain an initial low-quality photo alignment, only considering measured camera locations (Reference preselection mode). After that, photos with quality value <0.5, or out of focus, were excluded from further photogrammetric processing, as suggested in the user manual of the software (Agisoft 2018). The focal length and photo dimensions were automatically detected by the software, after the import of digital images and then used for the subsequent calibration of the intrinsic parameters of the camera (principal point coordinates, distortion coefficients). After this first quality check, we added Ground Control Points (GCPs) in all photos, where available, in order to: (i) scale and georeference the point cloud (and thus the resulting model); (ii) optimize extrinsic parameters, such as estimated camera locations and orientations; (iii) improve the accuracy of the final model. Using RTK/GNSS GPS information, we expect to obtain a model with centimetric accuracy, as already highlighted by several authors; for a more in-depth reading about the accuracy we suggested the work proposed from the following authors: Tonkin and Midgley (2016), Chandler and Buckley (2016), Ruzgiene et al. (2015) and Ouédraogo et al. (2014).

We then realigned the photos using high accuracy settings: as a result of this step, camera location and orientation were better established, and the sparse point cloud was computed by the software (e.g. Fig. 22.5c).

The next step consisted in reconstructing the dense point cloud (e.g. Fig. 22.5d), calculated from the sparse point cloud, using a Mild depth filtering and medium quality settings. The mesh was reconstructed considering a Height field surface, the dense cloud as source data for a Medium number of Face count. Finally, since the DSM could be generated both by the Mesh and by the dense cloud as source data, we preferred the latter to obtain a better resolution, referring the DSM to the geographic coordinate system (WGS84); the Orthomosaic was generated considering the DSM as surface (and thus the WGS84 system). In our models, the overall RMSE is lower than 0.10 m.

22.4 Results

22.4.1 Theistareykir Fissure Swarm, NE Iceland

In the first key area, we covered a surface of about 6.7 km^2 conducting 12 different surveys, at 90–100 m of height from the ground; the surveys were carried out during two campaigns during summer of 2017 and 2018 (Fig. 22.7), which enabled

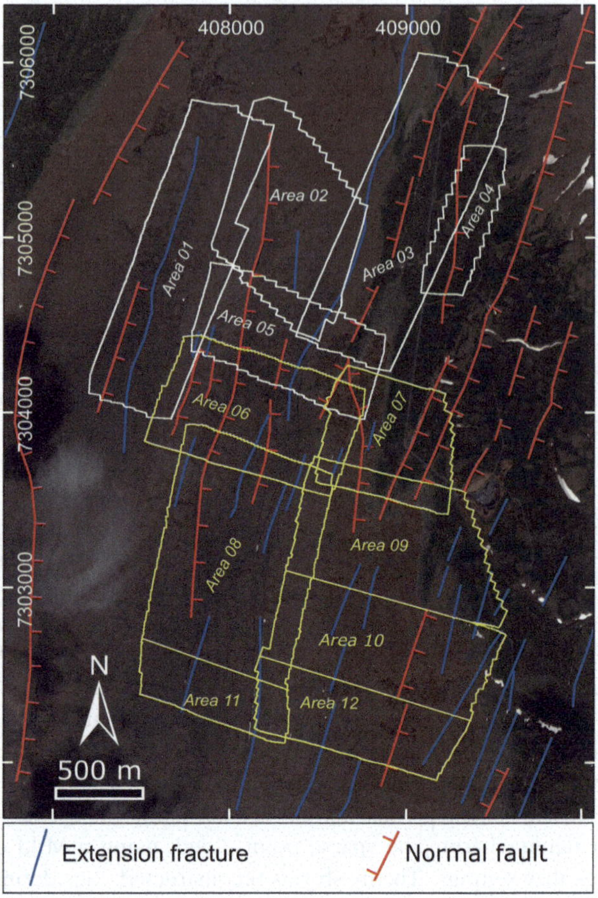

Fig. 22.7 Areas studied through UAV-based surveys in 2017 and 2018, respectively shown as white and yellow boxes. Extension fractures and normal faults are from Saemundsson et al. (2012). Reference system: WGS84/UTM zone 28 N

collecting a total of 8841 photos. Thanks to the above described workflow, the photos were used to reconstruct a high quality Orthomosaic and a DSM (Fig. 22.8a–b).

Due to the large amount of data and to the two different UAVs we used for the surveys, the two areas were processed separately. The area surveyed in 2017 resulted in 4652 UAV-collected photos, an Orthomosaic and a DSM characterized by a resolution of 4.17 and 17.6 cm/pixel, respectively, for an area of 3.04 km^2. The area surveyed in 2018 resulted in 4189 UAV-collected photos, an Orthomosaic and a DSM characterized by a resolution of 2.56 and 10.2 cm/pixel, respectively, for an area of 3.87 km^2. This area is important in order to understand rifting processes that have occurred in the last 8–10 ka BP. Thanks to the present work, we mapped a total of 546 fractures classified as normal faults (29) and extension fractures (517). They strike from N-S to NE-SW, whereby the strike NNE-SSW is the most represented, with 42% of the whole set between 11°N and 20°N, as shown in the rose diagram in Fig. 22.8c. The longest fracture reaches a length of 1173 m, whereas the overall

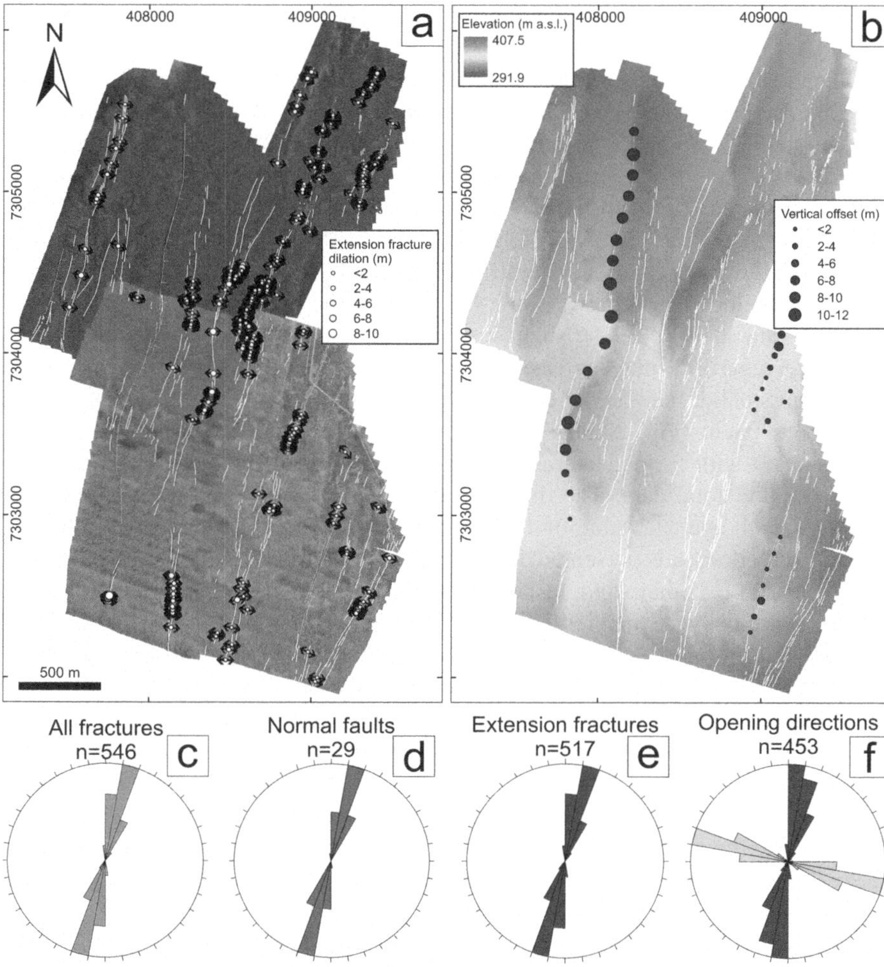

Fig. 22.8 (**a**) Orthomosaics of the surveyed areas represented in Fig. 22.7 with reported scaled dilation values of extension fractures as white circles and opening directions as black diverging arrows, (**b**) DSMs of the areas with scaled vertical offset. Faults and extension fractures are represented as white lines. Normal faults can be distinguished by the white short lines that indicate the downthrown block. Rose diagrams display the average strike of (**c**) all structures, (**d**) normal faults and (**e**) extension fractures, and (**f**) opening direction vs strike of the surveyed extension fractures. Reference system: WGS84/UTM zone 28 N

average length is 77.9 m. The number of mapped fractures is much greater than the 37 main fractures documented by Saemundsson et al. (2012) (Fig. 22.2), thanks to the high resolution of the new DSMs and Orthomosaics.

Regarding extension fractures, they mainly strike from N-S to NNE-SSW (Fig. 22.8e), the longest one reaches a length of 706 m and the overall average length is 70.3 m. Much more importantly, we gathered 1359 structural data at 453 structural stations, where we collected the values of strike, opening direction (Fig. 22.8f) and dilation.

Opening direction values range between 81.5°N and 148.8°N (Fig. 22.9g); more in detail, they range between 92.9°N and 148.8°N in the 2017 surveyed area, and between 81.5°N and 142.1°N in the 2018 surveyed area. We calculated an average value of 108.2°N in the area surveyed in 2017, with most of the values (48%) included between 101°N and 110°N and a value of 103.6°N in the area surveyed in 2018, with most of the values (36%) again between 101°N and 110°N. Considering the whole set we obtained an overall opening direction of 106.4°N, with 44% of the values included in the 101°–110°N interval. The average azimuth is 12.6°N (rose diagram in Fig. 22.8f), suggesting a slight overall right-lateral component of 4°. We regard as pure extensional fractures those which presented a lateral component smaller than 5°, whereas the others showed a predominant right or left lateral component (Fig. 22.9h): we thus counted 165 pure extensional fractures, 208 fractures with a right lateral and 80 fractures with a left lateral component, out of our 453 total data. The dilation range resulted between 0.29 and 9.56 m, with an overall average value of 1.73 m. Spatial relations of fracture azimuth and opening directions are presented in Fig. 22.9. Figure 22.9a, c show the relation between fracture azimuth and, respectively, easting and northing: the azimuth tends to rotate in a clockwise manner from NNW-SSE and N-S to NNE-SSW and NE-SW eastward. Furthermore, we noticed that average values are similar in the southern and in the northern sector, whereas a greater range of values can be observed in the central and southern sector. Figure 22.9b, d show the relation between opening direction and, respectively, easting and northing. As for fracture azimuth values, also opening direction values tend to increase going eastward, with maximum values of 140–145°N, and present a shorter range in the northern sector (Fig. 22.9b–d). Furthermore, the amount of opening is correlated with easting and northing (Fig. 22.9e–f): such values tend to decrease going eastward and northward, showing also a shorter range. These data will be commented in the discussion chapter.

In regard to normal faults, they mainly strike from N-S to NE-SW (Fig. 22.8d), showing an overall average length of 220 m. Since they represent a small percentage of the total number of fractures which affect the area, we did not focus our data collection on normal faults: along them we only measured vertical offset in 36 waypoints, as shown in Fig. 22.8b, obtaining a maximum value of 12 m. These offset values have been useful in order to improve our collection of data regarding extensional deformation, aimed at measuring the stretching ratio in the area. Thanks to the high resolution of the Orthomosaic and the DSM, therefore, we were able to collect the real dilation of extension fractures and the vertical offset of normal faults, which allowed us to calculate the corresponding dilatational component. We thus measured the total extensional component along two transects, with a 1.7 km length to the north and a 1.8 km length to the south, oriented in the given overall spreading direction, obtaining a total value of 24.33 m in the northern part of the area, and of 30.67 m in the southern part, which correspond respectively to a stretching ratio of 1.013 in the north and 1.017 in the south.

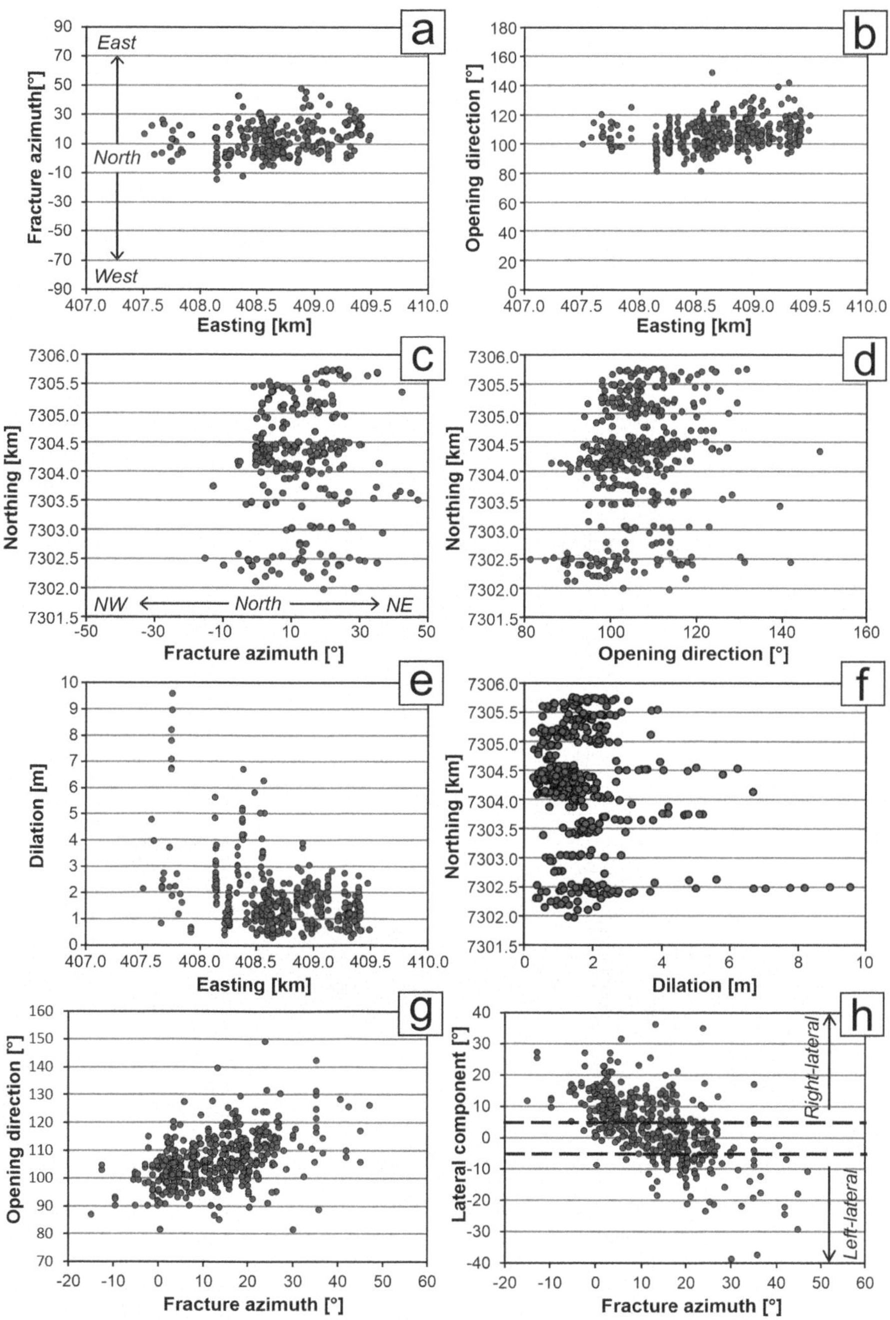

Fig. 22.9 Graphs show (**a–b**) fracture azimuth and opening direction vs. easting and (**c–d**) fracture azimuth and opening direction vs. northing, (**e–f**) fracture dilation vs easting and northing, (**g**) fracture azimuth vs opening direction and (**h**) vs lateral component of movement. Reference system: WGS84/UTM zone 28 N

22.4.2 Khoko Landslide

The second key area showcased in this work is the Khoko landslide, located near the Enguri Dam, in Georgia. This landslide covers an area of 1.2 km^2 and faces directly the Enguri Dam reservoir, in the southwestern sector of the Greater Caucasus (Fig. 22.4). A high resolution 3D model, an Orthomosaic with a resolution of 7.8 cm/pixel, and a DSM with a resolution of 15.6 cm/pixel, covering an area of 2.5 km^2 (Fig. 22.10) were obtained by processing the aerial photos with our workflow, allowing us to recognize the main geomorphological features of the landslide (Fig. 22.10b–c). Indeed, the slope shows several landforms associated with

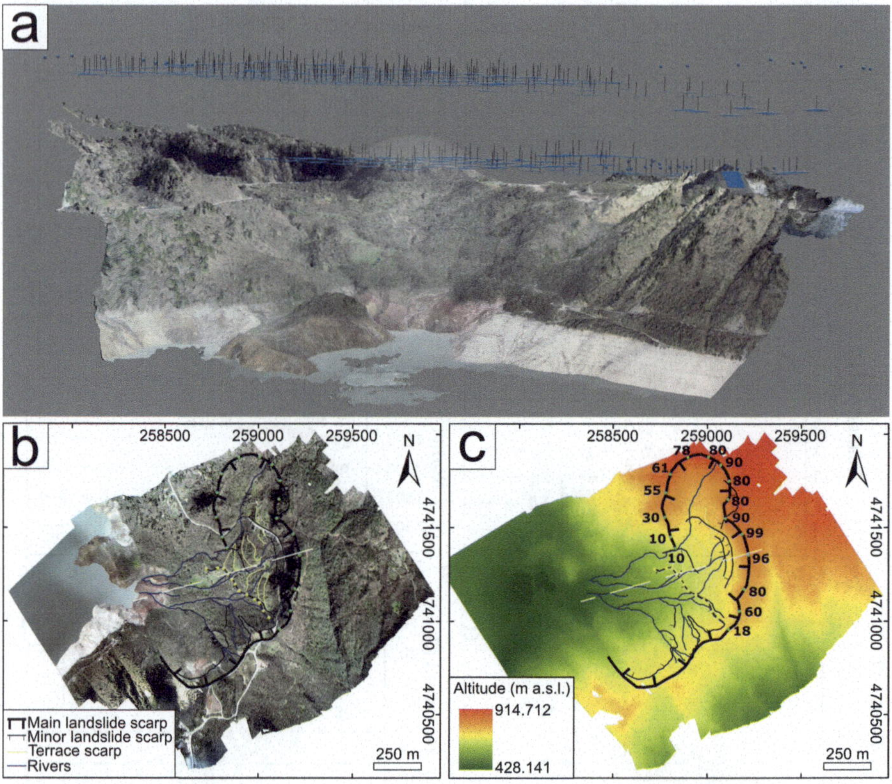

Fig. 22.10 (**a**) UAV-SfM-derived 3D model of the Khoko landslide. Computed camera positions are represented as blue rectangles, black lines show the pitch angle and camera orientation, (**b**) Resulting Orthomosaic and (**c**) DSM, with higher altitude values in red and lower in green. Coordinate values are referred to WGS 84 datum UTM zone 38 N. Interpreted Orthomosaic and DSM of the landslide, with all the main geomorphological features highlighted in different colors. Green and yellow dots in (**b–c**) represent the sites where the height of the main scarp (resulting values written aside) and of the minor scarp (values in Fig. 22.11b), respectively, were measured. White line represents the trace of the geological section provided in Fig. 22.11a. Reference system: WGS84/UTM zone 38 N

active/recent deformations: there are several discrete scarps facing westward, some of which affect the Jvari-Khaishi-Mestia road, with a concave sinuous shape in plain view, which form a continuous head scarp (Fig. 22.10b–c). We also measured the scarp height in correspondence of 16 waypoints: the resulting values range between 10 and 99 m. Besides the principal scarp, several other minor scarps were observed (Fig. 22.10b–c). Particularly, we took 12 measurements of a scarp which is almost 510 m long and cuts the landslide with a NW-SE strike, attesting to the strong instability of the slope. Results are displayed in Fig. 22.11b: the scarp reaches a maximum height of 32 m, showing greater values in the central part, which tend to decrease moving outward. The presence of this relevant scarp is highlighted also in the geological section in Fig. 22.11a (modified after Tibaldi et al. 2019): this secondary scarp and the principal one represent the emersion of the most relevant slip surfaces that affect the slope. Other minor scarps affect the area (Fig. 22.10b–c) and are characterized by a lower height. On the high-resolution 3D model and Orthomosaic it was possible to observe also some variations in inclination, which we classified as terrace scarps in Fig. 22.10b–c, according to Tibaldi et al. (2019), who checked them in the field to make sure they are not man-made features. These

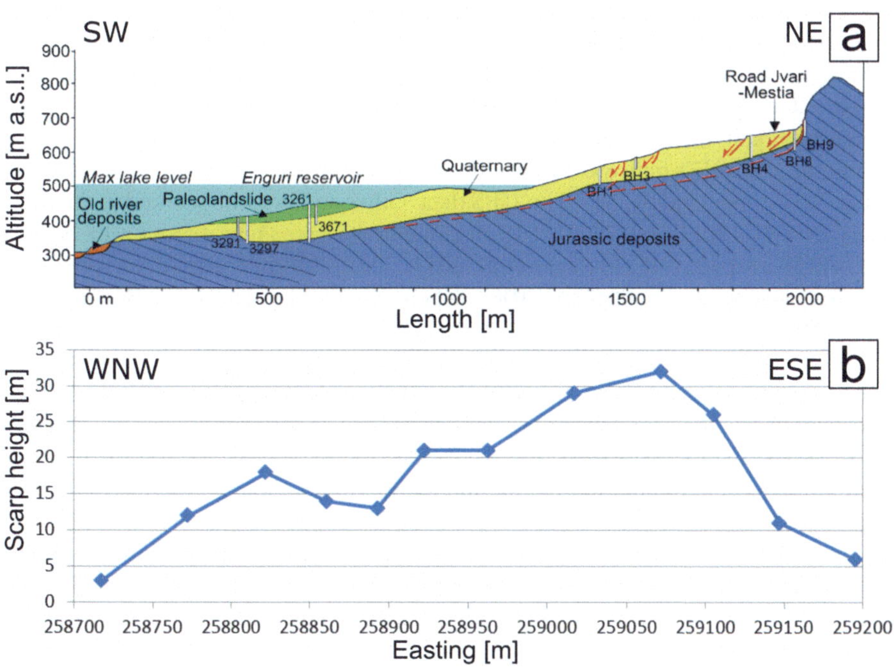

Fig. 22.11 (**a**) Geological section across the slope, modified after Tibaldi et al. (2019). White columns represent locations and depth of logs used to produce the cross section, from Tibaldi et al. (2019). Data for the submerged portion are derived from geological surveys performed in the Soviet era before construction of the Dam, (**b**) Profile representing the height of the minor scarp in Fig. 22.10c

scarps are mostly perpendicular to the slope dip and they are situated in the upper part of the slope; this suggests that they can represent secondary ruptures within the moving slope.

22.5 Discussion

22.5.1 Scientific Outcomes and Geohazard Implications for the Theistareykir Fissure Swarm

The 546 mapped structures, comprising both extension fractures (517) and normal faults (29), have azimuths between N-S and NE-SW, with the maximum frequency at NNE-SSW (rose diagram in Fig. 22.8c). The overall opening direction is about N106°; due to the fact that the surveyed area is characterized by lava deposits with an age of 8–10 ka, this value has not changed over the post-Last Glacial Maximum time interval and is almost identical to the data suggested by Hjartardóttir et al. (2012) regarding the NVZ. Other authors provided different values of spreading direction in the area through various methodologies. Hjartardóttir et al. (2012) suggested a spreading vector of N106° calculated from DeMets et al. (1994) who used geomagnetic data (up to 2.6 Ma – long-term data). DeMets et al. (2010) suggested a spreading vector of N104°, considering GPS and geological data (magnetic, bathymetric and earthquake data; up to 3.16 Ma – long-term data). Drouin et al. (2017) indicated a present-day spreading direction of N112° for the 2008–2014-time window, based on GPS data. Other authors who worked in the time intervals 1997–2011 and 2006–2010 documented a spreading direction of N109° and N115°, respectively (Metzger et al. 2011, 2012). More in detail, our spreading direction values are about N103–108°, when considering data from Bonali et al. (2019) (UAV survey of summer 2017) as a different set with respect to the summer 2018 survey. Taking into account that our data cover a time interval (Holocene times) that is much longer than GPS data but much shorter than the time span investigated by way of geomagnetic data, the range of N103–108° is very close to the previous, long-term N106° results (DeMets et al. 1994). The overall fracture azimuth is mostly orthogonal to the spreading vector, as also suggested for the active Krafla fissure swarm (Hjartardóttir et al. 2012), where fractures and eruptive fissures azimuth is usually N to NNE. Local perturbations in opening directions and the presence of lateral components can be related to possible dyke intrusions at a shallow depth, as shown by Ruch et al. (2016) for the 2014–2015 Bárðarbunga dyking event and by Eriksson et al. (2015) for the Neogene Alftafjordur dyke swarm. Furthermore, both fracture azimuths and opening directions in these areas do not seem to be influenced by strike-slip movements along the HFF, since they are farther away than 500 m. In fact, complex fractures geometry due to the interaction between the ThFS and the HFF have been detected at a distance <500 m by Tibaldi et al. (2016a). Furthermore, rose diagrams show that most opening directions occur perpendicular to fracture azimuth (Fig. 22.8f) although left and right-lateral components are also present, with a predominance of the latter (Fig. 22.9h). Our spreading direction can represent the combination of multiple incremental events of crustal tectonic stretching under plate spreading, in combination with events of shallow dyking. On the contrary, GPS-inferred spreading directions (N109–115°) cover just a few years/decades and represent deformation processes that are more limited in time.

In regard to the amount of opening, this increases westward and southward in the studied area. The westward increase can be explained based on the age of the extensional structures that increases in the same direction in the Theistareykir Fissure Swarm. The older faults, which are longer and have higher offsets, in fact are present in the westernmost part of this rift. Also, locally extension fractures show an increase in filling by debris and in vegetation cover westward, suggesting that their age increases in the same direction. For the southward increase in the amount of opening, this is coherent with the gradual closing of the ThFS towards the seacoast, i.e. northward. In this direction it is expected that normal fault profiles taper out and extension fissures decrease in number and opening. In regard to the cumulated amount of extension, the stretching ratio is 1.013–1.017. Dauteuil et al. (2001) estimated a stretching ratio of 1.009 along a transect located just north of the 1984 Krafla lava flow and a total extension of 30 m; this area provides a snapshot of the deformation accumulated over the last 10,000 years. Differently, Paquet et al. (2007) calculated a stretching ratio of 1.036–1.046 across the Tertiary Alftafjordur dyke swarm, but they described a deformation that lasted 1 Ma or more and analysed a swarm that emplaced at an original depth of 1 km. At very shallow/surface level, it is expected that the total thickness of emplaced dykes is lower than at depth due to dyke arrest. Our stretching ratio is thus more similar to the one estimated at Krafla during Holocene, an area where both tectonic and magmatic forces (dyking) contribute to surface deformation. The studied area has a great potential to be affected by both volcanic eruptions and earthquakes, also considering that: (i) post-LGM eruptions took place at the Stohriver volcanic centre; (ii) recent (post 2.4 ka BP) volcanic vents are reported by Pasquarè Mariotto et al. (2015), just north of the surveyed area and just south of the HFF, within the 2.4 ka-old unit (Fig. 22.2); (iii) the ThFS hosts the Holocene, Theistareykir central volcano, for which ground deformation data (InSAR and GPS) indicate an inflating shallow magma chamber throughout 2006–2008 (Metzger et al. 2011); (iv) historical, Magnitude > 6 earthquakes have been reported both along the ThFS and the HFF (Magnúsdóttir and Brandsdóttir 2011). These natural processes can affect the activity of the brand-new geothermal power plant, located 2 km away from the study area, and, consequently, the new companies that just began operating near Husavik town.

22.5.2 Scientific Outcomes and Geohazard Implications for the Khoko Landslide

The SfM resulting models for the Khoko landslide allowed us to gather several evidence of active slope deformation currently affecting the mountain ridge on its western flank. The identified scarps affect the present topographic surface and geological units up to Pleistocene and Holocene age. This observation, together with other evidence described in Tibaldi et al. (2019), such as offset road segments, GPS measures and extensometer data, indicate that the slope deformation is still active. The above presented data highlight the presence of a main head scarp (Fig. 22.10b–c), which runs east of the Jvari-Khaishi-Mestia road in the northern and in the central part of the area, and then crosses the road and runs west of it in the southern part. The scarp is composed of three segments with a concave shape facing west, interconnected with each other and forming a continuous scarp. This particular shape suggests that, originally, the landslide might have been composed by a number of discrete, smaller landslides, which have then been joined together by processes of slope instability. The strong instability of the landslide is manifested also by secondary scarps within the slope, reaching heights of several meters (Fig. 22.11b). The height of the main head scarp, instead, ranges between 10 and 99 m, indicating an important vertical slip in the head zone. The shape of this main scarp, obtained by analysing SfM derived models, is very similar to the one obtained through field survey by Tibaldi et al. (2019); regarding scarp height, instead, we obtained a greater range than the value of 20–70 m calculated in the field by the above-cited authors. Our greater range of values (up to 99 m) stems from the fact that the SfM-derived model enabled us to investigate sectors of the scarp that are difficult to access in the field, due to their steepness.

The main geomorphological features represented in Fig. 22.10b–c are similar to the ones surveyed in the field by Tibaldi et al. (2019), but we give more details and provide new data regarding the NW-SE striking scarp highlighted in Fig. 22.10b–c, which is almost 510 m long and cuts the slope, proving its strong instability. The height of this scarp has a maximum value of 32 m, with greater values in the centre of its profile (Fig. 22.11b). The slope is affected by other minor scarps, as highlighted in Fig. 22.10b–c, but none of them is as high as this scarp. The presence of this important scarp, which has been observed for the first time thanks to the UAV-SfM technique, suggests the existence of another important rupture surface, highlighted in Fig. 22.11a.

The extreme instability of the slope, confirmed by the above showcased data, is relevant for the assessment of the hydrogeological hazard in the area. In fact, the Khoko landslide faces directly the Enguri reservoir. Another relevant factor is represented by the high seismicity of the region, an area where earthquakes with Ms up to 7.0 were reported by Varazanashvili et al. (2011). As described by Tibaldi et al. (2019) two digital extensometers were set up in correspondence of the active head structures, together with benchmarks for GPS measurements; this monitoring system enabled documenting that the upper section of the slope moves continuously, as already observed elsewhere at Deep-Seated Gravitational Slope Deformations (DSGSDs) (Varnes et al. 1989, 1990, 2000). However, in some instances, the displacements at DSGSDs can be intermittent, with periods of slower deformation, and then with sudden movements, also in the order of meters, though without a complete failure (Tibaldi et al. 2004; Tibaldi and Pasquaré 2008; Pasquaré Mariotto and Tibaldi 2016). In the case of the Khoko landslide, in the worst-case scenario of a sudden displacement with a consistent seismic shaking, a huge rock volume might be involved, calculated by Tibaldi et al. (2019) in the order of $48 \pm 12*10^6$ m^3, with potential effects on the water reservoir. Moreover, the landslide represents also a threat for infrastructures, like the Jvari-Khaishi-Mestia road, and for an inhabited house situated in the middle of the landslide. In view of the above considerations regarding the hydrogeological and geomorphological hazard of the area, and considering the data presented in this paper, we suggest the repetition of systematic UAV surveys to monitor the slope movements (e.g. Turner et al. 2015) and the installation of more extensometers and benchmarks for GPS measurements, not only along the head of the main scarp, but also along the secondary scarp which cuts the slope, which we suggest could represent another important slip surface emerging across the slope. This additional monitoring effort would allow to: (i) individuate possible critical situations and to better understand the behaviour of the unstable slope; (ii) contribute to integrating the amount of data derived from the already positioned instruments (Tibaldi et al. 2019) along the main head scarp; (iii) take into account the effects caused by the presence of the secondary scarp, which had not been individuated before.

22.5.3 *Usefulness and Limitations of UAV-Based SfM for Geohazard Assessment*

The use of UAVs in geosciences has increased over the last decade (e.g. Niethammer et al. 2012; Lucieer et al. 2014; Clapuyt et al. 2016; Gomez et al. 2015; Jordan 2015; Stocker et al. 2015; Nishar et al. 2016; Chesley et al. 2017; Strick et al. 2017). Our work has enabled us to confirm that UAV-based SfM is a powerful tool to support classical fieldwork aimed at data collection for the assessment of geohazards associated with different kinds of geological phenomena (e.g. earthquakes and landslides). For example, Bonali et al. (2019) were able to collect 1098 measurements, comprising azimuth, dilation and opening vector (where detectable), at 387 sites by way of UAV-based SfM, hence providing a contribution, over a short time period, to reconstructing the tectonic evolution in an active rift zone in NE Iceland.

Furthermore, the SfM methodology, through the use of UAVs, becomes an alternative to more expensive techniques such as Terrestrial Laser Scanning (TLS) and Airborne Laser Scanning (ALS) in order to capture the micro-topographic variations and characteristics of the ground surface and outcrops. This is possible thanks to the capability of rapid data capturing, quick processing, and high accuracy of the output model; the SfM technique offers low-cost workflow that enables anyone to collect the data (Saputra et al. 2017), particularly for field research, even though TLS directly provides a point cloud that is immediately available for analysis (Nouwakpo et al. 2016). At a general level, the high quality Orthomosaics that can be obtained through SfM techniques open new possibilities for future applications, such as carrying out repeated UAV surveys across the same area at regular time intervals to determine the present-day deformation field and its evolution over time. Regarding Iceland, it would be possible to compare DSM and Orthomosaic before and after a rifting event (e.g. Krafla Fires or Bardarbunga rifting event) where the deformation is episodic and superficial deformation is larger than the present-day model accuracy. By the way, UAVs benefit from continuous advancements as regards flight stability and duration, camera quality in terms of resolution and sensors with the release of new products: this implies that the quality of future geological research will improve, and the extension of surveyed areas will be wider thanks to the possibility of flying UAVs at higher altitudes. A demonstration of this fast improvement is visible also in this work: in NE Iceland, in fact, we surveyed two adjacent areas with two different UAV models, the DJI Phantom 4 in 2017 and its upgrade, the DJI Phantom 4 Pro, in 2018, flying at the same height (90–100 m) but obtaining a higher resolution for both the Orthomosaic and the DSM of the 2018 surveyed area.

It is worth pointing out that this methodology presents also some limitations: (i) the battery life/flight time; (ii) weather conditions influencing the quality of the collected photos; (iii) legal flight restrictions that might limit the feasibility of UAV-based surveys; (iv) camera, sensor type and field of view strongly influence the resulting ground resolution; and (v) smallest tectonic features like striations on fault planes or slicken lines cannot be recognized even on high-resolution UAV photos (e.g. Vollgger and Cruden 2016). Furthermore, another limitation is represented by the time needed to place and/or mark GCPs, which are always needed for scaling and referencing the model. In physical geography, where SfM is used, it is recommended to use a minimum of three GCPs (or targets) with XYZ coordinates to scale and georeference the point cloud (Smith et al. 2016).

Javernick et al. (2014) decided to use a greater number of GCPs, even as many as 95, distributed across the study area in a 100 m grid, but with a lower image overlap of 70–60%. Other authors considered a lower number of GCPs but using a higher image overlap: (i) Vollgger and Cruden (2016) used 12 GCPs (randomly distributed) with an image overlap of 85–75%; (ii) Esposito et al. (2017) used 15 GCPs (uniformly distributed over the area) with 80–60% of overlap. Upcoming new UAVs equipped with RTK positioning system, better cameras and battery consumption combined with oblique imagery acquisition, could be part of the solution in order to decrease the number of GCPs and increase the accuracy of the resulting models. Another approach was proposed by James and Robson (2014) who suggested collecting also photos with oblique inclination with respect to the ground, in order to decrease the 'doming' effect on the DTM/DSM; it generally results by using the traditional parallel flight path with nadir imagery (Tonkin and Midgley

2016). In addition, if the geometry network results sufficiently strong, the number of GCPs could be substantially reduced (James et al. 2017).

Nevertheless, collecting photos with different orientation angle of the camera in the centre of the acquisition area affects battery power consumption especially for large areas, small UAVs and environment temperature. The flight strategy we used, considering the aim of this work, was selected in order to have a good compromise between survey time, battery consumption, area extension, data resolution and model accuracy. In our case, considering the time-intensive process of GCPs collection, we located only 5 GCPs per each mission, also using natural and ecological ones, that can be abandoned in the field (to speed up the process), but we further increased the photo overlap to 85–90% along flight path and 75–80% in lateral direction, thus obtaining a good match between field- and UAV-collected structural data. Our approach resulted sufficiently detailed for the survey of extension fractures, fault and landslide scarps, but further improvements may be needed to define the most efficient height for DSMs reconstruction, mainly devoted to measuring vertical offset at normal faults.

Regarding the size of the surveyed objects, we encountered very small features such as fractures with only few centimetres of opening, where dilation and opening direction can be precisely collected only in the field and not remotely collected on UAV-generated Orthomosaics. Bemis et al. (2014) applied the SfM in structural geology carrying out UAV-based surveys at flight height of 30–40 m, using a 15 Megapixel camera, which produces high resolution imagery with 1 pixel = 1 cm. They also indicated that sub-cm resolution may be reached at flight height of 15–20 m, though the benefits are countered by larger datasets and associated increase in data processing time.

To conclude, instead of time-expensive fieldwork necessary to collect data along active faults that can be tens of km long (e.g. Kozhurin et al. 2006; Pasquarè Mariotto et al. 2015), UAVs allow to cover wider areas in a faster way and to obtain better results than LiDAR and Terrestrial Laser Scanning (TLS) (Cawood et al. 2017; Lizarazo et al. 2017). Moreover, UAV-based surveys allow to investigate areas which are normally inaccessible due to their difficult logistic conditions, like active landslides (Tibaldi et al. 2015, 2019) or volcano edifices affected by lateral and/or caldera collapse (Tibaldi et al. 2008; Tortini et al. 2014; Russo et al. 2017). Finally, 3D models obtained by UAV-based SfM can be used for modern applications like the Immersive Virtual Reality, for research and teaching purposes (Gerloni et al. 2018; Bonali et al. 2019; Krokos et al. 2019).

22.5.4 Application of the Methodology to Other Regions in the Caucasus

UAV-based SfM techniques can be considered modern tools for geohazard assessment, making it possible to overcome difficult logistic conditions and the wideness of study areas, and allowing to collect several data on high resolution models, saving time and money by being integrated with field surveys.

Apart from the Khoko landslide in the western Caucasus, considered in this study, we suggest that this approach might be extended to other regions in the Caucasus, affected by different types of geohazards. In fact, the region, comprising territories of Georgia, Armenia, Azerbaijan, eastern Turkey and northwestern Iran, is geologically complex, being subject to seismic, volcanic, geomorphological and hydrological hazards. In particular, both strong seismicity and active volcanism represent a threat not only for the population and the environment, but also for strategically important infrastructures, such as the Caspian oil and gas pipelines. For example, the Baku-Tbilisi-Ceyhan (BTC) oil pipeline and the Baku-Tbilisi-Erzerum South Caucasian (SCP) natural gas pipeline were designed to sustain seismic events, but volcanic activity along their path was not considered (Pasquarè et al. 2011). Nevertheless, the BTC and SCP pipelines cross the northern part of the Javakheti recent volcanic province for 70 km, including the N-S trending Abuli Samsari Volcanic Ridge in Southern Georgia, with more than 20 volcanic centres, giving rise to a potential volcanic hazard (Lebedev et al. 2003). For this reason, Pasquaré et al. (2011) conducted a volcanic and seismic hazard assessment of this volcanic ridge, using geological data collected with field surveys. We thus suggest the use of the UAV-SfM techniques to monitor the topography of such areas, focusing on both slope stability aimed at landslide hazard assessment, and on the recognition and mapping of tectonic and volcano-tectonic features, like strike-slip and normal faults, fissures and associated Plio-Quaternary volcanoes. The use of aerial photogrammetry would allow to collect new data which could be integrated and compared with the work conducted by Pasquaré et al. (2011), in order to improve the geohazard assessment in such a strategically and economically important area.

Furthermore, the method adopted in this study could be extended also to the study of other landslides in the Caucasus region apart from the Khoko landslide, such as the Vokhchaberd – Garni landslide group in Armenia, whose origin could be related to the active Garni fault which is located only 1.5–2 km away, and that could have been reactivated by earthquakes triggered by the fault itself (Martino et al. 2007). Within this group, the giant Vokhchaberd landslide is the one posing a greater risk to the population, because it involves an area inhabited by about 1000 residents (Martino et al. 2007). In this case, the use of the UAV-SfM approach would allow to recognize the main geomorphological features of the slope, like terraces, scarps, bulging and eventual cracks, and to collect data along the main and minor scarps which might not be reachable in the field.

Another relevant geohazard that affects the Caucasus is related to the recent glacier retreat, glacier outburst floods and related debris flows, a worldwide phenomenon unequivocally recognized as a sign of the global climate change (Bogatikov et al. 2003; Kotlyakov et al. 2003; Seynova et al. 2003; Oerlemans 2005; Barry 2006; Oerlemans et al. 2009). In fact, in the Caucasus, glaciers occupy an area of about 1600 km^2, and a general retreat trend, accompanied by an increase in debris cover, has been observed in the late twentieth century (Stokes et al. 2006, 2007). The same authors carried out the observation and analysis of changes in glacial environment through field survey-derived data and remote satellite images. Nevertheless, such glacial studies can be hindered by glacier inaccessibility and hostile climate, thus not allowing a periodic monitoring of broad areas. For these reasons, satellite derived images were used to define glacial aerial extension, elevation changes and surface velocities (Gao and Liu 2001; Paul et al. 2015; Mouginot et al. 2017). Nevertheless, remote satellite images have limitations related to spatial and temporal resolution, which can be overcome by using UAVs, especially in the case of smaller glaciers which are not observable in detail using satellite images. The UAV-SfM approach would enable collecting high resolution Orthomosaics and DSMs to study glacier morphology, surface ablation, surficial displacements and debris coverage with a high resolution. Furthermore, UAV surveys can be carried out periodically to monitor changes in glacier features (Fugazza et al. 2015, 2018; Rossini et al. 2018). Limitations to the use of UAVs for glacier hazard assessment are imposed by battery life and altitude: the first does not affect the survey of smaller glaciers, whereby the use of multiple batteries is not required, whereas regarding altitude, the maximum reachable height above sea level depends on the used UAV model. Both the DJI Phantom 4 and Phantom 4 PRO models, which we used in our work, have a maximum service ceiling above sea level of 6000 m, which would thus allow the survey of Caucasian glaciers.

Furthermore, our UAV-SfM method can be applied to the easternmost areas of Azerbaijan, strongly affected by mud volcanism, whose formation is favoured by the high subsidence rate in the South Caspian Basin, which brings about high sediment accumulation rates (Katz et al. 2000); this condition is observed also in onshore Azerbaijan, and it leads to under compaction of sediments and high fluid pressures, both favouring the formation of mud volcanoes (Dimitrov 2002). Hundreds of mud volcanoes have been observed in the whole basin, which is defined as one of the areas with the highest density of such features in the world (Jakubov et al. 1971; Stewart and Davies 2006; Aliev et al. 2015). Several authors have studied this phenomenon in Azerbaijan: Hovland et al. (1997) studied the structure and geomorphology of the onshore Dashgil mud volcano, Planke et al. (2003) analysed mud and fluid migration in four onshore mud volcanoes, and Yusifov and Rabinowitz (2004) made a classification of offshore mud volcanoes based on characteristic seismic features. More recently, UAV images have been used to study these features both in Azerbaijan to document the deformation and growth of three onshore mud volcanoes (Odonne et al. 2019) and worldwide (e.g. Santagata 2017; Federico et al. 2019). The approach described in this study could be applied to study mud volcanoes in order to obtain 3D models and high resolution Orthomosaics and DSMs to monitor the morphological variations of the mud field and of the single edifices and to study the evolution of mud flows and volcanic cones. Moreover, a continuous aerial photogrammetric survey would allow to observe the vents migration, which could be related to other parameters like variations in gas output (e.g. Federico et al. 2019). Furthermore, as explained above, by using the UAV-SfM approach, more data would be collected in a very broad area, thus the eastern Azerbaijan region could be studied in a shorter time frame than pure field survey.

22.6 Conclusive Remarks

In the present work we applied the use of UAV-based SfM methodology to two different areas, both characterized by difficult logistic conditions and a great potential of occurrence of catastrophic natural phenomena like earthquakes, eruptions and landslides. At a general level, for both areas we obtained high-resolution Orthomosaics and DSMs that made it possible to collect a wealth of geological data, respect to only pure field survey, aimed at producing a larger base data set. Once one gathers an extensive, robust data set like the ones we presented, it is possible to carry out a more weighted analysis, aimed at better understanding the ongoing geological processes that lead to geohazard occurrence.

Regarding the NVZ, we collected a total of 453 quantitative measurements of the amount of opening and opening direction of Holocene extension fractures and 36 measurements of the height of fault scarps. These data allowed us to reconstruct, within the studied rift zone, an overall spreading direction of N106.4° during Holocene times, which has been compared with geodetic motion vectors, and a stretching ratio of 1.013–1.017 for 8–10 ka old lava units. We conclude that deformations in the area are related to both dyke intrusions and extensional tectonics.

In the Greater Caucasus, we applied the methodology to the recognition of the main geomorphological features associated with the active Khoko landslide, and to the measurement of the scarp height of the principal slip surfaces. Thanks to the high resolution of the resulting models we also noticed another main scarp, within the slope, which reaches heights of tens of meters and had never been noticed in the field, proving the instability of the slope and suggesting the presence of another important rupture surface. The use of the UAV-SfM techniques thus allowed us to enhance the geomorphological knowledge of the landslide, and to contribute to the assessment of the hydrogeological hazard of the area.

Based on the above, we suggest the use of the UAV-SfM techniques also in other regions of the Caucasus, in order to collect data which can integrate the results of field surveys and to better assess geohazards like the ones related to landslides, volcanism, seismicity and glacier retreat, considering also the presence in the region of important infrastructures, like strategic oil and gas lifelines.

Acknowledgments This work is a contribution to: NATO projects SfP G4934 and ARW G5566, the International Lithosphere Program – Task Force II, and has also been carried out in the framework of ESA Project Nr. 38829 and MIUR project ACPR15T4_00098 – Argo3D (http://

argo3d.unimib.it/). This effort is also an outcome of Project MIUR – Dipartimenti di Eccellenza 2018–2022. Agisoft Metashape is acknowledged for photogrammetric data processing.

References

Adamia SH, Lordkipanidze MB, Zakariadze GS (1977) Evolution of an active continental margins exemplified by the alpine history of the Caucasus. Tectonophysics 40:183–199

Adamia S, Alania V, Chabukiani A et al (2010) Evolution of the Late Cenozoic basins of Georgia (SW Caucasus): a review. In: Sosson M, Kaymakçı N, Stephenson R, Bergerat F (eds) Sedimentary basin tectonics from the Black Sea and Caucasus to the Arabian Platform, Special publication 340. Geological Society, London, pp 239–259

Adamia SH, Zakariadze G, Chkhotua T et al (2011) Geology of the Caucasus: a review. Turk J Earth Sci 20:489–544

Adamia SH, Alania V, Tsereteli N et al (2017) Post-collisional tectonics and seismicity of Georgia. In: Tectonic evolution and seismicity of south-west Asia. Geological Society of America (GSA) Special paper 525, pp 1–38

Alania V, Chabukiani A, Chagelishvili R et al (2016) Growth structures, piggyback basins and growth strata of Georgian part of Kura foreland fold and thrust belt: implication for Late Alpine kinematic evolution. In: Sosson M, Stephenson R, Adamia SH (eds) Tectonic evolution of the Eastern Black Sea and Caucasus, Special publications 428. Geological Society of London, London. https://doi.org/10.1144/SP428.5

Agisoft LLC (2018) PhotoScan user manual, professional edition. http://www.agisoft.com/pdf/photoscan-pro_1_4_en.pdf, version 1.4. Accessed date: 1 October 2018

Aliev AA, Guliev IS, Dadashov FH et al (2015) Atlas of the world mud volcanoes. Nafta Press Publishing House, Baku

Angster S, Wesnousky S, Huang WL et al (2016) Application of UAV photography to refining the slip rate on the Pyramid Lake fault zone, Nevada. Bull Seismol Soc Am 106(2):785–798

Antoniou V, Paraskevi N, Bardouli P et al (2019) The story map for Metaxa Mine (Santorini, Greece): a unique site where history and volcanology meet each other. GISTAM 2019. SciTePress

Avdeev B, Niemi NA (2011) Rapid Pliocene exhumation of the central Greater Caucasus constrained by low-temperature thermochronometry. Tectonics 30(2):TC2009. https://doi.org/10.1029/2010TC002808

Baiocchi V, Dominici D, Milone MV et al (2013) Development of a software to plan UAVs stereoscopic flight: an application on post-earthquake scenario in L'Aquila city. In: International conference on computational science and its applications. Springer, Berlin/Heidelberg, pp 150–165

Banks C, Robinson A, Williams M (1997) Structure and regional tectonics of the Achara-Trialeti fold belt and the adjacent Rioni and Kartli foreland basins, Republic of Georgia. In: Robinson AG (ed) Regional and petroleum geology of the Black Sea and Surrounding Region. American Association of Petroleum Geologists Memoir 68, pp 331–336

Barry RG (2006) The status of research on glaciers and global glacier recession: a review. Prog Phys Geogr 30(3):285–306

Bemis SP, Micklethwaite S, Turner D et al (2014) Ground-based and UAV-based photogrammetry: a multi-scale, high-resolution mapping tool for structural geology and paleoseismology. J Struct Geol 69:163–178

Benassi F, Dall'Asta E, Diotri F et al (2017) Testing accuracy and repeatability of UAV blocks oriented with gnss-supported aerial triangulation. Remote Sens 9(2):172

Bergerat F, Angelier J (2008) Immature and mature transform zones near a hot spot: the South Iceland Seismic Zone and the Tjörnes Fracture Zone (Iceland). Tectonophysics 447:142–154

Bi H, Zheng W, Ren Z et al (2017) Using an unmanned aerial vehicle for topography mapping of the fault zone based on structure from motion photogrammetry. Int J Remote Sens 38(8–10):2495–2510

Blistan P, Kovanic L, Patera M et al (2019) Monitoring of selected geohazards by using Unmanned Aerial Systems (UAS). CeON repository. https://depot.ceon.pl/handle/123456789/17036

Bogatikov OA, Gurbanov AG, Melekestsev IV et al (2003) Catastrophic rock-ice collapse and rapid shove of the Kukurtli Glacier (Elbrus Volcano, Northern Caucasus) in first and second centuries. Dokl Earth Sci 391(5):627–630

Bonali FL, Tibaldi A, Marchese F et al (2019) UAV-based surveying in volcano-tectonics: an example from the Iceland rift. J Struct Geol 121:46–64

Brook MS, Merkle J (2019) Monitoring active landslides in the Auckland region utilising UAV/structure-from-motion photogrammetry. Jpn Geotech Soc Spec Publ 6(2):1–6

Burns JHR, Delparte D (2017) Comparison of commercial structure-from-motion photogrammetry software used for underwater three-dimensional modeling of coral reef environments. Int Arch Photogramm Remote Sens Spat Inf Sci 42:127–131

Cawood AJ, Bond CE, Howell JA et al (2017) LiDAR, UAV or compass-clinometer? Accuracy, coverage and the effects on structural models. J Struct Geol 98:67–82

Cembrano J, Lara L (2009) The link between volcanism and tectonics in the southern volcanic zone of the Chilean Andes: a review. Tectonophysics 471(1–2):96–113

Chandler JH, Buckley SJ (2016) Structure from motion (SFM) photogrammetry vs terrestrial laser scanning. In: Uysal M, Toprak AS, Polat N (eds) Geoscience handbook 2016: AGI data sheets – DEM generation with UAV photogrammetry and accuracy analysis in Sahitler hill. Measurement 73. American Geosciences Institute, Alexandria. https://doi.org/10.1016/j.measurement.2015.06.010

Cheng KS, Wei C, Chang SC (2004) Locating landslides using multi-temporal satellite images. Adv Space Res 33:296–301

Cheng Y, He C, Rao G et al (2018) Geomorphological and structural characterization of the southern Weihe Graben, central China: implications for fault segmentation. Tectonophysics 722:11–24

Chesley JT, Leier AL, White S et al (2017) Using unmanned aerial vehicles and structure-from-motion photogrammetry to characterize sedimentary outcrops: an example from the Morrison Formation, Utah, USA. Sediment Geol 354:1–8

Cignetti M, Godone D, Wrzesniak A et al (2019) Structure from motion multisource application for landslide characterization and monitoring: the Champlas du Col case study, Sestriere, North-Western Italy. Sensors 19:2364

Clapuyt F, Vanacker V, Van Oost K (2016) Reproducibility of UAV-based earth topography reconstructions based on structure-from-motion algorithms. Geomorphology 260:4–15

Cook KL (2017) An evaluation of the effectiveness of low-cost UAVs and structure from motion for geomorphic change detection. Geomorphology 278:195–208

Darmawan H, Walter TR, Brotopuspito KS et al (2018) Morphological and structural changes at the Merapi lava dome monitored in 2012–15 using unmanned aerial vehicles (UAVs). J Volcanol Geotherm Res 349:256–267

Dauteuil O, Angelier J, Bergerat F et al (2001) Deformation partitioning inside a fissure swarm of the northern Icelandic rift. J Struct Geol 23:1359–1372

De Beni E, Cantarero M, Messina A (2019) UAVs for volcano monitoring: a new approach applied on an active lava flow on Mt. Etna (Italy), during the 27 February–02 March 2017 eruption. J Volcanol Geotherm Res 369:250–262

De Quay GS, Roberts GG, Rood DH et al (2019) Holocene uplift and rapid fluvial erosion of Iceland: a record of post-glacial landscape evolution. Earth Planet Sci Lett 505:118–130

Deffontaines B, Chang KJ, Champenois J et al (2016) Active interseismic shallow deformation of the Pingting terraces (Longitudinal Valley–Eastern Taiwan) from UAV high-resolution topographic data combined with InSAR time series. Geom Nat Haz Risk 8(1):120–136

DeMets C, Gordon RG, Argus DF et al (1994) Effect of recent revisions to the geomagnetic reversal time scale on estimates of current plate motions. Geophys Res Lett 21(20):2191–2194

DeMets C, Gordon RG, Argus DF (2010) Geologically current plate motions. Geophys J Int 181(1):1–80
Dimitrov LI (2002) Mud volcanoes – the most important pathway for degassing deeply buried sediments. Earth Sci Rev 59(1):49–76
Drouin V, Sigmundsson F, Ófeigsson BG et al (2017) Deformation in the Northern Volcanic Zone of Iceland 2008–2014: an interplay of tectonic, magmatic, and glacial isostatic deformation. J Geophys Res 122(4):3158–3178
Eriksson PI, Riishuus MS, Elming SÅ (2015) Magma flow and palaeo-stress deduced from magnetic fabric analysis of the Alftafjörður dyke swarm: implications for shallow crustal magma transport in Icelandic volcanic systems. Geol Soc Lond, Spec Publ 396(1):107–132
Esposito G, Mastrorocco G, Salvini R et al (2017) Application of UAV photogrammetry for the multi-temporal estimation of surface extent and volumetric excavation in the Sa Pigada Bianca open-pit mine, Sardinia, Italy. Environ Earth Sci 76(3):103
Fallati L, Polidori A, Salvatore C et al (2019) Anthropogenic Marine Debris assessment with unmanned aerial vehicle imagery and deep learning: a case study along the beaches of the Republic of Maldives. Sci Total Environ 693:133581
Favalli M, Fornaciai A, Nannipieri L et al (2018) UAV-based remote sensing surveys of lava flow fields: a case study from Etna's 1974 channel-fed lava flows. Bull Volcanol 80(3):29
Federico C, Liuzzo M, Giudice G et al (2019) Variations in CO_2 emissions at a mud volcano at the southern base of Mt Etna: are they due to volcanic activity interference or a geyser-like mechanism? Bull Volcanol 81(1):2
Fjäder K, Gudmundsson A, Forslund T (1994) Dikes, minor faults and mineral veins associated with a transform fault in North Iceland. J Struct Geol 16(1):109–119
Forte A, Cowgill E, Bernardin T et al (2010) Late Cenozoic deformation of Kura fold-thrust belt, southern Greater Caucasus. Geol Soc Am Bull 122:465–486
Freeman M, Vernon C, Berrett B et al (2019) Sequential earthquake damage assessment incorporating optimized sUAV remote sensing at Pescara del Tronto. Geosciences 9:332
Fugazza D, Senese A, Azzoni RS et al (2015) High-resolution mapping of glacier surface features. The UAV survey of the Forni Glacier (Stelvio National Park, Italy). Geogr Fis Din Quat 38(1):25–33
Fugazza D, Scaioni M, Corti M et al (2018) Combination of UAV and terrestrial photogrammetry to assess rapid glacier evolution and conditions of glacier hazards. Nat Hazards Earth Syst Sci 18:1055–1071
Gao J, Liu Y (2001) Applications of remote sensing, GIS and GPS in glaciology: a review. Prog Phys Geogr 25(4):520–540
Gao M, Xu X, Klinger Y et al (2017) High-resolution mapping based on an unmanned aerial vehicle (UAV) to capture paleoseismic offsets along the Altyn-Tagh fault, China. Sci Rep 7(1):1–11
Garcia S, Angelier J, Bergerat F et al (2002) Tectonic analysis of an oceanic transform fault zone revealed by fault slip data and earthquake focal mechanisms: the Husavik-Flatey Fault, Iceland. Tectonophysics 344:157–174
Gerloni IG, Carchiolo V, Vitello FR et al (2018) Immersive virtual reality for earth sciences. In: 2018 Federated Conference on Computer Science and Information Systems (FedCSIS) IEEE, pp 527–534
Gomez C, Hayakawa Y, Obanawa H (2015) A study of Japanese landscapes using structure from motion derived DSMs and DEMs based on historical aerial photographs: new opportunities for vegetation monitoring and diachronic geomorphology. Geomorphology 242:11–20
Gong JH, Wang DC, Li Y et al (2010) Earthquake induced geological hazard detection under hierarchical stripping classification framework in the Beichuan area. Landslides 7(2):181–189
Govedarica M, Jakovljević G, Álvarez-Taboada F (2018) Flood risk assessment based on LiDAR and UAV points clouds and DEM. In: Proceedings of remote sensing for agriculture, ecosystems, and hydrology XX, SPIE, vol 107830B
Gudmundsson A (1987) Geometry, formation and development of tectonic fractures on the Reykjanes Peninsula, southwest Iceland. Tectonophysics 139(3–4):295–308

Gudmundsson A, Brynjólfsson S, Jónsson MT (1993) Structural analysis of a transform fault-rift zone junction in North Iceland. Tectonophysics 220:205–221

Halldórsson P, Stefánsson R, Einarsson P et al (1984) Seismic hazard at Dysnes, Geldinganes, Helguvík, Vatnsleysuvík, Vogastapi and Þorlákshöfn. Ministry of Industry and Commerce report, p 34 (in Icelandic)

Hashemi-Beni L, Jones J, Thompson G et al (2018) Challenges and opportunities for UAV-based digital elevation model generation for flood-risk management: a case of Princeville, North Carolina. Sensors 18:3843

Hill AC, Limp F, Casana J, Laugier EJ, Williamson M (2019) A new era in spatial data recording: low-cost GNSS. Adv Archaeol Pract 7(2):169–177

Hjartardóttir ÁR, Einarsson P, Bramham E et al (2012) The Krafla fissure swarm, Iceland, and its formation by rifting events. Bull Volcanol 74(9):2139–2153

Hjartardóttir R, Einarsson P, Magnusdóttir S et al (2016) Fracture systems of the Northern Volcanic Rift Zone, Iceland: an onshore part of the Mid-Atlantic plate boundary. In: Wright TJ, Ayele A, Ferguson DJ, Kidane T, Vye-Brown C (eds) Magmatic rifting and active volcanism, Special publications 420. Geological Society, London, pp 297–314

Hovland M, Hill A, Stokes D (1997) The structure and geomorphology of the Dashgil mud volcano, Azerbaijan. Geomorphology 21(1):1–15

Hu JP, Wu WB, Tan QL (2012) Application of unmanned aerial vehicle remote sensing for geological disaster reconnaissance along transportation lines: a case study. Appl Mech Mater 226:2376–2379

Jakubov AA, Ali-Zade AA, Zeinalov MM (1971) Mud volcanoes of the Azerbaijan SSR. In: Atlas. Azerbaijan National Academy of Sciences, Baku

James MR, Robson S (2012) Straightforward reconstruction of 3D surfaces and topography with a camera: accuracy and geoscience application. J Geophys Res 117(F3):F03017

James MR, Robson S (2014) Mitigating systematic error in topographic models derived from UAV and ground-based image networks. Earth Surf Process Landf 39(10):1413–1420

James MR, Robson S, d'Oleire-Oltmanns S et al (2017) Optimising UAV topographic surveys processed with structure-from-motion: ground control quality, quantity and bundle adjustment. Geomorphology 280:51–66

Jancin M, Young KD, Voight B et al (1985) Stratigraphy and K/AR ages across the west flank of the northeast Iceland Axial Rift Zone, in relation to the 7 MA volcano-tectonic reorganization of Iceland. J Geophys Res Solid Earth 90(B12):9961–9985

Javernick L, Brasington J, Caruso B (2014) Modeling the topography of shallow braided rivers using structure-from-motion photogrammetry. Geomorphology 213:166–182

Jiao Q, Jiang W, Zhang J et al (2016) Identification of paleoearthquakes based on geomorphological evidence and their tectonic implications for the southern part of the active Anqiu–Juxian fault, eastern China. J Asian Earth Sci 132:1–8

Johnson K, Nissen E, Saripalli S et al (2014) Rapid mapping of ultrafine fault zone topography with structure from motion. Geosphere 10(5):969–986

Jordan BR (2015) A bird's-eye view of geology: the use of micro drones/UAVs in geologic fieldwork and education. GSA Today 25(7):50–52

Katz B, Richards D, Long D et al (2000) A new look at the components of the petroleum system of the South Caspian Basin. J Pet Sci Eng 28:161–182

Kotlyakov VM, Rototaeva LV, Desinov OV et al (2003) Causes and consequences of the catastrophic advance of the Kolka surging glacier in the central Caucasus. Dokl Earth Sci 389A(3):447–451

Kozhurin A, Acocella V, Kyle PR et al (2006) Trenching studies of active faults in Kamchatka, eastern Russia: palaeoseismic, tectonic and hazard implications. Tectonophysics 417(3–4):285–304

Krokos M, Bonali F, Vitello F et al (2019) Workflows for virtual reality visualisation and navigation scenarios in earth sciences. In: Proceedings of the 5th international conference on geographical information systems theory, applications and management, vol 1. SciTePress, pp 297–304

Langhammer J, Vackova T (2018) Detection and mapping of the geomorphic effects of flooding using UAV photogrammetry. Pure Appl Geophys 175(9):3223–3245. https://doi.org/10.1007/s00024-018-1874-1

Lebedev VA, Chernishev IV, Dudauri OZ et al (2003) The Samsari volcanic center as an example of recent volcanism in the Lesser Caucasus: K-Ar geochronological and Sr-Nd Isotopic data. Dokl Earth Sci 393A:1323–1328

Lizarazo I, Angulo V, Rodríguez J (2017) Automatic mapping of land surface elevation changes from UAV-based imagery. Int J Remote Sens 38(8–10):2603–2622

Lucieer A, Jong SMD, Turner D (2014) Mapping landslide displacements using structure from motion (SfM) and image correlation of multi-temporal UAV photography. Prog Phys Geogr 38(1):97–116

Magnúsdóttir S, Brandsdóttir B (2011) Tectonics of the Þeistareykir fissure swarm. Jökull 61:65–79

Marino CM, Tibaldi A (1988) Use of Landsat and Seasat data as a tool in kinematic analysis: the Tunisian atlas. Int J Remote Sens 9(10–11):1659–1673

Martino S, Paciello A, Sadoyan T et al (2007) Dynamic numerical analysis of the giant Vokhchaberd landslide (Armenia). In: Proceedings of the 4th international conference on earthquake geotechnical engineering, Tessaloniki, Greece, vol 6. Paper Springer, New York, pp 25–28

Metzger S, Jónsson S (2014) Plate boundary deformation in North Iceland during 1992-2009 revealed by InSAR time-series analysis and GPS. Tectonophysics 634:127–138

Metzger S, Jónsson S, Geirsson H (2011) Locking depth and slip-rate of the Húsavík Flatey fault, North Iceland, derived from continuous GPS data 2006-2010. Geophys J Int 187(2):564–576

Metzger S, Jónsson S, Danielsen G et al (2012) Present kinematics of the Tjörnes Fracture Zone, North Iceland, from campaign and continuous GPS measurements. Geophys J Int 192(2):441–455

Mosar J, Kangarli T, Bochud M et al (2010) Cenozoic–recent tectonics and uplift in the Greater Caucasus: a perspective from Azerbaijan. In: Sosson M, Kaymakçı N, Stephenson R, Bergerat F (eds) Sedimentary basin tectonics from the Black Sea and Caucasus to the Arabian Platform, Special publication 340. Geological Society, London, pp 261–279

Mouginot J, Rignot E, Scheuchl B et al (2017) Comprehensive annual ice sheet velocity mapping using Landsat-8, Sentinel-1, and RADARSAT-2 data. Remote Sens 9(4):364

Müller D, Walter TR, Schöpa A et al (2017) High-resolution digital elevation modeling from TLS and UAV campaign reveals structural complexity at the 2014/2015 Holuhraun eruption site, Iceland. Front Earth Sci 5:59

Nichol JE, Wong MS (2005) Satellite remote sensing for detailed landslide inventories using change detection and image fusion. Int J Remote Sens 26:1913–1926

Nichol JE, Shaker A, Wong MS (2006) Application of high-resolution stereo satellite images to detailed landslide hazard assessment. Geomorphology 76:68–75

Niethammer U, James MR, Rothmund S et al (2012) UAV-based remote sensing of the Super-Sauze landslide: evaluation and results. Eng Geol 128:2–11

Nishar A, Richards S, Breen D et al (2016) Thermal infrared imaging of geothermal environments and by an unmanned aerial vehicle (UAV): a case study of the Wairakei–Tauhara geothermal field, Taupo, New Zealand. Renew Energy 86:1256–1264

Nouwakpo SK, Weltz MA, McGwire K (2016) Assessing the performance of structure- from-motion photogrammetry and terrestrial LiDAR for reconstructing soil surface microtopography of naturally vegetated plots. Earth Surf Process Landf 41(3):308–322

Odonne F, Imbert P, Dupuis M et al (2019) Mud volcano growth by radial expansion: examples from onshore Azerbaijan. Mar Pet Geol 112:104051. https://doi.org/10.1016/j.marpetgeo.2019.104051

Oerlemans J (2005) Extracting a climate signal from 169 glacier records. Science 308(5722):675–677

Oerlemans J, Giesen RH, Van den Broeke MR (2009) Retreating alpine glaciers: increased melt rates due to accumulation of dust (Vadret da Morteratsch, Switzerland). J Glaciol 55(192):729–736

Ouédraogo MM, Degré A, Debouche C et al (2014) The evaluation of unmanned aerial system-based photogrammetry and terrestrial laser scanning to generate DEMs of agricultural watersheds. Geomorphology 214:339–355. https://doi.org/10.1016/j.geomorph.2014.02.016

Paquet F, Dauteuil O, Hallot E et al (2007) Tectonics and magma dynamics coupling in a dyke swarm of Iceland. J Struct Geol 29(9):1477–1493

Pasquarè Mariotto F, Tibaldi A (2016) Inversion kinematics at deep-seated gravity slope deformations revealed by trenching techniques. Nat Hazards Earth Syst Sci 16:663–674

Pasquaré Mariotto F, Bonali FL, Tibaldi A et al (2015) Holocene displacement field at an emerged oceanic transform-ridge junction: the Husavik-Flatey Fault-Gudfinnugja Fault system, North Iceland. J Struct Geol 75:118–134

Pasquarè F, Tormey D, Vezzoli L et al (2011) Mitigating the consequences of extreme events on strategic facilities: evaluation of volcanic and seismic risk affecting the Caspian oil and gas pipelines in the Republic of Georgia. J Environ Manag 92(7):1774–1782. https://doi.org/10.1016/j.jenvman.2011.02.003

Paul F, Bolch T, Kääb A et al (2015) The glaciers climate change initiative: methods for creating glacier area, elevation change and velocity products. Remote Sens Environ 162:408–426

Pesci A, Amoroso S, Teza G et al (2018) Characterization of soil deformation due to blast-induced liquefaction by UAV-based photogrammetry and terrestrial laser scanning. Int J Remote Sens 39(22):8317–8336

Planke S, Svensen H, Hovland M et al (2003) Mud and fluid migration in active mud volcanoes in Azerbaijan. Geo-Mar Lett 23:258–268. https://doi.org/10.1007/s00367-003-0152-z

Ramli MF, Yusof N, Yusoff MK et al (2010) Lineament mapping and its application in landslide hazard assessment: a review. Bull Eng Geol Environ 69(2):215–233

Rao G, He C, Cheng Y et al (2018) Active Normal Faulting along the Langshan Piedmont Fault, North China: implications for Slip Partitioning in the Western Hetao Graben. J Geol 126(1):99–118

Rathje EM, Franke K (2016) Remote sensing for geotechnical earthquake reconnaissance. Soil Dyn Earthq Eng 91:304–316

Rossini M, Di Mauro B, Garzonio R et al (2018) Rapid melting dynamics of an alpine glacier with repeated UAV photogrammetry. Geomorphology 304:159–172

Ruch J, Wang T, Xu W et al (2016) Oblique rift opening revealed by reoccurring magma injection in central Iceland. Nat Commun 7:12352

Russo E, Waite GP, Tibaldi A (2017) Evaluation of the evolving stress field of the Yellowstone volcanic plateau, 1988 to 2010, from earthquake first-motion inversions. Tectonophysics 700:80–91

Ruzgiene B, Berteška T, Gečyte S et al (2015) The surface modelling based on UAV photogrammetry and qualitative estimation. Measurement 73:619–627. https://doi.org/10.1016/j.measurement.2015.04.018

Saemundsson K (1974) Evolution of the axial rifting zone in northern Iceland and the Tjörnes fracture zone. Geol Soc Am Bull 85(4):495–504

Saemundsson K, Hjartarson A, Kaldal I (2012) Geological map of the Northern Volcanic Zone, Iceland. Northern part 1: 100.000. Reykjavik, Iceland GeoSurvey and Landsvirkjun

Sanhueza D, Picco L, Ruiz-Villanueva V et al (2019) Quantification of fluvial wood using UAVs and structure from motion. Geomorphology 345:106837. https://doi.org/10.1016/j.geomorph.2019.106837

Santagata T (2017) Monitoring of the Nirano Mud Volcanoes Regional Natural Reserve (North Italy) using unmanned aerial vehicles and terrestrial laser scanning. J Imaging 3(4):1–10

Saputra A, Rahardianto T, Gomez C (2017) The application of structure from motion (SfM) to identify the geological structure and outcrop studies. In: Proceedings of the AIP conference. AIP Publishing LLC 1857(1):030001

Seynova IB, Popovnin VV, Zolotaryov YA (2003) Intensification of glacial debris flows in the Gerkhozhan River basin, Caucasus, in the late 20th century. Landslides 14:39–43
Smith MW, Carrivick JL, Quincey DJ (2016) Structure from motion photogrammetry in physical geography. Prog Phys Geogr 40(2):247–275
Soeters A, Cornelis JW (1996) Slope instability recognition, analysis and zonation. In: Schuster RL, Turner AK (eds) Landslides investigation and mitigation. Transportation research board, special report 247. National Academy Press, Washington, DC, pp 129–177
Sosson M, Rolland Y, Danelian T et al (2010) Subductions, obduction and collision in the Lesser Caucasus (Armenia, Azerbaijan, Georgia), new insights. In: Sosson M, Kaymakci N, Stephanson R, Bergarat F, Storatchenoko V (eds) Sedimentary basin tectonics from the Black Sea and Caucasus to the Arabian Platform, Special publication 340. Geological Society, London, pp 329–352
Stal C, Bourgeois J, De Maeyer P et al (2012) Test case on the quality analysis of structure from motion in airborne applications. In: 32nd EARSeL Symposium: Advances in geosciences, European Association of Remote Sensing Laboratories (EARSeL)
Stewart SA, Davies RJ (2006) Structure and emplacement of mud volcano systems in the South Caspian Basin. Am Assoc Pet Geol Bull 90:771–786. https://doi.org/10.1306/11220505045
Stöcker C, Eltner A, Karrasch P (2015) Measuring gullies by synergetic application of UAV and close range photogrammetry – a case study from Andalusia, Spain. Catena 132:1–11
Stokes CR, Gurney SD, Shahgedanova M et al (2006) Late-20th-century changes in glacier extent in the Caucasus Mountains, Russia/Georgia. J Glaciol 52:99–109
Stokes CR, Popovnin V, Aleynikov A et al (2007) Recent glacier retreat in the Caucasus Mountains, Russia, and associated increase in supraglacial debris cover and supra–/proglacial lake development. Ann Glaciol 46:195–203
Strick R, Ashworth P, Best J et al (2017) The challenges in using UAV and plane imagery to quantify channel change in sandy braided rivers. In: Abstracts of EGU General Assembly conference 19: 13378
Tadini A, Bonali FL, Corazzato C et al (2014) Spatial distribution and structural analysis of vents in the Lunar Crater Volcanic Field (Nevada, USA). Bull Volcanol 76(11):1–15
Takasu T (2009) RTKLIB: open source program package for RTK-GPS, FOSS4G 2009 Tokyo, Japan
Themistocleous K, Danezis C (2018) Monitoring cultural heritage sites affected by geohazards. In: Proceedings of earth resources and environmental remote sensing/GIS applications IX, SPIE, vol 10790
Thoroddsen T (1925) Die Geschichte der isländischen Vulkane. Høst in Komm (in German)
Tibaldi A, Bonali FL (2018) Contemporary recent extension and compression in the central Andes. J Struct Geol 107:73–92
Tibaldi A, Ferrari L (1991) Multisource remotely-sensed data, field checks and seismology for the definition of active tectonics in the Ecuadorian Andes. Int J Remote Sens 12:2343–2358
Tibaldi A, Ferrari L (1992) Latest Pleistocene-Holocene tectonics of the Ecuadorian Andes. Tectonophysics 205(1–3):109–125
Tibaldi A, Pasquarè F (2008) Quaternary deformations along the 'Engadine–Gruf tectonic system', Swiss–Italian border. J Quat Sci 23(5):475–487
Tibaldi A, Rovida A, Corazzato C (2004) A giant deep-seated slope deformation in the Italian Alps studied by paleoseismological and morphometric techniques. Geomorphology 58:27–47
Tibaldi A, Pasquarè F, Papanikolaou D et al (2008) Discovery of a huge sector collapse at the Nisyros volcano, Greece, by on-land and offshore geological-structural data. J Volcanol Geotherm Res 177(2):485–499
Tibaldi A, Corazzato C, Rust D et al (2015) Tectonic and gravity-induced deformation along the active Talas–Fergana Fault, Tien Shan, Kyrgyzstan. Tectonophysics 657:38–62
Tibaldi A, Bonali FL, Pasquaré Mariotto F (2016a) Interaction between transform faults and rift systems: a combined field and experimental approach. Front Earth Sci 4:33

Tibaldi A, Bonali FL, Einarsson P et al (2016b) Partitioning of Holocene kinematics and interaction between the Theistareykir Fissure Swarm and the Husavik-Flatey Fault, North Iceland. J Struct Geol 83:134–155
Tibaldi A, Alania V, Bonali FL et al (2017a) Active inversion tectonics, simple shear folding and back-thrusting at Rioni Basin, Georgia. J Struct Geol 96:35–53
Tibaldi A, Russo E, Bonali FL et al (2017b) 3-D anatomy of an active fault-propagation fold: a multidisciplinary case study from Tsaishi (Georgia), western Caucasus. Tectonophysics 717:253–269
Tibaldi A, Oppizzi P, Gierke JS et al (2019) Landslides near Enguri dam (Caucasus, Georgia) and possible seismotectonic effects. Nat Hazards Earth Syst Sci 19:71–91
Tonkin TN, Midgley NG (2016) Ground-control networks for image based surface reconstruction: an investigation of optimum survey designs using UAV derived imagery and structure-from-motion photogrammetry. Remote Sens 8(9):16–19. https://doi.org/10.3390/rs8090786
Tortini R, Bonali FL, Corazzato C et al (2014) An innovative application of the Kinect in Earth sciences: quantifying deformation in analogue modelling of volcanoes. Terra Nova 26(4):273–281
Trippanera D, Ruch J, Passone L et al (2019) Structural mapping of dike-induced faulting in Harrat Lunayyir (Saudi Arabia) by using high resolution drone imagery. Front Earth Sci 7:168
Tsereteli N, Tibaldi A, Alania V et al (2016) Active tectonics of central-western Caucasus, Georgia. Tectonophysics 691:328–344
Turner D, Lucieer A, Watson C (2012) An automated technique for generating georectified mosaics from ultra-high resolution unmanned aerial vehicle (UAV) imagery, based on structure from motion (SfM) Point clouds. Remote Sens 4:1392–1410
Turner D, Lucieer A, De Jong SM (2015) Time series analysis of landslide dynamics using an unmanned aerial vehicle (UAV). Remote Sens 7(2):1736–1757
Varazanashvili O, Tsereteli N, Tsereteli E (2011) Historical earthquakes in Georgia (up to 1900): source analysis and catalogue compilation. Monograph Pub Hause MVP, Tbilisi
Varnes DJ, Radbruch-Hall DH, Savage WZ (1989) Topographic and structural conditions in areas of gravitational spreading of ridges in the western United States. USGS professional paper 1496
Varnes DJ, Radbruch-Hall DH, Varnes KL (1990) Measurement of ridge spreading movements (sackungen) at Bald Eagle Mountain, Lake County, Colorado, 1975–1989. USGS open file report, pp 90–543
Varnes DJ, Coe JA, Godt JW et al (2000) Measurement or ridge-spreading movements (sackungen) at Bald Eagle Mountain, Lake County, Colorado, II: continuation of the 1975–1989 measurements using a Global Positioning System in 1997 and 1999. USGS open file report 00–205, on-line edition
Vincent SJ, Morton AC, Carter A et al (2007) Oligocene uplift of the Western Greater Caucasus: an effect of initial Arabia-Eurasia collision. Terra Nova 19:160–166
Vincent SJ, Carter A, Lavrishchev VA et al (2011) The exhumation of the western Greater Caucasus: a thermochronometric study. Geol Mag 148:1–21
Vollgger SA, Cruden AR (2016) Mapping folds and fractures in basement and cover rocks using UAV photogrammetry, Cape Liptrap and Cape Paterson, Victoria, Australia. J Struct Geol 85:168–187
Walter TR, Jousset P, Allahbakhshi M et al (2018) Underwater and drone based photogrammetry reveals structural control at Geysir geothermal field in Iceland. J Volcanol Geotherm Res 391:106282. https://doi.org/10.1016/j.jvolgeores.2018.01.010
Warrick JA, Ritchie AC, Schmidt KM et al (2019) Characterizing the catastrophic 2017 Mud Creek landslide, California, using repeat structure-from-motion (SfM) photogrammetry. Landslides 16:1201–1219
Weismüller C, Urai JL, Kettermann M et al (2019) Structure of massively dilatant faults in Iceland: lessons learned from high-resolution unmanned aerial vehicle data. Solid Earth 10(5):1757–1784
Westoby MJ, Brasington J, Glasser NF et al (2012) "Structure-from-motion" photogrammetry: a low-cost, effective tool for geoscience applications. Geomorphology 179:300–314

Wise DU, Funiciello R, Parotto M et al (1985) Topographic lineament swarms: clues to their origin from domain analysis of Italy. Geol Soc Am Bull 96(7):952–967

Wu F, Ran Y, Xu L et al (2017) Paleoseismological study of the late quaternary slip-rate along the South Barkol Basin Fault and its tectonic implications, Eastern Tian Shan, Xinjiang. Acta Geol Sin-Engl 91(2):429–442

Yao Y, Chen J, Li T et al (2019) Soil liquefaction in seasonally frozen ground during the 2016 Mw 6.6 Akto earthquake. Soil Dyn Earthq Eng 117:138–148

Young KD, Jancin B, Orkan NI (1985) Transform deformation of tertiary rocks along the Tjörnes fracture zone, North Central Iceland. J Geophys Res 90:9986–10010

Yusifov M, Rabinowitz PD (2004) Classification of mud volcanoes in the south Caspian Basin, Offshore Azerbaijan. Mar Pet Geol 21:965–975. https://doi.org/10.1016/j.marpetgeo.2004.06.002

Zekkos D, Manousakis J, Athanasopoulos-Zekkos A et al (2018) Structure-from-motion based 3D mapping of landslides & fault rupture sites during 2016 Kaikoura earthquake reconnaissance. In: Proceedings of the 11th US national conference on earthquake engineering, integrating science, engineering & policy, Los Angeles, California

Chapter 23
Automatic Telemetric Monitoring/Early Warning Systems, with Multi-task Sensor, Applied to Mass Movements

Zurab Chelidze, Nodar Varamashvili, Tamaz Chelidze, Tengiz Kiria, and Tamar Tsamalashvili

Abstract The several geohazards, the costs for supporting personnel for long periods at potential mass movement sites, the growing number of exposed vulnerable objects and the limited resources of developing countries, the most affected by natural disasters, call for the development of cost-effective and at the same time, accurate, automatic monitoring/early warning telemetric systems.

We have developed the operating prototype of an original, cost-effective multi-sensor telemetric monitoring/early warning system with autonomous power source, aimed at forecasting the initiation of landslides/mudflows and sending alarms as far as 15–20 km by radio signals, or to a longer distances – by Internet or Mobile nets. Humidity, tilt and acceleration sensors, using modern high-tech elements (modified MEMS sensors and processing-diagnostic-remote transmitting ARDUINO blocks), are proposed. This ensures the cost-effectiveness of the system. The performance of our sensors has been compared with standard devices. Moreover, our system has been tested in laboratory conditions for defining optimal sensitivity, frequency and amplitude ranges. The EWS system, based on measurements of differential humidity, tilt and acceleration signals, was installed in October 2018 within the Gldani landslide (Tbilisi, Georgia). The EWS detected significant anomalies during and after the landslide event of 25 January 2019, as well as during intensive rainfall on 18 July 2019, which was not followed by significant landslide displacement.

The map of areas/objects affected by risk has been compiled for Georgia, in which the locations of dams, pipelines, roads, and high density of population are superimposed on a landslide hazard susceptibility map. This allows for the identification of areas, where the installation of EWS systems is especially recommended.

Z. Chelidze (✉) · N. Varamashvili · T. Kiria · T. Tsamalashvili
M. Nodia Institute of Geophysics, Tbilisi State University, Tbilisi, Georgia

T. Chelidze
M. Nodia Institute of Geophysics, Tbilisi State University, Tbilisi, Georgia

European Centre "Geodynamical Hazards of High Dams", Council of Europe, Tbilisi, Georgia

© Springer Nature B.V. 2021

F. L. Bonali et al. (eds.), *Building Knowledge for Geohazard Assessment and Management in the Caucasus and other Orogenic Regions*, NATO Science for Peace and Security Series C: Environmental Security,
https://doi.org/10.1007/978-94-024-2046-3_23

Keywords Early warning systems · Mass movement · Hazard · Sensor · Georgia

23.1 Introduction

Each year, landslides and debris-flows cause several disasters in mountainous areas all over the world. Several works, in the past, have documented this type of hazard, in Europe, the Caucasus and Central Asia (e.g. Tibaldi and Pasquaré 2007; Tibaldi et al. 2015, 2018; Pasquaré Mariotto and Tibaldi 2016). Thus, it is of great importance, for the purpose of Disaster Risk Reduction (DRR), to create reliable and cost-effective early warning systems (EWS) for monitoring mass movements in potentially hazardous areas. The key elements in the solution of the problem, as stressed in the Sendai Framework for Disaster Risk Reduction (UNISDR 2015), are science and technology. The Sendai Framework calls for "The improvement of preparedness in order to respond effectively to a disaster, by implementing a simple, low-cost EWS and improving the dissemination of information about EWS of natural disasters at local and national levels." In recent years, many works have been published, devoted to the development of modern EWS (Malet and Maquaire 2008; Intrieri et al. 2012; Anderson and Holcombe 2013; Stähli et al. 2015; Fathani et al. 2016; Calvello and Piciullo 2016; Wu et al. 2019); however, most are by no means cost-effective, as they propose the use of expensive equipment or complicated procedures for the prediction of landslide events.

The terrain of Georgia, due to the frequent occurrence of large-scale geological processes (National Disaster Risk Reduction Strategy of Georgia (2017), Towards a Multi-Hazard Early Warning System for Georgia, UNDP 2016), the steady growth of the population, the presence of vulnerable infrastructures and facilities, belongs to the most affected mountainous regions in the world.

The most recent geomorphological catastrophes were a mudflow that took place on 17 May 2014 in the Dariali Gorge (close to the Russia-Georgia border) and a major flashflood/mudflow that struck in Tbilisi on 13 June 2015. The Tbilisi event caused a series of disasters and 20 persons were killed; economic losses amounted to hundreds of millions of US dollars. According to the UNDP report (Towards a Multi-Hazard Early Warning System for Georgia, UNDP 2016) "the lack of adequate equipment, human resource and finances is an obstacle for availability of reliable and timely warnings".

Therefore, the problem of building up cost-effective, telemetric monitoring/early warning systems with multi-task sensors suitable for detecting and signaling the initiation of mass movements (debris flows/landslides) is of paramount importance for Georgia. It is worth pointing out that there are many institutions in the world developing such systems, based on geodesy, strainmeters, Global Positioning Systems (GPS), laser and radar interferometry, optical fiber techniques (OFT), and so on. The cost of such equipment is, as a rule, very high (of the order of hundreds of thousands of US dollars) and it is practically impossible for developing countries to purchase even one of such systems. Taking into account the huge number of mass movement sources – only in Georgia there are 40,000 potential debris flow/landslide sources – the financial expenses are unaffordable. That's why the development of cost-effective and, at the same time, accurate monitoring/early warning telemetric systems is urgently needed. The network approach is important, as the mass movement in the test area is as a rule very heterogeneous and, with a small number of sensors, the system might miss the initiation of dangerous events. The unexpensive equipment enables also: (i) to install ramified arrays of multi-sensors on the potential hazard source, which increases the reliability of the system; (ii) to avoid huge financial losses due to very probable destruction of EWSs by major mass movements, and (iii) facilitate the renewal of damaged EWS networks.

Our debris flow/landslide EWS project aims to develop a precise enough and cost-effective, telemetric geophysical monitoring system with an autonomous power source (solar batteries), suitable for signaling debris flow/landslide initiation using radio signals or internet connection. The above has been made possible, recently, thanks to availability of new cheap technologies, such as microelectromechanical sensors MEMS (Rai-Choudhury 2001), contactless soil moisture sensors and processing-diagnostic-remote transmitting ARDUINO blocks (Blum 2013), compatible with modern digital signal processing elements with low power consumption.

According to the widely accepted physical principles of mass movement formation, the main phases of such events are: (i) strong rainfalls; (ii) percolation of rainwater through the landslide rupture surface; (iii) small-scale, pre-slide mechanical displacements/tilts; (iv) onset of the massive landslide/debris-flow.

Accordingly, the suggested debris flow/landslide EWS network comprises the monitoring of two main factors, leading to a mass movement: (i) soil moisture (a relatively long-term precursor) and (ii) displacement/acceleration caused by mechanical motion (mainly short-term, but also mid-term precursor). Our device is a diagnostic system, which continuously acquires information, compares it with accepted criteria (critical values of soil humidity, tilts and differential acceleration) and works out a decision, after which the alarm signal is transmitted telemetrically.

23.2 Details on the Instrumentation

23.2.1 The Integrated System, Containing Multi-task Sensor, Data Acquisition and Analogue Processing Platform

In order to detect anomalous changes in humidity, tilt and acceleration, we hereby suggest a methodology, based on the measurement of differential signals related to humidity, tilt and acceleration, relative to the same signals at a reference station. The criteria for releasing different alarm signals (quiet, warning 1, warning 2, alarm 1, high level alarm 2) based on data on differential humidity, tilt and acceleration are elaborated.

As humidity sensor, we use a high-frequency (HF) 2.7 GHz micro-radar module. The HF electro-magnetic (EM) signal, emitted by micro-radar, is partially absorbed and reflected by the surrounding media.

The standard microelectromechanical sensors (MEMS) can register accelerations, but the output signal is relatively weak and contains a significant noise component. That is why, in our analogue processing system, we use amplifiers and filters to obtain the amplified and noise-filtered desired signal.

The input signals from the multi-task sensors are sent to the analogue processing platform, where the amplitudes, phases and spectral characteristics of signals obtained from each sensor are collected and their differences relative to the reference sensor located in a stable area are calculated. The difference between the signals from the reference sensor and the field sensors describes the variations of acceleration, tilt and mass movement events in the local area in a much better way than absolute signal values. At the same time, variations caused by earthquakes, industrial vibrations and other background noise influences can be suppressed. Differential line drivers are used for converting balanced signals to unbalanced ones. As mentioned, the signals from field sensors are compared to a reference signal and the result of the comparison is rectified by bipolar driver and bipolar detector circuit for both sensors and both directions of acceleration individually. Detected variations are applied to inputs of integrator circuits and summed up. For measuring tilt, the output signal from MEMS is filtered by low-pass filter in order to obtain the component, proportional to a slow geo-mechanical process – the inclination of the ground mass. The filter passes the low and infra-low components of the MEMS output signal in the range from DC to 1.0 Hz, which reflects a slow component of displacement– the tilt.

23.2.2 Power Source

As the mass movement sources can be located in unpopulated remote areas, we pay much attention to energy-saving in the electronic design of the EWS, as well as to the high reliability of the independent power source module (IPSM).

23.2.3 Telemetry System

For the transfer of information on the state of the landslide body to local and national DRR leaders, it is necessary to have a reliable and economic telemetry channel. One of the key elements of our EWS is the processing-diagnostic-remote transmitting block – Arduino Mega 2560: it receives permanently signals from the sensors (humidity, tilts, acceleration), performs the processing of data from sensors

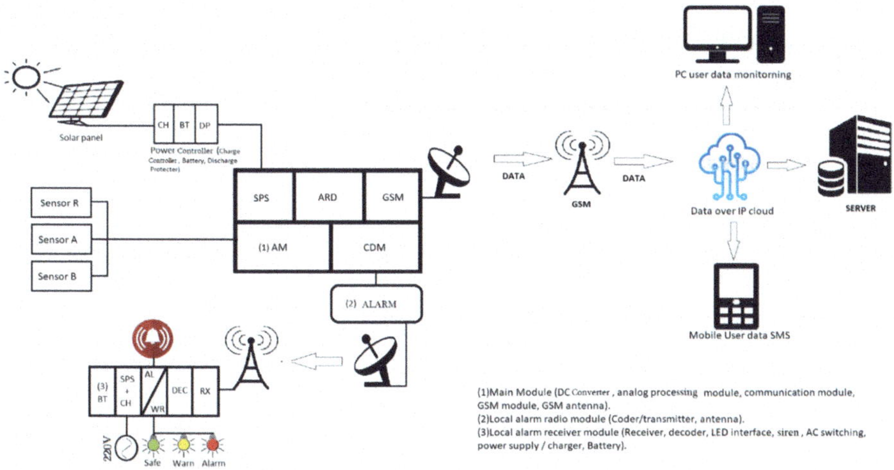

Fig. 23.1 The integrated block scheme of our EWS "Watchdog", including sensors, power source, acquisition, processing and telemetry blocks

and identifies the level of alarm by using programmed mathematical calculations and existing database. The information on the state of the landslide source (alarm level) is sent over GSM network regularly (every 30 min) or immediately when the anomalous signal is detected.

23.2.4 The Integrated EWS "Watchdog"

The integrated block scheme of our EWS, which we refer to as "Watchdog", including sensors, power source, acquisition, processing and telemetry blocks is shown in Fig. 23.1.

23.3 Methodology

The novelty of our research is that it involves the complex study of long- and short-term precursors of debris flow/landslides and the development of an integrated, low-cost early warning system based on high-tech elements. It is known that most mass movements are caused by intense percolation of rainwater into the soil (Guzzetti et al. 2008; Baum and Godt 2010), so one element of our sensors measures soil moisture, with the purpose of revealing a comparatively long-term precursor, as for the activation of a mass displacement, the rainwater should reach the landslide's rupture surface. At the same time, it is worth noting that the increased soil moisture, by itself, does not always lead to mass movement initiation. Therefore,

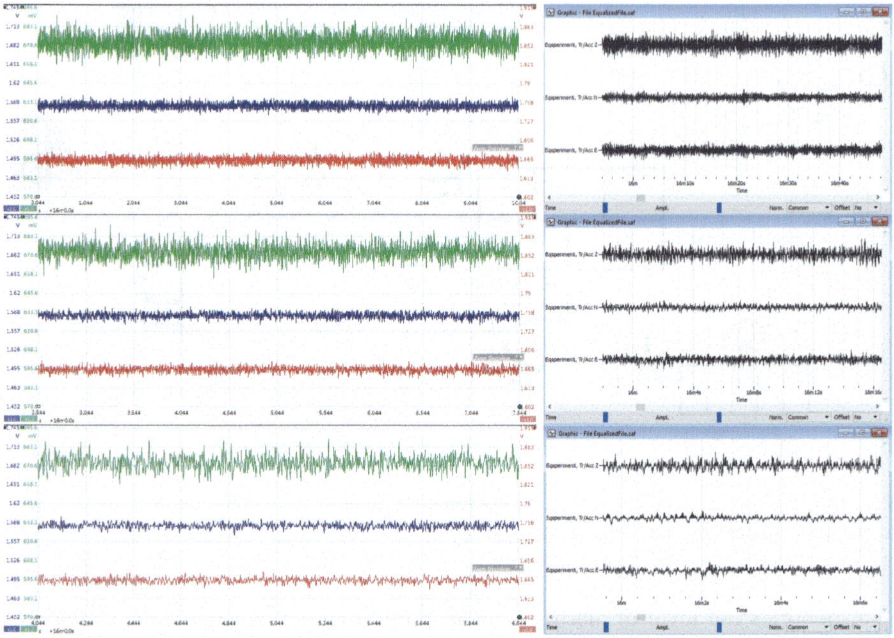

Fig. 23.2 Records of modified MEMS accelerometer (left side) and standard accelerometer TROMINO (Italy – right side) at various sampling rates

the system includes additionally (mechanical) tilt and acceleration sensors, which can provide both long- and, especially, short-term signals, generated by the initiation of mechanical displacements within the landslide body. Such an approach, based on the combination of multi-sensors, provides warnings/alarms of different levels, in a similar way as warning systems designed to address the occurrence of terrorist attacks.

23.3.1 Tests on Multi-sensors

The performance of the multi-sensors was tested in laboratory conditions, by comparing their outputs with standard serial devices. Figure 23.2 presents the records from the modified MEMS accelerometer and the standard TROMINO accelerometer (Italy), at various sampling rates. The records are very similar, as well as their spectra.

The data obtained by the tilt-measuring element of the MEMS accelerometer were compared to the tilt readings from the "Applied Geomechanics" 701 Platform Tiltmeter (Fig. 23.3a).

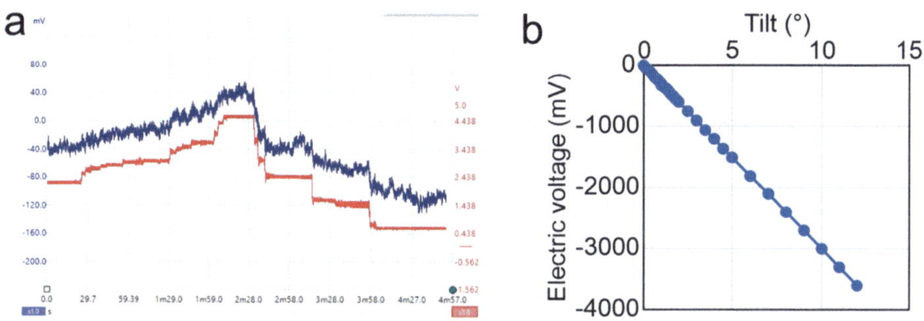

Fig. 23.3 (**a**) Comparison of the recordings of tilt-measuring element of MEMS (blue) with the tilt readings on the precision device "Applied Geomechanics" 701 Platform Tiltmeter (red). On the Y-axis the readings of the above devices in mV and on the X-axis the time in sec are represented; the tilt was changed in time during experiment; (**b**) Results of the calibration of MEMS tilt sensor on the special graduated table: on the Y-axis the readings of the MEMS tiltmeter in mV and on the X-axis the tilt of the supporting plate in degrees are represented (note the linearity of the graph)

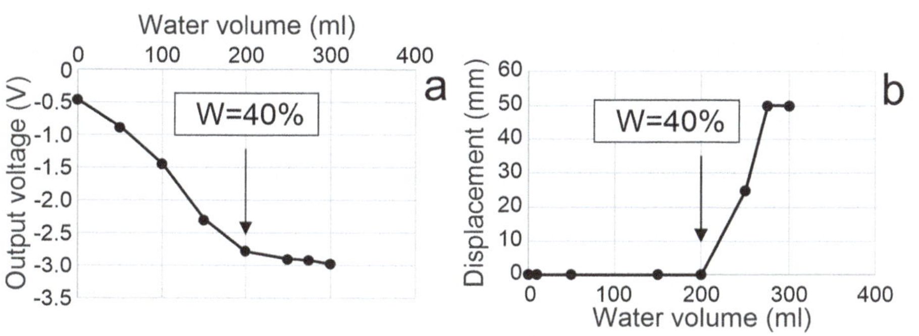

Fig. 23.4 (**a**) Results of the humidity sensor calibration: (**a**) output voltage of micro-radar versus water volume (ml) in the quartz sand sample; (**b**) Mass movement of sand at various humidity – the displacement of sand sample toe (mm) on the horizontal shear plane versus water volume (ml) in the sample. Note: significant changes in both micro-radar output and large mechanical displacement occur at the same humidity of the sample: W = 40%

MEMS tilt sensors were also calibrated using special graduated table: Figure 23.3b shows the graph of tilt in degrees versus the output voltage on the MEMS tiltmeter: the graph is practically linear.

Lastly, the humidity sensor was tested on the sand sample with different water content (Fig. 23.4).

It is clear that the radar signal strongly depends on the humidity of the sample as well as the mobility of water-containing sand. The arrow with the W = 40% label corresponds to the water volumetric percentage in the sample, at which both the radar signal and the displacement reveal drastic changes.

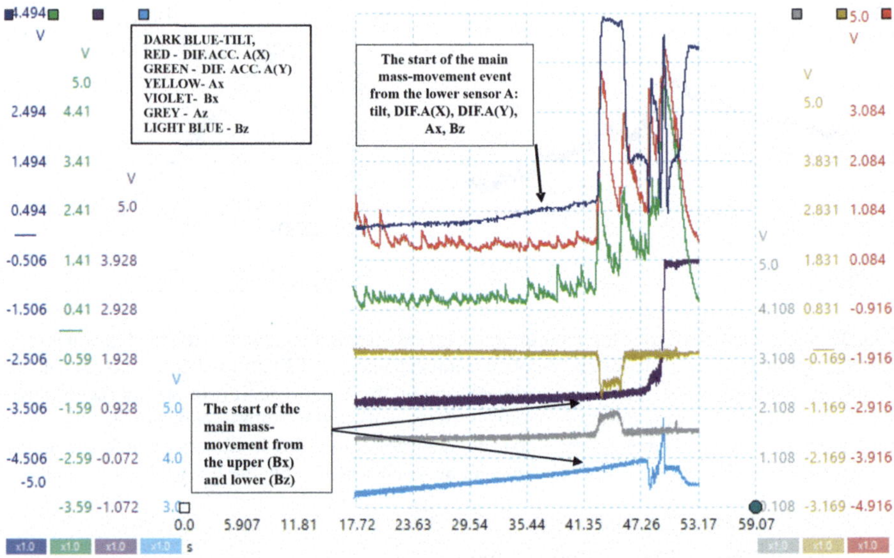

Fig. 23.5 Recordings from sensors into the modeled landslide, at increasing inclination values of the box containing the granular media (see text for details). The arrow shows the onset of the mass movement: the earlier signal arrived from the lower sensor (**a**) and the following one from the upper sensor (**b**). The mass movement began from the lower part of the model (close to the toe) and then propagated to the upper part

23.3.2 Tests of the EWS on the Laboratory Model of Landslide

In Fig. 23.5 we show the recording from sensors on the Picoscope, during the laboratory modeling of a landslide, obtained by changing the inclination of a large box (240 × 60 × 60 cm), partially filled with quartz sand at room humidity (60%). Two sensors, A and B, were inserted into the sand, and the immobile reference sensor was fixed on the floor. One side of the box was raised gently, until the final mass-movement occurred.

In Fig. 23.5, recordings correspond to: dark blue – tilt; red – differential signal of X-component of acceleration; green –the same of Y-component of acceleration; dark yellow – the direct output of x-component of sensor A; violet – the same for x-component of sensor B (both signals – without differentiation, filtering and amplification); grey – the same for z-component of sensor A; light blue – the same for z-component of sensor B. It is clear that, after reaching the critical inclination, the collapse of the sand mass occurs, accompanied by strong signals, first from the lowest sensor (A), close to the toe of the model mass, and then from the upper sensor (B), located close to the future crown. The results displayed in Fig. 23.5 confirm the benefit of using differential methods for monitoring sensor signals – the amplitudes of anomalies of differential signals (red and green) during both precursory and final

Fig. 23.6 (**a**) Gldani landslide; (**b**) The installed EWS, with reference multi-sensors, solar battery, acquisition unit and the processing-diagnostic-remote transmitting block – Arduino Mega 2560, assembled on the rod-post

stages of the sand mass collapse are much clearer than those of direct signals from sensors (yellow, violet, grey, light blue).

23.4 Field Tests of the EWS

The test area was the October 2018 landslide in Tbilisi. The landslide was of anthropogenic origin: in the 1970s, below a hillside, a tramline was built. As a result of this, the toe of the slope was cut and the mass movement began. In 1984–1989 interval, on the hill top, three 16-storey buildings were erected (Fig. 23.6a).

The landslide resumed activity and came as close as 17 m to one of the 16-storey buildings, putting several residents at risk. A number of preventive measures were put in place, and the landslide was stopped. In October 2018, our EWS was installed into the Gldani landslide body (Fig. 23.6b).

On 25 January 2019, a significant part of the relatively stable hill slope was reactivated, due to the cutting of the landslide toe. In Fig. 23.7, we have shown the configuration of our telemetric EWS with reference and field sensors as well as the contour of the excavated soil for the foundation of buildings, which, together with heavy rainfall, activated the (partial) slide event.

The recordings of humidity and tilt, before and after the 25 January 2019 event, are shown in Fig. 23.8: the strong effects on humidity and especially on tilt (with a sharp increase from 200 to 800 mV) are clearly visible. It is worth noting that in Figs. 23.8 and 23.9 the humidity sensor's signal is a reciprocal to the humidity value. It is interesting that, in January 2019, no heavy rain event was recorded, so the onset of the mass movement was due to the excavation of the building's foundations. Figure 23.9 displays humidity, tilt and accelerations (X, Z) recordings before and after the 18 July 2019 heavy rainfall episode, which lasted several hours: the recorded precipitation was around 50 mm.

Fig. 23.7 The Gldani landslide after 25 January 2019. Number 1 points to the station equipped with reference sensor; number 2 and 3 show the location of field sensors; the white dashed line shows the contour of the displaced mass; the magenta line indicates the contour of the soil, excavated for the building's foundations (this led to mass movement activation)

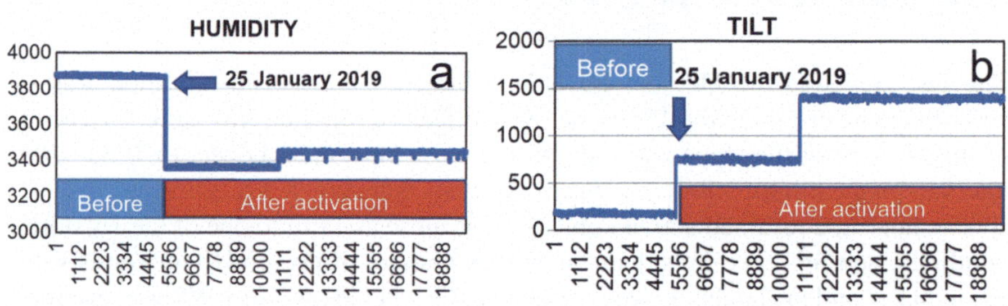

Fig. 23.8 The recordings of humidity (**a**) and tilt (**b**) before and after the 25 January event. The Y-axis shows the differential output signal of sensors (mV); on the X-axis, the numbers of sensor readouts are represented (recording IDs)

It is worth underscoring that the differential system of processing the sensors' signals is quite sensitive to (local) ground humidity and mechanical tilts/vibrations changes of the landslide mass, generated by heavy rain: it is possible to observe strong anomalies in all recordings, some time after the end of the rainfall episode. We believe this time was needed for allowing the percolation of water through the landslide slip surface. Our algorithm of alarm grade, with five levels (quiet, warning 1, warning 2, alarm 1, high level alarm 2), based on the combination of outputs from four sensors (Fig. 23.9), seems to be satisfactory, as the issued alarm level was "warning 2". Specifically, this alarm level corresponds to the absence of the X and Z components of acceleration, small anomalies in tilt and significant anomalies in

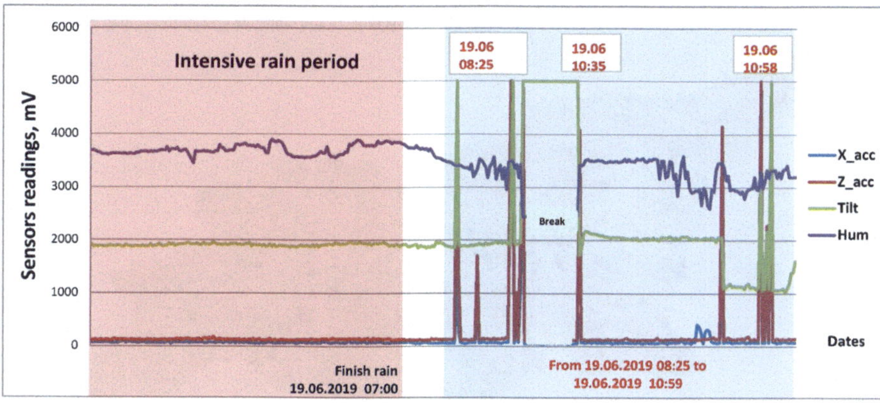

Fig. 23.9 Recordings of humidity (blue), tilt (green) and acceleration (X – light blue, Z – red) during and after the heavy rain event of 18 July 2019. There was a 40-min gap in data transmission, marked by an arrow. Note that strong anomalies in all recordings occurred approximately 1.5 h after the end of the rain event

humidity: in the joint line of telemetry signals, we have the following reading [0, 0, 1, 3], which corresponds to the absence of accelerations [0, 0], small change of tilt [1] and strong increase in humidity [3]. The EWS issued this warning signal with 1.5 h lag after rain termination, i.e. after 25 July 2019. We obtained the Georgian patent for the above-described EWS: # U 2019 2017 Y, 19 July 2018.

23.5 Choosing the Landslide High-Risk Locations for Optimal Placement of EWS Posts in Georgia

Needless to say, it is practically impossible to install EWSs at all localities potentially subject to landslide/debris-flow initiation. At present, across Georgia 50,000 landslide and 350 debris-flow sources have been individuated. In the high and medium mass movement risk zone, 1420 settlements are located. Therefore, monitoring all these sources involves major human and financial recourses, and this makes it impossible to assess the risk to which the whole territory of Georgia is subjected. Clearly, mass movement sources that are located far away from important structures or settlements, do not pose large risk to the population and infrastructures. That is why, in a first phase, we collected data on high-risk objects (dams, pipelines, international and national highways, settlements with population density larger than 300 units/km^2) and inserted their locations on the landslide/debris-flow susceptibility map of Georgia (Fig. 23.10). We regard as landslide high-risk objects the ones that are located in the high landslide susceptibility areas (red). All the results presented in this paper are discussed in detail in a recent volume (Chelidze et al. 2018).

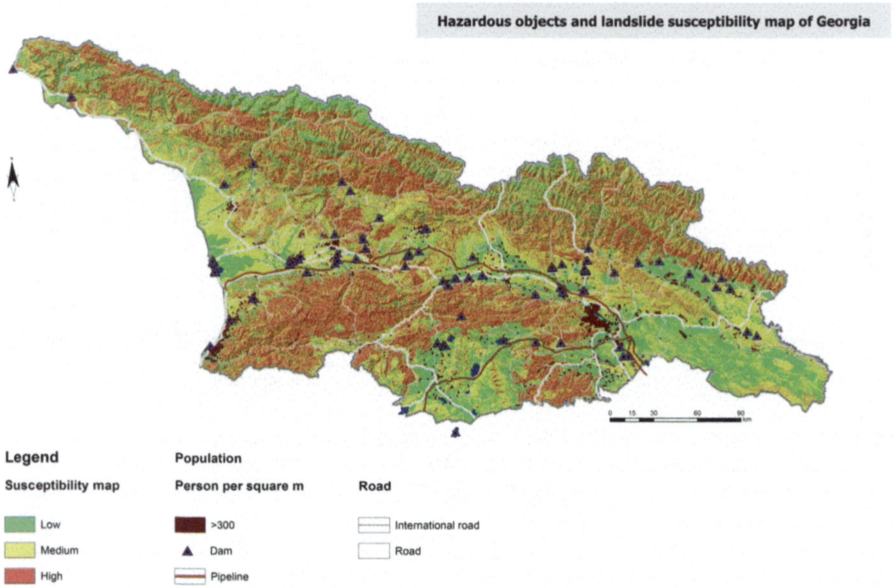

Fig. 23.10 The locations of high-risk objects, superimposed on the landslide hazard susceptibility map of Georgia

23.6 Conclusions

We have developed a prototype of an original, cost-effective multi-sensor telemetric monitoring/early warning system with autonomous power source, aimed at forecasting the initiation of a landslide/mudflow and transferring different-level alarms by internet or mobile networks.

We have manufactured humidity, tilt and acceleration sensors by using modern, high-tech devices (modified MEMS sensors and processing-diagnostic- remote transmitting ARDUINO blocks). This allows for the cost-effectiveness of the system (the price of the EWS hardware is around 600 US dollars; including the construction of the rod-post, the total cost is 1300 US dollars). We have compared the performance of our sensors with standard devices. The system has also been tested in laboratory conditions on a model- landslide, with the purpose of defining precursors and final sliding signatures in tilts and differential accelerations.

The criteria for emitting different alarm signals (quiet, warning 1, warning 2, alarm 1 and high level alarm 2) on the basis of information on humidity, differential tilt and acceleration have also been elaborated.

Our EWS system was installed into the Gldani landslide (Tbilisi) in October 2018, showing significant anomalies after man-induced activities (the landslide event of 25 January 2019) and the strong rainfall event of 18 July 2019.

Finally, as it is impossible to install EWSs at all mass movement sources, we have superimposed the locations of key, vulnerable areas/objects (dams, pipelines, roads, highly populated areas) on the landslide hazard susceptibility map of Georgia; this has enabled us to produce a high-landslide-risk map, with all the locations in which the installation of our EWS is recommended.

Acknowledgements This work was supported by a grant from the Shota Rustaveli National Scientific Foundation of Georgia # 216732 (2016-2018) and by the grant "Major Catastrophes Agreement of the Council of Europe" (Ref. No: CA/2017/08 FIMS PO No:537534). The Authors wish to acknowledge with gratitude the contributions provided by Prof. J.-P. Malet and Prof. E. Tsereteli. The support of Drs. N. Ghlonti, J. Kiria, technicians Z. Maghradze, L. Dvali and D. Amilakhvari was also very valuable.

References

Anderson M, Holcombe E (2013) Community-based landslide risk reduction: managing disasters in small steps. The World Bank, Washington, DC

Baum R, Godt J (2010) Early warning of rainfall-induced shallow landslides and debris flows in the USA. Landslides 7:259–272

Blum J (2013) Exploring Arduino: tools and techniques for engineering wizardry. Wiley, Indianopolis

Calvello M, Piciullo L (2016) Assessing the performance of regional landslide early warning models: the EDuMaP method. Nat Hazards Earth Syst Sci 16:103–122

Chelidze T, Varamashvili N, Chelidze Z et al (2018) Cost-effective telemetric monitoring and early warning systems for signaling landslide initiation, Tbilisi (in Georgian)

Fathani TF, Karnawati D, Wilopo W (2016) An integrated methodology to develop a standard for landslide early warning system. Nat Hazards Earth Syst Sci 16:2123–2135

Guzzetti F, Peruccacci S, Rossi M et al (2008) The rainfall intensity–duration control of shallow landslides and debris flows: an update. Landslides 5(1):3–17

Intrieri E, Gigli G, Mugnai F et al (2012) Design and implementation of a landslide early warning system. Eng Geol 147:124–136

Malet JP, Maquaire O (2008) Risk assessment methods of landslides. Ramsoil, risk assessment methodologies for soil threats. Sixth framework programme, project report, 2

National Disaster Risk Reduction Strategy of Georgia (2017). www.preventionweb.net/files/54533_drrstrategy2017annex1eng.pdf

Pasquaré Mariotto F, Tibaldi A (2016) Inversion kinematics at deep-seated gravity slope deformations revealed by trenching techniques. Nat Hazards Earth Syst Sci 16:663–674

Rai-Choudhury P (2001) MEMS and MOEMS technology and applications. SPIE Press, Washington

Stähli M, Sättele M, Huggel C et al (2015) Monitoring and prediction in early warning systems for rapid mass movements. Nat Hazards Earth Syst Sci 15:905–917

Tibaldi A, Pasquaré F (2007) Quaternary deformations along the Engadine-Gruf tectonic system, Swiss-Italian Alps. J Quat Sci 23:475–487

Tibaldi A, Corazzato C, Rust D et al (2015) Tectonic and gravity-induced deformation along the active Talas-Fergana Fault, Tien Shan, Kyrgyzstan. Tectonophysics 657:38–62

Tibaldi A, Korzhenkov AM, Pasquaré Mariotto F et al (2018) NATO and earth scientists: an ongoing collaboration to assess geohazards and contribute to societal security in Central Asia and the Caucasus. Episodes 41:193–205

UNDP (2016) Towards a multi-hazard early warning system for Georgia. United Nations development programme – Georgia, p 25

UNISDR U (2015) Sendai framework for disaster risk reduction 2015–2030. In: Proceedings of the 3rd United Nations world conference on DRR, Sendai, Japan, March, pp 14–18

Wu Y, Niu R, Lu Z (2019) A fast monitor and real time early warning system for landslides in the Baige landslide damming event, Tibet, China Nat Hazards Earth Syst Sci Discuss. https://doi.org/10.5194/nhess-2019-48

Chapter 24
Extensometer-Based Monitoring of Active Deformation at the Khoko Landslide (Jivari, Georgia)

Federico Pasquaré Mariotto and Paolo Oppizzi

Abstract There are several ways to monitor the evolution of a mass movement in mountainous areas, such as a landslide: from inclinometers, helpful in measuring deep deformation, to piezometers, for determining water level within the unstable mass, to electronic and manual extensometers, suitable for assessing local deformation. Here, we document the results of an extensometer-based campaign of measurements carried out in Georgia from 2016 to 2019, at the giant Khoko landslide, located on the southern slope of the Greater Caucasus. The campaign was conducted in the framework of a NATO-funded project, aimed at identifying geohazards affecting the Enguri artificial reservoir and the related hydroelectrical plant. Our results, which are meant to be integrated by the more accurate GPS-measurements performed during the NATO-supported project, indicate that the Khoko landslide is indeed active, and its displacements appear to be controlled by variations in hydraulic load, in turn induced by lake oscillations and, although to a minor extent, by rainfall.

Keywords NATO · Enguri · Dam · Extensometer · Landslide

24.1 Introduction

In November 2015, an international team of scientists, under the aegis of NATO, started the first investigations focused on the wide gamut of natural hazards affecting the Enguri water reservoir, situated on the southern slope of the Greater Caucaus, Georgia (Fig. 24.1a). The related hydroelectrical plant, which was built during the

F. Pasquaré Mariotto
Department of Human and Innovation Sciences, Insubria University, Como, Italy

P. Oppizzi (✉)
Geolog.ch, Bellinzona, Switzerland
e-mail: paolo.oppizzi@geolog.ch

© Springer Nature B.V. 2021

F. L. Bonali et al. (eds.), *Building Knowledge for Geohazard Assessment and Management in the Caucasus and other Orogenic Regions*, NATO Science for Peace and Security Series C: Environmental Security,
https://doi.org/10.1007/978-94-024-2046-3_24

Fig. 24.1 (**a**) Location of the Enguri dam and Enguri reservoir in the Greater Caucasus; Country borders are from GADM data, © 2018 GADM available at https://gadm.org/, version 3.6 (released on 6 May 2018); yellow dashed lines represent the main active faults in the region, from the EMME Project (Gulen et al. 2011). Reference system: WGS84/geographic coordinates. (**b**) The upper portion of the active Khoko landslide. (**c**) Panoramic view of the Enguri dam

Soviet era together with the Enguri Dam (Fig. 24.1c), is responsible for up to 55% of the energy supply to the country (Tibaldi et al. 2018, 2019). A thorough inspection of the slopes along the eastern flank of the water reservoir, aimed at evaluating the presence of mass movements and individuating the types of landslides, revealed the existence of a number of active landslides. Most attention was paid to the huge Khoko landslide (Fig. 24.1b and 24.2a), thoroughly described in a recent paper by Tibaldi et al. (2019). The NATO-funded team set out to investigate the hydrogeological hazard posed by the landslide, also in terms of the possible role played by active faults in triggering earthquakes in the region surrounding the dam. Indeed, although this particular issue is outside the scope of the present paper, it is worth pointing out that the study area is affected by still active mountain building processes, which led to the formation of the Greater and Lesser Caucasus, and resulting from the continent–continent collision between the Eurasian and African–Arabian plates (Reilinger et al. 1997, 2006; Koçyigit et al. 2001; Pasquaré et al. 2011). Such processes are accompanied by diffuse seismicity, reaching Ms of 6–7 (Tsereteli et al. 2016) and macroseismic intensities as great as 10 (Varazanashvili et al. 2018). Active compressional tectonics, both at the core of the Greater Caucasus and along its southern border, have been documented in a number of research

Fig. 24.2 (**a**) Black lines show the main head scarps of the unstable slope area facing the Enguri reservoir. (**b**) and (**c**) Photos of one of the several slip planes that offset the Jvari-Khaishi-Mestia road, taken in November 2015 and May 2016, respectively. Note the substantial increase in offset. (Modified after Tibaldi et al. (2019), photographs courtesy of the Authors)

efforts (Tsereteli et al. 2016; Tibaldi et al. 2017a, b); in the study area, this tectonic regime is expressed by a zone of active thrusting along fault planes dipping towards NNE (Tibaldi et al. 2019). As has been known for decades, there is a possible linkage between major faults and the development of landslides and deep-seated gravity-induced slope deformations (DGSDs). This linkage has been thoroughly investigated, in the past, at mountainous locations in the Alps as well as in Central Asia (e.g. Tibaldi et al. 2004, 2015; Tibaldi and Pasquaré 2008; Pasquaré Mariotto and Tibaldi 2016).

Back to the main topic of this work, it is worth highlighting that the uppermost portion of the Khoko landslide is cut, across along a 2-km-long stretch, by the Zugdidi-Jvari-Mestia-Lasdili road, at about 700 m a.s.l. Several clues to active displacements were observed, by the NATO-funded team, along two road stretches, about 150–200 m long, represented by differential failure of the road surface (Figs. 24.2b–c), opening of holes, deformation of walls and road drainage systems. Long before the beginning of our investigations, along a few road stretches that had been subject to the greatest loading, the road surface (asphalt paving) had periodically been replaced every 2–5 years.

Upslope of the active Khoko landslide, deep holes (up to 4 m deep) and natural trenches (up to 20 m long) were discovered, affecting the slope. The trenches are aligned with escarpments facing down towards the artificial lake; at the foot of these scarps, large blocks (up to 5 m in size) were found: this suggests a previous collapse of the slope. The presence of holes, trenches and scarps along the upper part of a slope can be interpreted as clear evidence of gravitational instability. Only small trees are located within the holes and trenches, whereas larger and older trees are located outside the area affected by these gravity features. This, in turn, indicates that the holes and trenches are quite recent. In view of the recognized critical situation, in 2016 it was agreed upon the need to install horizontal extensometers within the Khoko landslide, aimed at monitoring, in real time, the deformation across its upper slope as well as the displacements affecting the Zugdidi-Jvari-Mestia-Lasdili road.

24.2 Methodology and Instrumentation

In May, 2016, two trenches, each suitable for hosting a horizontal, digital extensometer (Wire Linear Potentiometric Transducer, SF500), were excavated; the location of the sites selected for trench opening is indicated in Fig. 24.3.

The two trenches were opened perpendicularly to the roadway, within which the instrumentation protection system was placed. The opening was performed in two stages, to enable vehicles to drive through the area along alternating lanes. The protection for the measurement stations consists of a channel in reinforced concrete, buried down to a depth of at least 50 cm (Fig. 24.4).

The wire was inserted into a pipe, laid down horizontally and protected with sand. At both ends, steel pipes were positioned, aimed at securing the measurement wire and the electronic instrumentation. Each vertical tube was equipped with a steel cover and gasket. The two covers were buried underneath a 15 cm-thick soil layer. The operations for putting in place the wires were made more difficult, at station n.

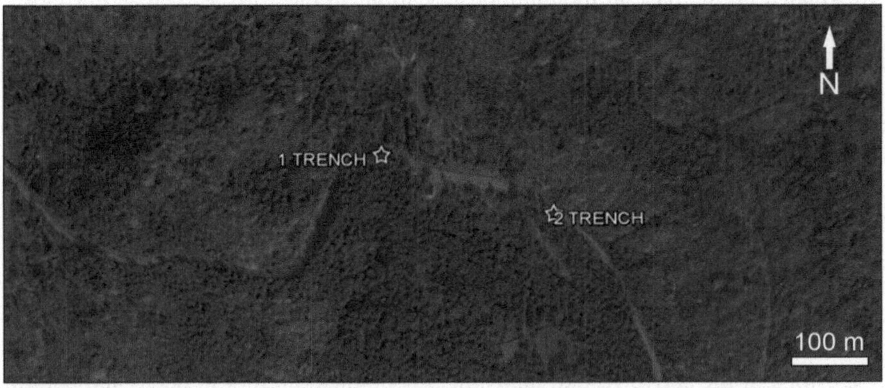

Fig. 24.3 Location of the two measurement stations along the road Zugdidi-Jvari-Mestia-Lasdili, spaced 200 m from each other. See Fig. 24.2 for its location

Fig. 24.4 Opening of the first trench and installation of the protection for the extensometer

1, by the presence of a pavement in concrete beneath the present-day asphalt layer; at station n. 2, trenching was made difficult by the presence of remnants of the old road, made of about 40 cm of concrete, found at a depth of 1.20 m.

The first extensometer was put in operation in November 2016, whereas the second one was put in place in the spring of 2017. Accordingly, data collection from measurement station n. 1 started in October 2016, and it began at station n. 2 in May 2017.

24.3 Results and Discussion

In Fig. 24.5, we have shown the readings collected over a 29-month interval, between the beginning of May 2016 and the beginning of October 2019, by the extensometer placed at station n. 1, together with lake level variations and rainfall.

As can be clearly observed, deformation peaked from 16 May 2017 to 8 August 2017, with a total extension of 52 mm, corresponding to a rate of 0.61 mm/day. For comparison, the overall extension recorded during the 29-month period was

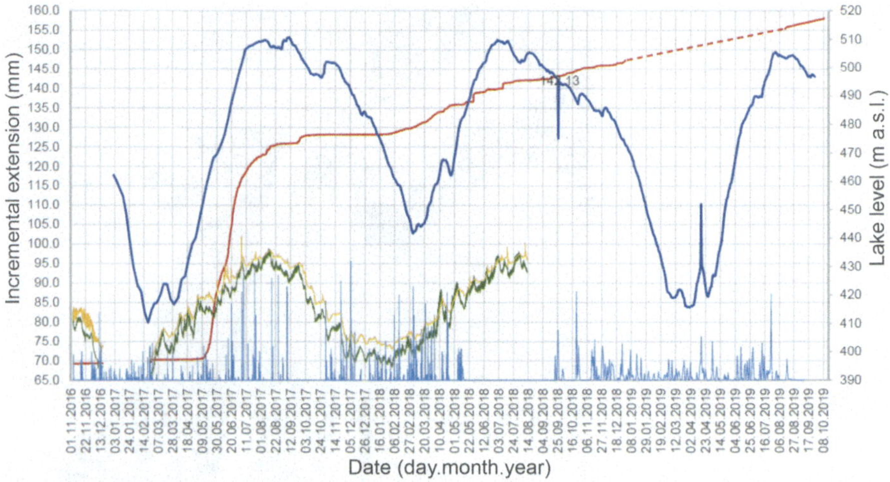

Fig. 24.5 Graph showing the readings of the incremental extension (in mm), associated with landslide surface displacement, recorded from May 2017 to October 2019, at station n.1. The red line shows the evolution of extension; in blue lake level oscillations are represented. Light blue vertical segments show rainfall amounts

0.1 mm/day. The beginning of the measurements at station n. 1 coincided with an exceptionally low level of the reservoir, which had been partly emptied to carry out major maintenance work. During the low-level phase, a section of the toe of the landslide was abnormally desaturated with respect to the normal high-and-low cycles of the lake. The subsequent infilling of the basin to a "normal level" coincided with the documented acceleration in the movement, marked also by the opening of fractures on the road surface. As mentioned, this interval of extension increase followed the almost complete drawdown of the lake (21 February 2017) and the successive period of lake infilling, although a delay of about two months can be observed. Another interval of increase, although much less pronounced than the previous one, can be documented after 6 March 2018, in concomitance with another infilling of the lake. These acceleration phases have been superimposed over the general trend of extension, which amounts to 88 mm in 29 months, equal to 3 mm/month on average. This slower, creep-like movement is attested by changes of slope inclination, sinkholes and development of fractured areas within the landslide body. As regards the measurements at station n. 2, data are available over a 15-month interval (after 19 February 2019, the extensometer ceased to function due to a technical inconvenience). Here, deformation developed with a total of 20 mm in about 15 months (Fig. 24.6). Also in this case, the extension recorded by the instrument is only partially consistent with rainfall; in fact, the first peak of extension occurred during a sustained rainy period but, later on, as can be seen from the graph in Fig. 24.6, deformation was more influenced by reservoir filling operations.

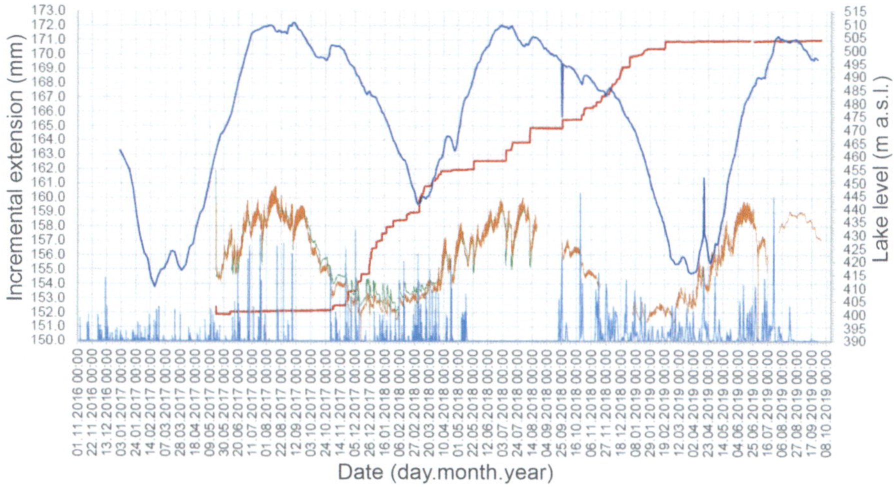

Fig. 24.6 Graph showing the readings of the incremental extension (in mm), associated with landslide surface displacement, recorded from the end of October 2017 to February 2019 at station n. 2. The red line shows the evolution of extension; in blue lake level oscillations are represented. Light blue vertical segments show rainfall amounts

24.4 Conclusions

Extensometer measurements represent a direct and inexpensive way for controlling deformation at specific locations in geohazard-prone areas. In the case of the major Khoko landslide, the results of an almost 3-year long campaign of measurement testify to the activity of the mass movement, at a rate of about 3.6 cm/year; deformation at the Khoko landslide appears to have been controlled, in the investigated period, by variations in hydraulic load, induced mainly by lake oscillations and, to a lesser extent, by rainfall. It is worth highlighting that a much more accurate way of assessing the movement of the landslide is represented by geodetic techniques, which have also been applied successfully during the 3-year NATO project. This overall monitoring effort will help individuate possible future accelerations of deformation at the unstable mass overlooking the Enguri artificial reservoir.

References

Gulen L, EMME WP2 Team (2011) Active faults and seismic sources of the Middle East region: earthquake model of the Middle East (EMME) project. In: Abstracts of the AGU fall meeting, San Francisco, California, 5–9 December 2011

Koçyigit A, Yılmaz A, Adamia S et al (2001) Neotectonics of East Anatolia Plateau (Turkey) and Lesser Caucasus: implication for transition from thrusting to strike-slip faulting. Geodin Acta 14:177–195

Pasquaré F, Tormey D, Vezzoli L et al (2011) Mitigating the consequences of extreme events on strategic facilities: evaluation of volcanic and seismic risk affecting the Caspian oil and gas pipelines in the Republic of Georgia. J Environ Manag 92:1774–1782

Pasquaré Mariotto F, Tibaldi A (2016) Inversion kinematics at deep-seated gravity slope deformations revealed by trenching techniques. Nat Hazards Earth Syst Sci 16:663–674

Reilinger RE, McClusky SC, Oral MB et al (1997) Global positioning system measurements of present-day crustal movements in the Arabia-Africa-Eurasia plate collision zone. J Geophys Res 102:9983–9999

Reilinger RE, McClusky SC, Vernant P et al (2006) GPS constraints on continental deformation in the Africa-Arabia-Eurasia continental collision zone and implications for the dynamics of plate interactions. J Geophys Res 111:B05411. https://doi.org/10.1029/2005JB004051

Tibaldi A, Pasquaré F (2008) Quaternary deformations along the "Engadine–Gruf tectonic system", Swiss–Italian border. J Quat Sci 23:475–487

Tibaldi A, Rovida A, Corazzato C (2004) A giant deep-seated slope deformation in the Italian Alps studied by paleoseismological and morphometric techniques. Geomorphology 58:27–47

Tibaldi A, Corazzato C, Rust D et al (2015) Tectonic and gravity-induced deformation along the active Talas–Fergana Fault, Tien Shan, Kyrgyzstan. Tectonophysics 657:38–62

Tibaldi A, Alania V, Bonali FL et al (2017a) Active inversion tectonics, simple shear folding and back-thrusting at Rioni Basin, Georgia. J Struct Geol 96:35–53

Tibaldi A, Russo E, Bonali FL et al (2017b) 3-D anatomy of an active fault-propagation fold: a multidisciplinary case study from Tsaishi (Georgia), western Caucasus. Tectonophysics 717:253–269

Tibaldi A, Korzhenkov AM, Pasquaré Mariotto F et al (2018) NATO and earth scientists: an ongoing collaboration to assess geohazards and contribute to societal security in Central Asia and the Caucasus. Episodes 41:193–205

Tibaldi A, Oppizzi P, Gierke JS, Oommen T, Tsereteli N, Gogoladze Z (2019) Landslides near Enguri dam (Caucasus, Georgia) and possible seismotectonic effects. Nat Hazards Earth Syst Sci 19:71

Tsereteli N, Tibaldi A, Alania V et al (2016) Active tectonics of Central-Western Caucasus, Georgia. Tectonophysics 691:328–344

Varazanashvili O, Tsereteli N, Bonali FL et al (2018) GeoInt: the first macroseismic intensity database for the Republic of Georgia. J Seismol 22(3):625–667. https://doi.org/10.1007/s10950-017-9726-5

Chapter 25
Aspects of Risk Management and Vulnerability Assessment of Buildings in the Republic of Georgia

Teimuraz Melkadze

Abstract This manuscript presents an ad-hoc methodology for the assessment of the vulnerability of buildings to be applied in Georgia where such assessment is included in the management risk approach designed to work at the national level.

The risk management approach is based on the determination of several factors: specifically, it considers the severity and probability of occurrence of adverse events, the multiplicity and complexity of their impacts (anthropological, economic, environmental, political and social), and the determination of physical and social vulnerability. The physical vulnerability assessment of buildings is here generally defined against all events that produce mechanical effects on buildings, with special focus on natural hazards.

Keywords Natural hazards · Risk assessment · Physical vulnerability of the buildings · Natural disasters · Vulnerability classes of buildings · Risk matrix · Natural hazards in the Caucasus region

25.1 Introduction

Natural disasters such as earthquakes and floods, as well as man-made catastrophes like the Chernobyl and Fukushima nuclear accidents (Steinhauser et al. 2014), have increasingly led to disruption of the ecological balance, leaving the world as a whole, including the Southern Caucasus region, under an increased exposure to risks and to a variety of consequent emergency situations (e.g. Kellenberg and Mobarak 2008).

Georgia presents both a variable climate and a mountainous territory, and these factors imply the possibility of abundant rainfall and highly energetic water discharge (e.g. Tatishvili et al. 2016), with negative feedbacks in terms of floods

T. Melkadze (✉)
Department of Civil Engineering, Georgian Technical University, Tbilisi, Georgia

© Springer Nature B.V. 2021

F. L. Bonali et al. (eds.), *Building Knowledge for Geohazard Assessment and Management in the Caucasus and other Orogenic Regions*, NATO Science for Peace and Security Series C: Environmental Security,
https://doi.org/10.1007/978-94-024-2046-3_25

and landslide and positive feedbacks in terms of hydropower energy production (Chomakhidze 2007). The Georgian territory is also affected by strong earthquakes (Tsereteli et al. 2016; Varazanashvili et al. 2018), as well as volcanoes with Holocene activity are present (e.g. Pasquarè et al. 2011). Consequently, this country is prone to possible crises, due to earthquakes, landslides, mudflows, floods, storms, droughts and more. In addition to the above, the geopolitical location and the current state of industrial processes cause a high level of exposure to environmental catastrophes. The death toll during recent natural disasters that struck Georgia confirms the extent of the risk; among the most recent, it is worth mentioning the mudflow event in the Kazbegi Municipality, which hit on May 17, 2014 (Gaprindashvili and Gaprindashvili 2015), with 9 casualties; the recent flooding in Tbilisi, which took place on June 14, 2015 (Gaprindashvili et al. 2016), with 22 casualties 2 missing people; the wildfire in the Borjomi Municipality, that broke out on August 20, 2017 (https://www.nato.int/cps/en/natohq/news_146447.htm?) and lasted 3 weeks, a time span during which nearly 1299 hectares of forest were incinerated.

The global response to risks induced by natural phenomena was provided during the UN Third World Conference on Disaster Risk Reduction; this event, held on March 14–18, 2015 in the city of Sendai (Japan), was aimed at discussing the international strategy on disaster risk reduction (Aitsi-Selmi and Murray 2015). The main outcome of this Conference was the "Sendai Framework for Disaster Risk Reduction 2015–2030" (UNISDR 2015). This international document is designed to help countries to consider multiple hazards for the emergency management at all levels, and in all fields. The implementation of the Sendai Risk Reduction Framework should be based on national peculiarities, internal state laws and international obligations.

Accordingly, legislative changes were implemented in Georgia during 2014–2018, resulting in disaster risk management obligations for civil building construction. For instance, according to part 1, article 8 of the "Georgian space planning, architectural and construction activities code", space and urban planning must be based on the principle of sustainable development, which ensures that land use for economic and social purposes is in compliance with emergency management risk reduction requirements.

The purposes of this work are: *(i)* to present a methodology for assessing certain components of risk, particularly the physical vulnerability of buildings, in normative-technical regulations of building safety and risk management; *(ii)* to determine the basic parameter for physical vulnerability assessment.

The following is a risk assessment methodology that is currently being implemented in Georgia and accounts for the specificity of the structural condition of existing buildings and their vulnerability, also taking into account that most buildings were built with standards that are now obsolete. Consequently, in the near future, it is planned to develop nationwide risk management plans based on the methodology presented in this article.

25.2 Risk Management Approach

The risk management approach is here presented through the risk matrix designed at the national level (Melkadze 2017). The emergency situation risk matrix (R) is obtained by multiplying the risk intensity matrix (F), the probability of event occurrence matrix (Z) and the event's impact matrix (W): R = F × Z × W.

The probability of event occurrence matrix (Z) is determined by the repetition of a particular event over a given time period, while the risk intensity matrix (F) is determined according to the intensity or strength of the exposure. Thus, their magnitude is determined on any given territory through methods established by scientific observations and based on the probability theory.

Finally, the third parameter needed to determine the risk matrix, which also represents the main parameter, is the emergency situation impact (W), which depends on the Vulnerability Matrix. The Vulnerability assessment requires the identification of the social impact of the event and of the resistance level of the buildings facing the hazard. There are, in fact, two types of vulnerability – physical and social.

The calculation of the emergency situation impact matrix (W) is necessary for assessing the quantitative impact of the hazard, and it is given by the ratio of the sum of all categories impact of the probable hazard (V) and the endurance/resilience matrix (P) (Eq. 25.1):

$$\boldsymbol{W} = \frac{\mathbf{0.4}\boldsymbol{V}_{\text{Anthropogenic}} + \mathbf{0.2}\boldsymbol{V}_{\text{economic}} + \mathbf{0.2}\boldsymbol{V}_{\text{social-political}} + \mathbf{0.2}\boldsymbol{V}_{\text{ecological}}}{\mathbf{P}} \tag{25.1}$$

The impact of each category of emergency is determined by the vulnerability matrix, and depends on different parameters for each category, according to the formulas provided in Eq. 25.2:

$$\begin{array}{ll} \boldsymbol{V}_{\text{Anthropogenic}} = \left[\sum_{k=1}^{n}\mathbf{M}_{\boldsymbol{k}}\right]; & \boldsymbol{V}_{\text{ecological}} = \left[\sum_{k=1}^{n}\mathbf{U}_{\boldsymbol{k}}\right]; \\ \boldsymbol{V}_{\text{economic}} = \left[\sum_{k=1}^{n}\mathbf{E}_{\boldsymbol{k}}\right]; & \boldsymbol{V}_{\text{social-political}} = \left[\sum_{k=1}^{n}\mathbf{I}_{\boldsymbol{k}}\right]; \end{array} \tag{25.2}$$

Where:

$M_k = f(S_{physical}, S_{social})$; $U_k = f(S_{physical}, S_{social})$; $W_k = f(S_{physical}, S_{social})$ and $I_k = f(S_{physical}, S_{social})$ are identifiers of anthropogenic, environmental/ecological, economic and socio-political emergency impact categories, all functions of the vulnerabilities ($S_{physical} = f(H)$) $\boldsymbol{S}_{physical}(\boldsymbol{H})$ and ($S_{social} = f(H)$).

Their values shall be determined by referring to predetermined limit values at a particular risk, "n" and "k" represent the number and counter of the parameters used to define the V values. The more parameters are provided, the greater the accuracy of the calculated risk value. The parameters set for each emergency impact category in the case of Georgia are presented in Table 25.1.

Table 25.1 Parameters of the Emergency Impact Categories

M_1	Number of people killed
M_2	Number of people injured
M_3	Number of people who need help more than 14 days
M_4	Number of people who need help less than 14 days
U_1	Deterioration in protected areas
U_2	Water resources pollution and deterioration
U_3	Groundwater pollution
U_4	Damage to agricultural areas
E_1	Physical damage (number of direct material damage)
E_2	Effective damage (indirect material damage)
E_3	Interruption / disruption of economic activity
E_4	Stopping or delaying the profitability of an object of the economy
I_1	Interruption / disruption of water supply
I_2	Interruption / disruption of power supply
I_3	Interruption / disruption of gas supply
I_4	Interruption / disruption of telecommunications
I_5	Impact on public order and security
I_6	Damage to cultural heritage sites

In this work, only the physical vulnerability ($S_{physical}$) assessment is discussed (see Sect. 25.3). In fact, the social vulnerability (S_{social}) relates to specific approaches, which are the focus of separate research, outside the scope of the present work. Its definition must take into account the population size and density, the socio-economic status of residents, their ability to cope with the expected emergency situation and their civil safety awareness. As it is not possible to quantify it, its impact on the risk calculation algorithm may not be considered; as a consequence, the accuracy of the calculation will be lower.

Finally, the Endurance/Resilience Matrix (P) evaluation involves the assessment of the infrastructure and material resources, and the capacity to cope with the consequences of emergencies.

25.3 Physical Vulnerability Assessment of the Buildings

The physical vulnerability ($S_{physical}(H)$) is defined by the endurance of buildings and the sustainable functioning of associated assurance systems and it depends only on the characteristics of the exposed objects. Its assessment is based on the determination of vulnerability parameters, as defined by the formulas in Eq. 25.3:

$$\boldsymbol{S_{physical}(H) = 1/S}, \tag{25.3}$$

where, $\boldsymbol{S = 0.25\,S_1 + 0.15\,S_2 + 0.15\,S_3 + 0.45\,S_4}$

All the parameters are presented hereunder.

Parameter $\mathbf{S_1}$ (Resistance) is determined by the analysis of the expected damage on buildings as brought about by the adverse event, based on experimental research, and it is expressed through the vulnerability classification. This approach is based on a building type classification scale that is harmonized with European standards: according to it, the vulnerability of buildings is divided into 6 classes (Table 25.2), with decreasing vulnerability from A (highest) to F (lowest), as shown in Table 25.3. This method is based on the European Macroseismic Scale 1998 (EMS-98) intensity definition (Grünthal 1998), modified according to the number of floors, and with the addition of typological classes to better describe the characteristics of the buildings in Georgia. The modified vulnerability classes and the characteristics typological classes are highlighted (in grey) in Table 25.2.

Parameter $\mathbf{S_2}$ (compliance with the requirements of technical regulations) is determined by the conditions in which the spatial planning and construction requirements, defined by the relevant technical regulations, are implemented.

The above-mentioned parameter is determined in terms of building certificates and the initial design assignments, and it can be given the following numerical values:

- 1.0, when it fully complies with the requirements of the technical regulations of the building.
- 0.9, partial non-compliance with the requirements of the technical regulations of the building.
- 0.85, when it fully complies with the requirements of the technical regulations of the building but encloses (non-load-bearing) structures which are made of burnable and rot-susceptible materials.
- 0.75, when it fully complies with the requirements of the technical regulations of the building, but encloses load-bearing structures which are made of burnable and rot-susceptible materials.
- 0.5, for buildings and facilities that are in complete non-compliance with the technical regulations.

If there are several instances of discrepancy at the same time, the minimum value will be considered (Development of Rules for Preparation of Emergency Risk Management Plan 2017).

Parameter $\mathbf{S_3}$ (compliance with current technical conditions) is based on the technical conditions of the structure and on external signs of the construction of the building, seen during visual inspections, according to Table 25.4.

Hereunder are listed all S_3 categories:

- Normal Condition – There is virtually no damage.
- Fully Satisfactory Condition – There is moderate damage (load-bearing building structures are lightly damaged, and non-bearing building structures are moderately damaged).

Table 25.2 Vulnerability classes of buildings according to the structural types of buildings. Symbols: white circle, the most vulnerable class; continuous line, available range; dashed line, less probability range (rare case)

№	Constructive types of buildings	Floor*	Vulnerability classes					
			A	B	C	D	E	F
Masonry building types								
1	Rubble or fieldstone masonry buildings	1	○					
2	Adobe (earth brick) masonry buildings	1	○─	─┤				
3	Simple stone masonry buildings	2-3	├┄┄	┄○				
4	Whole brick and whole concrete and natural stones (Including massive stones) masonry buildings	3-4		├─	─○┄	┄┤		
5	Unreinforced, with manufactured stone units	3-4		├─	─○┄	┄┤		
6	Reinforced of confined masonry buildings	4			├┄	┄○─	─┤	
7	Same, regular complex arranged concrete system with increased strength	4			├─	─○─	─┄	┄┤
8	Buildings constructed of large concrete or vibrator blocks; also, diverse walls of different types of internal monolithic reinforced concrete	5			├─	─○─	─┤	
9	Voluminous concrete block buildings	5			├─	─	─○┄	┄┤
Reinforced concrete buildings								
10	Monolithic reinforced concrete structure buildings	Up to 24		├┄	┄─	─○─	─┄	┄┤
11	Unaccustomed large-scale buildings	Up to 9			├─	─○─	─┄	┄┤
12	Frame-type reinforced concrete buildings	Up to 5		├┄	┄─	─○┄	┄┤	
13	Connections, diaphragm or reinforced framed concrete buildings (hard drive and etc), including high-rise buildings **	Up to 9			├─	─	─○─	─┤
Buildings made of metal, wood and others, including constructions equipped with special systems								
14	In terms of arctic-constructive solution, wide range of steel framed buildings (flexible buildings, modern high-rise buildings **)	-	├┄	┄	┄─	─	─○─	─┤
15	Large spatial structures, pneumatic (inflatable) structures, etc.	1			├─	─	─○─	─┤
16	Buildings with special systems of seismic isolation and damping of oscillations, which regulates the reaction of the building	-					├─	─○
17	Wooden frame structures	1-2		├┄	┄─	─○┄	┄┤	
18	Buildings whose load-bearing structures are made of wooden beams and logs	1-2			├┄	┄─	─○─	─┤

* In case of more than the specified floor, vulnerability of the appropriate type of building is recommended to increase by one class

** The height of the buildings discussed in the above line should not exceed 40 meters, whereas its ratio with the smallest transverse dimension should not exceed 6:1 range

Table 25.3 S_1 parameter values based on building vulnerability class

Vulnerability class	A	B	C	D	E	F
$\mathbf{S_1}$	0.3	0.5	0.7	0.9	1	1.75

Table 25.4 S_3 parameter based on building compliance

Category of technical condition		S_3
1	Normal condition	1
2	Fully satisfactory condition	0.95
3	Partially satisfactory condition	0.85
4	Unsatisfactory condition	0.75
5	State of emergency condition	0.65

- Partially Satisfactory Condition – Damage ranges from significant to severe (load-bearing building structures are moderately damaged, and non-bearing building structures are strongly damaged).
- Unsatisfactory Condition – Very severe damage is observed (load-bearing building structures are strongly damaged, and non-bearing building structures are very strongly damaged).
- State of Emergency Condition – Extreme damage (particular load-bearing elements are destroyed; the whole building is on the verge of collapse).

Parameter $\mathbf{S_4}$ (compliance with the characteristics of the analysis based on the structural calculation) is determined by a ratio, in which the numerator corresponds to existing factual values, and the denominator is determined by the conditions in case of a guaranteed required capacity, which allows for normal operation. The numerical value of this parameter, which should not exceed 1.2, is determined by the precise instrumental evaluation of the premises according to set parameters. The following types of operations are performed to determine $\mathbf{S_4}$:

- Geological, seismic, meteorological, climatic, ecological and other types of analysis and assessment of the local site.
- Calculation of the impact caused by damage, in case of possible emergency situations, to the building-planning and selection of appropriate reporting schemes.
- Numerical modelling of the design structure of the building.
- Direct calculation with the analysis of the results.

Therefore, the numerical value of the $\mathbf{S_4}$ coefficient is determined by the precise instrumental evaluation of the engineering structure according to the specified parameters.

Finally, in case of non-availability of one of the above-explained structural vulnerability parameters (S_1, S_2, S_3, S_4), the quantitative share in the total value is equally distributed among all the coefficients.

25.4 Final Remarks

This article presents some suggestions for the improvement and development of the already existing disaster risk management methodology in Georgia. Indeed, the amendments to the current technical regulations are aimed at applying the recommendations of the Sendai Disaster Risk Reduction Framework Program 2015–2030 and complying with EU civil protection procedures.

It has been discussed the main components of risk assessment, the method for evaluating the physical vulnerability of buildings as well as the methods to determine the main relevant criteria. It also has been specified the coefficients that determine vulnerability, by considering different parameters. It is foreseen that the present methodology will be implemented and applied to Georgian buildings, so as to allow disaster risk mitigation challenges in civil engineering to be easily identified, and design the most appropriate preventive measures.

References

Aitsi-Selmi A, Murray V (2015) The Sendai framework: disaster risk reduction through a health lens. Bull World Health Organ 93(6):362

Chomakhidze D (2007) Georgia: natural energy resources. Central Asia and the Caucasus (4):26–34

Decree No. 254 of the Government of Georgia (6 October 2017) Development of Rules for Preparation of Emergency Risk Management Plan

Gaprindashvili G, Gaprindashvili M (2015) Catastrophic debrisflow in Dariali (Georgia) in the year 2014. Nat Sci 7(07):379–389

Gaprindashvili G, Gaprindashvili M, Tsereteli E (2016) Natural disaster in Tbilisi City (Riv. Vere Basin) in the year 2015. Int J Geosci 7(09):1074–1087

Grünthal G (1998) European macroseismic scale 1998 (EMS-98). European Seismological Commission (ESC)

Kellenberg DK, Mobarak AM (2008) Does rising income increase or decrease damage risk from natural disasters? J Urban Econ 63(3):788–802

Melkadze T (2017) Manage emergency situations. Georgian Technical University, Republic of Georgia

Pasquarè FA, Tormey D, Vezzoli L et al (2011) Mitigating the consequences of extreme events on strategic facilities: evaluation of volcanic and seismic risk affecting the Caspian oil and gas pipelines in the republic of Georgia. J Environ Manag 92(7):1774–1782

Steinhauser G, Brandl A, Johnson TE (2014) Comparison of the Chernobyl and Fukushima nuclear accidents: a review of the environmental impacts. Sci Total Environ 470:800–817

Tatishvili MR, Kartvelishvili LG, Mkurnalidze IP (2016) Thunderstorm and hail processes over Georgian territory against global climate change background. J Georgian Geophys Soc 19:22

Tsereteli N, Tibaldi A, Alania V et al (2016) Active tectonics of central-western Caucasus, Georgia. Tectonophysics 691:328–344

UNISDR U (2015) Sendai framework for disaster risk reduction 2015–2030. Proceedings of the 3rd United Nations world conference on DRR, Sendai, Japan, 2015

Varazanashvili O, Tsereteli N, Bonali FL et al (2018) GeoInt: the first macroseismic intensity database for the Republic of Georgia. J Seismol 22(3):625–667

The manufacturer's authorised representative in the EU is Springer
Nature Customer Service Centre GmbH, Europaplatz 3, 69115 Heidelberg,
Germany. If you have any concerns regarding our products, please
contact ProductSafety@springernature.com

Printed and bound by CPI Group (UK) Ltd, Croydon, CR0 4YY
29/09/2026
02225004-0001